Novel Plant Protein Processing

Proteins serve as an important nutritional as well as structural component of foods. Not only do they provide an array of amino acids necessary for maintaining human health but also act as thickening, stabilizing, emulsifying, foaming, gelling, and binding agents. The ability of a protein to possess and demonstrate such unique functional properties depends largely on its inherent structure, configuration, and how it interacts with other food constituents like polysaccharides, lipids, and polyphenolic compounds. Proteins from animal sources have superior functionality, higher digestibility, and lower antinutrient components than plant proteins. However, consumer preferences are evolving worldwide for ethically and sustainably sourced, clean, cruelty-free, vegan, or vegetarian plant-based food products. Unlike proteins from animal sources, plant proteins are more versatile and religiously and culturally acceptable among vegetarian and vegan consumers and associated with lower food-processing waste, water, and soil requirement.

Thus, the processing and utilization of plant proteins have gained worldwide attention, and as such numerous scientific studies are focusing on enhancing the utilization of plant proteins in food and pharmaceutical products through various processing and modification techniques to improve their techno-functional properties, bioactivity, bioavailability, and digestibility. *Novel Plant Protein Processing: Developing the Foods of the Future* presents a roadmap for plant protein science and technology which will focus on plant protein ingredient development, plant protein modification, and the creation of plant protein-based novel foods.

KEY FEATURES

- Includes complete information about novel plant protein processing to be used in future foods
- Presents a roadmap to upscale the meat analog technological processes
- Discusses marketing limitations of plant-based proteins and future opportunities

This book highlights the important scientific, technological advancements that are being deployed in the future foods using plant proteins, concerns, opportunities, and challenges and as an alternative to maintaining a healthy and sustainable modern food supply. It covers the most recent research related to the plant protein-based future foods which include their extraction, isolation, modification, characterization, development, and final applications. It also covers the formulation and challenges: emphasis on the modification for a specific use, legal aspects, business perspective, and future challenges. This book is useful for researchers, readers, scientists, and industrial people to find information easily.

Novel Plant Protein Processing

Developing the Foods of the Future

Edited by
Zakir Showkat Khan,
Sajad Ahmad Wani and
Shemilah Fayaz

CRC Press
Taylor & Francis Group
Boca Raton London New York

CRC Press is an imprint of the
Taylor & Francis Group, an **informa** business

Cover image: Shutterstock 2116460495

First edition published 2024
by CRC Press
2385 NW Executive Center Drive, Suite 320, Boca Raton FL 33431

and by CRC Press
4 Park Square, Milton Park, Abingdon, Oxon, OX14 4RN

CRC Press is an imprint of Taylor & Francis Group, LLC

© 2024 selection and editorial matter, Zakir Showkat Khan, Sajad Ahmad Wani, and Shemilah
Fayaz; individual chapters, the contributors

ISBN: 978-1-032-43816-0 (hbk)
ISBN: 978-1-032-43992-1 (pbk)
ISBN: 978-1-003-36979-0 (ebk)

DOI: 10.1201/9781003369790

Typeset in Times
by Apex CoVantage, LLC

Contents

Chapter 3   Modification Methods of Plant-Based Proteins: An Overview...................57

Priti Sharad Mali and Pradyuman Kumar

Chapter 4 Biological Modification of Plant-Based Proteins 76

Subhasri Dharmaraj and Pintu Choudhary

Chapter 5 Thermal Modification of Plant-Based Proteins 95

Kanchan Suri, Seerat Bhinder, Seeratpreet Kaur and Mehak Katyal

Preface

Proteins are the most important macronutrients required for human survival; however, protein production is a concern nowadays because conventional animal protein production requires large amounts of resources. Plant-based proteins are promising alternatives due to the popularity of vegan/vegetarian diets, affordability, health consciousness, ethical issues, etc. Studies suggest that plant-based proteins might overtake animal-based proteins in food systems by 2054. The major disadvantage associated with plant-based proteins is a lower functional profile which is defined by poor solubility, emulsifying and gelling properties, and thus it can limit their use in the food industry. In comparison to animal proteins, there is a growing interest in food and pharmaceutical industries toward plant protein technology. Therefore, collecting information in the form of a book can be a great achievement.

This book presents a roadmap for plant protein science and technology which will focus on plant protein ingredient development, plant protein modification, and the creation of plant protein-based novel foods. It highlights the important scientific, technological advancements that are being deployed in the future foods using plant proteins, with their concerns, opportunities, and challenges. Plant protein-based future foods are an alternative to maintaining a healthy and sustainable modern food supply. The book covers the most recent research related to the plant protein-based future foods, which include their extraction, isolation, modification, characterization, development, and final applications. It presents the complete direction for the development of plant protein–based future foods and the potential role of plant protein–based future foods in the current scenario. It also covers the formulation and challenges: emphasis on the modification for a specific use, legal aspects, business perspective, and future challenges. Such information available in the form of a book can help researchers, readers, scientists, and industrial people locate useful information easily.

About the Editors

Zakir Showkat Khan is currently working as Assistant Professor TEQIP III in the Department of Food Technology, Islamic University of Science and Technology, Awantipora, Jammu and Kashmir, India. He has completed his Master's Degree in Food Technology from Sant Longowal Institute of Engineering and Technology, Punjab, India, and currently pursuing his PhD from Guru Nanak Dev University, Amritsar, Punjab, India. He has qualified Graduate Aptitude Test in Engineering (2011) conducted by the Ministry of Human Resource Development India (MHRD) in February 2011. He has published more than 18 research/review articles and 2 book chapters. He has attended ten national and five international conferences, seminars, and workshops throughout the world. Dr. Zakir Showkat Khan has participated in more than ten faculty development programs. He is Potential Peer Reviewer of reputed international journals related to Food Science and Technology, which belong to popular publishing houses, viz. Elsevier, Taylor & Francis, Wiley, Springer, etc. He is also working as a member of various associations both national and international. Dr. Zakir Showkat Khan is the recipient of prestigious research grants like "Collaborative research scheme TEQIP III" funded by MHRD (2019–2021), and "Entrepreneurship development scheme (2019)" funded by Government of India.

Dr. Sajad Ahmad Wani is currently working as D.S. Kothari Post-doctoral Fellow in the Department of Food Science and Technology, SKUAST-Kashmir, Jammu and Kashmir, India. He has earned his Master's Degree in Food Technology from IUST, Awantipora, Jammu and Kashmir, India, and PhD from Sant Longowal Institute of Engineering and Technology, Punjab, India. He has qualified National Eligibility Test (2015) (NET) conducted by Indian Council of Agricultural Research (ICAR) in December 2015. He has published more than 40 research/review articles, 10 book chapters, and 7 books. He is also writing articles for the popular magazine *Food & Beverage News* (India's first magazine for food and beverage industry). He has attended 50 national and 20 international conferences, seminars, and workshops throughout the world. Dr. Wani has participated in various faculty development programs and Dr. Wani serves as Honorary Associate Editor of esteemed *International Journal of Food Science and Technology* (Wiley). He is Potential Peer Reviewer of reputed international journals related to Food Science and Technology, which belong to popular publishing houses, viz., Elsevier, Taylor & Francis, Wiley, Springer, etc. He is also working as member of various associations like IFT, IFERP, International Association for Agricultural Sustainability, AFSTI, Asia Society of Researchers, and Asian Council of Science Editors. Dr. Wani is the recipient of prestigious "D.S. Kothari Post-doctoral Fellowship" and "Maulana Azad National Fellowship (MANF-2013–14)" from University Grant Commission, New Delhi, India.

Shemilah Fayaz is currently working as Assistant Professor TEQIP III in the Department of Food Technology, Islamic University of Science and Technology, Awantipora, Jammu and Kashmir, India. She has completed his master's degree in Food Technology from National Dairy Research Institute, Karnal, India, and is currently pursuing her PhD from Guru Nanak Dev University, Amritsar, Punjab, India. She has qualified Graduate Aptitude Test (GATE) in Engineering (2013) conducted by the Ministry of Human Resource Development India (MHRD) in February 2013. She has published more than six research/review articles and two book chapters. She has attended four national and three international conferences, seminars, and workshops throughout the world. Ms. Shemilah Fayaz has participated in more than six faculty development programs. She is Potential Peer Reviewer of reputed international journals related to Food Science and Technology, which belong to popular publishing houses, viz., Taylor & Francis, Wiley, Springer, etc. She is also working as a member of various associations, both national and international. She is the recipient of prestigious research grants like "Collaborative research scheme TEQIP III" funded by MHRD (2019–2021), Government of India.

Contributors

Bisma Amir
Department of Nutrition
School of Health Sciences
University of Management and Technology
Lahore, Pakistan

Saira Amir
Department of Nutrition
School of Health Sciences
University of Management and Technology
Lahore, Pakistan

Suhail Anees
Department of Clinical Biochemistry
University of Kashmir
Srinagar, Jammu and Kashmir, India

Riya Barthwal
Department of Food Technology
Govind Ballabh Pant University of Agriculture
    and Technology
Pantnagar, Uttarakhand, India

Kanchan Bhatt
Department of Food Science and Technology
Dr. Yashwant Singh Parmar University of
    Horticulture & Forestry
Nauni, Solan, Himachal Pradesh, India

Seerat Bhinder
Department of Biotechnology Engineering
    and Food Technology
Chandigarh University
Mohali, Punjab, India

Fatma Boukid
ClonBio Group Limited
Dublin, Ireland

Priyanka Chakraborty
Department of Pharmacology
BCDA College of Pharmacy &
    Technology
Hridaypur, West Bengal, India

Pintu Choudhary
National Institute of Food Technology
Entrepreneurship and Management
Thanjavur, Tamil Nadu, India

Sailee Chowdhury
Department of Pharmaceutical Chemistry
BCDA College of Pharmacy & Technology
Hridaypur, West Bengal, India

Subhasri Dharmaraj
National Institute of Food Technology
Entrepreneurship and Management
Thanjavur, Tamil Nadu, India

Hao Feng
Department of Family and Consumer Sciences
North Carolina A&T State University
Greensboro, North Carolina, United States
Department of Food Science and Human
    Nutrition
University of Illinois Urbana-Champaign
Urbana, Illinois, United States

Showkat Ahmad Ganie
Department of Clinical Biochemistry
University of Kashmir
Srinagar, Jammu and Kashmir, India

Rukiye Gundogan
Department of Food Engineering
Faculty of Chemical and Metallurgical
    Engineering
Istanbul Technical University
Maslak, Istanbul, Türkiye

Rabia Hamid
Department of Nanotechnology
University of Kashmir
Srinagar, Jammu and Kashmir, India

Ozan Kahraman
Applied Food Sciences Inc.
Austin, Texas, United States

Ragya Kapoor
Department of Food Science and Human
  Nutrition
University of Illinois Urbana-Champaign
Urbana, Illinois, United States

Koyel Kar
Department of Pharmaceutical Chemistry
BCDA College of Pharmacy & Technology
Hridaypur, West Bengal, India

Gulsah Karabulut
Department of Food Engineering
Sakarya University
Sakarya, Türkiye
Department of Food Science and Human
  Nutrition
University of Illinois at Urbana-Champaign
Urbana, Illinois, United States

Asli Can Karaca
Department of Food Engineering
Faculty of Chemical and Metallurgical
  Engineering
Istanbul Technical University
Maslak, Istanbul, Türkiye

Atefeh Karimidastjerd
Department of Food Engineering
Faculty of Chemical and Metallurgical
  Engineering
Istanbul Technical University
Maslak, Istanbul, Türkiye

Deepika Kathuria
Dairy Chemistry Division
NDRI, Karnal, Haryana, India

Mehak Katyal
Department of Nutrition and Dietetics (FAHS)
Manav Rachna International Institute of
  Research and Studies
Faridabad, Haryana, India

Seeratpreet Kaur
PG Department of Food Science and
  Technology
Khalsa College
Amritsar, Punjab, India

Zakir Showkat Khan
Guru Nanak Dev University
Amritsar, Punjab, India

Pradyuman Kumar
Department of Food Engineering and
  Technology
Sant Longowal Institute of Engineering and
  Technology (Deemed-to-be University)
Longowal, Punjab, India

Supriya Kumari
Guru Nanak Dev University
Amritsar, Punjab, India

Priti Sharad Mali
Department of Food Engineering and
  Technology
Sant Longowal Institute of Engineering
  and Technology (Deemed-to-be
  University)
Longowal, Punjab, India

Shreya Mandal
Division of Biochemistry
ICAR-Indian Agricultural Research
  Institute
New Delhi, India

Kamalika Mazumder
Department of Pharmaceutical Chemistry
BCDA College of Pharmacy & Technology
Hridaypur, West Bengal, India

Anna Aleena Paul
Department of Food Processing Technology
St. Teresa's College (Autonomous)
Ernakulam, Kerala, India

Srutee Rout
Department of Agricultural and Food
  Engineering
IIT Kharagpur, Kharagpur
West Bengal, India

Nilushni Sivapragasam
College of Agriculture and Veterinary
  Sciences
United Arab Emirates University, UAE

Prem Prakash Srivastav
Department of Agricultural and Food
    Engineering
IIT Kharagpur, Kharagpur
West Bengal, India

Kanchan Suri
PG Department of Agriculture
Khalsa College
Amritsar, Punjab, India

Shweta Suri
Amity Institute of Food Technology (AIFT)
Amity University
Noida, Uttar Pradesh, India

Priyanka Suthar
Department of Food Science and Technology
Dr. Yashwant Singh Parmar University of
    Horticulture & Forestry
Nauni, Solan, Himachal Pradesh, India

Hafsa Tahir
Department of Nutrition
School of Health Sciences
University of Management and Technology
Lahore, Pakistan

Gizem Sevval Tomar
Department of Food Engineering
Faculty of Chemical and Metallurgical
    Engineering
Istanbul Technical University
Maslak, Istanbul, Türkiye

Gulcin Yildiz
Department of Food Engineering
Igdir University
Igdir, Türkiye

1 Introduction to Plant Protein-Based Future Foods

Saira Amir, Bisma Amir and Hafsa Tahir

1.1 INTRODUCTION

As the global population is expected to rise above 9 billion people by the year 2050, the world is facing the challenge to make enough food accessible for the entire world (FAO). Aside from aiding the total calorie requirement, protein is the second-most essential micronutrient needed for the growth and development of the human body. Traditionally, the production of animal protein requires a considerable amount of resources, i.e., land, water, and harvest, and their consumption is not deemed safe for humans due to several diseases in animals (Pojić et al., 2018; Sun-Waterhouse et al., 2014). Plant-based proteins are a promising solution to this issue due to the long history of crop production, low cost, environmental stability, and accessibility in various parts of the world.

Plant-based proteins are included in food that is vegan. They offer a sufficient amount of amino acids, are efficiently utilized by humans, and aid in curing many diseases. Further, plant-based protein is also a good source of fiber, oligosaccharides, polyunsaturated fatty acids, and carbohydrates. These proteins are strongly linked with reducing low-density lipoprotein cholesterol (LDL), diabetes mellitus type 2, cardiovascular diseases, and diabetes (Guasch-Ferré et al., 2019). There are many different sources of plant-based proteins such as cereals, pseudocereals, legumes, nuts, and seeds (Lonnie et al., 2020). However, It is known that plant proteins have lower quality, poor functionality, reduced solubility, gelling, emulsifying, and foaming properties that somehow limit their use in the food industry. To bridge this gap, advancement in the development of plant protein ingredients and knowledge to make plant-based foods are surely needed.

1.1.1 PLANT-BASED SOURCES

Plant-based sources of protein dominate the protein supply among all the existing sources of protein in the human diet (57%), while the 34% remaining is meat (18%), dairy products (10%), other animal products (9%), and fish and shellfish (6%) (McGuire, 2015). Several different sources of proteins from plants have been explored to tackle the existing challenges of feeding the increasing population (Day & Technology, 2013; Hughes et al., 2014; X. Wang et al., 2010). Depending on the sources, plant-based protein might be lacking some of the essential amino acids. For example, legumes contain lower sulfur-containing amino acids, i.e., methionine and cysteine, and cereals are deficient in lysine (Nosworthy et al., 2017). However, the pseudocereals like amaranth and quinoa contain a good amount of lysine. Sometimes, the factors like soil diversity, precipitation levels, altitude, climate conditions, and agricultural practices can result in the same plants containing different nutrients (Goldflus et al., 2006; K.-l. Liu et al., 2017).

1.1.1.1 Legumes

Legume-rich diet has many health benefits for humans (Frías et al., 2011). They are the best among other dietary options for high energy, carbohydrates, protein, vitamins, minerals, and fiber content. Common legumes include chickpeas, lupins, peas, soybeans, and common beans. In developing countries, common beans are the major source of vegan protein (Espinosa-Páez et al., 2017). Peas have been used in various food products to improve human dietary protein intake. Another major

DOI: 10.1201/9781003369790-1

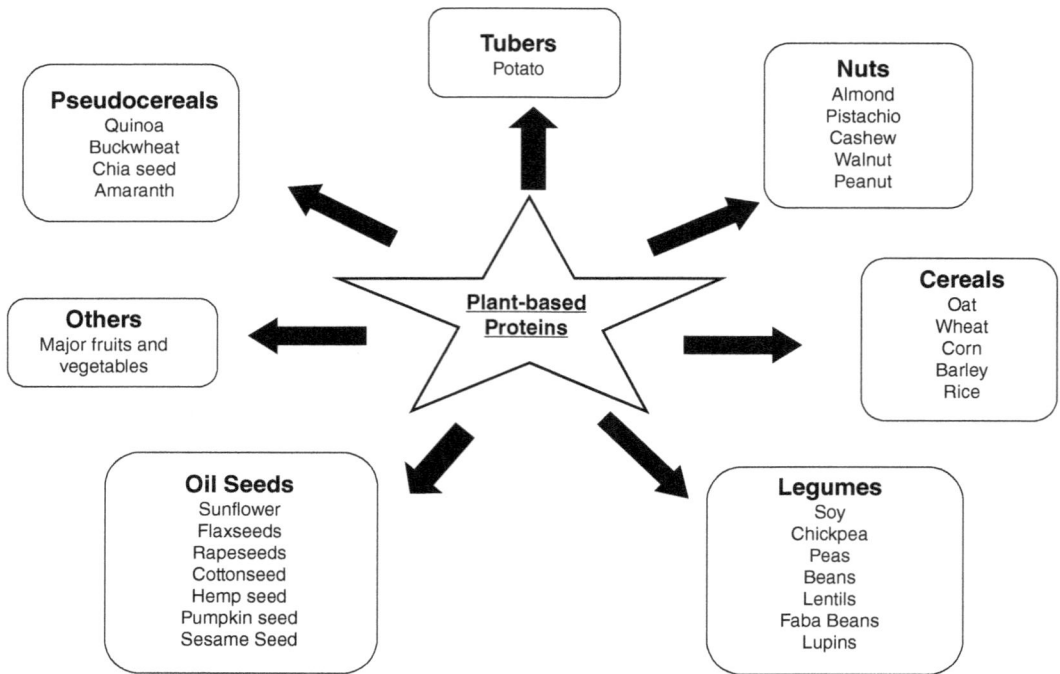

FIGURE 1.1 Plant-based sources of protein.

source of high dietary protein is food products containing chickpeas. Pigeon peas are also a rich source of sulfur-containing amino acids (Adenekan et al., 2018).

1.1.1.2 Cereals

Cereals including corn, barley, wheat, and rice are the staple food all around the globe (Amagliani et al., 2017). In both developed as well as developing countries, rice is one of the most consumed cereals. In rice, the highest lysine content is present in albumin, and globulin is higher in sulfur-containing amino acids. Various isolation techniques have been studied to improve the quality of rice protein (L. Yang et al., 2012). Millet is rich in essential amino acids such as lysine. Cereal-based proteins are widely used in the food and bakery industry. Many studies have shown that the use of legumes with cereals results in improved nutritional quality (Coda et al., 2017).

1.1.1.3 Pseudocereals

Pseudocereals are plants having dicotyledons and are thought to be false cereals, such as buckwheat, quinoa, and amaranth (Alvarez-Jubete et al., 2010). Recently, pseudocereals with proteins like quinoa and amaranth have been used to meet the high protein demand. They are rich in protein, fiber, unsaturated fatty acids, minerals, and vitamins. Being gluten free, they are excellent to be introduced into the diet for celiac patients (López et al., 2018).

1.1.1.4 Seeds

Seeds are an excellent source of good-quality nutrition (Amagliani et al., 2017). Flaxseeds are rich in high-quality protein, fiber, essential amino acids, and phenolic compounds but may have limited lysine content (Anaya et al., 2015). Watermelon seeds have a good proportion of arginine and leucine (Kaul, 2011), while the chia seeds lack lysine (Olivos-Lugo et al., 2010). Paprika seeds are found to be rich in aromatic amino acids such as threonine, tryptophan, and lysine, but they contain low amounts of isoleucine and sulfur-containing amino acids (El-Adawy & Taha, 2001).

1.1.1.5 Nuts

Nuts have high-quality protein, lipids, and fatty acids. Nontraditional almonds like baru and pequi have a complete amino acid profile (de Oliveira Sousa et al., 2011). Baru almonds have all the essential amino acids, while the pequi and cashew nuts are rich in sulfur amino acids, but they lack the lysine. Peanuts are considered a lower source of protein as they are limited in lysine and valine.

1.1.1.6 Meat Analogs

Plant-based meat analogs (also known as meat substitutes, faux meat, or mock meat) are plant-based products that look, taste, and feel like animal meat (V. Joshi & Kumar, 2015; Kumar et al., 2017). For the creation of a wide range of meat-free goods including burger patties, nuggets, and sausages, several non-animal protein sources (e.g., cereals, legumes, vegetables, fungi, and microalgae) are employed to replace animal proteins (Caporgno et al., 2020; Curtain & Grafenauer, 2019; Fresán et al., 2019). Table 1.1 provides a summary of the main elements in fibrous meat analogs, which include plant proteins (20–50%), polysaccharides (2–30%), vegetal lipids (0–5%), and additional components to create a meat-like experience (Hsieh, 2014).

Due to their technological functionalities (solubility, viscosity, emulsification, foaming, solubility, film formation, flavor binding, and gelling), plant proteins perform a variety of roles in the structure, texture, color, and flavor of meat analogs (Chiang et al., 2019; Mattice & Marangoni, 2020; Schreuders et al., 2019a). Soy protein has been the most commonly utilized protein in meat analog products due to its strong functional characteristics and balanced amino acid content (protein digestibility-corrected amino acid score (PDCAAS) of 1.00 as compared to meat) (Kumar et al., 2017; Yuliarti et al., 2021). At a reasonable cost, soy proteins guarantee a dual function as extenders and binders. Wheat protein has a long history of usage in meat analogs because of its rheological and viscoelastic qualities, which enable the production of a fibrous-like texture in end meat products (Chiang et al., 2019; Samard et al., 2019). Particularly, gluten can act as a binder and extender, minimizing cooking losses during processing. Pea protein is a popular substitute for soy protein because it is highly adaptable, hypoallergenic, and functional. It is typically combined with additional sources (such as gluten) to enhance the nutritional and sensory qualities of

TABLE 1.1
Ingredients of Plant-based Meat Analogs

Ingredient	Role	Sources	Reference
Non-animal protein	Nutrition, texture, color, structure, and flavor	Soy, legumes, lupin, wheat, potato, rice, and pea	Azzollini et al., 2019; Bohrer, 2019; Smetana et al., 2019
Polysaccharides	Water binding and consistency	Flours, fibers, native starches	S. Lin et al., 2002; Yao et al., 2004
Lipids	Texture, mouth feel, and flavor	Cocoa butter and coconut oil (saturated fatty acids) Sunflower oil, sesame oil, canola oil, and avocado oil (unsaturated fatty acids)	Dekkers et al., 2018; Emin et al., 2017; Pietsch et al., 2019
Coloring agents	Meat color	Beet juice extract, lycopene, or leghemoglobin	Fraser et al., 2018; Oreopoulou & Tzia, 2007; Rayner et al., 2018; Schreuders et al., 2019a
Flavoring agents	Flavor	Herbs and spices, savory yeast extract, and paprika	Wi et al., 2020
Fortification agents	Nutritional value	Sodium ascorbate, zinc gluconate, tocopherols, thiamine hydrochloride	Caporgno et al., 2020; Wi et al., 2020

meat substitutes. Alternative proteins derived from rice and potatoes are becoming more popular, particularly for the creation of meat substitutes free from gluten (Avebe, 2020). The source and make-up of fatty acids play a crucial role in stimulating the flavor of meat that results from lipid oxidation and the volatiles produced by the Maillard reaction during thermal processing (Diez-Simon et al., 2019). To strengthen the flavor as well as to enhance texture and mouthfeel, lipids high in saturated fatty acids (like cocoa butter and coconut oil) and unsaturated fatty acids (e.g., sunflower oil, avocado oil, sesame oil, and canola oil) are utilized (Emin et al., 2017; Pietsch et al., 2019). Polysaccharides play significant structural and functional roles in shaping meat analogs because of their thickening and emulsifying capabilities, which are typically needed to increase consistency and water binding (S. Lin et al., 2000). The primary usage of native starches and flours (such as potato, corn, wheat, cassava, pea, and rice) is to enhance the texture and consistency of flours (Krintiras et al., 2015; Yuliarti et al., 2021). Because of their high water-holding capacity and ability to produce stable oil/water emulsions, fibers from various sources, such as peas, oat, potatoes, soybean, citrus, bamboo, and apple, can thicken products and minimize cooking loss (J. Zhang et al., 2020). In order to simulate the red color of beef, coloring agents including leghemoglobin, lycopene, beet juice extract, and annatto extracts (E 160b) are utilized, while titanium dioxide is used to simulate the color of chicken (Fraser et al., 2018; Oreopoulou & Tzia, 2007; Schreuders et al., 2019a). Plant-based proteins must undergo a series of changes from their natural state (globular shape) to get textured protein (linear shape) by using various processing techniques in order to resemble the fibrousness of meat muscle. These processes include extrusion, wet or electro-spinning, high-temperature conical simple shearing, freeze structuring, blend proteins, hydrocolloids, and 3D printing.

1.1.1.7 Milk Analogs

Plant-based milk analogs often have somewhat similar compositions and structures because they are typically designed to have physicochemical and sensory properties similar to those of cow's milk (D. J. McClements et al., 2019). The majority of milk substitutes are colloidal dispersions made of oil bodies, fat droplets, protein aggregates, and/or fragments of plant tissues dispersed in an aqueous solution with dissolved sugars and salts. There are two alternative ways to make these colloidal suspensions (D. J. McClements, 2020). Initially, they can be created through the mechanical or enzymatic disruption of whole plant materials like soybeans, almonds, coconut flesh, rice, oats, or hazelnuts. They can also be produced by mixing plant-based oils (such as soy, sunflower, flaxseed, canola, corn, or olive oil) with water in the presence of plant-based emulsifiers like proteins, polysaccharides, phospholipids, or surfactants. In both situations, the final composition and structure need to be precisely managed to produce a finished product with the necessary physicochemical, sensory, and shelf-life properties. The content and structural organization of plant-based milks determines their physicochemical and sensory properties (D. J. McClements et al., 2019). The creamy appearance of milk analogs is caused by the scattering of light waves by colloidal particles such as oil bodies, lipid droplets, protein aggregates, or plant-tissue fragments. Usually, as the concentration and contrast of the refractive indices of these colloidal particles increase, so does the lightness of a plant-based milk analog.

Table 1.2 compares the nutritional characteristics of many plant-based milk substitutes to those of cow's milk. There are significant differences in the nutrient contents of various types of milk analogs, as well as between them and actual milk. The protein level of the milk analog is lower than that of actual milk, with the exception of soymilk. The total and saturated fat contents of the milk analogs are lower than those of whole milk but higher than skim milk. Compared to actual milk, the milk analogs don't have any cholesterol. Although the milk analogs' sugar concentrations are lower than those of cow's milk, they mostly include sucrose whereas cow's milk also contains lactose. Due to their calcium carbonate fortification, all of the milk substitutes have higher calcium contents than actual milk. Also, they are enriched with a number of oil-soluble vitamins, such as vitamins A, D, and E.

TABLE 1.2

Ingredients of Plant-based Milk Analogs

	Whole Milk	Skim Milk	Soy	Almond	Coconut	Oat
Calories/240 mL (kcal)	150	80	110	60	70	90
Percentage content (w/w)						
Total Fat	3.3	0.0	1.9	1.0	1.9	1.5
Total carbohydrate	5.0	5.4	3.8	3.3	2.1	5.4
Sugars	5.0	5.0	2.5	2.9	2.1	2.1
Dietary fiber	0.0	0.0	0.8	0.0	0.0	0.0
Protein	3.3	3.3	3.3	0.4	0.0	0.4
Calcium	0.125	0.129	0.188	0.188	0.192	0.192
Sodium	0.052	0.052	0.038	0.063	0.027	0.040
Cholesterol	0.015	0.002	0.0	0.0	0.0	0.0

1.1.2 ANIMAL VERSUS PLANT-BASED PROTEINS

Both plants and animals can provide dietary proteins. Animal protein is often regarded as being less environmentally sustainable, although having a bigger demand. To maintain the stability of environment, moral principles, and affordability of food; improve food safety; satisfy increased buyer demand; and combat protein-energy malnutrition, a gradual shift from animal to plant-based protein diet may be preferable. Since the previous 20 years, private businesses and scientific research teams have concentrated mostly on algae, earthworms or earthworm meal, insects, and other invertebrates as alternate sources of protein (Bessada et al., 2019). Food obtained through plant sources has a major role in the human diet as a vital source of bioactive peptides, vitamins, and phenolic compounds. These bioactive components aid humans against many pathogens (Karaś et al., 2017). As a substitute for animal proteins, the proteins generated from plant-based foods are being employed more and more in the human diet as a cost-effective and health-promoting alternative. Since it is also difficult and costly to extract a sufficient quantity of animal proteins, plant proteins are a good alternative for enhancing human nutrition.

A total of 80% of the protein consumed worldwide comes from plants, like cereal grains, soy, beans, nuts, pulses, fruits, and vegetables, while the remaining 20% comes from animals, such as eggs, dairy products, fish, meat, and cheese. Plant-based proteins are higher in fiber, oligosaccharides, polyunsaturated fatty acids, and carbohydrates than animal-based proteins (Guasch-Ferré et al., 2019). Increased urbanization and economic growth have caused a number of changes in the eating habits of people in low-income and middle-income countries, including the demand for animal-based diets, which was observed in developing nations. Typically, animal-based meals supply the daily requirement for protein. However, the acceptance of substitute sources of proteins for human consumption was prompted by changes in consumer demands. Therefore, there is a greater need to study new aspects of animal proteins, such as population increase, climate change, and the generation of economically and environmentally sustainable protein sources. This research should primarily focus on high-content plant proteins that are climatically robust and provide a balanced diet for people.

1.2 PROTEIN STRUCTURE

Proteins are macromolecules made up of linear polymers of amino acid residues connected by peptide bonds. They offer a variety of structural, nutritional, and functional properties which are valuable to the food industry for the development of food products (Arif & Pauls, 2018). Given the increased popularity of plant-based diets, it's critical to pick a plant protein that can match animal proteins while still being functional. Emerging plant protein sources can benefit from understanding the functional and structural characteristics of animal-based proteins, with a particular emphasis on the creation of plant-derived protein components and their application in prepared diets. Proteins are the most important functional element in a food system due to their structuring, emulsifying, texturizing, hydration, nutritional and foaming qualities (Wilding et al., 1984). One of the main issues in this area is how to alter or create plant-based proteins in configurations that provide functional as well as physicochemical qualities similar to those provided by animal based proteins. According to their structural characteristics, proteins are broadly categorized as globular (albumins, hemoglobin, globulins, phosphoproteins, glycoproteins, glutelins, nucleoproteins, and prolamins), fibrous (myosin, collagen, and keratin fibrin), or flexible (casein), based on their physico-chemical characteristics as the sequence and quantity of amino acid residues on the polymer chain. In contrast to other globular forms, which are typically water-insoluble, fibrous structures are typically water-soluble (Grossmann & Weiss, 2021; D. J. McClements & Grossmann, 2021).

1.2.1 GLOBULAR PROTEINS

Plant-based proteins are primarily composed of globular proteins, which are classified as albumins, globulins, prolamins, and glutelins (Grossmann & Weiss, 2021). These proteins are typically found as multimers that are covalently linked together. Albumin and globulins are primarily found in all pulses (>50%) (Boye et al., 2010) and in a few pseudocereals (amaranth and quinoa), whereas glutelins (wheat) and prolamins (barley, maize, wheat, and rye) account for 85% of the protein in the family of cereal (Veraverbeke & Delcour, 2002) and pseudocereal. The polypeptide chains that make up these globular proteins fold into a tightly packed shape as a result of hydrogen bonds, hydrophobic effects, electrostatic forces, disulfide bonds, and van der Waals forces. The amounts of proline and glutamines in each structure, as well as the amino acid sequences, are quite comparable across prolamins and glutelins; nevertheless, they are different in terms of intramolecular, inter-molecular structures, and molar mass (González-Pérez & Arellano, 2009).

Factors impacting globular protein's solubility are:

- Differences in molecular weight
- Flexibility of molecules
- Protein surface containing acidic, basic, and hydrophobic structures
- Association and dissociation of subunits (Lakemond et al., 2000; Molina et al., 2004)

Globulins can create more robust interfacial films, which is probably because of how the larger subunits interact and unfold. Foaming and emulsifying properties can change on the basis of which kind of globulin fraction is employed. In the course of food processing, each subunit of both 7S and 11S globulins go through unfolding, dissociation, and, then, reaggregation under the proper ionic strength and heating conditions, which increase their functionalities about emulsification, gelation, solubility, and foaming (Nishinari et al., 2018). For example, the following is how the functionality of 7S and 11S globulins differ: While soybean and pea 7S globulins are effective emulsifiers (M. Chen et al., 2019; Liang & Tang, 2013), soybean and pea 11S globulins are effective at stabilizing foam (Ruíz-Henestrosa et al., 2007). Greater contact at the interface with the larger subunits is the cause of this. In contrast, 7S globulins in fava beans exhibit lesser emulsification as compared to 11S globulins (Kimura et al., 2008). Low water solubility and higher molecular weight components of glutelins and prolamins at neutral pH have an impact on the proteins' capacity to foam and emulsify. The native prolamin found in wheat, however, is a powerful foaming agent, especially in the alkaline pH (Thewissen et al., 2011). Whey (lactoglobulins, lactalbumins) and albumins (egg white) are two ingredients of animal-based globular proteins utilized in the food business. These proteins are water soluble and have good gelling, emulsifying, and foaming capabilities (Dissanayake &

Vasiljevic, 2009; Patel & Kilara, 1990). Despite being mostly globular (soy, mung bean, potato, pea, and rice proteins), plant proteins differ from animal-based globular proteins because of molecular features and functionalities. However, a wide range of plant-based proteins denature only when they reach higher temperatures (e.g., soy glycinin denatures at around 90°C in contrast to 63–93°C which works for egg albumins) (D. J. McClements & Grossmann, 2021). This is despite the fact that pea and soy protein have emulsifying properties and solubility similar to the protein of egg. Because of this, achieving the same structural development and textural qualities as actual eggs frequently necessitates higher temperatures or longer heating durations. Additionally, isolated and extracted plant globular proteins are used in food applications. The functionality of the plant proteins may be harmed as a result of denaturation and aggregation.

1.2.2 Fibrous Proteins

The three main types of fibrous or meat proteins are sarcoplasmic, stromal (elastin and collagen), and myofibrillar (troponins, tropomyosin, actin, and myosin). Fibrous proteins have a complicated hierarchical structure of fibrous protein bundles contained in connective tissue made of collagen triple helices. One muscle fiber (or cell), which varies in size from 1 to 40 millimeters (mm) based on the species of animal and type of muscle, weighs between 20 and 100 millimeters (mm). Animal tissues that are composed of linked polypeptide chains are the source of collagen. Foods' textural properties are influenced by the arrangement of myosin and actin molecules in muscular tissues, which resemble fibers. The development and production of fake meat products have a long history, and a significant percentage of the studies have been on simulating the fibrous structure from plant-based proteins. Out of all, only wheat glutenin possesses viscoelastic and cohesive properties, allowing it to create proteinaceous fibrous networks that are commonly used in meat substitutes when comparing the fibrous qualities of animal to plant protein. Due to the nature of other plant-based proteins, making products that resemble meat is difficult. Extrusion that aids in replicating a fibrous structure to that of globular plant proteins has been used in the food business for a vast timeline to produce textured vegetable proteins (TVPs). Fibrous proteins that resemble flesh can be created by modifying or treating plant proteins.

1.2.3 Flexible Proteins

The architectures of filamentous proteins are flexible and disordered; casein, for example, contains a coil shape having regions of both hydrophobic as well as hydrophilic amino acids (Farrell Jr et al., 2002). When combined with calcium phosphate, casein produces giant colloidal particles known as casein micelles (Bhat et al., 2016; Broyard & Gaucheron, 2015). The high amounts of prolyl residues that result in open and flexible conformations and the random coil configurations with hydrophobic, hydrophilic, and phosphate groups in them give casein outstanding surface-active and stabilizing capabilities [29]. Although highly stable at a pH of 7, caseins can coagulate when exposed to an acid or an enzyme. The manufacture and quality of cheeses and yogurts are significantly influenced by the way that caseins join to make networks that are three-dimensional, giving them their distinctive texture characteristics. These molecular characteristics are very different from that of globular plant proteins. Due to the non-polar as well as polar surfaces of the caseins, they are effective emulsifiers. Additionally, because they contain a large number of anionic phosphate groups, calcium ions can bind them all together, which is essential for the formation of gel. Collagen, a superhelical protein composed of three parallel alpha chains, is the building block of flexible proteins like gelatin. It melts when heated, but when it is chilled, it makes helical sections that crosslink with each other using hydrogen bonds, giving it cold-set gelation qualities (L. Lin et al., 2017). Since the majority of naturally occurring plant proteins do not possess micellar or flexible random coil structures, simulating the properties of casein and gelatin has proven to be quite difficult. So a major issue in the research is how to assemble plant proteins into these superstructures. Although there are some polysaccharides that can mimic the characteristics of these flexible proteins, more focus may be placed on engineering these structures into plant-based proteins. It might take the form of adding random coils to globular plant protein structures or putting different plant-based proteins together that resemble casein micellar structures.

1.3 FUNCTIONAL PROFILING OF PLANT PROTEINS

1.3.1 WATER- AND OIL-HOLDING PROPERTIES

Proteins' respective WHCs and OHCs, which stand for "water and oil holding capacities," quantify how much liquid they hold per unit of mass. These characteristics are necessary for various food applications, including the syneresis of plant-based yogurts and the characteristic juiciness of plant-based meats and cookability. The usual procedures for determining these characteristics are those given by Beuchat (1977) and M. J.-Y. Lin et al. (1974) which entail suspending a known mass (g/g) of protein in vegetable oil distilled water or distilled water and hard shaking. After that, the formed slur mixture is centrifuged to eliminate any extra oil or water. Calculating the amount of oil or water the protein can hold (reported as g water/oil per g protein) involves comparing the mass of the sample before centrifugation and after. It should be noted that this strategy is not always applied uniformly. Comparing the outcomes of various research can be challenging since different protein concentrations, mixing durations, and centrifugation settings are employed. For instance, the movement of the liquid into the powder and the dispersion of the protein powder in the surrounding liquid both require time; therefore, the outcomes are dependent on the amount of mixing time. Additionally, mixing temperature and the pH of the protein solution have an effect on how well the proteins retain water. For instance, soy protein retains water more readily between pH 6 and 8 and between 40°C and 70°C (C. Wang & Zayas, 1991). Even while results for water retention showed no difference between incubation times of 10 and 30 min, a longer incubation time might have an impact. The speed of centrifugation and timeframes utilized in the analysis vary from study to study, from 1,600 to 16,000 g and from 10 to 30 min, respectively. The water-holding capacity has been shown to enhance with centrifugation speed and time; hence, it would be good to standardize these settings. Some scientists reversed the centrifuge tubes after decanting the supernatant to drain any extra water or oil; however, this practice resulted in some sample losses that affected the findings. For soluble proteins, the WHC can be calculated after the protein mixture has transformed into a gel state, such as by heating, the addition of cross-linking agents, or cooling.

The OHC and WHC can be determined after a protein element is added to a food item – for example, a vegan equivalent of meat, cheese, fish, or eggs. WHC and OHC values of plant protein isolate typically rise as protein content does. The magnitude of the rise, nevertheless, often varies between the WHC and OHC readings. For example, with increasing pea protein concentration, the increase in WHC is more significant than the increase in OHC. The types of plant proteins (soy, chickpea, pea, and lentil protein) also affect the WHC values; this may be because the hydrophobicity of their surfaces varies. Legume protein isolate's WHC and OHC vary from 1.8 to 6.8 g/g and from 3.5 to 6.8 g/g, respectively (Gundogan & Karaca, 2020; Keskin et al., 2022; Lafarga et al., 2020). In addition, the effects of heat treatment and high-pressure processing on the structure and functionality of pulse (lentil, pea, and faba bean) protein were assessed by Hall & Moraru (2021). In comparison to untreated controls, both treatments increased the samples' ability to store water. The potential of pea protein isolates to boost the water-holding capacity in a dose-dependent manner has led to their use as meat extenders in chicken nuggets (Shoaib et al., 2018). When pea protein isolate was introduced, the overall product cook loss was decreased, going from 12.4% to 5.0%. The plant proteins in the product have the effect of retaining more water and oil. Adding more than 3% pea protein isolate reduced the overall moisture level of the chicken nuggets even though the cooking loss was reduced; this could affect the nuggets' desired sensory qualities. Due to their superior water-holding capacity, plant protein concentrates and isolates have also been employed as texturized vegetable proteins (TVPs) in meat substitutes. The WHC affects the extruded TVPs' porosity and air cell size (Samard & Ryu, 2019). In the past, soy protein isolates were used to make TVPs, but these days, pea, mung bean, and peanut proteins are also used in the production of TVPs. High (55%) or low (26–35%) moisture extrusion can be used to make pea-based TVPs (Schreuders et al., 2019b). In comparison to mung bean, peanut, and gluten TVP, pea protein TVP has been found to have a better capacity to hold water and a higher capacity to hold oil (Samard, Ryu, et al., 2019). The cooking loss and shrinkage of chicken sausage analogs have been reported to be reduced by a plant protein-based formulation including a blend of SPI, gluten, and chickpea flour. These findings imply that plant protein combinations may be employed to enhance the water- or oil-holding capacity of plant-based meat substitutes.

1.3.2 GELLING PROPERTIES

Plant proteins' ability to gel is crucial for food applications that call for a semi-solid structure, such as products made from plant-based versions of dairy, meat, fish, and eggs. The least gelation concentration (LGC), which is the protein concentration at which the protein solution forms a gel that does not slip from a test tube when it is inverted, is the basis for the most popular technique for evaluating the gelling capabilities of proteins (Sathe et al., 1982). Plant protein solutions are available with protein of about 2–20%. To encourage the aggregation thermal denaturation and thermal denaturation of the proteins, they are heated at roughly 100°C for a set period (e.g., 60 min). Following heating, the sample is given a set amount of time to cool before the tubes are turned upside down for visual inspection. The LGC is the lowest protein concentration found in the inverted tube with the protein sample. The ability of plant proteins to create gels is usefully revealed by this method, but it does not reveal anything about the characteristics of the gels that are generated, such as their hardness or brittleness. As a result, numerous researchers employ additional techniques to evaluate the gels' textural attributes. Compression tests, in which stress–strain correlations are recorded as a sample is compressed/decompressed at a set rate, are the most used technique for assessing the textural characteristics of gels made from plant proteins (Shand et al., 2007). For instance texture profile analysis (TPA) can assess the firmness, adhesion, springiness, cohesion, gumminess, and resilience of gels. It has been noted that using this approach, lupine protein gels exhibited higher hardness than gels made from pea or faba bean proteins (K. K. Ma, 2020). Gel characteristics can also be described using dynamic shear rheology data, notably as a function of temperature. Using dynamic oscillatory measurements, Langton and colleagues examined the gelation of faba bean protein mixtures at pH 5 and 7 as a function of temperature (25–95°C) (Langton et al., 2020). In comparison to pH 7 gels, they found that pH 5 gels had a higher storage modulus (G') at a lower temperature. Other studies demonstrated that gels made from kidney bean protein were stronger and more thermally stable than gels made from pea protein (Shevkani et al., 2015). These techniques can also establish the gelation temperature and whether a gel is thermally reversible or irreversible.

The type of plant proteins affects how well they gel. Most plant proteins have a concentration range between 10% and 18% where they can gel the least; however, some can gel at much lower levels. For instance the LGC value of chickpea proteins is between 5% and 7%. It should be noted that the quoted LGC values rely on a variety of factors, including protein type, the presence of other ingredients, and gelation parameters including pH, ionic strength, and heating conditions. This highlights the significance of standardizing settings when comparing various protein sources and how the same protein may have varied LGC values, depending on the conditions employed. To enhance the textural characteristics of meat products, plant-based proteins are frequently added for gelling. For example, it has been claimed that adding lentil and chickpea flour to beef burgers increased their toughness (Motamedi et al., 2015). Similarly to this, making sausages using chickpea protein concentrate enhanced the strength of gel in the finished product (Ghribi et al., 2018). In a different investigation, it was discovered that adding 20% or 60% chicken meat to soy-based sausage did not affect their gel strength or other textural qualities, including cohesiveness, chewiness, stiffness, adhesiveness, and gumminess (Kamani et al., 2019). In contrast, the hybrid sausages had larger water content than the chicken-meat-free variant of the sausage, which may have contributed to their lower gel strength. Therefore, there is a lot of room for improvement in the use of plant proteins in the production of hybrid meat products and meat-free goods. Authors have analyzed how the textural characteristics of beef patties are affected by soy, pea, lentil, and bean proteins used as meat extenders (Žugčić et al., 2018). The maximum hardness, gumminess, and chewiness were seen in the beef patties incorporating soy protein. Due to their reduced protein content (55–60%) compared to the soy protein ingredient utilized (90%), the beef patties using pulse proteins may have had less desirable texture characteristics. Yogurt and tofu substitutes have been made with the use of faba bean flour (Jiang et al., 2020). A tofu imitation was produced by removing the starch from faba bean flour, and it had a firmer texture and a higher capacity to hold water. This was mostly due to the higher protein content.

1.3.3 Protein Solubility

Protein solubility influences the transport of plant proteins to the oil–water or air–water interface, which regulates their emulsification and foaming capabilities (Johnston et al., 2015). An approach used to assess the solubility of proteins involves mixing a measured amount of protein powder with a buffer solution, then adjusting the pH by mixing in 0.1 M NaOH or HCl (Morr et al., 1985). The resulting solution is then centrifuged, and kept under controlled circumstances for a predetermined amount of time, and the supernatant is gathered to determine its protein concentration. Researchers have applied this approach in a few distinct ways. For instance, the protein is dissolved in the aqueous solution either before or after the pH is adjusted to the desired result. Furthermore, the amount of time needed for the protein to disperse and dissolve in the aqueous solution ranges from 30 min to overnight. Additionally, the incubation temperature and stirring parameters may change from study to study – for example, from room temperature to refrigeration.

For instance, it has been found that several plant proteins are more soluble when incubated at 50°C as opposed to 25°C. On the other hand, Chao and Aluko (2018) examined the impact of laboratory heat treatment (50–100°C) on the solubility of pea protein isolate protein. They concluded that the protein values of the heated and unheated samples were comparable, indicating that the heat treatment did not result in severe aggregation that would have reduced water contact. The significance of standardizing test settings is once again highlighted by these findings. For instance, it was noted that one group's study (Karaca et al., 2011) found that the protein solubility of soybean, faba bean, and pea protein isolates produced by IEP was higher at pH 7 than that of another group's study (Fernández-Quintela et al., 1997). This discrepancy may be due to the fact that the proteins in the solutions were incubated for only 30 min in the earlier investigation versus overnight in the later study. The Kjeldahl, Dumas, Bradford assay and Lowry procedures are among the analytical techniques that have been employed to measure the protein concentration in the supernatant throughout the research. As a result, it would be beneficial to do protein solubility analyses under uniform circumstances so that findings from various research could be directly compared (Khan et al., 2021) Plant proteins have the lowest water solubility (20%) in the pH range of 4 to 6 because their isoelectric points are in this range. Since the protein molecules have a low electrostatic attraction toward one another as a result, they can easily form van der Waals, hydrophobic, or hydrogen bonds with one another. In contrast, plant proteins are typically more soluble when the pH is lower than when they reach their isoelectric point because this lowers their charge and increases the electrostatic attraction between them. For instance, the water solubility of soy, chickpea, faba bean, pea, and lentil proteins is relatively high (>80%) at pH 8 and fairly high (40–60%) at pH 3. To maximize protein solubility, it is therefore advised to employ pH levels of 8 or above, but this is not always feasible. Meat products like sausages and hamburgers often have pH levels between 5 and 7 based on the used type of meat, which is near the isoelectric points of the plant proteins. For instance, the pH of chorizo sausage that contained

3% plant proteins as meat extenders (soy, bean, lentil, or broad bean proteins) was about 5.8, which is close to the isoelectric point of these proteins (Thirumdas et al., 2018). Using only plant proteins, a meat analog was created in a different study where the pH was approximately 7, making the plant proteins more soluble (Kamani et al., 2019). It should be highlighted that both soluble and insoluble proteins may be advantageous to attain the desired texture and other quality qualities in plant-based meat analogs.

1.3.4 EMULSIFYING PROPERTIES

Proteins' capacity to create and maintain emulsions is typically used to describe their emulsifying capabilities (Yasumatsu et al., 1972). The size, flexibility, shape, charge, hydrophobicity, and aggregation state of proteins are some of the many variables that determine their ability to emulsify. As a result, methodologies for describing their emulsifying properties are crucial. For assessing the emulsifying abilities of proteins, a number of techniques have been suggested. The emulsion activity index (EAI) and the emulsion activity (EA) are two of the most widely used methods (Pearce & Kinsella, 1978; Yasumatsu et al., 1972). Despite the fact that these techniques have significant drawbacks, they are used because of their ability to be carried out using basic equipment that is frequently found in laboratories. In these procedures, an oil-in-water emulsion is made by weighing out a protein dose, adding it to a buffer solution, and mixing it with a predetermined quantity of vegetable oil in a high-shear mixer. However, there are also significant disparities among them. The EA method entails generating an oil-in-water emulsion, centrifuging it under controlled circumstances (duration and speed), and then measuring the volumes of the top "emulsion" layer (VE) and the entire sample (VT) (Yasumatsu et al., 1972). Then, the emulsion activity is computed using the formula EA = 100 × VE/VT. According to the EAI approach, a predetermined quantity of protein is added to the aqueous phase while creating a 25% oil-in-water emulsion under regulated blending parameters. Following the addition of a determined amount of surfactant solution (0.1% SDS) to the emulsion to dissolve any flocs, the turbidity of the diluted solution is measured at 550 nm. From the turbidity and droplet concentration, an equation is then employed to get the EAI value (Eq. 1.1). The emulsion's turbidity and droplet size have a direct correlation, which is the basis for this approach. It has been noted (D. J. McClements, 2007) that both procedures are significantly influenced by the type of blender and blending parameters employed in the test, as this results in emulsions with varying droplet sizes, making it challenging to compare results between experiments. This is due to the fact that both the oil–water interfacial area and the oil concentration affect how much emulsifier is needed to stabilize the emulsion. The difference in homogenizing rotational speed can have an impact on the size of the emulsion's particles since higher rotational speeds produce stronger shear forces, which in turn reduce the size of the droplets (Cui et al., 2014). However, these techniques are helpful for contrasting the effectiveness of various protein emulsifiers under comparable experimental circumstances.

$$\text{EAI}\left(m^2/g\right) = \frac{2 \times 2.303 \times A0 \times DF}{c \times \varphi \times \left(1-\theta\right) \times 1000},$$

$$\text{ESI}\left(\min\right) = \frac{A0}{A0-A10} \times 10 \qquad \text{Eq. (1.1)}$$

Equation (1.1) depicts the equations for the emulsifying activity index (EAI) and the emulsifying stability index (ESI) (where DF is the dilution factor, c is the initial protein content (g/mL), φ shows the optical path, θ indicates the oil fraction utilized to make the emulsion, and A0 and A10 demonstrate the absorbance of diluted emulsions at 0 and 10 min, respectively.)

Another straightforward test for emulsion stability has been suggested (Yasumatsu et al., 1972). To speed up the breakdown of an emulsion, it is first incubated at 80°C for 30 min. After that, it is centrifuged, and the volume of the cream layer is then calculated. Next, the emulsion stability is computed using the emulsion layer's volume at the tube's top after centrifugation:

$$100 \times V_{E,H}/V_{E,I} \qquad \text{Eq. (1.2)}$$

where $V_{E,H}$ and $V_{E,I}$ are the emulsion layer volumes with and without heat treatment, respectively.

Another proposed emulsion stability test (Pearce et al., 1978) involves heating an aliquot of emulsion to 80°C for 30 min, diluting it with a 0.1% SDS solution, and measuring the turbidity at 550 nanometers. This technique is based on changes in the emulsion's droplet size upon heating, which results in variations in turbidity. Then, using the formulae provided in Equation (1.1), the emulsion activity index and emulsion stability index can be determined. The aforementioned techniques are extremely simplistic despite being frequently utilized, and more sophisticated techniques have taken their place (D. J. McClements, 2007). For instance, evaluating the mean particle diameter versus protein content when subjected to regulated homogenization conditions allows researchers to determine if a protein can form emulsions. The minimal achievable droplet diameter ($d$min) and the minimum protein concentration (Cmin) needed to generate small droplets may then be determined. Additionally, it is occasionally feasible to figure out the surface load of the emulsifier (mg/m^2), which establishes the quantity of emulsifier necessary to create an emulsion with a particular droplet concentration and size. When samples are incubated under controlled circumstances, such as ionic strength (0–500 mM NaCl), pH (2–8), and temperature (30–90°C), for a predetermined amount of time, their particle size distribution, creaming stability, and microstructure are measured (Gumus et al., 2017). Researchers can also conduct zeta-potential, interfacial tension, surface hydrophobicity, and rheology studies to learn more about how well plant protein emulsifiers work. It should be mentioned that direct comparison of the functional performance of plant proteins is challenging due to the employment of different techniques and operating settings to test the emulsifying capabilities of plant proteins by different studies.

Table 1.3 lists three of the more conventional techniques (Pearce et al., 1978; Yasumatsu et al., 1972) that were used to determine the emulsifying qualities of plant proteins. According

TABLE 1.3
Plant Proteins' Ability to Emulsify Documented in Published Studies Utilizing a Variety of Techniques

Protein Isolate Type	Reference	Protein Content (%)	Emulsifying Activity	Emulsifying Stability (%)
Mung Bean	Branch & Maria, 2017	81.5	63.2	62.8
Soybean	Branch & Maria, 2017	86.0	74.5	81.2
Pea	Butt & Rizwana, 2010	83.6	21.0	43.2
Mung bean	Butt & Rizwana, 2010	85.5	41.1	45.5
Green mung bean	X. Tang et al., 2021	84.6	62.0	53.0
Pigeon pea	X. Tang et al., 2021	86.9	73.0	71.0
Yellow lentil	Morr et al., 1985	87.8	72.0	64.0
Soy	Morr et al., 1985	92.4	71.0	70.0
Commercial soy	Morr et al., 1985	88.6	54.0	49.0
White lentil	Morr et al., 1985	91.2	68.0	67.0
Chickpea	Morr et al., 1985	89.1	66.0	53.0
Cowpea	Morr et al., 1985	91.0	69.0	61.0
Pea	Morr et al., 1985	89.2	76.0	62.0
Yellow mung bean	X. Tang et al., 2021	90.0	62.0	53.0
Grass pea	Feyzi et al., 2018	87.5	87.5	29.8

to these findings, the source and concentration of plant proteins affect how well they can emulsify. Despite using the identical EA approach across investigations, there is a significant difference in emulsifying activity (Pearce et al., 1978). Even within the same plant protein, there are variations: emulsifying activity in pea protein isolates ranges from 21% to 76%, depending on the study (Fernández-Quintela et al., 1997; Lafarga et al., 2020). Since soybean protein isolate primarily has a high protein content and EA (Karaca et al., 2011; Lafarga et al., 2020), it can be the best choice for applications where emulsification is crucial. But there isn't much of a difference between soy and other plant-based proteins. For instance, cowpea protein isolate and white lentil protein isolate both contain significant EA, at 68% and 69%, respectively. Fatty meat products like patties and frankfurters can be emulsified and bound using surface active plant proteins. One example is the claim that beef sausage's emulsion stability was improved when lupin flour was added (Leonard et al., 2019). As the amount of lupin flour used increased, less fluid and fat were released from the sausages, increasing the cooking yield. The egg yolk in salad dressings has also been replaced by pulse proteins as an emulsifier (Z. Ma et al., 2016). The scientists demonstrated that salad dressings made from pea, lentil, and chickpea protein isolates might have physical characteristics that were comparable to those of conventional egg-based salad dressings.

1.3.5 FOAMING PROPERTIES

By adhering to the air–water interface and forming a barrier around the air bubbles, plant-based proteins can be employed to stabilize foams as well. This is significant when it comes to food applications that call for a creamy or fluffy texture—for instance, ice cream, cakes, and cream (whipped). By testing a protein's foaming stability and capacity, foaming qualities can be identified. The capacity of foaming measures the amount of foam that can be produced by vigorously mixing a protein solution, whereas foam stability measures how long it takes a protein to stabilize the foam before it collapses. The typical way to make foams is to blend or homogenize a protein solution. When the foam is produced, its volume at the start is measured by pouring it right away into a graduated cylinder, which enables the foaming capacity to be calculated. In order to evaluate the foam stability, it is also noted how the foam's volume changes over time. It is challenging to directly compare different researches because of the wide variability in mixing times and speeds that have been reported in various investigations. For instance, mixing for a longer period of time or at a faster rate can produce more foam, which can change how the foaming capacity and stability are calculated.

$FC = 100 \times (V_{2-1})/V_1$, where V_1 and V_2 are the volumes of the protein solution before and after whipping, respectively, can be used to calculate the foaming capacity. Foaming stability can be determined using the formula $FS = 100 \times V_t/V_0$, where V_t is the foam's volume at time t (often considered to be 30 min post whipping) and V_0 is the foam's baseline volume (immediately after whipping). Table 1.4 reports on the foaming characteristics of a few plant proteins. A generous range of foaming abilities has been observed in various investigations, which can be related to various foam-creation blending techniques. Green lentils, kidney beans, and pea proteins all have exhibited foaming stability levels higher than 90%, with soy having the highest level. Indicating that a larger concentration of protein promotes foam stability, the higher foaming stability values were seen for proteins with comparably high protein contents (>90%). The ratio of the volume of whipped protein solution to its weight, or specific volume (mL/g), has been utilized in several experiments to gauge foaming capabilities (Gupta et al., 2018).

1.4 BIOACTIVE PROPERTIES OF PLANT PROTEINS

Secondary metabolites are plant-derived bioactive components produced from metabolism and have promising therapeutic capabilities, particularly antioxidant properties characteristics. Carotenoids and phenolics are regarded as the principal bioactive or phytochemical substances that can aid to improve human health (Singh et al., 2015). Carotenoids, which are lipophilic chemicals, are

TABLE 1.4

Published Foaming Properties of Plant Proteins Using Various Methods

	Protein Type	Reference	Protein Content (%)	Foaming Capacity (%)	Foaming Stability (%)
Concentrates	Faba bean	Fernández-Quintela et al., 1997	81.2	15.0	77.0
	Soy bean	Fernández-Quintela et al., 1997	82.20	22	93
	Pea	Fernández-Quintela et al., 1997	84.90	15.0	94.0
	Green Lentil	Aydemir & Yemenicioğlu, 2013	87.00–95.00	34.8	96.7
	Mung bean	Branch & Maria, 2017	81.53	89.7	78.3
	Soybean	Branch & Maria, 2017	86.00	68.7	100.0
	Pea	Butt & Rizwana, 2010	83.60	78	N/A
	Mung bean	Butt & Rizwana, 2010	85.46	110	N/A
Flour	Chickpea	Aydemir et al., 2013	71.00–77.00	43.9	64.8
	Soybean	Aydemir et al., 2013	70.00	32.0	43.7
	Chickpea	Kaur & Singh, 2007	89.90–94.40	30.4–44.3	N/A
Isolates	Soybean	Aydemir et al., 2013	92.0	36.0	88.0
	Bombay bean	Gundogan & Karaca, 2020	N/A	83	75
	Akkus bean	Gundogan & Karaca, 2020	N/A	91	72
	Hinis bean	Gundogan & Karaca, 2020	N/A	72	80
	Sumav bean	Gundogan & Karaca, 2020	N/A	81	71
	Gembos bean	Gundogan & Karaca, 2020	N/A	76	82
	Pea	Stone et al., 2015	80.60–89.00	81.1	27.1
	Grass pea	Feyzi et al., 2018	92.5	41	100
	Faba bean	Singhal et al., 2016	92.14–99.36	143.3–183.3	55.9–71.59
	Pea	Shevkani et al., 2015	92.8	87.0–132.0	94.0–96.0
	Kidney bean	C.-H. Tang et al., 2009	92.5	244.9	87.8

mostly found in abundance in orange and yellow fruits and vegetables (Ahmed et al., 2014). These substances are highly helpful for the food industry in creating colors and dietary agents that are good for your health. Additionally, carotenoids are gaining popularity due to their significant anti-oxidative activity, which can assist to minimize the risk of certain chronic diseases (Gürbüz et al., 2018). The majority of polyphenols, which are natural antioxidants, come from food and medicinal plants such as vegetables, cereals, fruits, medicinal herbs, spices, drinks, and mushrooms. There are various classes of polyphenols, including phenolic acids, anthocyanins, and flavonoids. Many biological characteristics, including anti-cancerous, anti-inflammatory, and anti-aging activities, have been linked to natural antioxidants, particularly carotenoids, and polyphenols (Xu et al., 2017). Fruits including blueberries, apples, plums, kiwis, cherries, etc. are high in hydroxycinnamic acids (HCAs), which have values of 0.5–2 g/kg fresh weight. One of the most prevalent phenolic acids is caffeic acid, and in some fruits, it makes up between 75% and 100% of all HCAs (Manach et al., 2004). However, in cereal grains, ferulic acid is the most prevalent phenolic acid and accounts for roughly 90% of the total polyphenol content of wheat grain (Xu et al., 2017). Anthocyanins are commonly employed as dyes because of their vibrant hues, but they can also be helpful in controlling diabetes, inflammation, cancer, cardiovascular disease, and a number of other human conditions. Anthocyanins have considerable antioxidative potential, which mostly accounts for their health-promoting qualities (Casati et al., 2016; Khoo et al., 2017). Consuming fruits and vegetables is advised since they are a good source of natural antioxidants including vitamin E (tocopherols) and vitamin C (ascorbic acid). Antioxidants have various health-promoting properties, such as regulating immunological function, lowering DNA damage, and improving lipid peroxidation. A naturally occurring flavanone called naringenin has been shown to have the ability to regulate cognitive functioning and maintain insulin signaling in the brain. It is primarily found in grapes and citrus fruits (Ghofrani et al., 2015). Organosulfur compounds are yet another form of plant-derived bioactive (OSCs). Onion (Allium) has high levels of OSCs, and frequent consumption of Allium has been linked to the prevention of a variety of chronic diseases, including diabetes, cardiovascular disease, obesity, and metabolic disorders (Moreno-Ortega et al., 2021).

1.4.1 Cardiovascular and Metabolic Disorders

Numerous studies have suggested that dietary proteins derived from plants may have protective effects against the factors of cardio-metabolic risk. In 2017, the first study on the production and consumption of plant-based proteins as a replacement for animal based proteins was conducted. In this research, the researchers examined and presented the cardiovascular disease biomarkers derived from dietary plant-based proteins. They also looked into blood lipid levels and discovered a drop in apolipoprotein B, non-high-density lipoprotein, and low-density lipoprotein cholesterol. Additionally, the researchers carried out randomized experiments that demonstrated plant protein's efficiency in lowering the risk factors linked to adult cardiovascular illnesses. Another study indicated that plant proteins, primarily soy products, had a greater impact on hypercholesterolemic patients in terms of decreasing their lipid profiles than animal proteins (H. Zhao et al., 2020). Most of the advantages of plant-based proteins and metabolic health issues have been examined in populations in the adolescent period. In light of the fact that metabolic syndrome, obesity, and weight management are major and increasingly prevalent health problems among teenagers worldwide, several studies have taken place to explore the advantages of consuming plant-based proteins. Nevertheless, the management of protein intake is essential to many physiological processes and development. Hence, the use of plant-based proteins as substitute to animal-based proteins in adolescent diets aids in the management of cardio-metabolic variables and obesity (Y. Lin et al., 2015). A number of research studies have concluded that increasing the amount of plant-based proteins in the diet of people will lower their chance of developing cardiovascular disease and other risk factors (Campbell, 2019). Additionally, it was discovered that consuming plant-based proteins rather than animal proteins reduces blood pressure in hypertensive individuals, especially senior people (Tielemans et al., 2014). Most studies also found a connection between mortality and dietary intake of plant protein sources. Researchers also noted the impact of dietary protein preference on mortality in a cohort study from the NIH-AARP Diet and Health Study. From 1996 through December 2011, more than six lakh Americans in the 50- to 71-year-old age range were monitored for this study. The death rate from cardiovascular disease, stroke, and both male and female mortality rates have been found to be negatively correlated with dietary intake of plant proteins. In both men and women, they saw 10% reduction in the risk of overall mortality when only 3% animal-based protein was substituted with plant-based protein (Huang et al., 2020). As a result, in terms of longevity and mortality, replacing animal proteins with plant proteins in the diet is advantageous. A newly published evaluation of 32 cohort studies concluded that a diet containing plant-based protein reduces the risk of cardiovascular-related and all-cause mortality. Using plant protein in place of animal protein extends life (Naghshi et al., 2020)

1.4.2 Diabetes

It is yet unclear if switching from animal to plant-based proteins lowers the risk of diabetes in the general population, despite the fact that plant-based diets are mostly connected with lowering diabetes risk (Tonstad et al., 2013). After researching and analyzing the Nurses' Health Study II dataset study, it was found that a 5% replacement of vegetable protein for animal protein was associated with a 23% reduced risk of type 2 diabetes (V. S. Malik et al., 2016). In a meta-analysis, animal-based protein sources were substituted with plant-based protein for 35% of dietary protein intake in eight-week randomized controlled trials. According to the findings of this study, patients with diabetes—those who have both type 1 and 2 diabetes—have had considerable improvements in their fasting glucose, fasting insulin, and HbA1c levels (Viguiliouk et al., 2015). In a cohort research, participants were given a protein-rich diet, and it was discovered that increased protein consumption is linked to a decreased chance of developing diabetes and pre-diabetes occurrences, with plant-based proteins being the key driver (Sluik et al., 2019). Additionally, compared to processed animal products, a diet containing plant-based protein has a number of bioactive elements that have positive impacts on health. Another randomized crossover trial found that replacing red meat with legumes (peas, chickpeas, beans, and lentils) significantly reduced fasting blood glucose, insulin, and triglyceride levels in people with type 2 diabetes, pointing to the potential benefit of plant-based proteins over animal-based proteins (Hosseinpour-Niazi et al., 2015).

1.4.3 CANCER

A wide range of elements including environmental, genetic, nutritional, and other ingrained characteristics influences the development of cancer. With the use of gene–environment interaction analysis and other factors, including cancer risk factors, lifestyle, and genetics, a study analyzed and examined the risk factor of colorectal cancer in individuals. The researchers examined the relationship between colorectal cancer and the genetic variety of fatty acid metabolism, which is mostly linked to a larger intake of meat, and came to the conclusion that those who eat a heavy meat diet are at an increased risk of developing colorectal cancer (Andersen et al., 2019). Consequently, replacing animal protein with plant-based protein is a superior method of lowering the risk of colon cancer in people having specific gene variants.

1.4.4 RENOPROTECTIVE EFFECT

Trials have been conducted on the diet to determine the distinctions that aid in the treatment of chronic diseases, particularly chronic kidney disease (CKD), and it has low amounts of vegetables, fruits, healthy oils, and dairy foods. It is higher in total grains, total protein foods, saturated fats, added sugar, and sodium (Smyth et al., 2016). According to recent research, the origin of the protein—for instance, whether it is plant or animal—might be a significant factor influencing how well the kidneys work, in addition to the amount of protein consumed (Bernier-Jean et al., 2021). A significant 23% decreased rate of mortality was noted for people with chronic renal disease who consumed plant-based proteins (X. Chen et al., 2016). The plant-based proteins, which are primarily derived from rice endosperm and soybean, have also demonstrated kidney protective activity in diabetic rat models (Kubota et al., 2016). Additionally, other elements, including phytochemicals and fiber, also significantly contributed to kidney protection through the consumption of whole foods from plant-based diets and other parts of plants. Therefore, adding high-quality plant proteins is advised for their renoprotective properties.

1.5 PLANT PROTEIN EXTRACTION

Growth of the market was aided by developments in technologies for protein synthesis, such as upstream cultivation optimization (e.g., nutritional, physical parameters, and bioreactor design), expression hosts engineering, and the creation of protein extraction techniques (Pojić et al., 2018; Schillberg et al., 2019). The extracted protein yield and its nutritive and functional qualities can be enhanced with the use of protein extraction technology. Consequently, it is important to choose an appropriate sort of protein extraction technique.

1.5.1 DRY PROTEIN EXTRACTION

To create fiber- or protein-rich fractions, air classification and/or sieving techniques have been widely used. These approaches are primarily a part of innovative techniques for the dry extraction

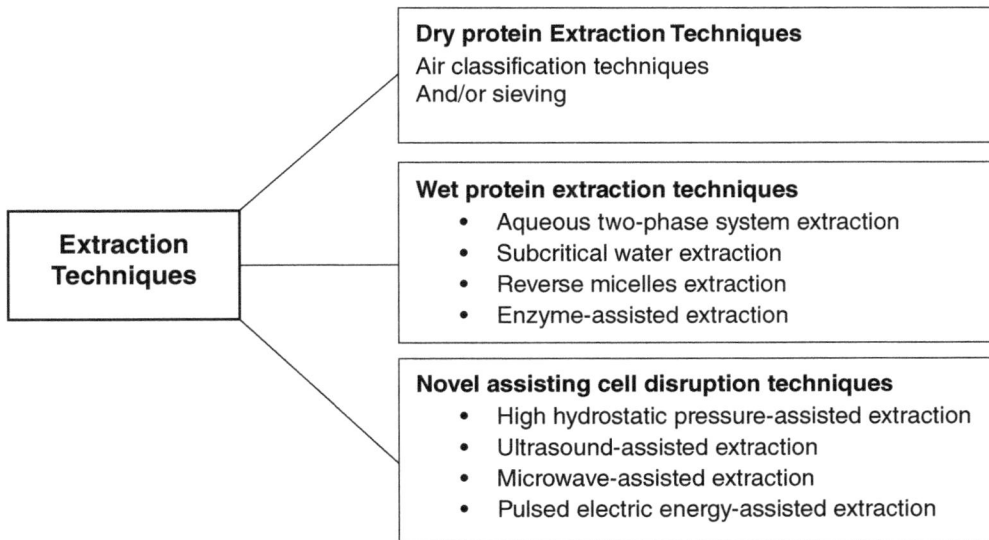

FIGURE 1.2 Plant protein extraction techniques.

of protein. Despite producing a high yield of protein, it required more energy as compared to wet techniques of protein extraction. Additionally, the presence of impurities and particle agglomeration is a drawback of these methods (Pojić et al., 2018).

1.5.2   WET PROTEIN EXTRACTION

In wet protein extraction methods, the protein first solubilizes in a medium with a pH that is far from the isoelectric point, and it subsequently precipitates in a medium with a pH that is close to the isoelectric point. There are several methods for extracting proteins using acidic and alkaline solutions (Pojić et al., 2018).

1.5.2.1   Enzyme-Assisted Extraction

This approach was found on the idea of disrupting cell walls using certain enzymes that break down pectin, cellulose, and/or hemicelluloses, as well as proteases that aid in the hydrolysis of protein to increase solubility. The release of protein bodies is made possible by the breakdown of cell walls. This procedure demands a longer processing time, expensive equipment, increased energy consumption, and acceptable environmental conditions like pH and temperature. The majority of the time, however, this strategy produces superior items for human use with a reduced negative influence on the environment (Lu et al., 2016; Pojić et al., 2018).

1.5.2.2   Subcritical Water Extraction

In order to keep the water in a liquid form, this approach uses hot water that has a temperature range of 100–374°C and high pressure. By adopting this technique, biomaterials including proteins and carbohydrates have been hydrolyzed without the use of any additional catalysts. For instance, using this technique greatly boosted the yield of soy protein extraction by 59.3% when soy meals were heat denatured (Lu et al., 2016).

1.5.2.3   Reverse Micelles Extraction

This technique uses reverse micelle, a nanometer-sized surfactant molecule aggregate that often comprises inner water molecule cores inside non-polar liquids. Reverse micelles include polar water molecules that aid in the solubilization of proteins and other hydrophilic macromolecules. Reverse micelles were created to create a three-phase system known as a water–surfactant–organic solvent system, which prevents the protein denaturation using organic solvents inside pools of polar water (Pojić et al., 2018).

1.5.2.4   Aqueous Two-Phase Systems' Extraction

This extraction process is created by mixing two polymers, such as two salts or one salt and one polymer, at a specific temperature. This technique has been regarded as the most environmentally benign protein extraction technique. It extracts proteins using an ionic liquid aqueous two-phase system, with a 99.6% yield (Pojić et al., 2018).

1.5.3   NOVEL-ASSISTING CELL DISRUPTION TECHNIQUES

Cell disruption is the first step in both dry and wet protein extraction procedures, and it aids in the release of protein from the protein-containing bodies. Disruption of cells used to be achieved mechanically (milling, grinding, etc.), chemically, or thermally.

1.5.3.1   Microwave-Assisted Extraction

In order to disrupt hydrogen bonds, facilitate the migration of dissolved ions, and increase the porosity of the matrix (biological)—all of that contributing to the protein's extraction—this method makes use of electromagnetic radiation with a frequency ranging from 300 MHz to 300 GHz. For instance, one study reported using this method for rice bran protein extraction (Phongthai et al., 2016).

1.5.3.2   Ultrasound-Assisted Extraction

This approach makes use of sound waves that produce cavitation at a frequency of 20 kHz, increasing the porosity of the matrix and enhancing solvent permeation. This process benefits from efficient lower extraction temperatures, selective extraction, quick energy transfer, mixing and thermal gradients, quick response times, fewer equipment requirements, and higher output. However, reports of protein denaturation and a disturbance of its functional characteristics exist (Pojić et al., 2018).

1.5.3.3   Pulsed Electric Energy-Assisted Extraction

There are now numerous ways for extracting proteins using pulsed electric energy. In order to induce structural changes in the target compound, this technique employs electric pulses with short durations (ranging from a few nanoseconds to a few milliseconds) and high pulse amplitudes (ranging from 100–300 V/cm to 10–50 kV/cm). The food industry has made extensive use of a variety of PEE techniques, including pulsed high-voltage electrical discharges (HVED), pulsed electric fields (PEF), and ohmic heating (POH) (Pojić et al., 2018).

1.5.3.4   High Hydrostatic Pressure-Assisted Extraction

In the food business, large-scale microbial cell disruption, emulsification, and meat tenderization are the main uses of high hydrostatic pressure-assisted protein extraction. Instead of proteins, this approach only works with bioactive substances. However, the cell wall swelling, increase in the dynamic viscosity, and particle size have resulted in a reduction in the efficiency of extraction and separation yield with the application of multiple cycles of HHP (Pojić et al., 2018).

1.6   MODIFICATION OF PLANT PROTEINS

Protein modification is the process of changing the chemical groups or molecular structure of a protein using particular techniques to increase its bioactivity and usefulness. They can create food products with several uses, thanks to plant-based proteins that have been modified. Table 1.5 provides a brief overview of the several categories of protein modification, including biological, physical, chemical, and other innovative strategies. The methods of physical modification include

TABLE 1.5
Different Methods of Modification

Methods of Modification			Reference	Applications
Physical Modification	Heat Treatment	Infrared Irradiation	Ogundele & Kayitesi, 2019	Decrease the Antinutritional Factors Enhance Digestibility
		Microwave heating	Xiang et al., 2020	Improves gelling, digestibility and emulsifying properties Splits disulfide and hydrogen bonds to induce protein folding Modulates protein while preserving its structure Enhances the enzymatic modification efficiency
		Conventional thermal treatment	Mir et al., 2020	Used in food and pharmaceutical industry Enhances the gelling properties and thermal stability Improves nutritional, emulsifying properties and digestibility Reduces negative impacts of antinutritional components
		Radio frequency treatment	Han et al., 2018	Increases the hydrophobicity of surface Improves emulsifying and oil-holding capacity
		Ohmic heating	Pereira et al., 2018	Milk pasteurization Improves emulsifying properties and reduces the heating time
	High pressure treatment	Dynamic high-pressure fluidization	Doost et al., 2018	Microbial cell inactivation Improves digestibility, emulsifying ability, functionality and versatility Enhances the foaming and emulsifying properties Decreases the size of nanoparticles
		High hydrostatic pressure	Lee et al., 2016	Microbial inactivation, changes in emulsification and textures Increases nutritional value and protein hydrophobicity Improves gelation, antioxidant activity, techno-functional and emulsifying properties, water- and oil-holding capacity.
	Gamma irradiation		M. A. Malik et al., 2017	Extends the shelf life Enhances the hydrophobicity of surface, thermal stability, antioxidant ability, digestibility, oil-holding capacity, foaming and emulsifying properties Reduces the water-binding capacity
	Ultraviolet radiation		Panozzo et al., 2016	Reduces the allergenicity and immunoreactivity of plant proteins Enhances the mechanical properties of films Increases the solubility, sulfhydryl content, antioxidant activity, surface hydrophobicity, foaming and emulsifying properties
	Ultrafiltration		Eckert et al., 2019	Enhances the hydrophobicity of surface, foaming, emulsifying and oil-holding capacity Reduces antinutrient components

(Continued)

TABLE 1.5 *(Continued)*
Different Methods of Modification

Methods of Modification		Reference	Applications
	Electron beam irradiation	X. Zhang et al., 2020	Assists in sterilizing the food and the process of extraction Reduces surface hydrophobicity Enhances the nutritional value, functional properties, solubility, and emulsifying activity
	Sonication	Gharibzahedi & Smith, 2020	Increases gelling, emulsifying properties, water- and oil-holding capacity, solubility, antioxidant ability, hydrophobicity, and digestibility Reduces the antinutrients, foaming stability, and allergenicity
	Ball mill treatment	Ramadhan & Foster, 2018	Improves the gelling properties and solubility
	Extrusion	Doost et al., 2019	Inactivate enzymes, microbes, and naturally found toxic substances Reduces the antinutrient compounds Enhances digestibility, and techno-functionality, also helps in the generation of meat-like texture
	Cold atmospheric plasma processing	Tolouie et al., 2018	Improves the surface hydrophobicity, antioxidant activity, foaming and gelling properties, emulsifying properties, solubility, techno-functional properties, and surface activity Inactivates the viruses, spores, and microbes found on food surface
Chemical Modification	Cationization	Nesterenko et al., 2014	Modifies the techno-functional properties Enhances encapsulation, solubility, and emulsifying properties
	Glycation	Nasrabadi et al., 2015	Improves the solubility, immunomodulatory properties, protein functionality, foaming and emulsifying ability, flavor profile, and thermal stability
	Acylation	Y. Zhao et al., 2017	Improves emulsifying, functional and foaming properties, solubility, water-holding capacity, hydrophobicity, gelling properties, emulsion and thermal stability
	Phosphorylation	Y. Liu et al., 2020	Maintains bioavailability of nutrients Increases the solubility, viscosity, in-vitro digestibility, thermal aggregation, thermal stability, foaming and emulsifying ability
	Deamidation	W. He et al., 2019	Reduces allergenicity, grittiness, beany flavor, and lumpiness Increases the water-holding capacity, emulsifying and foaming properties, solubility, and techno-functional properties
	pH-shifting treatment	Yildiz et al., 2017	Improves antioxidant activity, protein reactivity, extensibility, solubility, foaming and emulsifying ability Modifies the protein's functional and structural properties

TABLE 1.5 *(Continued)*
Different Methods of Modification

Methods of Modification			Reference	Applications
Biological modification	Fermentation		Schlegel et al., 2019	Improves digestibility, nutritional value, antioxidant properties, solubility, foaming, functional properties, oil- and water-holding capacity
				Decreases antinutrient compounds, allergens, and beany, bitter flavors
	Enzymatic modification		Nivala et al., 2017	Reduces the bitterness
				Enhances emulsifying and foaming ability, solubility, techno-functionality, plant protein's bioactivity, and interfacial properties
Others	Amyloid fibrillization		Smith et al., 2019	Improves hydrophobicity, stability, and functionality
				Enhances the gels, rheological properties, degradable films, foam and water purification filters
	Complexation	Protein–protein	Zheng et al., 2020	Improves water solubility and techno-functionality
		Protein–surfactant	S.-R. Dong et al., 2017	Increases encapsulation, stability, physiochemical properties, water dispersibility, thermal stability, PH, foaming and emulsifying properties
		Protein-polysaccharide	Fan et al., 2020	Improves physical stability, techno-functionality, solubility, susceptibility, foaming and emulsifying properties
				Reduces the bitterness in potato protein
		Protein-phenolic	Hu et al., 2018	Exhibits activities like antioxidant, anticancer, antimicrobial, anti-inflammatory and antiallergenic

heat treatment (such as infrared irradiation, microwave heating, ohmic heating, radio frequency treatment, and conventional thermal treatment), pulsed-electric field, ultraviolet radiation, electron beam irradiation, gamma irradiation, high-pressure treatment (such as high hydrostatic pressure, dynamic high-pressure fluidization), ultrafiltration, ball mill use, extrusion, sonication, and cold atmospheric plasma processing. The methods for chemical modification involve deamidation, acylation, phosphorylation, glycation, pH-shifting treatment, and cationization. Enzymatic modification and fermentation are two methods used in biological modification. Other modification methods, such as complexation (such as protein phenolic, protein-protein, protein-polysaccharide, and protein-surfactant) and amyloid fibrillization as shown in Table 1.5, have been discovered in addition to physical, chemical, and biological alterations (Nasrabadi et al., 2021).

1.7 CREATING FUTURE FOODS USING PLANT PROTEINS

1.7.1 PROTEIN–POLYSACCHARIDE INTERACTIONS

The majority of foods are an intricate blend of different ingredients. The main constituent in the majority of plant-based products, aside from proteins, is polysaccharides. Starch, agars, pectins, cellulose, gums, alginates, and carrageenans are only a few examples of the wide family of polysaccharides, which are sugar polymers joined with glycosidic bonds. Less-refined plant ingredients could be used by capitalizing on the polysaccharides present naturally in many sources of plant protein sources because they serve as the main structural and stabilizing components of food products through their gelling, emulsifying, and thickening capabilities (Le et al., 2017). Their functionality can be increased when combined with proteins due to interactions between the two biopolymers (Schmitt et al., 1998). Understanding and managing protein–polysaccharide interactions are therefore of tremendous relevance for the development of plant-based foods such as plant-based desserts, milk, and ice cream. Because plant proteins are not very soluble, polysaccharides have been added to biopolymers to increase their solubility, sometimes in conjunction with a modification step. The following are a few examples: simple complexation (Lan et al., 2018; S. Liu et al., 2010), sonication (C. Li et al., 2014; Yildiz et al., 2018), and conjugation (Qu et al., 2018; Saatchi et al., 2019; Zhang et al., 2012). Particularly noteworthy is that some researchers claim that the solubility of biopolymer increased near the isoelectric point of protein (Yildiz et al., 2018), whereas the minimum protein solubility shifted toward more of the acidic regions (Willett et al., 2019). That is probably caused by a shift in net charges of biopolymer surface following complexation and, consequently, modification. For the purpose of reducing the precipitation of plant proteins, the change in the biopolymer isoelectric point that is apparent will be helpful in the development of high-protein-containing acidic beverages. This tactic merits additional investigation, such as incorporating different kinds of processing techniques.

Polysaccharides enhance a number of plant proteins' characteristics in addition to solubility, including viscosity (Cai et al., 2021), foaming (D. J. McClements & Grossmann, 2021; Mohanan et al., 2020; Naderi et al., 2020), emulsifying (D. Dong & Hua, 2018; Feng et al., 2021; Shen & Li, 2021; H. Yang et al., 2020; Zha et al., 2019), and gelling (Z. He et al., 2021; Sim et al., 2019; Sim & Moraru, 2020; Y.-R. Wang et al., 2021; Wee et al., 2017). The altering of biopolymer interfacial properties would obviously affect emulsifying and foaming properties, but an intriguing strategy is to take advantage of plant proteins' low solubility to produce insoluble protein–polysaccharide particles of plant proteins as Pickering emulsion stabilizers (Feng et al., 2021; H. Yang et al., 2020). Processing has a crucial function, which should also be noted. For instance, high-pressure processing caused granules of starch to stay ungelatinized and intact, employed as a filler in the pressure-induced gel matrix of protein (Sim & Moraru, 2020), as opposed to heat processing of plant protein–starch mixes, which resulted in a mixed protein–starch gel network (M. Joshi et al., 2014; J.-Y. Li et al., 2007). Polysaccharides do not necessarily enhance the performance of plant proteins (G.-Y. Li et al., 2021). In other circumstances, it may even make their characteristics worse (e.g., lowering foaming solubility and solubility because of the development of electrostatic compounds that are insoluble [Stone et al., 2014]). Further research is required to comprehend exactly how these circumstances arise.

1.7.2 Flavors Obtained from Plant Proteins

An essential sensory characteristic that influences a buyer's food consumption and purchase choice is flavor. Various volatile molecules from meat or relevant model systems including meat ingredients have been found through numerous investigations on the chemistry of meaty flavors. There is great anticipation in creating this meaty flavor from non-meat sources, for instance, plant proteins, because alternative proteins are becoming more popular. By combining free amino compounds (such as amino acids or peptides) with reducing sugars (such as hexoses or pentoses) in a precise way to form melanoidins, the Maillard reaction can provide these meaty flavors (Van Ba et al., 2012). These meaty flavors are produced by enzymatic hydrolysis, which breaks down the plant proteins into amino acids and peptides. In the Maillard process, diketones, ketones, aliphatic aldehydes, and lower fatty acids are the taste chemicals that are produced in the greatest amounts (Reineccius, 2005). In contrast, heterocyclic molecules containing sulfur, nitrogen, oxygen, or combinations of these elements are significantly more prevalent and have a big impact on how food flavors emerge when they are heated. When sulfur-containing amino acids (like cysteine) and reducing sugars are combined, it frequently results in the creation of a meaty flavor; pentoses like ribose or xylose are preferred for this reaction (Kerler et al., 2010). Most food products are thought to produce their meaty flavor primarily through a chemical interaction between cysteine and reducing sugars. The Strecker degradation of cysteine is catalyzed by the dicarbonyl molecules produced during the Maillard process, and the main degradation products are hydrogen sulfide, acetaldehyde, and mercaptoacetaldehyde (De Roos, 1992). Following a sequence of events, these Strecker degradation by-products result in the creation of molecules with a meaty flavor. Several plant proteins, including flaxseed protein, quinoa protein, pea protein, soybean protein, and others, have been used in published studies to create meaty flavors. Several fragrance compounds from the Maillard reaction products, including ketones, pyrazines, furans, aldehydes, and others, were identified using a gas chromatography–mass spectroscopy study (MRPs). According to studies (Fadel et al., 2015; Wei et al., 2019), 2 methyl 3 furanthiol, which is an odorant molecule having a sweet, sweet, and sulfurous scent in the MRPs (Van Ba et al., 2012), has been identified. In a model system, cysteine and reducing sugar underwent the Maillard process to create this molecule. In addition to the protein hydrolysates and reducing sugars, these researchers also reported adding sulfur-containing substances including thiamine, taurine, and cysteine during the heat treatment for the Maillard reaction (Lotfy et al., 2021; Wei et al., 2019; Yu et al., 2018; Zhou et al., 2021). By using solely the free peptides from the hydrolyzed plant proteins to have a reaction with the reducing sugars, further research might be done to prevent these sulfur-containing chemicals.

1.8 NUTRITION AS A FOCUS

Nutritional researchers have asserted the presence of a "protein gap" since the 1970s. Providing enough protein for human health continues to be a significant issue, given the predicted growth in population. The need for plant-based proteins is growing, so it's critical to advance science and technology that will keep plant proteins' nutritional value and functional properties intact. In the past, either humans or animals have been employed to assess the quality of protein. Net protein utilization (NPU), for instance, is the measurement of the retained protein as a percentage of the consumed protein. NPU is a single metric that combines biological value (BV) and digestibility (D), two factors that determine the quality of proteins. Because the technique is onerous and time consuming, the digestible indispensable amino acid score (DIAAS) and protein digestibility corrected amino acid score (PDCAAS) are now commonly used. Based on the requirements for human amino acids as well as the quality of the protein that can be digested, PDCAAS and DIAAS are employed to calculate the protein's level of digestibility. A dietary protein's PDCAAS value is calculated by multiplying its real fecal nitrogen digestibility (the amount of nitrogen expelled in feces in relation to the amount consumed) by its amino acid score. To find out the true ileal digestibility of some essentially required amino acids, the DIAAS, as compared to the PDCAAS, uses the ileal digestibility coefficients of every single amino acid rather than the protein's actual fecal nitrogen digestibility (Marinangeli & House, 2017). Animal and plant protein's nutritional quality is influenced by the quantity and quality of their essential amino acids, their ability to be digested, how much of them are consumed, how valuable they are biologically, and how well they can be digested when the essential amino acids are rectified (Organization, 1991). When contrasted to plant sources, almost all animal proteins are thought to have greater protein quality, making them frequently advised for satisfying dietary protein requirements at a moderate calorie load. In contrast to plant proteins, which are typically lacking or have restricted amounts of one or more important nutrients, animal proteins typically include a healthy mix of both non-essential and essential amino acids (Berrazaga et al., 2019). Plant-based proteins are thought to be nutritionally deficient because they lack several essential amino acids and are less gastrointestinally accessible or digested than animal proteins. Most grain proteins, for instance, have less lysine content, while the pulse proteins have lower tryptophan, cysteine, and methionine levels. As indicated in Table 1.6, animal proteins typically have

TABLE 1.6

PDCAAS and DIAAS for Food and Protein Fractions

Food	PDCAAS	DIAAS
Whole milk	1.00	1.14
Soy protein isolate	0.98	0.90
Chickpeas	0.74	0.83
Pea protein isolate	0.89	0.82
Cooked peas	0.60	0.58
Tofu	0.56	0.52
Cooked rice	0.62	0.59
Rice protein concentrate	0.42	0.37
Almonds	0.39	0.40

higher PDCAAS/DIAAS scores (>0.9), while plant proteins frequently have lesser scores (0.4–0.9) (Berrazaga et al., 2019; Phillips, 2017).

The reduction in plant proteins digestibility is caused by three main factors:

(1) Antinutrients including phytates and trypsin inhibitors, which prevent proteins from being digested and absorbed, are present (Gilani et al., 2005).

(2) Animal and plant proteins differ structurally in that plant proteins have more β-sheet structures and fewer α-helixes, which make it easier for proteins to aggregate (Carbonaro et al., 2012).

(3) Decreased proteolytic digestibility in plants due to the presence of dietary fiber (Duodu et al., 2003).

Few options exist to enhance the quality of plant-based proteins for the purpose of human consumption.

• Antinutrients from the plant-based proteins (isolated) are eliminated in order to raise the nutritional quality of protein to achieve PDCAAS value of 1.00, which is comparable to diets originating from animals (e.g., soy).

• It is projected that enhancing the amount of plant-based proteins taken in each meal will efficiently make up for the low anabolic response they produce when compared to animal-based proteins (Norton et al., 2009). However, when using these strategies, plant proteins would not produce an anabolic response comparable to that of animal sources since leucine is a key necessary amino acid that may not be increased by eliminating antinutrients from plant-based proteins (Boye et al., 2010; Gilani et al., 2005). As a result, it has been demonstrated that consuming wheat protein with a leucine content that is equal to that of whey protein increases postprandial rates of muscle protein synthesis (Gorissen et al., 2016).

• Another tactic to enhance the nutritional profile of the plant-based protein is to fortify it with the amino acids needed for protein synthesis. According to a study, adding leucine, isoleucine, and valine to soy proteins improved whole-body protein synthesis (Engelen et al., 2007).

• The nutritional value of plant proteins can be increased by combining the correct kind of proteins that are plant-based to make a complex source of essential amino acids (Hertzler et al., 2020). The separate PDCASS values of peas and rice are modest, but when they are combined, they can raise the PDCASS value to 1.00 (Day, 2013).

REFERENCES

Adenekan, M. K., Fadimu, G. J., Odunmbaku, L. A., Oke, E. K. (2018). Effect of isolation techniques on the characteristics of pigeon pea (*Cajanus cajan*) protein isolates. *Food Science and Nutrition*, 6(1), 146–152.

Ahmed, F., Fanning, K., Netzel, M., Turner, W., Li, Y., & Schenk, P. M. (2014). Profiling of carotenoids and antioxidant capacity of microalgae from subtropical coastal and brackish waters. *Food Chemistry*, 165, 300–306.

Alvarez-Jubete, L., Arendt, E. K., Gallagher, E. J. (2010). Nutritive value of pseudocereals and their increasing use as functional gluten-free ingredients. *Trends in Food Science and Technology*, 21(2), 106–113.

Amagliani, L., O'Regan, J., Kelly, A. L., & O'Mahony, J. A. (2017). The composition, extraction, functionality and applications of rice proteins: A review. *Trends in Food Science and Technology*, 64, 1–12.

Anaya, K., Cruz, A. C., Cunha, D. C., Monteiro, S. M., & Dos Santos, E. A. J. P. (2015). Growth impairment caused by raw linseed consumption: Can trypsin inhibitors be harmful for health? *Plant Foods for Human Nutrition*, 70(3), 338–343.

Andersen, V., Halekoh, U., Tjønneland, A., Vogel, U., & Kopp, T. I. J. I. J. OMS. (2019). Intake of red and processed meat, use of non-steroid anti-inflammatory drugs, genetic variants and risk of colorectal cancer: A prospective study of the Danish "diet, cancer and health" cohort. *International Journal of Molecular Sciences*, 20(5), 1121.

Arif, M., & Pauls, K. P. J. P. B. (2018). *Properties of plant proteins*. In *Plant bioproducts* (pp. 121–142).

Aydemir, L. Y., Yemenicioğlu, A. (2013). Potential of Turkish Kabuli type chickpea and green and red lentil cultivars as source of soy and animal origin functional protein alternatives— *LWT - Food Science and Technology*, 50(2), 686–694.

Azzollini, D., Wibisaphira, T., Lakemond, C. M. M., & Fogliano, V. (2019). Toward the design of insect-based meat analogue: The role of calcium and temperature in coagulation behavior of *Alphitobius diaperinus* proteins. *LWT*, 100, 75–82.

Bernier-Jean, A., Prince, R. L., Lewis, J. R., Craig, J. C., Hodgson, J. M., Lim, W. H., . . . Wong, G. J. N. D. T. (2021). Dietary plant and animal protein intake and decline in estimated glomerular filtration rate among elderly women: A 10-year longitudinal cohort study. *Nephrology, Dialysis, Transplantation*, 36(9), 1640–1647.

Berrazaga, I., Micard, V., Gueugneau, M., & Walrand, S. J. N. (2019). The role of the anabolic properties of plant-versus animal-based protein sources in supporting muscle mass maintenance: A critical review. *Nutrients*, 11(8), 1825.

Bessada, S. M., Barreira, J. C., Oliveira, M. B. P. J. (2019). Pulses and food security: Dietary protein, digestibility, bioactive and functional properties. *Trends in Food Science & Technology*, 93, 53–68.

Beuchat, L. R. (1977). Functional and electrophoretic characteristics of succinylated peanut flour protein. *Journal of Agricultural and Food Chemistry*, 25(2), 258–261.

Bhat, M. Y., Dar, T. A., & Singh, L. R. J. (2016). Casein proteins: Structural and functional aspects. *Milk Proteins - From Structure to Biological Properties and Health Aspects*, 10, 64187.

Bohrer, B. M. (2019). An investigation of the formulation and nutritional composition of modern meat analogue products. *Food Science and Human Wellness*, 8(4), 320–329.

Boye, J., Zare, F., & Pletch, A. J. (2010). Pulse proteins: Processing, characterization, functional properties and applications in food and feed. *Food Research International*, 43(2), 414–431.

Branch, S., & Maria, S. J. (2017). Evaluation of the functional properties of mung bean protein isolate for development of textured vegetable protein. *International Food Research Journal*, 24(4), 1595–1605.

Broyard, C., & Gaucheron, F. J. (2015). Modifications of structures and functions of caseins: A scientific and technological challenge. *Dairy Science & Technology*, 95, 831–862.

Butt, M. S., & Rizwana, B. J. (2010). Nutritional and functional properties of some promising legumes protein isolates. *Pakistan Journal of Nutrition*, 9(4), 373–379.

Cai, Y., Huang, L., Tao, X., Su, J., Chen, B., Zhou, F., . . . Van der Meeren, P. J. (2021). Effect of pH on okara protein-carboxymethyl cellulose interactions in aqueous solution and at oil-water interface. *Food Hydrocolloids*, 113, 106529.

Campbell, W. W. (2019). Animal-based and plant-based protein-rich foods and cardiovascular health: A complex conundrum. *The American Journal of Clinical Nutrition*, 110, 8–9.

Caporgno, M. P., Böcker, L., Müssner, C., Stirnemann, E., Haberkorn, I., Adelmann, H., . . . Mathys, A. (2020). Extruded meat analogues based on yellow, heterotrophically cultivated *Auxenochlorella protothecoides* microalgae. *Innovative Food Science & Emerging Technologies*, 59, 102275.

Carbonaro, M., Maselli, P., & Nucara, A. J. (2012). Relationship between digestibility and secondary structure of raw and thermally treated legume proteins: A Fourier transform infrared (FT-IR) spectroscopic study. *Amino Acids*, 43, 911–921.

Casati, L., Pagani, F., Braga, P. C., Scalzo, R. L., & Sibilia, V. (2016). Nasunin, a new player in the field of osteoblast protection against oxidative stress. *Journal of Functional Foods*, *23*, 474–484.

Chao, D., & Aluko, R. E. (2018). Modification of the structural, emulsifying, and foaming properties of an isolated pea protein by thermal pretreatment. *CyTA – Journal of Food*, *16*(1), 357–366.

Chen, M., Lu, J., Liu, F., Nsor-Atindana, J., Xu, F., Goff, H. D., . . . Zhong, F. J. (2019). Study on the emulsifying stability and interfacial adsorption of pea proteins. *Food Hydrocolloids*, *88*, 247–255.

Chen, X., Wei, G., Jalili, T., Metos, J., Giri, A., Cho, M. E., . . . Beddhu, S. J. (2016). The associations of plant protein intake with all-cause mortality in CKD. *American Journal of Kidney Diseases*, *67*(3), 423–430.

Chiang, J. H., Loveday, S. M., Hardacre, A. K., & Parker, M. E. (2019). Effects of soy protein to wheat gluten ratio on the physicochemical properties of extruded meat analogues. *Food Structure*, *19*, 100102.

Coda, R., Varis, J., Verni, M., Rizzello, C. G., & Katina, K. (2017). Improvement of the protein quality of wheat bread through faba bean sourdough addition— *LWT - Food Science and Technology*, *82*, 296–302.

Cui, Z., Chen, Y., Kong, X., Zhang, C., & Hua, Y. (2014). Emulsifying properties and oil/water (O/W) interface adsorption behavior of heated soy proteins: Effects of heating concentration, homogenizer rotating speed, and salt addition level. *Journal of Agricultural and Food Chemistry*, *62*(7), 1634–1642.

Curtain, F., & Grafenauer, S. (2019). Plant-based meat substitutes in the flexitarian age: An audit of products on supermarket shelves. *Nutrients*, *11*(11), 2603.

Day, L. J. (2013). Proteins from land plants–potential resources for human nutrition and food security. *Trends in Food Science & Technology*, *32*(1), 25–42.

de Oliveira Sousa, A. G., Fernandes, D. C., Alves, A. M., De Freitas, J. B., & Naves, M. M. V. J. (2011). Nutritional quality and protein value of exotic almonds and nut from the Brazilian Savanna compared to peanut. *Food Research International*, *44*(7), 2319–2325.

De Roos, K. (1992). Meat flavor generation from cysteine and sugars. In *Flavor precursors*, (pp. 203–216). ACS Publications.

Dekkers, B. L., Emin, M. A., Boom, R. M., & van der Goot, A. J. (2018). The phase properties of soy protein and wheat gluten in a blend for fibrous structure formation. *Food Hydrocolloids*, *79*, 273–281.

Diez-Simon, C., Mumm, R., & Hall, R. D. (2019). Mass spectrometry-based metabolomics of volatiles as a new tool for understanding aroma and flavour chemistry in processed food products. *Metabolomics*, *15*, 1–20.

Dissanayake, M., & Vasiljevic, T. J. (2009). Functional properties of whey proteins affected by heat treatment and hydrodynamic high-pressure shearing. *Journal of Dairy Science*, *92*(4), 1387–1397.

Dong, D., & Hua, Y. J. (2018). Emulsifying behaviors and interfacial properties of different protein/gum arabic complexes: Effect of pH. *Food Hydrocolloids*, *74*, 289–295.

Dong, S.-R., Xu, H.-H., Tan, J.-Y., Xie, M.-M., & Yu, G.-P. (2017). The structure and amphipathy characteristics of modified γ-zeins by SDS or alkali in conjunction with heating treatment. *Food Chemistry*, *233*, 361–368.

Doost, A. S., Dewettinck, K., Devlieghere, F., & Van der Meeren, P. J. (2018). Influence of non-ionic emulsifier type on the stability of cinnamaldehyde nanoemulsions: A comparison of polysorbate 80 and hydrophobically modified inulin. *Food Chemistry*, *258*, 237–244.

Doost, A. S., Nasrabadi, M. N., Wu, J., A'yun, Q., Van der Meeren, P. J. (2019). Maillard conjugation as an approach to improve whey proteins functionality: A review of conventional and novel preparation techniques. *Trends in Food Science & Technology*, *91*, 1–11.

Duodu, K., Taylor, J., Belton, P., & Hamaker, B. J. (2003). Factors affecting sorghum protein digestibility. *Journal of Cereal Science*, *38*(2), 117–131.

Eckert, E., Han, J., Swallow, K., Tian, Z., Jarpa-Parra, M., & Chen, L. J. (2019). Effects of enzymatic hydrolysis and ultrafiltration on physicochemical and functional properties of faba bean protein. *Cereal Chemistry*, *96*(4), 725–741.

El-Adawy, T. A., & Taha, K. M. J. (2001). Characteristics and composition of different seed oils and flours. *Food Chemistry*, *74*(1), 47–54.

Emin, M., Quevedo, M., Wilhelm, M., & Karbstein, H. (2017). Analysis of the reaction behavior of highly concentrated plant proteins in extrusion-like conditions. *Innovative Food Science & Emerging Technologies*, *44*, 15–20.

Engelen, M. P., Rutten, E. P., De Castro, C. L., Wouters, E. F., Schols, A. M., & Deutz, N. E. J. T. (2007). Supplementation of soy protein with branched-chain amino acids alters protein metabolism in healthy elderly and even more in patients with chronic obstructive pulmonary disease. *American Journal of Clinical Nutrition*, *85*(2), 431–439.

Espinosa-Páez, E., Alanis-Guzmán, M. G., Hernández-Luna, C. E., Báez-González, J. G., Amaya-Guerra, C. A., & Andrés-Grau, A. M. J. (2017). Increasing antioxidant activity and protein digestibility in *Phaseolus vulgaris* and *Avena sativa* by fermentation with the *Pleurotus ostreatus* Fungus. *Molecules*, *22*(12), 2275.

Fadel, H. H., Samad, A. A., Kobeasy, M., Mageed, M. A. A., & Lotfy, S. N. (2015). Flavour quality and stability of an encapsulated meat-like process flavouring prepared from soybean based acid hydrolyzed protein. *International Journal of Food Processing Technology*, *2*(1), 17–25.

Fan, R., Zhang, T., Tai, K., & Yuan, F. J. (2020). Surface properties and adsorption of lactoferrin-xanthan complex in the oil-water interface. *Journal of Dispersion Science and Technology*, *41*(7), 1037–1044.

Farrell Jr, H., Qi, P., Brown, E., Cooke, P., Tunick, M., Wickham, E., & Unruh, J. J. (2002). Molten globule structures in milk proteins: Implications for potential new structure-function relationships. *Journal of Dairy Science*, *85*(3), 459–471.

Feng, T., Wang, X., Wang, X., Zhang, X., Gu, Y., Xia, S., & Huang, Q. J. (2021). High internal phase Pickering emulsions stabilized by pea protein isolate-high methoxyl pectin-EGCG complex: Interfacial properties and microstructure. *Food Chemistry*, *350*, 129251.

Fernández-Quintela, A., Macarulla, M., Del Barrio, A., & Martínez, J. J. (1997). Composition and functional properties of protein isolates obtained from commercial legumes grown in northern Spain. *Plant Foods for Human Nutrition*, *51*, 331–341.

Feyzi, S., Milani, E., & Golimovahhed, Q. A. J. (2018). Grass pea (Lathyrus sativus L.) protein isolate: The effect of extraction optimization and drying methods on the structure and functional properties. *Food Hydrocolloids*, *74*, 187–196.

Fraser, R. Z., Shitut, M., Agrawal, P., Mendes, O., & Klapholz, S. (2018). Safety evaluation of soy leghemoglobin protein preparation derived from Pichia pastoris, intended for use as a flavor catalyst in plant-based meat. *International Journal of Toxicology*, *37*(3), 241–262.

Fresán, U., Mejia, M. A., Craig, W. J., Jaceldo-Siegl, K., & Sabaté, J. (2019). Meat analogs from different protein sources: A comparison of their sustainability and nutritional content. *Sustainability*, *11*(12), 3231.

Frías, J., Giacomino, S., Peñas, E., Pellegrino, N., Ferreyra, V., Apro, N., et al. (2011). Assessment of the nutritional quality of raw and extruded Pisum sativum L. var. laguna seeds. *LWT – Food Science and Technology*, *44*(5), 1303–1308.

Gharibzahedi, S. M. T., Smith, B. J. (2020). The functional modification of legume proteins by ultrasonication: A review. *Trends in Food Science & Technology*, *98*, 107–116.

Ghofrani, S., Joghataei, M.-T., Mohseni, S., Baluchnejadmojarad, T., Bagheri, M., Khamse, S., & Roghani, M. (2015). Naringenin improves learning and memory in an Alzheimer's disease rat model: Insights into the underlying mechanisms. *European Journal of Pharmacology*, *764*, 195–201.

Ghribi, A. M., Amira, A. B., Gafsi, I. M., Lahiani, M., Bejar, M., Triki, M., . . . Besbes, S. J. (2018). Toward the enhancement of sensory profile of sausage "Merguez" with chickpea protein concentrate. *Meat Science*, *143*, 74–80.

Gilani, G. S., Cockell, K. A., & Sepehr, E. J. (2005). Effects of antinutritional factors on protein digestibility and amino acid availability in foods. *Journal of AOAC International*, *88*(3), 967–987.

Goldflus, F., Ceccantini, M., & Santos, W. J. (2006). Amino acid content of soybean samples collected in different Brazilian states: Harvest 2003/2004. *Brazilian Journal of Poultry Science*, *8*, 105–111.

González-Pérez, S., & Arellano, J. B. (2009). Vegetable protein isolates. In *Handbook of hydrocolloids* (pp. 383–419). Elsevier.

Gorissen, S. H., Horstman, A. M., Franssen, R., Crombag, J. J., Langer, H., Bierau, J., . . . Van Loon, L. J. J. T. (2016). Ingestion of wheat protein increases in vivo muscle protein synthesis rates in healthy older men in a randomized trial. *Journal of Nutrition*, *146*(9), 1651–1659.

Grossmann, L., & Weiss, J. J. (2021). Alternative protein sources as techno-functional food ingredients. *Annual Review of Food Science and Technology*, *12*, 93–117.

Guasch-Ferré, M., Zong, G., Willett, W. C., Zock, P. L., Wanders, A. J., Hu, F. B., & Sun, Q. J. (2019). Associations of monounsaturated fatty acids from plant and animal sources with total and cause-specific mortality in two US prospective cohort studies. *Circulation Research*, *124*(8), 1266–1275.

Gumus, C. E., Decker, E. A., & McClements, D. J. J. (2017). Formation and stability of ω-3 oil emulsion-based delivery systems using plant proteins as emulsifiers: Lentil, pea, and faba bean proteins. *Food Biophysics*, *12*, 186–197.

Gundogan, R., & Karaca, A. C. J. (2020). Physicochemical and functional properties of proteins isolated from local beans of Turkey. *LWT*, *130*, 109609.

Gupta, S., Chhabra, G. S., Liu, C., Bakshi, J. S., & Sathe, S. K. J. (2018). Functional properties of select dry bean seeds and flours. *Journal of Food Science*, *83*(8), 2052–2061.

Gürbüz, N., Uluişik, S., Frary, A., Frary, A., & Doğanlar, S. (2018). Health benefits and bioactive compounds of eggplant. *Food Chemistry*, *268*, 602–610.

Hall, A. E., & Moraru, C. I. J. (2021). Structure and function of pea, lentil and faba bean proteins treated by high pressure processing and heat treatment. *LWT*, *152*, 112349.

Han, Z., Cai, M.-j., Cheng, J.-H., & Sun, D.-W. J. (2018). Effects of electric fields and electromagnetic wave on food protein structure and functionality: A review. *Trends in Food Science and Technology*, 75, 1–9.

He, W., Yang, R., & Zhao, W. J. (2019). Effect of acid deamidation-alcalase hydrolysis induced modification on functional and bitter-masking properties of wheat gluten hydrolysates. *Food Chemistry*, 277, 655–663.

He, Z., Liu, C., Zhao, J., Li, W., & Wang, Y. J. (2021). Physicochemical properties of a ginkgo seed protein-pectin composite gel. *Food Hydrocolloids*, 118, 106781.

Hertzler, S. R., Lieblein-Boff, J. C., Weiler, M., & Allgeier, C. J. (2020). Plant proteins: Assessing their nutritional quality and effects on health and physical function. *Nutrients*, 12(12), 3704.

Hosseinpour-Niazi, S., Mirmiran, P., Hedayati, M., & Azizi, F. J. (2015). Substitution of red meat with legumes in the therapeutic lifestyle change diet based on dietary advice improves cardiometabolic risk factors in overweight type 2 diabetes patients: A cross-over randomized clinical trial. *European Journal of Clinical Nutrition*, 69(5), 592–597.

Hsieh, F. H. (2014). *Physicochemical properties of soy-and pea-based imitation sausage patties* (Doctoral dissertation, University of Missouri).

Hu, B., Shen, Y., Adamcik, J., Fischer, P., Schneider, M., Loessner, M. J., & Mezzenga, R. J. (2018). Polyphenol-binding amyloid fibrils self-assemble into reversible hydrogels with antibacterial activity. *ACS Nano*, 12(4), 3385–3396.

Huang, J., Liao, L. M., Weinstein, S. J., Sinha, R., Graubard, B. I., & Albanes, D. J. (2020). Association between plant and animal protein intake and overall and cause-specific mortality. *JAMA Internal Medicine*, 180(9), 1173–1184.

Hughes, G. J., Kress, K. S., Armbrecht, E. S., Mukherjea, R., Mattfeldt-Beman, M. J. (2014). Initial investigation of dietitian perception of plant-based protein quality. *Food Science and Nutrition*, 2(4), 371–379.

Jiang, Z.-Q., Wang, J., Stoddard, F., Salovaara, H., & Sontag-Strohm, T. J. (2020). Preparation and characterization of emulsion gels from whole faba bean flour. *Foods*, 9(6), 755.

Johnston, S. P., Nickerson, M. T., & Low, N. H. J. (2015). The physicochemical properties of legume protein isolates and their ability to stabilize oil-in-water emulsions with and without genipin. *Journal of Food Science and Technology*, 52, 4135–4145.

Joshi, M., Aldred, P., Panozzo, J., Kasapis, S., & Adhikari, B. J. (2014). Rheological and microstructural characteristics of lentil starch–lentil protein composite pastes and gels. *Food Hydrocolloids*, 35, 226–237.

Joshi, V., & Kumar, S. (2015). Meat analogues: Plant based alternatives to meat products-A review. *International Journal of Food and Fermentation Technology*, 5(2), 107–119.

Kamani, M. H., Meera, M. S., Bhaskar, N., & Modi, V. K. J. (2019). Partial and total replacement of meat by plant-based proteins in chicken sausage: Evaluation of mechanical, physico-chemical and sensory characteristics. *Journal of Food Science and Technology*, 56, 2660–2669.

Karaca, A. C., Low, N., & Nickerson, M. J. (2011). Emulsifying properties of chickpea, faba bean, lentil and pea proteins produced by isoelectric precipitation and salt extraction. *Food Research International*, 44(9), 2742–2750.

Karaś, M., Jakubczyk, A., Szymanowska, U., Złotek, U., & Zielińska, E. J. (2017). Digestion and bioavailability of bioactive phytochemicals. *International Journal of Food Science & Technology*, 52(2), 291–305.

Kaul, P. J. (2011). Nutritional potential, bioaccessibility of minerals and functionality of watermelon (*Citrullus vulgaris*) seeds. *LWT - Food Science and Technology*, 44(8), 1821–1826.

Kaur, M., & Singh, N. J. (2007). Characterization of protein isolates from different Indian chickpea (*Cicer arietinum* L.) cultivars. *Food Chemistry*, 102(1), 366–374.

Kerler, J., Winkel, C., Davidek, T., & Blank, I. J. (2010). Basic chemistry and process conditions for reaction flavours with particular focus on Maillard-type reactions. *Food Flavour Technology*, 2, 51–88.

Keskin, S. O., Ali, T. M., Ahmed, J., Shaikh, M., Siddiq, M., & Uebersax, M. A. J. (2022). Physico-chemical and functional properties of legume protein, starch, and dietary fiber—A review. *Legume Science*, 4(1), e117.

Khan, Z. S., Sodhi, N. S., Dhillon, B., Dar, B., Bakshi, R. A., & Shah, S. F. (2021). Seabuckthorn (Hippophae rhamnoides L.), a novel seed protein concentrate: Isolation and modification by high power ultrasound and characterization for its functional and structural properties. *Journal of Food Measurement and Characterization*, 15(5), 4371–4379.

Khoo, H. E., Azlan, A., Tang, S. T., & Lim, S. M. (2017). Anthocyanidins and anthocyanins: Colored pigments as food, pharmaceutical ingredients, and the potential health benefits. *Food & Nutrition Research*, 61(1), 1361779.

Kimura, A., Fukuda, T., Zhang, M., Motoyama, S., Maruyama, N., Utsumi, S. J. (2008). Comparison of physicochemical properties of 7S and 11S globulins from pea, fava bean, cowpea, and French bean with those of soybean: French bean 7S globulin exhibits excellent properties. *Journal of Agricultural and Food Chemistry*, 56(21), 10273–10279.

Krintiras, G. A., Göbel, J., Van der Goot, A. J., & Stefanidis, G. D. (2015). Production of structured soy-based meat analogues using simple shear and heat in a Couette Cell. *Journal of Food Engineering*, *160*, 34–41.

Kubota, M., Watanabe, R., Yamaguchi, M., Hosojima, M., Saito, A., Fujii, M., . . . Kadowaki, M. J. (2016). Rice endosperm protein slows progression of fatty liver and diabetic nephropathy in Zucker diabetic fatty rats. *British Journal of Nutrition*, *116*(8), 1326–1335.

Kumar, P., Chatli, M., Mehta, N., Singh, P., Malav, O., & Verma, A. K. (2017). Meat analogues: Health promising sustainable meat substitutes. *Critical Reviews in Food Science and Nutrition*, *57*(5), 923–932.

Lafarga, T., Álvarez, C., Villaró, S., Bobo, G., & Aguiló-Aguayo, I. J. (2020). Potential of pulse-derived proteins for developing novel vegan edible foams and emulsions. *International Journal of Food Science & Technology*, *55*(2), 475–481.

Lakemond, C. M., de Jongh, H. H., Hessing, M., Gruppen, H., & Voragen, A. G. J. (2000). Soy glycinin: Influence of pH and ionic strength on solubility and molecular structure at ambient temperatures. *Journal of Agricultural and Food Chemistry*, *48*(6), 1985–1990.

Lan, Y., Chen, B., & Rao, J. J. (2018). Pea protein isolate–high methoxyl pectin soluble complexes for improving pea protein functionality: Effect of pH, biopolymer ratio and concentrations. *Food Hydrocolloids*, *80*, 245–253.

Langton, M., Ehsanzamir, S., Karkehabadi, S., Feng, X., Johansson, M., & Johansson, D. P. J. (2020). Gelation of faba bean proteins-Effect of extraction method, pH and NaCl. *Food Hydrocolloids*, *103*, 105622.

Le, X. T., Rioux, L.-E., Turgeon, S. L. J. A. i. C., & Science, I. (2017). Formation and functional properties of protein–polysaccharide electrostatic hydrogels in comparison to protein or polysaccharide hydrogels. *Advances in Colloid and Interface Science*, *239*, 127–135.

Lee, H., Yildiz, G., Dos Santos, L., Jiang, S., Andrade, J., Engeseth, N., & Feng, H. J. (2016). Soy protein nano-aggregates with improved functional properties prepared by sequential pH treatment and ultrasonication. *Food Hydrocolloids*, *55*, 200–209.

Leonard, W., Hutchings, S. C., Warner, R. D., & Fang, Z. J. (2019). Effects of incorporating roasted lupin (Lupinus angustifolius) flour on the physicochemical and sensory attributes of beef sausage. *International Journal of Food Science & Technology*, *54*(5), 1849–1857.

Li, C., Xue, H., Chen, Z., Ding, Q., & Wang, X. J. (2014). Comparative studies on the physicochemical properties of peanut protein isolate–polysaccharide conjugates prepared by ultrasonic treatment or classical heating. *Food Research International*, *57*, 1–7.

Li, G.-Y., Chen, Q.-H., Su, C.-R., Wang, H., He, S., Liu, J., et al. (2021). Soy protein-polysaccharide complex coacervate under physical treatment: Effects of pH, ionic strength and polysaccharide type. *Innovative Food Science and Emerging Technologies*, *68*, 102612.

Li, J.-Y., Yeh, A.-I., & Fan, K.-L. J. (2007). Gelation characteristics and morphology of corn starch/soy protein concentrate composites during heating. *Journal of Food Engineering*, *78*(4), 1240–1247.

Liang, H.-N., & Tang, C.-H. J. (2013). pH-dependent emulsifying properties of pea [Pisum sativum (L.)] proteins. *Food Hydrocolloids*, *33*(2), 309–319.

Lin, L., Regenstein, J. M., Lv, S., Lu, J., & Jiang, S. J. (2017). An overview of gelatin derived from aquatic animals: Properties and modification. *Trends in Food Science & Technology*, *68*, 102–112.

Lin, M. J.-Y., Humbert, E., & Sosulski, F. J. (1974). Certain functional properties of sunflower meal products. *Journal of Food Science*, *39*(2), 368–370.

Lin, S., Huff, H., & Hsieh, F. (2000). Texture and chemical characteristics of soy protein meat analog extruded at high moisture. *Journal of Food Science*, *65*(2), 264–269.

Lin, S., Huff, H., & Hsieh, F. (2002). Extrusion process parameters, sensory characteristics, and structural properties of a high moisture soy protein meat analog. *Journal of Food Science*, *67*(3), 1066–1072.

Lin, Y., Mouratidou, T., Vereecken, C., Kersting, M., Bolca, S., de Moraes, A. C. F., . . . Valtueña, J. J. (2015). Dietary animal and plant protein intakes and their associations with obesity and cardio-metabolic indicators in European adolescents: The HELENA cross-sectional study. *Nutrition Journal*, *14*(1), 1–11.

Liu, K.-L., Zheng, J.-B., & Chen, F.-S. J. (2017). Relationships between degree of milling and loss of Vitamin B, minerals, and change in amino acid composition of brown rice— *LWT - Food Science and Technology*, *82*, 429–436.

Liu, S., Elmer, C., Low, N., & Nickerson, M. J. (2010). Effect of pH on the functional behaviour of pea protein isolate–gum Arabic complexes. *Food Research International*, *43*(2), 489–495.

Liu, Y., Wang, D., Wang, J., Yang, Y., Zhang, L., Li, J., et al. (2020). Functional properties and structural characteristics of phosphorylated pea protein isolate. *International Journal of Food Science and Technology*, *55*(5), 2002–2010.

Lonnie, M., Laurie, I., Myers, M., Horgan, G., Russell, W. R., & Johnstone, A. M. J. N. (2020). Exploring health-promoting attributes of plant proteins as a functional ingredient for the food sector: A systematic review of human interventional studies. *Nutrients, 12*(8), 2291.

López, D. N., Galante, M., Robson, M., Boeris, V., & Spelzini, D. J. I. j. o. b. m. (2018). Amaranth, quinoa and chia protein isolates: Physicochemical and structural properties. *International Journal of Biological Macromolecules, 109*, 152–159.

Lotfy, S. N., Saad, R., El-Massrey, K. F., & Fadel, H. H. J. (2021). Effects of pH on headspace volatiles and properties of Maillard reaction products derived from enzymatically hydrolyzed quinoa protein-xylose model system. *LWT, 145*, 111328.

Lu, W., Chen, X.-W., Wang, J.-M., Yang, X.-Q., & Qi, J.-R. J. (2016). Enzyme-assisted subcritical water extraction and characterization of soy protein from heat-denatured meal. *Journal of Food Engineering, 169*, 250–258.

Ma, K. K. (2020). *The comparison of functional and physical properties of commercial pulse proteins to soy protein.* Future Foods, 6, 100155.

Ma, Z., Boye, J. I., & Simpson, B. K. J. (2016). Preparation of salad dressing emulsions using lentil, chickpea and pea protein isolates: A response surface methodology study. *Journal of Food Quality, 39*(4), 274–291.

Malik, M. A., Sharma, H. K., & Saini, C. S. J. (2017). Effect of gamma irradiation on structural, molecular, thermal and rheological properties of sunflower protein isolate. *Food Hydrocolloids, 72*, 312–322.

Malik, V. S., Li, Y., Tobias, D. K., Pan, A., & Hu, F. B. J. (2016). Dietary protein intake and risk of type 2 diabetes in US men and women. *American Journal of Epidemiology, 183*(8), 715–728.

Manach, C., Scalbert, A., Morand, C., Rémésy, C., & Jiménez, L. (2004). Polyphenols: Food sources and bioavailability. *The American Journal of Clinical Nutrition, 79*(5), 727–747.

Marinangeli, C. P., & House, J. D. J. (2017). Potential impact of the digestible indispensable amino acid score as a measure of protein quality on dietary regulations and health. *Nutrition Reviews, 75*(8), 658–667.

Mattice, K. D., & Marangoni, A. G. (2020). Comparing methods to produce fibrous material from zein. *Food Research International, 128*, 108804.

McClements, D. J. (2007). Critical review of techniques and methodologies for characterization of emulsion stability. *Critical Reviews in Food Science and Nutrition, 47*(7), 611–649.

McClements, D. J. (2020). Development of next-generation nutritionally fortified plant-based milk substitutes: Structural design principles. *Foods, 9*(4), 421.

McClements, D. J., & Grossmann, L. J. (2021). A brief review of the science behind the design of healthy and sustainable plant-based foods. *NPJ Science of Food, 5*(1), 17.

McClements, D. J., & Grossmann, L. J. (2021). The science of plant-based foods: Constructing next-generation meat, fish, milk, and egg analogs. *Comprehensive Reviews in Food Science and Food Safety, 20*(4), 4049–4100.

McClements, D. J., Newman, E., & McClements, I. F. (2019). Plant-based milks: A review of the science underpinning their design, fabrication, and performance. *Comprehensive Reviews in Food Science and Food Safety, 18*(6), 2047–2067.

McGuire, S. (2015). FAO, IFAD, and WFP. The state of food insecurity in the world 2015: Meeting the 2015 international hunger targets: Taking stock of uneven progress. Rome: FAO, 2015. *Advances in Nutrition, 6*(5), 623–624.

Mir, N. A., Riar, C. S., & Singh, S. J. (2020). Structural modification in album (Chenopodium album) protein isolates due to controlled thermal modification and its relationship with protein digestibility and functionality. *Food Hydrocolloids, 103*, 105708.

Mohanan, A., Nickerson, M. T., & Ghosh, S. J. (2020). Utilization of pulse protein-xanthan gum complexes for foam stabilization: The effect of protein concentrate and isolate at various pH. *Food Chemistry, 316*, 126282.

Molina, M. I., Petruccelli, S., Añón, M. C. J. (2004). Effect of pH and ionic strength modifications on thermal denaturation of the 11S globulin of sunflower (Helianthus annuus). *Journal of Agricultural and Food Chemistry, 52*(19), 6023–6029.

Moreno-Ortega, A., Ordóñez, J. L., Moreno-Rojas, R., Moreno-Rojas, J. M., & Pereira-Caro, G. (2021). Changes in the organosulfur and polyphenol compound profiles of Black and fresh onion during simulated gastrointestinal digestion. *Foods, 10*(2), 337.

Morr, C., German, B., Kinsella, J., Regenstein, J., Buren, J. V., Kilara, A., . . . Mangino, M. J. (1985). A collaborative study to develop a standardized food protein solubility procedure. *Journal of Food Science, 50*(6), 1715–1718.

Motamedi, A., Vahdani, M., Baghaei, H., & Borghei, M. A. (2015). Considering the physicochemical and sensorial properties of momtaze hamburgers containing lentil and chickpea seed flour. *Nutrition and Food Sciences Research, 2*(3), 55–62.

Naderi, B., Keramat, J., Nasirpour, A., & Aminifar, M. J. (2020). Complex coacervation between oak protein isolate and gum Arabic: Optimization & functional characterization. *International Journal of Food Properties*, *23*(1), 1854–1873.

Naghshi, S., Sadeghi, O., Willett, W. C., & Esmaillzadeh, A. J. (2020). Dietary intake of total, animal, and plant proteins and risk of all cause, cardiovascular, and cancer mortality: Systematic review and dose-response meta-analysis of prospective cohort studies. *BMJ*, *370*.

Nasrabadi, M. N., Doost, A. S., & Mezzenga, R. J. F. H. (2021). Modification approaches of plant-based proteins to improve their techno-functionality and use in food products. *Food Hydrocolloids*, *118*, 106789.

Nasrabadi, M. N., Goli, S. A. H., & Nasirpour, A. J. (2015). Evaluation of biopolymer-based emulsion for delivering conjugated linoleic acid (CLA) as a functional ingredient in beverages. *Journal of Dispersion Science and Technology*, *36*(6), 778–788.

Nesterenko, A., Alric, I., Silvestre, F., & Durrieu, V. J. (2014). Comparative study of encapsulation of vitamins with native and modified soy protein. *Food Hydrocolloids*, *38*, 172–179.

Nishinari, K., Fang, Y., Nagano, T., Guo, S., & Wang, R. (2018). Soy as a food ingredient. In *Proteins in food processing* (pp. 149–186). Elsevier.

Nivala, O., Mäkinen, O. E., Kruus, K., Nordlund, E., & Ercili-Cura, D. J. (2017). Structuring colloidal oat and faba bean protein particles via enzymatic modification. *Food Chemistry*, *231*, 87–95.

Norton, L. E., Layman, D. K., Bunpo, P., Anthony, T. G., Brana, D. V., & Garlick, P. J. J. T. (2009). The leucine content of a complete meal directs peak activation but not duration of skeletal muscle protein synthesis and mammalian target of rapamycin signaling in rats. *Journal of Nutrition*, *139*(6), 1103–1109.

Nosworthy, M. G., Neufeld, J., Frohlich, P., Young, G., Malcolmson, L., & House, J. D. J. (2017). Determination of the protein quality of cooked Canadian pulses. *Food Science & Nutrition*, *5*(4), 896–903.

Ogundele, O. M., & Kayitesi, E. J. (2019). Influence of infrared heating processing technology on the cooking characteristics and functionality of African legumes: A review. *International Journal of Food Science & Technology*, *56*, 1669–1682.

Olivos-Lugo, B. L., Valdivia-López, M. Á., & Tecante, A. J. (2010). Thermal and physicochemical properties and nutritional value of the protein fraction of Mexican chia seed (Salvia hispanica L.). *Food Science and Technology International*, *16*(1), 89–96.

Oreopoulou, V., & Tzia, C. (2007). *Utilization of plant by-products for the recovery of proteins, dietary fibers, antioxidants, and colorants*. Paper presented at the Utilization of by-products and treatment of waste in the food industry.

Organization, W. H. (1991). *Protein quality evaluation: Report of the joint FAO/WHO expert consultation, Bethesda, MD, USA 4–8 December 1989* (Vol. 51). Food & Agriculture Org.

Panozzo, A., Manzocco, L., Lippe, G., & Nicoli, M. C. J. (2016). Effect of pulsed light on structure and immunoreactivity of gluten. *Food Chemistry*, *194*, 366–372.

Patel, M. T., & Kilara, A. J. (1990). Studies on whey protein concentrates. 2. Foaming and emulsifying properties and their relationships with physicochemical properties. *Journal of Dairy Science*, *73*(10), 2731–2740.

Pearce, K. N., & Kinsella, J. E. J. (1978). Emulsifying properties of proteins: Evaluation of a turbidimetric technique. *Journal of Agricultural and Food Chemistry*, *26*(3), 716–723.

Pereira, R. N., Teixeira, J. A., Vicente, A. A., Cappato, L. P., da Silva Ferreira, M. V., da Silva Rocha, R., & da Cruz, A. G. J. (2018). Ohmic heating for the dairy industry: A potential technology to develop probiotic dairy foods in association with modifications of whey protein structure. *Current Opinion in Food Science*, *22*, 95–101.

Phillips, S. M. J. (2017). Current concepts and unresolved questions in dietary protein requirements and supplements in adults. *Frontiers in Nutrition*, 13.

Phongthai, S., Lim, S.-T., & Rawdkuen, S. J. (2016). Optimization of microwave-assisted extraction of rice bran protein and its hydrolysates properties. *Journal of Cereal Science*, *70*, 146–154.

Pietsch, V. L., Bühler, J. M., Karbstein, H. P., & Emin, M. A. (2019). High moisture extrusion of soy protein concentrate: Influence of thermomechanical treatment on protein-protein interactions and rheological properties. *Journal of Food Engineering*, *251*, 11–18.

Plant-based Meat - Avebe. (2020). Retrieved from https://www.avebe.com/markets/food/plant-based-meat/

Pojić, M., Mišan, A., & Tiwari, B. J. (2018). Eco-innovative technologies for extraction of proteins for human consumption from renewable protein sources of plant origin. *Trends in Food Science & Technology*, *75*, 93–104.

Qu, W., Zhang, X., Han, X., Wang, Z., He, R., & Ma, H. J. (2018). Structure and functional characteristics of rapeseed protein isolate-dextran conjugates. *Food Hydrocolloids*, *82*, 329–337.

Ramadhan, K., & Foster, T. J. J. (2018). Effects of ball milling on the structural, thermal, and rheological properties of oat bran protein flour. *Journal of Food Engineering*, *229*, 50–56.

Rayner, M. G., Rayner, J. L., & Miller, R. (2018). *Pseudo-loaf food compositions*. Google Patents.

Reineccius, G. (2005). *Flavor chemistry and technology*. CRC Press.

Ruíz-Henestrosa, V. P., Sánchez, C. C., Escobar, M. D. M. Y., Jiménez, J. J. P., Rodríguez, F. M., & Patino, J. M. R. J. (2007). Interfacial and foaming characteristics of soy globulins as a function of pH and ionic strength. *Colloids and Surfaces A: Physicochemical and Engineering Aspects*, 309(1–3), 202–215.

Saatchi, A., Kiani, H., & Labbafi, M. J. (2019). A new functional protein-polysaccharide conjugate based on protein concentrate from sesame processing by-products: Functional and physico-chemical properties. *International Journal of Biological Macromolecules*, 122, 659–666.

Samard, S., Gu, B. Y., & Ryu, G. H. (2019). Effects of extrusion types, screw speed and addition of wheat gluten on physicochemical characteristics and cooking stability of meat analogues. *Journal of the Science of Food and Agriculture*, 99(11), 4922–4931.

Samard, S., & Ryu, G. H. J. (2019). Physicochemical and functional characteristics of plant protein-based meat analogs. *Journal of Food Processing and Preservation*, 43(10), e14123.

Sathe, S., Deshpande, S., & Salunkhe, D. J. (1982). Functional properties of lupin seed (Lupinus mutabilis) proteins and protein concentrates. *Journal of Food Science*, 47(2), 491–497.

Schillberg, S., Raven, N., Spiegel, H., Rasche, S., & Buntru, M. J. (2019). Critical analysis of the commercial potential of plants for the production of recombinant proteins. *Frontiers in Plant Science*, 10, 720.

Schlegel, K., Sontheimer, K., Hickisch, A., Wani, A. A., Eisner, P., Schweiggert-Weisz, U. J. F. s., & nutrition. (2019). Enzymatic hydrolysis of lupin protein isolates—Changes in the molecular weight distribution, techno-functional characteristics, and sensory attributes. *Food Science & Nutrition*, 7(8), 2747–2759.

Schmitt, C., Sanchez, C., Desobry-Banon, S., & Hardy, J. J. (1998). Structure and technofunctional properties of protein–polysaccharide complexes: A review. *Critical Reviews in Food Science and Nutrition*, 38(8), 689–753.

Schreuders, F. K., Dekkers, B. L., Bodnár, I., Erni, P., Boom, R. M., & van der Goot, A. J. (2019a). Comparing structuring potential of pea and soy protein with gluten for meat analogue preparation. *Journal of Food Engineering*, 261, 32–39.

Schreuders, F. K., Dekkers, B. L., Bodnár, I., Erni, P., Boom, R. M., & van der Goot, A. J. J. (2019b). Comparing structuring potential of pea and soy protein with gluten for meat analogue preparation. *Journal of Food Engineering*, 261, 32–39.

Shand, P., Ya, H., Pietrasik, Z., & Wanasundara, P. J. (2007). Physicochemical and textural properties of heat-induced pea protein isolate gels. *Food Chemistry*, 102(4), 1119–1130.

Shen, Y., & Li, Y. J. (2021). Acylation modification and/or guar gum conjugation enhanced functional properties of pea protein isolate. *Food Hydrocolloids*, 117, 106686.

Shevkani, K., Singh, N., Kaur, A., & Rana, J. C. J. (2015). Structural and functional characterization of kidney bean and field pea protein isolates: A comparative study. *Food Hydrocolloids*, 43, 679–689.

Shoaib, A., Sahar, A., Sameen, A., Saleem, A., Tahir, A. T. J. (2018). Use of pea and rice protein isolates as source of meat extenders in the development of chicken nuggets. *Journal of Food Processing and Preservation*, 42(9), e13763.

Sim, S. Y., Karwe, M. V., & Moraru, C. I. J. (2019). High pressure structuring of pea protein concentrates. *Journal of Food Process Engineering*, 42(7), e13261.

Sim, S. Y., & Moraru, C. I. J. (2020). High-pressure processing of pea protein–starch mixed systems: Effect of starch on structure formation. *Journal of Food Process Engineering*, 43(2), e13352.

Singh, J. P., Kaur, A., Shevkani, K., & Singh, N. (2015). Influence of jambolan (S yzygium cumini) and xanthan gum incorporation on the physicochemical, antioxidant and sensory properties of gluten-free eggless rice muffins. *International Journal of Food Science & Technology*, 50(5), 1190–1197.

Singhal, A., Stone, A. K., Vandenberg, A., Tyler, R., & Nickerson, M. T. J. (2016). Effect of genotype on the physicochemical and functional attributes of faba bean (*Vicia faba* L.) protein isolates. *Food Science and Biotechnology*, 25, 1513–1522.

Sluik, D., Brouwer-Brolsma, E. M., Berendsen, A. A., Mikkilä, V., Poppitt, S. D., Silvestre, M. P., . . . Raben, A. J. T. (2019). Protein intake and the incidence of pre-diabetes and diabetes in 4 population-based studies: The PREVIEW project. *American Journal of Clinical Nutrition*, 109(5), 1310–1318.

Smetana, S., Pernutz, C., Toepfl, S., Heinz, V., & Van Campenhout, L. (2019). High-moisture extrusion with insect and soy protein concentrates: Cutting properties of meat analogues under insect content and barrel temperature variations. *Journal of Insects as Food and Feed*, 5(1), 29–34.

Smith, K. B., Fernandez-Rodriguez, M. Á., Isa, L., & Mezzenga, R. J. (2019). Creating gradients of amyloid fibrils from the liquid–liquid interface. *Soft Matter*, 15(42), 8437–8440.

Smyth, A., Griffin, M., Yusuf, S., Mann, J. F., Reddan, D., Canavan, M., . . . O'Donnell, M. J. (2016). Diet and major renal outcomes: A prospective cohort study. The NIH-AARP diet and health study. *Journal of Renal Nutrition*, 26(5), 288–298.

Stone, A. K., Avarmenko, N. A., Warkentin, T. D., & Nickerson, M. T. J. (2015). Functional properties of protein isolates from different pea cultivars. *Food Science and Biotechnology*, 24, 827–833.

Stone, A. K., Teymurova, A., Dang, Q., Abeysekara, S., Karalash, A., & Nickerson, M. T. J. (2014). Formation and functional attributes of electrostatic complexes involving Napin protein isolate and anionic polysaccharides. *European Food Research and Technology*, 238, 773–780.

Sun-Waterhouse, D., Zhao, M., & Waterhouse, G. I. J. (2014). Protein modification during ingredient preparation and food processing: Approaches to improve food processability and nutrition. *Food and Bioprocess Technology*, 7, 1853–1893.

Tang, C.-H., Wang, X.-Y., Yang, X.-Q., & Li, L. J. (2009). Formation of soluble aggregates from insoluble commercial soy protein isolate by means of ultrasonic treatment and their gelling properties. *Journal of Food Engineering*, 92(4), 432–437.

Tang, X., Shen, Y., Zhang, Y., Schilling, M. W., & Li, Y. (2021). Parallel comparison of functional and physicochemical properties of common pulse proteins. *LWT*, 146, 111594.

Thewissen, B. G., Celus, I., Brijs, K., & Delcour, J. A. J. (2011). Foaming properties of wheat gliadin. *Journal of Agricultural and Food Chemistry*, 59(4), 1370–1375.

Thirumdas, R., Brnčić, M., Brnčić, S. R., Barba, F. J., Gálvez, F., Zamuz, S., Lacomba, R., & Lorenzo, J. M. (2018). Evaluating the impact of vegetal and microalgae protein sources on proximate composition, amino acid profile, and physicochemical properties of fermented Spanish "chorizo" sausages. *Journal of Food Processing and Preservation*, 42(11), e13817.

Tielemans, S. M., Kromhout, D., Altorf-van der Kuil, W., & Geleijnse, J. M. J. N. (2014). Associations of plant and animal protein intake with 5-year changes in blood pressure: The Zutphen Elderly Study. *Metabolism, and Diseases, C*, 24(11), 1228–1233.

Tolouie, H., Mohammadifar, M. A., Ghomi, H., Hashemi, M. J. (2018). Cold atmospheric plasma manipulation of proteins in food systems. *Critical Reviews in Food Science and Nutrition*, 58(15), 2583–2597.

Tonstad, S., Stewart, K., Oda, K., Batech, M., Herring, R., Fraser, G. J. (2013). Vegetarian diets and incidence of diabetes in the Adventist Health Study-2. *Nutrition, Metabolism, and Cardiovascular Diseases*, 23(4), 292–299.

Van Ba, H., Hwang, I., Jeong, D., & Touseef, A. J. (2012). Principle of meat aroma flavors and future prospect. *Latest Research into Quality Control*, 2, 145–176.

Veraverbeke, W. S., & Delcour, J. A. J. (2002). Wheat protein composition and properties of wheat glutenin in relation to breadmaking functionality. *Critical Reviews in Food Science and Nutrition*, 42(3), 179–208.

Viguiliouk, E., Stewart, S. E., Jayalath, V. H., Ng, A. P., Mirrahimi, A., De Souza, R. J., . . . Leiter, L. A. J. (2015). Effect of replacing animal protein with plant protein on glycemic control in diabetes: A systematic review and meta-analysis of randomized controlled trials. *Nutrients*, 7(12), 9804–9824.

Wang, C., & Zayas, J. J. (1991). Water retention and solubility of soy proteins and corn germ proteins in a model system. *Journal of Food Science*, 56(2), 455–458.

Wang, X., Gao, W., Zhang, J., Zhang, H., Li, J., He, X., & Ma, H. J. (2010). Subunit, amino acid composition and in vitro digestibility of protein isolates from Chinese Kabuli and desi chickpea (*Cicer arietinum* L.) cultivars. *Food Research International*, 43(2), 567–572.

Wang, Y.-R., Yang, Q., Li-Sha, Y.-J., & Chen, H.-Q. J. (2021). Structural, gelation properties and microstructure of rice glutelin/sugar beet pectin composite gels: Effects of ionic strengths. *Food Chemistry*, 346, 128956.

Wee, M., Yusoff, R., Lin, L., & Xu, Y. J. (2017). Effect of polysaccharide concentration and charge density on acid-induced soy protein isolate-polysaccharide gels using HCl. *Food Structure*, 13, 45–55.

Wei, C.-K., Ni, Z.-J., Thakur, K., Liao, A.-M., Huang, J.-H., & Wei, Z.-J. J. (2019). Color and flavor of flaxseed protein hydrolysates Maillard reaction products: Effect of cysteine, initial pH, and thermal treatment. *International Journal of Food Properties*, 22(1), 84–99.

Wi, G., Bae, J., Kim, H., Cho, Y., & Choi, M.-J. (2020). Evaluation of the physicochemical and structural properties and the sensory characteristics of meat analogues prepared with various non-animal based liquid additives. *Foods*, 9(4), 461.

Wilding, P., Lillford, P. J., & Regenstein, J. M. J. (1984). Functional properties of proteins in foods. *Journal of Chemical Technology and Biotechnology. Biotechnology*, 34(3), 182–189.

Willett, W., Rockström, J., Loken, B., Springmann, M., Lang, T., Vermeulen, S., . . . Wood, A. J. (2019). Food in the Anthropocene: The EAT–Lancet Commission on healthy diets from sustainable food systems. *The Lancet*, 393(10170), 447–492.

Xiang, S., Zou, H., Liu, Y., Ruan, R. J. (2020). Effects of microwave heating on the protein structure, digestion properties and Maillard products of gluten. *Journal of Food Science & Technology*, 57, 2139–2149.

Xu, D.-P., Li, Y., Meng, X., Zhou, T., Zhou, Y., Zheng, J., . . . Li, H.-B. (2017). Natural antioxidants in foods and medicinal plants: Extraction, assessment and resources. *International Journal of Molecular Sciences*, 18(1), 96.

Yang, H., Su, Z., Meng, X., Zhang, X., Kennedy, J. F., & Liu, B. J. (2020). Fabrication and characterization of Pickering emulsion stabilized by soy protein isolate-chitosan nanoparticles. *Carbohydrate Polymers*, *247*, 116712.

Yang, L., Chen, J.-H., Zhang, H., Qiu, W., Liu, Q.-H., Peng, X., . . . Yang, H.-K. J. (2012). Alkali treatment affects in vitro digestibility and bile acid binding activity of rice protein due to varying its ratio of arginine to lysine. *Food Chemistry*, *132*(2), 925–930.

Yao, G., Liu, K., & Hsieh, F. (2004). A new method for characterizing fiber formation in meat analogs during high-moisture extrusion. *Journal of Food Science*, *69*(7), 303–307.

Yasumatsu, K., Sawada, K., Moritaka, S., Misaki, M., Toda, J., Wada, T., & Ishii, K. (1972). Whipping and emulsifying properties of soybean products. *Agricultural and Biological Chemistry*, *36*(5), 719–727.

Yildiz, G., Andrade, J., Engeseth, N. E., & Feng, H. J. (2017). Functionalizing soy protein nano-aggregates with pH-shifting and mano-thermo-sonication. *Journal of Colloid and Interface Science*, *505*, 836–846.

Yildiz, G., Ding, J., Andrade, J., Engeseth, N. J., & Feng, H. (2018). Effect of plant protein-polysaccharide complexes produced by mano-thermo-sonication and pH-shifting on the structure and stability of oil-in-water emulsions. *Innovative Food Science & Emerging Technologies*, *47*, 317–325.

Yu, M., He, S., Tang, M., Zhang, Z., Zhu, Y., & Sun, H. J. (2018). Antioxidant activity and sensory characteristics of Maillard reaction products derived from different peptide fractions of soybean meal hydrolysate. *Food Chemistry*, *243*, 249–257.

Yuliarti, O., Kovis, T. J. K., & Yi, N. J. (2021). Structuring the meat analogue by using plant-based derived composites. *Journal of Food Engineering*, *288*, 110138.

Zha, F., Dong, S., Rao, J., & Chen, B. J. (2019). Pea protein isolate-gum Arabic Maillard conjugates improves physical and oxidative stability of oil-in-water emulsions. *Food Chemistry*, *285*, 130–138.

Zhang, J., Liu, L., Jiang, Y., Shah, F., Xu, Y., & Wang, Q. (2020). High-moisture extrusion of peanut protein-/carrageenan/sodium alginate/wheat starch mixtures: Effect of different exogenous polysaccharides on the process forming a fibrous structure. *Food Hydrocolloids*, *99*, 105311.

Zhang, X., Qi, J.-R., Li, K.-K., Yin, S.-W., Wang, J.-M., Zhu, J.-H., & Yang, X.-Q. J. (2012). Characterization of soy β-conglycinin–dextran conjugate prepared by Maillard reaction in crowded liquid system. *Food Research International*, *49*(2), 648–654.

Zhang, X., Wang, L., Chen, Z., Li, Y., Luo, X., & Li, Y. J. (2020). Effect of high energy electron beam on proteolysis and antioxidant activity of rice proteins. *Food & Function*, *11*(1), 871–882.

Zhao, H., Song, A., Zheng, C., Wang, M., & Song, G. J. (2020). Effects of plant protein and animal protein on lipid profile, body weight and body mass index on patients with hypercholesterolemia: A systematic review and meta-analysis. *Acta Diabetologica*, *57*, 1169–1180.

Zhao, Y., Li, X., Liu, Y., Zhang, L., Wang, F., & Lu, Y. (2017). High performance surface-enhanced Raman scattering sensing based on Au nanoparticle-monolayer graphene-Ag nanostar array hybrid system. *Journal of Chemical Sciences*, *247*, 850–857.

Zheng, J., Gao, Q., Tang, C.-h., Ge, G., Zhao, M., & Sun, W. J. (2020). Heteroprotein complex formation of soy protein isolate and lactoferrin: Thermodynamic formation mechanism and morphologic structure. *Food Hydrocolloids*, *100*, 105415.

Zhou, X., Cui, H., Zhang, Q., Hayat, K., Yu, J., Hussain, S., . . . Ho, C.-T. J. (2021). Taste improvement of Maillard reaction intermediates derived from enzymatic hydrolysates of pea protein. *Food Research International*, *140*, 109985.

Žugčić, T., Abdelkebir, R., Barba, F. J., Rezek-Jambrak, A., Gálvez, F., Zamuz, S., Granato, D., & Lorenzo, J. M. (2018). Effects of pulses and microalgal proteins on quality traits of beef patties. *Journal of Food Science and Technology*, *55*, 4544–4553.

2 Challenges in Designing of Plant Protein-Based Future Foods

Shreya Mandal

2.1 INTRODUCTION

The agricultural and food sectors have produced an abundant supply of nutritious foods during the past century or more, which has contributed to a decrease in hunger and malnutrition worldwide (WRI, 2019). By 2050, the United Nations projects that there will be 9.7 billion people on earth (United Nations, 2019). Proteins are the most important source of amino acids in the human diet, and they frequently encode peptides that are involved in biological processes (Corredig et al., 2020). The market for meat products has risen at an unbelievable rate due to the rapid growth of the people worldwide and the economic growth of developing nations. The demand for meat worldwide has increased by 58% during the last two decades (Whitnal et al., 2019). However, there are a variety of constraints on the production and consumption of meat, such as high resource use, pollution, residue from animal antibiotics, and zoonotic diseases (Zhang et al., 2022). The production of meat is being challenged by problems including climate change, water and land scarcity, and animal welfare. The animal protein-based diet has also come under fire for having a high carbon footprint; ineffective production methods; a possible lack of nutritional balance; and lifestyle disorders including type 2 diabetes, cardiovascular disease, and cancer (Hu et al., 2019). Considering these issues and the rising demands for meat products, research has turned to the production of meat substitutes, particularly plant-based meat substitutes. Since 2010, a multitude of alternative protein (AP) products have been produced, including plant-based meat that resembles the appearance ("the bleeding burger"), texture (mouthfeel), flavour (meaty/savory), and nutritional profiles (iron, vitamins, etc.) of animal goods (Andreani et al., 2023; Zhang et al., 2022). The plant protein-based foods (PPBFs) were traditionally made with soy and wheat proteins. Proteins from different oilseeds (such as canola and sunflower), cereals (such as rice), legumes, and pulses (such as lupin, beans, peas, and lentils) are also used as sources of protein (Arrutia et al., 2020). These products tend to be more processed than traditional meat products. By considering all the crucial elements of developing novel food items of PPBF, this chapter attempts to highlight the major hurdles associated with the manufacturing and commercialization PPBF products. In fact, the author discusses the nutritional profiles of PPBFs, their impact on health, role in environmental sustainability, limitations in production technology as well as customer preferences. Insights of other opinions by critically evaluating recent studies from various disciplines to highlight debate and hurdles on this topic from an interdisciplinary perspective are staged. Each of the seven sections will represent a brief overview of the current situation. Figure 2.1 shows major challenges in designing plant protein-based future foods.

2.2 POTENTIAL NUTRITIVE VALUE

Functional characterization, which evaluates various protein quality parameters such as amino acid profile (most crucial), thermal stability, molecular weight distribution, structural morphology, and food functional properties (cookability, water-holding capacity, oil-holding capacity, emulsification activity, protein stability, protein solubility), and in-vitro digestibility, enables us to determine the

DOI: 10.1201/9781003369790-2

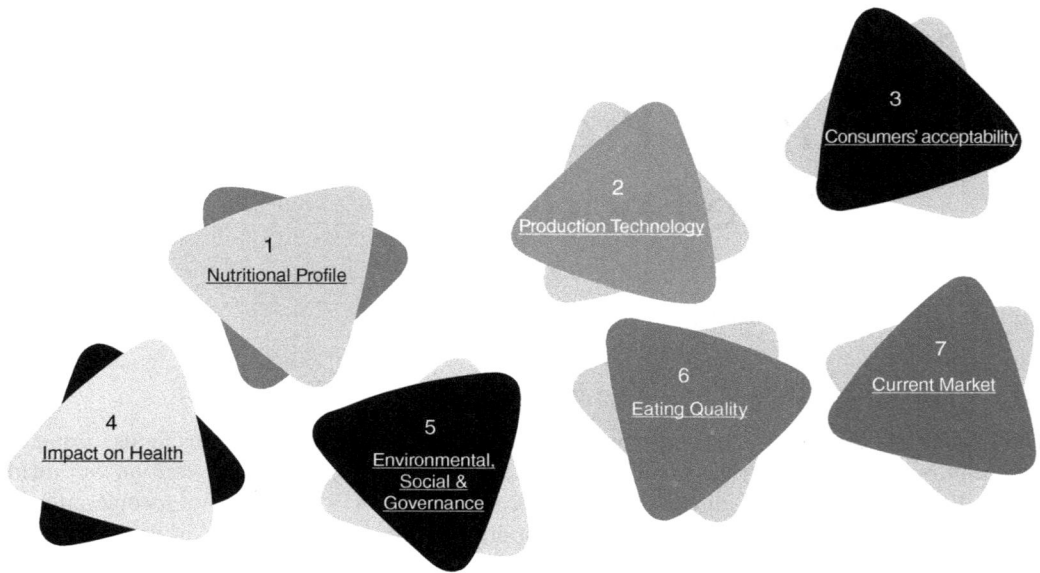

FIGURE 2.1 Major challenges in designing plant protein-based future foods (PPBFs).

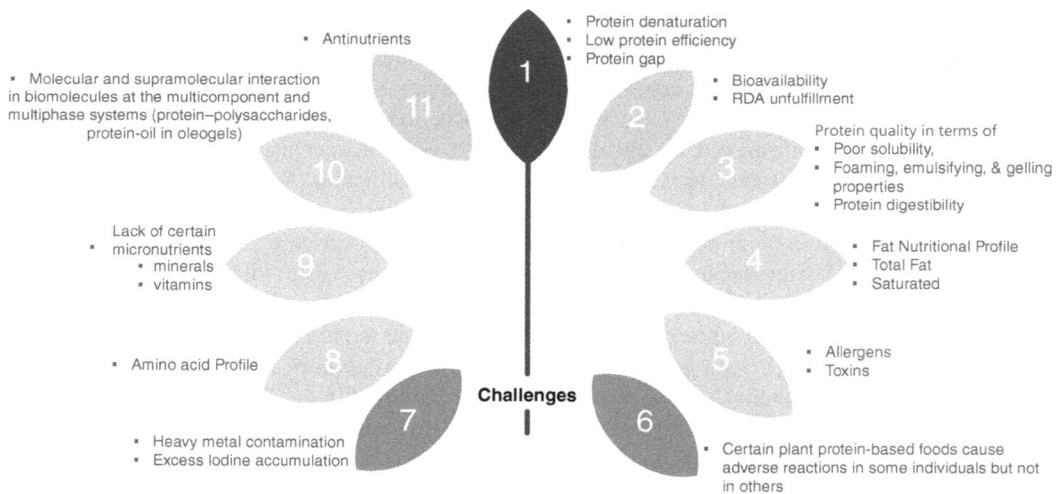

FIGURE 2.2 Unbalanced nutritional profile of plant protein-based future foods (PPBFs).

precise application of plant-derived proteins in food systems (Kumar et al., 2022). Various new sources of high-quality plant proteins are present, but their use as dietary supplements is restricted to their accessibility, purity, and consumer preferences. Figure 2.2 shows unbalanced nutritional profile of plant protein-based future food.

2.2.1 Protein Quality

Animal proteins have a balance of both essential and non-essential amino acids, but plant proteins are thought to be nutritionally deficient because they lack some essential amino acids

(Berrazaga et al., 2019). For instance, most cereal proteins are low in lysine, while pulse proteins are deficient in methionine, cysteine, and tryptophan. Application of plant-based proteins in PPBF items is restricted by their low protein concentration and absence of key amino acids. In this regard, a recent study by Cutroneo et al. (2022) examined the nutritional value of 269 commercial meat substitutes currently available on the Italian market by extracting information from their food labels. Burgers and meatballs made from plants contained less protein than their animal equivalents.

2.2.2 ANTINUTRIENTS

High quantities of phytic acid in PPBF can limit the absorption of trace elements and minerals, including iron, zinc, calcium, and manganese (Schonfeldt and Gibson Hall, 2012).

2.2.3 DIGESTIBILITY AND BIOAVAILABILITY

Protein availability for intestinal absorption is determined by in-vitro digestibility. Plant proteins are less digestive than animal proteins. Phillips et al. (2017) reported that animal proteins have the greatest PDCAAS/DIAAS scores (>0.9), while plant proteins frequently have lower values (0.4–0.9). The poor digestibility of plant proteins is caused due to three factors primarily:

(i) presence of *anti-nutrients* like phytates and trypsin inhibitors that prevent the digestion and absorption of proteins (Gilani et al., 2005)
(ii) structural variations i.e. plant proteins contain more sheet structures and fewer helixes than animal proteins, which promote protein aggregates (Carbonaro et al., 2012)
(iii) presence of *dietary fibre* that reduces plant proteolytic digestibility (Duodu et al., 2003)

2.2.4 TOXICITY AND ALLERGENICITY

Toxicity is another issue in using plant proteins, which can be caused by substances like gossypol, which is harmful in cottonseed protein. Cheney et al. (2016) reported many hazardous chemicals, including aplysiatoxin, debromoaplysiatoxin, prostaglandins, polycavernosides, and the diethyl peroxides caulerpin and caulerpicin, in protein isolate from natural seaweeds.

2.2.5 FLAVORING

The Maillard reaction may be used to impart a meaty taste to popular plant-based proteins. Corredig et al. (2020) reported that the development of Maillard products leads to a loss of critical lysine. More research might be done to exclude sulphur-containing molecules (such as cysteine, taurine, and thiamine) from the Millard reaction and instead employ only the free amino acids or peptides from the plant protein hydrolysates to react with the reducing sugars (Yong et al., 2021).

2.2.6 COLORING AGENTS

Hu et al. (2019) raised concern about the inclusion of LegH as coloring agent in PPBF products, indicating that prolonged heme iron intake increased diabetes risk.

2.2.7 FUNCTIONAL PROPERTIES

The functional characteristics of a protein that affect how it behaves during storage, processing, preparation, and consumption are foaming, emulsification, gelation, etc. (Kumar et al., 2022). Since proteins do not readily form networks in oil, it is difficult to use protein as oil structurants.

More research is required on protein-based oleogels, particularly those containing plant proteins (Feichtinger et al., 2020).

2.2.8 DOSAGE AND PRODUCT LABELLING

Phytoestrogens (anti-estrogenic/endocrine disruptors) are frequently cited as a negative regulator of phytoestrogen ingestion. ANSES (National Agency for Food, Environmental and Occupational Health Safety; previously AFSSA, 2007) proposed the following labelling: (i) for soy foods (tonyu, miso, tofu, yoghurt, and soy desserts): "Contains Xmg of isoflavones" (phytoestrogens); (ii) moderate consumption (1 mg/kg of body weight daily maximum); (iii) not suggested for youngsters younger than three; (iv) "The recommended daily portion of soybeans should not result in a daily consumption of isoflavones exceeding 1 mg/kg body weight" (expressed as aglycone of the main component). The label must advise against usage by women with a personal or familial history of breast cancer; (v) due to the presence of coumestrol, coumarin, and alkaloids, alfalfa packaging must include a warning against its use by women with a personal or familial history of breast cancer (Elakovich and Hampton, 1984).

2.2.9 ANABOLIC RESPONSE

Eliminating the anti-nutrients or increasing the consumption of plant-based proteins may not enhance the required amino acid content (leucine), because plant proteins would not induce an equivalent anabolic response, unlike animal proteins (Boye et al., 2010). Van Vliet et al. (2015) has reported that, compared to their animal counterparts, plant-based proteins are less effective at boosting postprandial anabolic rates, *i.e.* postprandial muscle protein synthesis. This raises the question of whether the chronic consumption of plant-based proteins versus animal-based proteins would result in different phenotypic results, notably disparities in muscle mass.

The obstacle for the food industry to generate PPBF products is robust supply of their major ingredient i.e., high-quality plant protein with acceptable flavour, texture, colour, mouthfeel, and availability or affordability since the performances of plant-based proteins are constrained owing to their subpar functioning. The nutritional profiles of today's plant-based meat alternatives are more processed (additives, preservatives such as antimicrobial and antioxidants, buffer, crosslinking agents) than those of traditional meat alternatives, according to a comparison of front-of-package labels and online information (Bohrer, 2019). Sodium and saturated fat levels in many vegetarian and vegan meat substitutes are also greater than in meat (Tso and Forde, 2021). Very few businesses disclose their products' iron and vitamin B12 levels, as well as the presence or absence of preservatives (Lacy-Nichols et al., 2021). Traditional plant-based diet food preparation techniques are vastly dissimilar from those employed in the creation of refined plant-based components in formulations and the fabrication of plant protein-based food (PPBF) products. Investigating whether or if the long-term health impacts of eating PPBF items differ from those of eating traditional plant-based diets is important (Toh et al., 2022). Therefore, inclusive research is needed on the effects of changes in plant protein-derived amino acid profiles, protein digestion, absorption, and allergenicity on human health and well-being (McClements, 2021).

2.3   PRODUCTION TECHNOLOGY

Proteins in food function as both nutrients and structural constituents. The structural property usually modulates protein's techno-functionality (Foegeding, 2015). The design of processes, upscaling strategies, low-cost manufacturing, choice of appropriate plant protein sources, and food safety are all in need of comprehensive research, which is limited till date. Additionally, intensive research is required to improve PPBF products' quality, assure the sustainability of their production, and

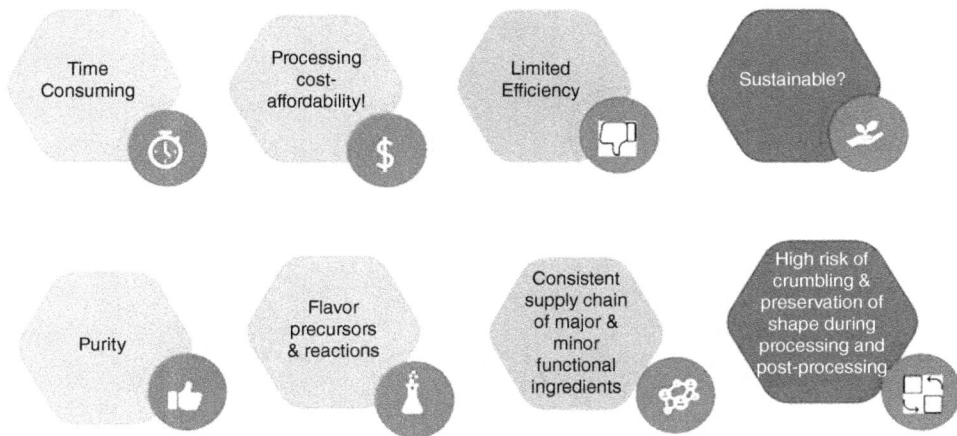

FIGURE 2.3 Challenges of plant protein-based foods' (PPBFs') production.

build a system for evaluating food safety (Zhang et al., 2022). Figure 2.3 shows challenges of plant protein-based food (PPBF) production.

2.3.1 SOURCE OF PLANT PROTEIN

Reconstructing the globular molecular structure of plant protein in the fibrous structure of muscle proteins is the most difficult aspect of PPBF production along with the colour, flavour, and aroma of real animal protein (Zhang et al., 2022). The difficulties vary depending on the kind of raw material used (Corredig et al., 2020). The lower sensory quality of plant-based alternatives to meat is viewed to be the most significant impediment to their broad adoption. Various plant-derived proteins are currently employed in the production of meat replacements. The selection of protein sources has a substantial effect on the perceived flavour characteristics of the finished product. To further examine the origin and identity of off-flavor chemicals from various plant protein sources, more systematic investigations focusing on their origin are required (Wang et al., 2022). Green sources with drastically differing proteins and physiological and biochemical barriers to their efficient isolation include leafy greens, grass, and algae. There are issues with yields, solubility, technological qualities, taste, and digestibility due to the cell architecture, intertwined fibres, strong cell walls, and phenolics found in the raw material (Tenorio et al., 2018). Interestingly, fat is an essential component of meat products because it makes them juicy, tender, and tasty (López-Pedrouso et al., 2021). Nonetheless, high fat content is deleterious to the creation of fibre structure, assembly of macromolecules, and during extrusion (Kyriakopoulou et al., 2019). An overview of these serious hurdles is presented in the subsequent sections.

2.3.2 SAFETY

Storage life of plant protein-based food items and food safety of plant-based meat alternatives with longer shelf-life are critical in the manufacturing of PPBF products (Tyndall et al., 2022). Despite extensive research on the microbiological and chemical safety of PPBF products, there is a dearth of scientific evidence about the safety of PBMA. PPBF items are prone to microbial contamination due to high moisture with neutral pH (He and Evans, 2020). Although the primary functional ingredients are healthy, there is a concern that nutritive value may be lost during processing (Choudhury et al., 2020). Nevertheless, the long-term health implications of other additives are yet unknown.

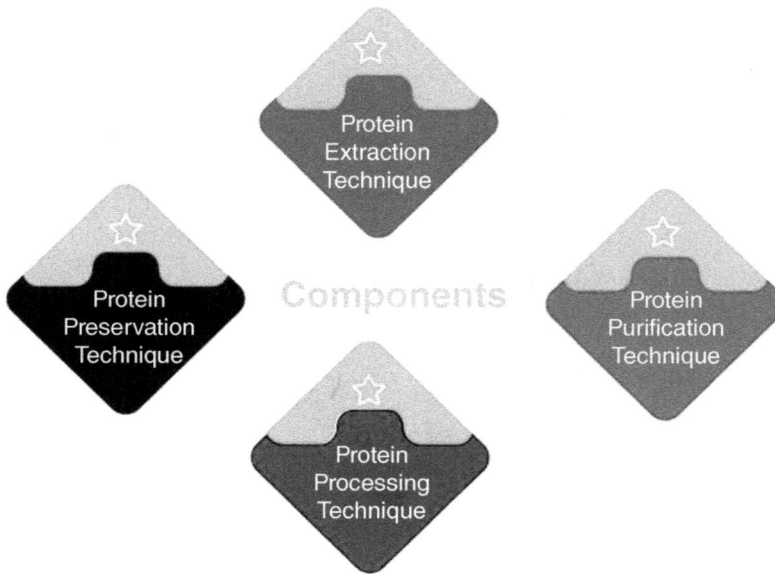

FIGURE 2.4 4Ps of production technology of plant protein-based foods (PPBFs).

Therefore, specifications for the usage of its components in PPBF products must be observed. Figure 2.4 shows 4Ps of production technology of plant protein-sourced foods.

2.3.3 SUPPLY CHAIN

Despite technology advancements in recent decades, establishing robust supply chains remains a challenge (Corredig et al., 2020).

2.3.4 PURITY

Plant protein components frequently contain contaminants that alter their functional properties, including dietary fibres, carbohydrates, other proteins, lipids, phenolic chemicals, and minerals (McClements, 2021).

2.3.5 INCONSISTENCY

Functional performance of commercial plant protein components frequently varies from batch to batch due to variations in their composition, denaturation state, aggregation state, and contaminants (McClements, 2021).

2.3.6 PROTEIN EXTRACTION

Numerous plant proteins are far less hydrophilic than animal proteins and cannot be readily isolated from plant material in their original state (Li and Vries, 2018). The poor solubility and higher molecular weight of plant protein sources make it difficult to identify suitable solvents for processing plant proteins in solution, thereby adding extra cost in recovery and extraction (Day et al., 2013). When considering the utilization of plant protein–polysaccharide complexes for interfacial stability, this presents a new set of issues (Li and Vries, 2018). Current research is still largely empirical, and most published articles are focused on single specific uses for a single plant protein/polysaccharide combination (Li and Vries, 2018). Although heating and acid precipitation offer great recovery rates, protein functionality is compromised due to its denaturation. Depending on the extraction procedure, the presence of various antinutritional metabolites glycoalkaloids (solanine, chaconine) can further deteriorate the quality of potato protein isolate (Nielsen et al., 2020).

2.3.7 PROTEIN PURIFICATION

Following extraction, plant proteins are purified and isolated from the extraction solvent. Ultrafiltration is a better choice to separate protein isolates since isoelectric precipitation process has negative impacts on protein techno-functional attributes (Yong et al., 2021).

2.3.8 PROTEIN PROCESSING

In *extrusion technology (a top down approach of structuring techniques)*, forecasting the thermo-mechanical dynamics during the material's flow through the extrusion barrel in order to maximize the material's structural and technological qualities is difficult (Corredig et al., 2020; Emin et al., 2017). While *bottom-up approaches* to structuring phenomena in multiphase systems of PPBF food production instrumentation have the potential to resemble the fibrous structure of meat, top-down approaches are better scalable and more efficient in their use of resources (Dekkers et al., 2018). In contrast, the top-down approach, while being effective, is more difficult to implement and can only produce the desired structure at larger length scales. Dekkers et al. (2018) also highlighted that in-situ analysis techniques to evaluate the structure's development during processing are urgently needed. Both *mechanical techniques of texturization, Warner-Bratzler and Kramer Shear Cell*, do not assess a single mechanical property which is a limitation of these methods (Xiong et al., 2006). To compare different products produced from plant protein sources as raw ingredients, it is crucial to have standardized texturization procedures for all mechanical techniques. Thereafter, *imaging techniques* for the characterization of PPBF products, such as *TEM and AFM*, have only been applied to meat to date (Schreuders et al., 2021). *Spectroscopy* is used to gain insight into the local composition, intermolecular interaction, and anisotropy of PPBF products like meat and meat analogues, but it is also quite costly (Schreuders et al., 2021). Large-scale production of *plant protein glycation* (non-enzymatic glycosylation happens at the beginning of the *Maillard reaction*) to improve the technological and functional properties of proteins is not yet possible (Kutzli et al., 2021). The expensive freeze-drying stage is the main barrier to a practical deployment of the *dry-state heating technique* for the manufacture of glycated proteins on an industrial scale (Zhu et al., 2010), and the *wet-state technique* only results in low yields (Zhu et al., 2008). Therefore, it is not yet viable to decipher the results of the study on how glycation affects the allergenic potential of soy proteins (Kutzli et al., 2021). Polysaccharides in *protein–polysaccharides molecular interaction* may even degrade techno-functional qualities of plant protein by decreasing solubility and foaming ability (Stone et al., 2014). More research is required to comprehend how these occurrences occur. Besides, *molecular and supramolecular interactions* within mixed systems remain mostly unexplored. Without these investigations at different length scales, it is difficult to predict, control, and enhance their behaviours in response to formulation or processing alterations (Jim et al., 2021; Alves and Tavares, 2019). The rheological characteristics of plant proteins relevant to the PPBF product-specific processing circumstances are shown systematically in *texture maps* (Schreuders et al., 2021). In *protein blends*, extremely system-specific synergistic or antagonistic properties have been described, indicating the significance of understanding the details of the interactions, as a function of processing conditions, and demonstrating that understanding the fundamentals of protein structure and function in general is insufficient (Jim et al., 2021; Alves and Tavares, 2019). In protein beverages, the use of high-temperature treatments such as UHT with or without steam injection can cause heat-induced aggregation events, the formation of off flavors, and a reduction in protein nutritional quality (Rivera del Rio et al., 2019). However, research on the influence of processing to improve the solubility of plant protein blends is limited. The molecular intricacies of heat-induced protein–protein interactions are lacking, and there are no reports on how the nutritional quality of proteins may be changed by these processing methods (Jim et al., 2021). For most plant proteins, as well as plant–protein blends, the ideal circumstances and material composition required to produce the desired structure remain unclear (Jim et al., 2021). As the details of the *molecular and supramolecular interactions* as a function of other components and environmental and processing conditions are unknown, it is necessary to investigate the nutritional and technological functionalities of blends compared to the proteins in isolation, particularly in mixed matrices. Protein blend is highly reliant on processing circumstances, as well as on the diversity of the source, the purity of the ingredients, the ratio and concentration of the blends, and their interactions with other system components (Jim et al., 2021). Soon, research into the impact of processing on the creation of distinct *supramolecular structures* and the disruption of these structures in the gastrointestinal tract will provide the essential information required to build the next generation of plant protein-based diets (Jim et al., 2021). Besides, the underlying principles driving structural changes and interactions in protein systems are well understood, but the knowledge of complicated matrices is inadequate. Corretedig et al. (2020) highlighted that there is a need of studying molecular and supramolecular interactions in *multi-component, multi-phase systems*, as well as the need for a better understanding of the effect of processing history on the functionality of new developing climate-friendly chemicals.

2.3.9 PROTEIN PRESERVATION AND ADDITIVES

Unprocessed plant–protein mixtures prepared from soy or pea protein frequently have an unpleasant bitter taste caused by saponins and volatile off-flavours owing to by-products of lipid oxidation such alcohols, aldehydes, and ketones (Roland et al., 2017). The use of flavoring agents and other additives, such as preservatives and texturizing agents, combined with high levels of saturated fat, poses issues with regard to nutritional benefits on consumer's health, "clean" label, food safety, and willingness to pay (Tuccillo et al., 2022). Numerous plant-based lipids include relatively large quantities of unsaturated fatty acids, which presents a problem in this area.

2.4 CUSTOMER ACCEPTABILITY

Stakeholders in the plant protein-based foods (PPBF) industry who are hoping to expand their market share are especially interested in knowing how well these products are received by consumers. According to published polls that used consumers to directly analyze consumer sentiments about plant protein-based meals (PPBM), most plant protein-based foods (PPBF) remain unacceptably low such as unsaturated fatty acids are susceptible to lipid oxidation during storage and processing, resulting in the development of rancid reaction in products that consumers find unappealing and that may be hazardous (He and Evans, 2020). Organoleptic quality (i.e. taste, flavour, texture, color, and appearance), affordability, convenience, health and wellness, safety, environmental sustainability, and unfamiliarity with the product are the main challenges to consumer adoption of plant protein-based foods (Tyndall et al., 2022; Boukid et al., 2021). Neophobia, societal conventions and rituals, varying eating goals of consumers (Jahn et al., 2021), and their inability to adhere to plant-based diets are all factors that prevent many consumers from choosing meals with plant proteins (Graça et al., 2019). Approximately 65% of buyers would continue to buy traditional meat products, while only 21% would pick plant protein-based foods (PPBFs) in an experiment on the preference of consumer foods conducted by Slade (2018). There is a shortage of knowledge regarding consumer views of various plant protein-based foods (PPBF) types and the factors that influence their consumption, so these are explained in the following section.

2.4.1 ORGANOLEPTIC QUALITY (TASTE, FLAVOUR, TEXTURE, COLOR, AND APPEARANCE)

2.4.1.1 Taste
The taste is the one of main reasons why people didn't eat more plant proteins. People in the western countries don't prefer seaweed as a suitable source of PPBF very often because the taste is often mistaken for "fishy" and "sea taste."

2.4.1.2 Texture
First generation of PPBF products, such as those made from mycoproteins, were poorly received in terms of taste and texture. As a result, meat eaters are less likely to consider such goods as actual meat alternative (Andreani et al., 2023; Hashempour-Baltork et al., 2020).

2.4.1.3 Flavour
It is challenging to recapitulate the flavour of animal meat in PPBF products yet averting undesirable flavours (such as bitter, burned, earthy, green, mushroomy) generated by the high percentage of legume protein (Giacalone et al., 2022; McClements 2021).

2.4.1.4 Appearance
It is typical to evaluate a product's look first because it is a key factor in determining whether consumers would accept the food. The appearance and texture of complete pieces of meat (e.g. steak) are more difficult to replicate (Rubio et al., 2020).

2.4.1.5 Color
Plant-based products may lose their colour when exposed to light or oxygen, or the taste may change when lipids oxidize, both of which are not good (Fiorentini et al., 2020). The final PPBF product's colour is significantly influenced by the colour of the seaweed used. In a trial on the preparation of new generation of PPBF products, e.g. sausages with sea tangle powder, colour acceptability was quite low (Raja et al., 2022; Cofrades et al., 2017). In a sensory panel study reported by Schouteten et al. (2016), which compared animal, plant, and insect-based burgers, emotional adjectives "contented, happy, and pleasant" were linked to animal burgers, while the words "disappointed, distrust, and discontented" were linked to plant burgers.

2.4.2 PROCESSED AND UNHEALTHY FOOD

During the development stage, the use of various chemicals is required to replicate the qualities of meat. Consequently, the product packaging for PPBF products frequently includes a lengthy list of unfamiliar ingredients, which may give customers the impression that they are purchasing highly processed and extremely unhealthy food (Hartmann et al., 2022). When compared to traditional sources of animal protein, PPBF can be considered "highly processed," which might scare away "clean label" consumers from consuming, who are suspicious of "unnatural" approaches to the manufacturing of food (Asioli et al., 2017). Research of European consumers conducted in 2017 indicated that the lack of naturalness associated with cultured meat was a factor that hindered acceptance of the product, even when they knew it could be better for the environment and animals (Siegrist and Sütterlin, 2017).

2.4.3 CONSUMERS' PERCEPTION

One of the most critical deciding factors in consumer acceptance of a product is how little the consumer knows about it, including how to prepare or cook it. This could make it harder to reach more people in the general consumer market. To completely evaluate consumers' acceptance of

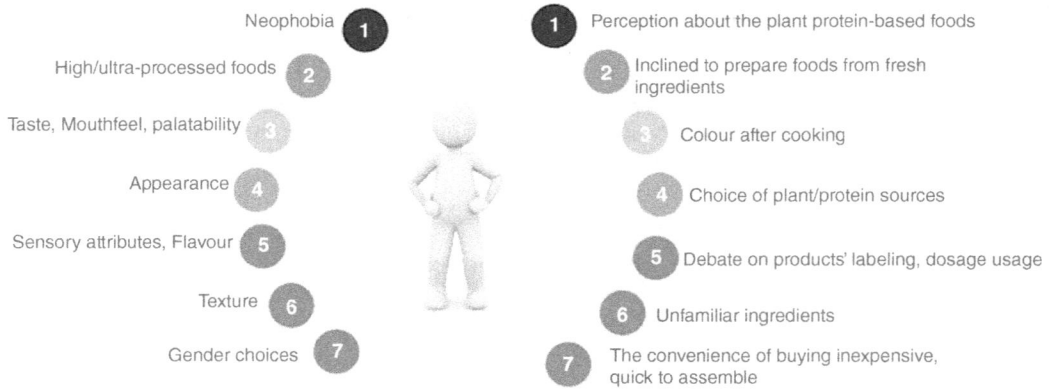

Neophobia 1

High/ultra-processed foods 2

Taste, Mouthfeel, palatability 3

Appearance 4

Sensory attributes, Flavour 5

Texture 6

Gender choices 7

1 Perception about the plant protein-based foods

2 Inclined to prepare foods from fresh ingredients

3 Colour after cooking

4 Choice of plant/protein sources

5 Debate on products' labeling, dosage usage

6 Unfamiliar ingredients

7 The convenience of buying inexpensive, quick to assemble

FIGURE 2.5 Factors influencing a consumer's attitude for purchasing.

PPBF, individuals should have first-hand experience (Andreani et al., 2023). The findings repeatedly show that respondents preferred traditional meat products to plant-based alternatives (Andreani et al., 2023). In terms of demographics, habits, and attitudes, consumers who are health-conscious, environmental enthusiasts, and young are more likely to embrace PPBFs (Giacalone et al., 2022). Consumers' views on animal protein alternatives and, more generally, their views on food innovation may also play a role. However, the biggest personal barriers to acceptability are connected to people's aversion to new foods and food technologies, their emotional attachment to meat, and the social contexts in which they are less likely to choose to use plant protein sources (Michel and Siegrist, 2019). Multiple studies have indicated that compared to flexitarians, people who regularly consume large amounts of animal protein may be less likely to switch to plant-based protein sources (Andreani et al., 2023). On the other side, vegetarian or vegan consumers are not on the lookout for the animal-meat-like sensory qualities in plant-based food products (Kerslake et al., 2022). In the subsequent research work, consumers were asked to evaluate combinations of animal-protein-derived food items, and the results suggested that the context of the meal, such as the type of dish, works as a substantial influence on the adoption of meat alternatives (He and Evans, 2020). The selection of plant/protein sources has a significant impact on sensory qualities and customer acceptance (Fiorentini et al., 2020). How meat alternatives are made from plant protein sources may also affect how people think about it. When a panel of people were asked how they felt about "animal-free", "clean", and "cultured", the "lab-grown," "animal-free", and "clean" terms made them feel better than "lab-grown." Besides, words that are often used to describe them are "artificial," "cell-based," "cultivated," "in vitro," and "synthetic" (Bryant and Barnett, 2019). Figure 2.5 shows factors influencing consumer's attitude for purchasing.

2.4.4 COST AND CONVENIENCE

Cost is a significant element influencing consumers' readiness to buy. To encourage wider adoption of PPBFs as a lifestyle, they must provide more direct benefits to the customer, such as being less expensive and easier to use (Datar et al., 2010). As reported by Caputo et al. (2022) in a choice experiment of sensory investigation between two groups (i) blind-informed and (ii) informed group, the beef burger earned the highest willingness to pay (WTP) in comparison to two Plant-Based Meet Alternatives (PBMAs) and one hybrid burger (70% beef and 30% mushrooms). This was because beef burger was the most familiar of the three options. Additionally, they observed that the informed group's preference and WTP for the plant-based patty labelled as "produced with animal-like protein" outweighed those for the hybrid burger and the plant-based burger "made with pea protein." Though vegetarians and vegans are both inclined to agree with the prospective benefits of PPBFs but less willing to try them (Wilks and Phillips, 2017). Many people believe that vegetarian diets are expensive compared to their current eating habits (Pohjolainen et al., 2015). Moreover, people are dissuaded from adopting a plant-based diet for a variety of practical reasons, including the difficulty of producing excellent vegetarian cuisine and the lack of choices available at restaurants (Lea et al., 2006). People prefer the ease of purchasing inexpensive, easy-to-prepare meals made from fresh ingredients. The Food & Health Survey conducted by the IFIC Foundation (n = 1,000) in 2020 revealed that Americans prioritize affordability, taste, and convenience above health and environmental sustainability as factors that affect their purchase and eating decisions (Kraak, 2022; International Food Information Council Foundation, Food & Health Survey, 2020).

2.4.5 GENDER CHOICES AND SOCIAL BARRIER

Recent survey conducted by Raja et al. (2022) suggests that young men liked seaweed-based (PPBFs) snacks and fast food more than young women. There may be a social barrier stopping individuals from altering their diets, since they may be unable to forego meat when most of their relatives and acquaintances are strict meat eaters (Lea et al., 2006).

2.5 SCOPE OF MARKETING

To meet future needs, alternative plant protein sources with more desired functional characteristics must be explored. Current plant protein procurement and restructuring techniques for the creation of plant-based meat are mostly guided by trial-and-error methods and practical experience (Li, 2020). PBMAs are gaining popularity among customers globally; yet, the small, young sector is only worth $939 million and accounts for less than 1% of total meat sales in the United States in 2019 (Good Food Institute, 2019). The traditional meat industry is litigating the plant-based meat sector for regulatory and legal classification. When producing these types of alternative food items, a variety of challenges impacting the marketing and regulation of plant protein-based food products must also be addressed. Figure 2.6 represents marketing challenges to promote plant protein-based food products.

2.5.1 MARKET OPPORTUNITY

Lack of understanding about the relative benefits and drawbacks of various plant-derived ingredients and production methods, as well as concerning safety issues (including allergenicity), international laws, and supply chain problems, are preventing the marketing of plant-based foods (McClements, 2021). The global market revenue for plant-based meat substitutes is anticipated to be worth USD 33.99 billion in 2027 (Global: Meat Substitutes Market Revenue 2016–2027|Statista, 2022), in contrast to the USD 1354 billion expected for the meat industry by that year (Global Meat Industry Value Projection, 2021–2027|Statista, 2022) (Andreani et al., 2023). As a result, PPBFs' market share is expected to continue being much lesser compared to that of the meat industry (Andreani et al., 2023). This analysis projects that plant protein-based foods are usually used to supplement animal protein industry rather than to replace it completely (Neuhofer et al., 2022). Using Mintel's Global New Product Database (GNPD), Andreani et al. (2023) translated that 120 distinct claims were found in the packaging's informational materials. Out of 4,965 goods, 2,849 (57%) made the claim that they were "Vegan/No Animal Ingredients," and 2,099 (42%) said they were "Plant Based." Additionally, in keeping with this, the statement "High/Added Protein" was the most frequent nutrition claim (n = 1,616; 33%).

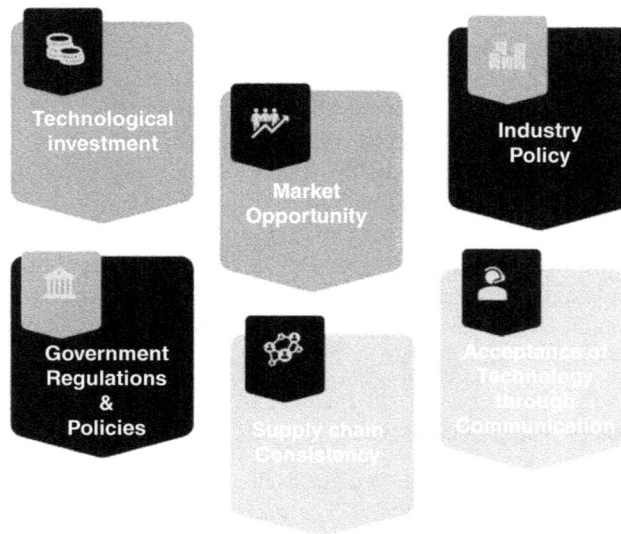

FIGURE 2.6 Marketing challenges to promote plant protein-based food products (PPBFs).

2.5.2 TECHNOLOGICAL INVESTMENT

To address dietary needs, it's critical to create new plant-based meat products with equivalent and even more desired sensory, nutritional, and nutrient properties. This process will need expanding scientific research in R&D sector into the functional minor and major ingredients of plant protein-based foods as well as production techniques (Li, 2020). The investment in plant-based meat alternatives' start-ups by business luminaries such as Bill Gates and big food firms such as Tyson Foods (United States) and Cargill (United States) has drawn the attention of other investors. Tyson Foods partnered with Beyond Meat in 2016, and the company collected more than $200 million in funding over the next two years. Beyond Meat filed for an initial public offering (IPO) in 2019, with a valuation of $1.5 billion. Impossible Foods acquired US$1.2 billion in investment after finishing a series F funding round of US$500 million in March 2020 (McClements, 2021).

2.5.3 PATENT STATISTICS

This quickly growing acceptance is supported by a patent landscape review, which offers technical and business insights on such topics as which products or processes are fundamental in the field, what "white space" is still available for innovation, the companies operating in the market, and their paperwork and commercialization activities (Tyndall et al., 2022). Using a mix of keywords, International Patent Classifications (IPC), and/or Cooperative Patent Classifications, the patent landscape for plant-based meat imitators was searched for in the ORBIT and Google Patents databases (CPC). The search concentrated on the following feature subcategories: (i) Juiciness, water-holding capacity, lubricity, feel, and consistency; (ii) flavor, including Maillard reaction, cooked flavour profile, flavour intensity, flavour balance, and aroma; (iii) cooking behaviour such as sizzle, shrinkage, and cooked chemical changes; and (iv) processing employing extrusion, Couette shear cell technology, and centrifugal spinning (Tyndall et al., 2022). For plant protein-based foods, 918 potentially pertinent patent families (patent applications deriving from a first filing) were found. Five groups of results were then defined by limiting the search results to each of the afore-mentioned feature subcategories. The search results were restricted to 506 patent families with a priority application (a first filed application in any country) in the last 20 years and 339 patent families with a priority application (a first filed application in any part of the world) in the last 10 years at the time the search was conducted due to the high number of results identified in some of the subsets till August 2020 (Tyndall et al., 2022).

2.5.4 ACCEPTANCE OF TECHNOLOGY

Many plants protein-based foods are highly processed and have a lot of added ingredients, which many people do not like. Because of this, more research is needed to find ways to make processed plant-based foods with fewer ingredients and less processing. Also, people often think that foods made from plants are healthier than foods made from animals. But this is not always true. More research is needed to make sure that plant-based foods are carefully made so that they have good nutrient profiles, and the nutrients are in a form that the body can use (McClements, 2021). Using an online survey of 602 US consumers, Lang (2020) investigated the acceptability of plant-based ingredients (mushrooms) combined with meat. The author examined acceptance of blending (evaluating it favourably, liking it, being interested in it, wanting to know more, and intending to consume it in the near future), ranking of blending benefits (blending would taste better, be healthier, be more environmentally sustainable, cost less, and be novel and interesting), and four lifestyle factors related to blending: food knowledge, food involvement, cooking habits, and healthy eating. The results indicated that the top three reasons for consuming blended products are all health-related (adding more vegetables to the diet, consuming meat in a healthier manner, and avoiding things that are bad for health), followed by taste and price benefits, with the environmental benefit receiving the lowest score. In terms of format preference, burgers ranked #1. The most influential acceptability factors were food innovation, food engagement, and healthy eating. Consumers viewed blended products as sustainable, novel, and intriguing, as well as healthy, but had moderate expectations that they would be less expensive and have a superior flavour (Grasso and Goksen, 2023).

2.5.5 COMMUNICATION

Raising awareness among consumers, especially through the creation of a curriculum-friendly project for use in classrooms, can help start a productive dialogue about the benefits of PPBF diets and their place in the larger ecosystems of sustainable food systems (Tyndall et al., 2022). A study conducted in Switzerland in 2018 found that sharing details about the manufacturing process with consumers did not improve their acceptability, suggesting that messaging focused on the finished product instead of the technical processes would be more effective (Siegrist et al., 2018). In a similar vein, a study conducted in the Netherlands in 2020 found that informing customers of the individual and societal benefits increased their level of support for the idea (Rolland et al., 2020). Banovic et al. (2022) surveyed a total of 1,958 buyers from across three countries to look at the difference in attitude between consumers with a focus on their own self-interest (health) and those with a focus on a self-transcendent goal (the environment), as well as those with an independent (self) versus interdependent (family) self-construal. According to the findings, environmental communications performed better when they addressed members of the target audience's immediate family, while health-focused messages performed better when they addressed the recipients' individual well-being. The authors found that to effectively communicate the benefits of eating PPBFs, it is important to consider not just the consumers' intentions, but also their own sense of identity (Banovic et al., 2022).

2.5.6 SUPPLY CHAIN

As new firms enter the field, money is made available to do research on the numerous unexplored areas throughout the supply, processing, and formulation chain for products and processes (Tyndall et al., 2022). To facilitate innovation, Tyndall et al. (2022) advocated for the development of national and international collaborations to facilitate robust supply chain for securing plant protein and other plant-derived ingredient supplies. Based on how they are sold, the market for plant-based meat alternatives can be split into two groups: B2C (grocery stores, online retail, supermarkets, and specialty food and drink stores) and B2B (hotels and restaurants). Strangely, Impossible Foods became better known and more visible by using social media and celebrity chefs to get people's attention. This is why they chose a B2B model (i.e. partnerships with companies like Burger King) instead of a B2C model. Impossible Foods started selling in stores after they got FDA approval for their key ingredient, "heme." Beyond Meat company's products reached the most people directly by following grocery stores (B2C). It's starting to show up on menus at fast food places like Carl's Jr., Del Taco, Subway, and others. This shows the importance of using both business-to-business (B2B) and business-to-consumer (B2C) models to sell as many plant-based meat alternatives as possible (Choudhury et al., 2020).

2.5.7 Government Regulations and Policies

Food inspection, labelling, packaging, imports, and facility safety are all under the control of the Food and Drug Administration (FDA) and Centre for Food Safety and Applied Nutrition (CFSAN). Current EU policy and regulation encourage innovation and financial investment in alternative proteins. An "EU Protein Plan" was unveiled by the European Commission in 2018 to promote the production of alternative proteins for human consumption (Rubio et al., 2020). The plan also identified current EU policy tools that offer alternatives for increasing the development of EU-grown plant proteins. New Zealand, Australia, and Canada have all proposed legislation to direct control of innovative foods. Food labelling also needs to be under government regulation (Rubio et al., 2020). However, federal government agencies have not given customers enough information or clear labelling to help them understand how PPBF products might support wholesome, sustainable diets. Federal regulatory and policy responses to the variety of PPBF products are sluggish (Kraak, 2022). Livestock and dairy companies are utilizing lawsuits to challenge and restrict the labelling and commercialization of plant protein-based goods in states because there is no government regulatory framework for PPBF products (Gencarella and Costea, 2020). The labelling scenario has been referred to as being "volatile" by constitutional lawyers (Gencarella and Costea, 2020). To forbid the marketing of PPBF products in several jurisdictions, trade unions are influencing state legislators and courts who decide product labelling regulations through lawsuits (Gencarella and Costea, 2020). Thirty US states contemplated passing legislation between 2018 and 2020 to restrict how plant-based protein products might be advertised as "sausage, burger, and bacon" (Skadden et al., 2021). The first rule that forbade PPBF products' firms from advertising a product as meat if it was not made from harvested production animals or poultry was passed in Missouri in 2018. Violations of this law might result in a $1,000 fine and a year in jail (Troitino, 2021). Unless manufacturers provide disclosures, flaunting PPBF products' labelling that implies animal products is prohibited by Oklahoma's Meat Consumer Protection Act (2020) and Louisiana's Truth in Labelling of Food Products Act (2019) (Gencarella and Costea, 2020). In the state legislatures of New York and Texas, measures were presented in 2021, which define meat and milk products as coming from livestock or mammals and forbid the use of these terminology to refer to PPBF products (Watson, 2021). PPBF products are not mentioned in the USDA's DGA 2020–2025 or MyPyramid, which advise Americans to consume "lean protein—choose protein sources including beans, fish, lean meats, and nuts" (USDA, MyPlate tools, 2021).

2.5.8 Industry

There is a limited amount of information available to both large and small enterprises about the business panorama of plant-based foods, such as the affordability and relative costs of actual chemicals and processes, regulatory frameworks, supply chain challenges, and safety concerns. The availability of this information would ease enterprises' entry into the plant-based food sector encouraging economic development, wealth generation, and jobs (McClements, 2021). In a list of the top ten retailers by sales across world, Amy's Kitchen is at the top, followed by Beyond Meat, which is the youngest. The oldest store on the list is Morningstar Farms, which is owned by Kellogg (Choudhury et al., 2020). Many plant-based food firms are reformulating their goods to lower the overall number of ingredients and to use mainly natural ingredients that consumers see as label-friendly. The labelling or branding of plant protein-based food products is an additional matter that requires attention. In certain nations, regulators prohibit the use of names like "meat," "milk," and "egg" on plant-based food substitutes, to avoid any product confusion and stating that these marketing terms should be kept for actual animal-derived foods (McClements, 2021; Alcorta et al., 2021). The traditional farming industry, whose livelihoods and markets are endangered by the rising popularity of plant-based foods, has been a major opponent of these phrases (McClements, 2021).

2.6 ENVIRONMENTAL SUSTAINABILITY

It is still necessary to evaluate how PBMAs will affect the environment (Andreani et al., 2023). The life cycle assessment (LCA) method has been used in this regard. It is a methodology used in a variety of situations to evaluate a product's environmental performance based on the ISO 14040 and ISO 14044 standards (Froldi et al., 2022).

2.6.1 WATER USAGE

The source of the primary protein has a significant impact on the amount of water that is required to produce PPBFs. Mycoprotein-based goods have higher water requirements than gluten-based products (0.954 kg/kg) and soy-based products (0.73 kg/kg), according to the results of a life cycle assessment (LCA) (Rubio et al., 2020; Smetana et al., 2015). Another LCA study found that a tonne of PPBF products required an average of 3,800 m³ of water (Rubio et al., 2020; Tuomisto et al., 2011). Additionally, proper guidelines are needed for water modelling so that inappropriate comparisons do not generate incorrect conclusions (Andreani et al., 2023).

2.6.2 DEFORESTATION

Even if environmental policies stop trees from being cut down, there could be more land that can be used to grow food to meet the growing demand for alternatives to meat. One example could be the growth of palm plantations in tropical regions, since coconut oil is used more and more in PPBF products (Goldstein et al., 2017).

2.6.3 ENERGY CONSUMPTION

In-depth analyses are required because proxy processes for the energy sources produce disparate data on how much energy is consumed (Bryant, 2022). The most important parts of the PPBF industry are production, the supply chain, and storage (Godfray et al., 2018). Because fossil fuels are used to make electricity, 80% of the damage to the environment is done during the production stage (Goldstein et al., 2017).

2.6.4 GREENHOUSE GAS (GHG) EMISSIONS

Farming is responsible for 9.9% of the world's greenhouse gas emissions (Bager et al., 2021).

2.7 PUBLIC HEALTH

Few researchers have compared the effects of vegetarian/vegan diets to those of omnivore diets; however, trials focusing especially on PPBF-based diets are still limited (Oussalah et al., 2020). Nonetheless, based on the publication of study protocols in clinical trial registries (e.g., ClinicalTrials.gov), Andreani et al. (2023) have anticipated that the execution and publication of trials investigating the effects of PPBF-based diets on nutritional and health-related factors in the near future will be funded. Crimarco et al. (2022) recently evaluated the impact of plant protein-based meats on biomarkers of inflammation using a secondary analysis of the Study With Appetizing Plant food—Meat Eating Alternatives Trial (SWAP-MEAT) (Andreani et al., 2023). Contrary to predictions, no improvement in inflammatory biomarkers was observed following the intake of plant-based meat (Andreani et al., 2023). Additional extensive research focusing on health markers is required.

2.7.1 PLANT PROTEIN-BASED FOODS, E.G. MILK ANALOGUES

Several plant protein-based milk alternatives are compared to cow's milk in terms of their nutritional value (McClements, 2021). Except for soymilk, milk substitutes do not have as much protein as real milk. The milk substitutes have more total fat and saturated fat than skim milk, but they are less than that of whole milk (McClements, 2021). It is important to keep in mind that there are different types of milk alternatives in each category. For example there are original, sweetened, flavoured, and low-calorie versions of soymilk, which affect how healthy they are. As the authors have already talked about other plant-based foods, milk analogues will have different metabolic and physiological effects than cow's milk, which go beyond the differences in their macronutrient content. This is because they digest and absorb at different rates and to different degrees. Overall, the reported differences in the nutritional profiles of milk analogues and milks may have some effect on human health, but this is not known now and will depend on the type and number of products consumed.

2.7.2 PLANT PROTEIN-BASED FOODS, E.G. MEAT ANALOGUES

Comparing the nutritional characteristics of commercially available plant protein-based meat substitutes and real meat, McClements (2021) showed that the Beyond Meat burger contains more calories, total fat, saturated fat, and salt than the traditional beef burger, as well as less protein. The Impossible burger contains significantly more saturated fat, carbohydrates, and sodium than a traditional beef burger, as well as less protein. The black bean burger contains far less total and saturated fat than the beef burger but significantly more carbohydrates and sodium (McClements, 2021). As the author previously discussed, animal proteins contain a more balanced composition of essential amino acids than plant-based proteins. These nutritional analyses demonstrate that there are significant disparities across plant-based products and that a product's composition of plant-derived ingredients does not automatically make it healthier. McClements (2021) clearly advocated that additional effort is necessary to improve the nutritional profiles of plant protein-based meat substitutes while maintaining their affordability, sustainability, and flavour. Furthermore, it will be essential to identify variances in their absorption and bioavailability, as these parameters significantly influence human health and well-being.

2.7.3 PLANT PROTEIN-BASED FOODS, E.G. EGG ANALOGUES

Comparing the nutritional characteristics of plant protein-based egg substitutes available on the market and real eggs, McClements (2021) highlighted that plant protein-based egg substitutes include more calories, total fat, polyunsaturated fat, sodium, and carbohydrates than real eggs but less saturated fat. The total protein content of each product was comparable. Although the nutritional content of egg proteins is superior to that of plant proteins, comparatively the mung bean proteins are preferred because they include the full complement of necessary amino acids, so this mung bean protein can be utilized in egg analogues (McClements, 2021). As discussed previously, plant protein-based egg substitutes may also have distinct metabolic and physiological reactions to hen's eggs in addition to variances in their macronutrient content, which must be considered. The severity of these consequences is yet unknown.

2.7.4 ASSOCIATED RISKS

Consumers are growing concerned about the allergenicity of wheat and soy products, as well as GMO soy, which, together with pea proteins and their blends, have been the primary protein components used in plant-based meats so far (Li, 2020). The possibility of vitamin B12 was underlined, particularly when novel PPBF products were introduced (Tso and Forde, 2021). This is a big worry for older people, because poor nutrition has been linked to a higher risk of becoming frail (Rodríguez-Mañas, 2020). People who eat primarily a plant-based diet may develop diseases linked to vitamin or mineral deficiency—for example anaemia due to the fact that iron deficiency is common in rural India where the diet is primarily based on cereals (Taneja et al., 2020). There is also some worry that a vegetarian or vegan diet, which lacks enough of the vitamins and minerals necessary for healthy growth, may have a negative impact on a growing fetus (Biesalski and Kalhoff, 2020). This was also shown by the EPIC-Oxford cohort, which found that vegans and vegetarians were more likely to break bones than people who eat meat (Tong et al., 2020). To assess the relationship between the effectiveness of a plant-based diet and the risk of frailty, a healthy plant-based diet index (hPDI), which is defined by healthy plant foods, and an unhealthful plant-based diet index (uPDI), which is defined by less healthy plant foods, were both produced by Sotos-Prieto (2022). The study backs up the idea that not all plant-based diets are healthy because it found a correlation between a diet high in juices, refined carbohydrates, or sugar-sweetened beverages with a higher risk of frailty. Refined carbs, beverages with added sugar, and processed meats are some of the items in uPDI that cause inflammation (Sotos-Prieto, 2022). Seaweeds are rich in iodine; however, excessive ingestion may have negative health implications. Dietary-induced iodine caused hyperthyroidism in a 20-year-old Japanese woman who consumed iodine-rich seaweed sweets, and similar symptoms were observed in a 71-year-old Japanese woman who consumed seaweed that had been cooked. The symptoms disappeared as soon as seaweed was stopped being given in the diet (Raja et al., 2022). These findings once again emphasize the significance of considering the nutritional value of PPBFs when converting to plant protein-based diets that omit animal items (Andreani et al., 2023).

2.8 EATING QUALITY

The biggest challenge in creating plant protein-based meat alternatives that are good to eat is making products that look, feel, taste, smell, and have the same nutritional profile as meat. Also, the raw product must be apt to be dealt with like meat and browned in a way that is like how meat is cooked so that the finished product has the same taste and texture as meat. The new generation of PPBF products must deal with the challenges of making them taste and smell the way people want and making them healthier (Tyndall et al., 2022).

2.9 REQUIREMENT OF AGRICULTURAL CONDITIONS

In conclusion, the climate and available land provide difficulties for the cultivation of plant-based diets. For example, in some parts of the world, like steep or rocky hills, it is not possible to grow crops economically, but it is still possible to raise animals for food (Loveday, 2019).

REFERENCES

AFSSA (French Agency for Food Safety). (2007). *Apport en protéines: consommation, qualité, besoins et recommandations*. Report of the working group, pp. 461.

Alcorta, A., Porta, A., Tarrega, A., Alvarez, M. D., Vaquero, M. P. (2021). Foods for plant-based diets: Challenges and innovations. *Foods*. 10(2), 1–23.

Alves, A. C., Tavares, G. M. (2019). Mixing animal and plant proteins: Is this a way to improve protein techno-functionalities? *Food Hydrocoll*. 97, 105–171.

Andreani, G., Sogari, G., Marti, A., Froldi, F., Dagevos, H., Martini, D. (2023). Plant-based meat alternatives: Technological, nutritional, environmental, market, and social challenges and opportunities. *Nutrients*. 15, 452.

Arrutia, F., Binner, E., Williams, P., Waldron, K. W. (2020). Oilseeds beyond oil: Press cakes and meals supplying global protein requirements. *Trends Food Sci. Technol*. 100, 88–102.

Asioli, D., Aschemann-Witzel, J., Caputo, V., Vecchio, R., Annunziata, A., Næs, T., Varela, P. (2017). Making sense of the "clean label" trends: A review of consumer food choice behavior and discussion of industry implications. *Food Res. Int.* (Ottawa, Ont.), 99(Pt 1), 58–71.

Bager, S. L., Persson, U. M., Dos Reis, T. N. P. (2021). Eighty-six EU policy options for reducing imported deforestation. *One Earth*. 4, 289–306.

Banovic, M., Barone, A. M., Asioli, D., Grasso, S. (2022). Enabling sustainable plant forward transition: European consumer attitudes and intention to buy hybrid products. *Food Qual. Prefer*. 96, Article 104440.

Berrazaga, I., Micard, V., Gueugneau, M., Walrand, S. (2019) The role of the anabolic properties of plant- versus animal-based protein sources in supporting muscle mass maintenance: A critical review. *Nutrients*. 11, 1825.

Biesalski, H. K., Kalhoff, H. (2020). Contra vegan diet during childhood growth and development—A commentary from the nutritional medicine perspective. *Aktuelle Ernahrungsmedizin*. 45(2), 104–113.

Bohrer, B. M. (2019). An investigation of the formulation and nutritional composition of modern meat analogue products. *Food Sci. Hum. Wellness*. 8(4), 320–329.

Boukid, F. (2021). Plant-based meat analogues: From niche to mainstream. *Eur. Food Res. Technol*. 247(2), 297–308.

Boye, J., Zare, F., Pletch, A. (2010). Pulse proteins: Processing, characterization, functional properties and applications in food and feed. *Food Res. Int*. 43, 414–431.

Bryant, C. J. (2022). Plant-based animal product alternatives are healthier and more environmentally sustainable than animal products. *Futur. Foods*. 6, 100174.

Bryant, C. J., Barnett, J. C. (2019). What's in a name? Consumer perceptions of in vitro meat under different names. *Appetite*. 137, 104–113.

Caputo, V., Sogari, G., Van Loo, E. J. (2022). Do plant-based and blend meat alternatives taste like meat? A combined sensory and choice experiment study. *Appl. Econ. Perspect. Policy*, 1–20.

Carbonaro, M., Maselli, P., Nucara, A. (2012). Relationship between digestibility and secondary structure of raw and thermally treated legume proteins: A Fourier transform infrared (FT-IR) spectroscopic study. *Amino Acids*. 43, 911–921.

Cheney, D. (2016). Toxic and harmful seaweeds. Seaweed in health and disease prevention. *Elsevier Sci*. Netherlands, pp. 407–421.

Choudhury, D., Singh, S., Si, J., Seah, H., Chen, D., Yeo, L., Tan, L. P. (2020). Trends in plant science & society commercialization of plant-based meat alternatives trends in plant science. *Trends Plant Sci*. xx(xx), 1–4.

Cofrades, S., Benedí, J., Garcimartin, A., Sánchez-Muniz, F. J., Jimenez-Colmenero, F. (2017). A comprehensive approach to formulation of seaweed-enriched meat products: From technological development to assessment of healthy properties. *Food Res. Int*. 99, 1084–1094.

Corredig, M., Young, N., Dalsgaard, T. K. (2020). ScienceDirect food proteins: Processing solutions and challenges. *Curr. Opin. Food Sci*. 35, 49–53.

Crimarco, A., Landry, M. J., Carter, M. M., Gardner, C. D. (2022). Assessing the effects of alternative plant-based meats v. Animal meats on biomarkers of inflammation: A secondary analysis of the SWAP-MEAT randomized crossover trial. *J. Nutr. Sci*. 11, e82.

Cutroneo, S., Angelino, D., Tedeschi, T., Pellegrini, N., Martini, D., Sinu Young Working Group., Dall'Asta, M., Russo, M. D., Nucci, D., Moccia, S. (2022). Nutritional quality of meat analogues: Results from the food labelling of Italian products (FLIP) project. *Front. Nutr*. 9, 852831.

Datar, I., Betti, M. (2010). Possibilities for an in vitro meat production system. *Innov. Food Sci. Emerg. Technol*. 11, 13–22.

Day, L. (2013). Proteins from land plants-Potential resources for human nutrition and food security. *Trends Food Sci. Technol.* 32, 25–42.

Dekkers, B. L., Boom, R. M., Goot, A. J. Van Der. (2018). Trends in food science & technology structuring processes for meat analogues. *Trends Food Sci. Technol.* 81, 25–36.

Duodu, K., Taylor, J., Belton, P., Hamaker, B. (2003). Factors affecting sorghum protein digestibility. *J. Cereal Sci.* 38, 117–131.

Elakovich, S. D., Hampton, J. M. (1984). Analysis of coumestrol, a phytoestrogen, in alfalfa tablets sold for human consumption. *J. Agric. Food Chem.* 32(1), 173–175. doi:10.1021/jf00121a041.

Emin, M. A., Quevedo, M., Wilhelm, M., Karbstein, H. P. (2017). Analysis of the reaction behavior of highly concentrated plant proteins in extrusion-like conditions. *Innov. Food Sci. Emerg. Technol.* 44, 15–20. ISSN 1466–8564.

Feichtinger, A., Scholten, E. (2020). Preparation of protein oleogels: Effect on structure and functionality. *Foods.* 9, 1745.

Fiorentini, M., Kinchla, A. J., Nolden, A. A. (2020). Role of sensory evaluation in consumer acceptance of plant-based meat analogs and meat extenders: A scoping review. *Foods.* 9, 1334.

Foegeding, E. A. (2015). Food protein functionality: A new model. *J. Food Sci.* 80, C2670–C2677.

Froldi, F., Lamastra, L., Trevisan, M., Mambretti, D., Moschini, M. (2022). Environmental impacts of cow's milk in Northern Italy: Effects of farming performance. *J. Clean. Prod.* 363, 132600.

Gencarella, N. R., Costea, F. A. (2020). *The complex labeling landscape for plant-based meat alternatives.* [Internet]. Jones Day White Paper. Cleveland (OH): Jones Day.

Giacalone, D., Clausen, P. M., Jaeger, R. S. (2022). Understanding barriers to consumption of plant-based foods and beverages: Insights from sensory and consumer science. *Curr. Opin. Food Sci.* 48, 100919. ISSN 2214-7993.

Gilani, G. S., Cockell, K. A., Sepehr, E. (2005). Effects of antinutritional factors on protein digestibility and amino acid availability in foods. *J. AOAC Int.* 88, 967–987.

Godfray, H. C. J., Aveyard, P., Garnett, T., Hall, J. W., Key, T. J., Lorimer, J., Pierrehumbert, R. T., Scarborough, P., Springmann, M., Jebb, S. A. (2018). Meat consumption, health, and the environment. *Science.* 361, eaam5324.

Goldstein, B., Moses, R., Sammons, N., Birkved, M. (2017). Potential to curb the environmental burdens of American beef consumption using a novel plant-based beef substitute. *PLOS ONE.* 12, e0189029.

The Good Food Institute. (2019). *Plant-based market overview.* GFI. www.gfi.org/marketresearch

Graça, J., Godinho, C. A., Truninger, M. (2019). Reducing meat consumption and following plant-based diets: Current evidence and future directions to inform integrated transitions. *Trends Food Sci. Technol.* 91, 380–390.

Grasso, S., Goksen, G. (2023). The best of both worlds? Challenges and opportunities in the development of hybrid meat products from the last 3 years. *LWT.* 173, 114235.

Hartmann, C., Furtwaengler, P., Siegrist, M. (2022). Consumers' evaluation of the environmental friendliness, healthiness and naturalness of meat, meat substitutes, and other protein-rich foods. *Food Qual. Prefer.* 97, 104486.

Hashempour-Baltork, F., Khosravi-Darani, K., Hosseini, H., Farshi, P., Reihani, S. F. S. (2020). Mycoproteins as safe meat substitutes. *J. Clean. Prod.* 253, 119958.

He, J., Evans, N. M. (2020). A review of research on plant-based meat alternatives: Driving forces, history, manufacturing, and consumer attitudes. *Compr. Rev. Food Sci. Food Saf.* 1–18.

Hu, F. B., Otis, B. O., McCarthy, G. (2019). Can plant-based meat alternatives be part of a healthy and sustainable diet? *JAMA*, 1–3.

International Food Information Council Foundation, Food & Health Survey. (2020). https://ific.org/ & https://foodinsight.org/2020-food-and-health-survey/

Jahn, S., Furchheim, P., Strässner, A. M. (2021). Plant-based meat alternatives: Motivational adoption barriers and solutions. *Sustainability.* 13(23), 13271.

Jim, L. M., Tavares, G. M., Corredig, M. (2021). Design future foods using plant protein blends for best nutritional and technological functionality. *Trends Food Sci. Technol.* 113, 139–150.

Kerslake, E., Kemper, J. A., Conroy, D. (2022). What's your beef with meat substitutes exploring barriers and facilitators for meat substitutes in omnivores, vegetarians, and vegans? *Appetite.* 170, 105864.

Kraak, V. I. (2022). Perspective: Unpacking the wicked challenges for alternative proteins in the United States: Can highly processed plant-based and cell-cultured food and beverage products support healthy and sustainable diets and food systems? *Adv. Nutr.* 2020(4), 38–47.

Kumar, M., Tomar, M., Potkule, J., Punia, S., Dhakane-lad, J., Singh, S., Dhumal, S., Chandra, P., Bhushan, B., Anitha, T., Alajil, O., Alhariri, A., Amarowicz, R., Kennedy, J. F. (2022). Food hydrocolloids functional characterization of plant-based protein to determine its quality for food applications. *Food Hydrocoll.* 123, 106986.

Kutzli, I., Weiss, J., Gibis, M. (2021). Glycation of plant proteins via Maillard reaction: Reaction chemistry, techno-functional properties, and potential food application. *Foods.* 10, 376.

Kyriakopoulou, K., Dekkers, B., Goot, A. T. J. (2019). *Chapter 6—plant-based meat analogues. Sustainable meat production and processing.* Academic Press. 103–126. ISBN 9780128148747.

Lacy-Nichols, J., Hattersley, L., Scrinis, G. (2021). Nutritional marketing of plant-based meat-analogue products: An exploratory study of front-of-pack and website claims in the USA. *Public Health Nutri.* 24(14), 4430–441.

Lang, M. (2020). Consumer acceptance of blending plant-based ingredients into traditional meat-based foods: Evidence from the meat-mushroom blend. *Food Qual. Prefer.* 79, 103758. ISSN 0950-3293.

Lea, E. J., Crawford, D., Worsley, A. (2006). Public views of the benefits and barriers to the consumption of a plant-based diet. *Eur. J. Clin. Nutr.* 60(7), 828–837.

Li, X., Vries, R. De. (2018). ScienceDirect interfacial stabilization using complexes of plant proteins and polysaccharides. *Curr. Opin. Food Sci.* 21, 51–56.

Li, Y. (2020). Feeding the future: Plant-based meat for global food security and environmental sustainability. *Cereal Foods World.* 65(4), 8–11.

López-Pedrouso, M., Lorenzo, J. M., Campagnol, P. C. B., Franco, D. (2021). Novel strategy for developing healthy meat products replacing saturated fat with oleogels. *Curr. Opin. Food Sci.* 40, 40–45. ISSN 2214–7993.

Louisiana's Truth in Labelling of Food Products Act. (2019). https://lawreview.law.lsu.edu/2020/01/22/pickin-on-veggies-louisianas-truth-in-labeling-of-food-products-act/ (http://legis.la.gov/legis/ViewDocument.aspx?d=1144034)

Loveday, S. M. (2019). Food proteins: Technological, nutritional, and sustainability attributes of traditional and emerging proteins. *Annu. Rev. Food Sci. Technol.* 10, 311–339.

McClements, D. J. (2021). The science of plant-based foods: Constructing next-generation meat, fish, milk, and egg analogs. *Compr. Rev. Food Sci. Food Saf.* 1–52.

Michel, F., Siegrist, M. (2019). How should importance of naturalness be measured? A comparison of different scales. *Appetite.* 140, 298–304. ISSN 0195-6663.

Neuhofer, Z. T., Lusk, J. L. (2022). Most plant-based meat alternative buyers also buy meat: An analysis of household demographics, habit formation, and buying behavior among meat alternative buyers. *Sci. Rep.* 12, 13062.

Nielsen, S. D., Schmidt, J. M., Kristiansen, G. H., Dalsgaard, T. K., Larsen, L. B. (2020). Liquid chromatography mass spectrometry quantification of α-solanine, α-chaconine, and solanidine in potato protein isolates. *Foods.* 9(4), 416.

Oklahoma's Meat Consumer Protection Act. (2020). https://oksenate.gov/press-releases/oklahoma-meat-consumer-protection-act-heads-governors-desk

Oussalah, A., Levy, J., Berthezène, C., Alpers, D. H., Guéant, J. L. (2020). Health outcomes associated with vegetarian diets: An umbrella review of systematic reviews and meta-analyses. *Clin. Nutr.* 39, 3283–3307.

Phillips, S. M. (2017). Current concepts and unresolved questions in dietary protein requirements and supplements in adults. *Front. Nutr.* 4, 13.

Pohjolainen, P., Vinnari, M., Jokinen, P. (2015). Consumers' perceived barriers to following a plant-based diet. *Br. Food J.* 117, 1150–1167.

Raja, K., Kadirvel, V., Subramaniyan, T. (2022). Seaweeds, an aquatic plant-based protein for sustainable nutrition—A review. *Future Foods.* 100142.

Rivera del Rio, A., Opazo-Navarrete, M., Cepero-Betancourt, Y., Tabilo-Muzinga, G., Boom, R. M., Janssen, A. E. M. (2019). Heat-induced changes in microstructure of spray-dried plant protein isolates and its implications on in vitro gastric digestion. *LWT- Food Sci & Tech.* 118, 108795.

Rodríguez-Mañas, L. (2020). Impact of nutritional status according to GLIM criteria on the risk of incident frailty and mortality in community-dwelling older adults. *Clin. Nutr.* 40(3), 1192–1198.

Roland, W. S. U., Pouvreau, L., Curran, J., Van de Velde, F., de Kok, P. M. T. (2017). Flavor aspects of pulse ingredients. *Cereal Chem.* 94, 58–65.

Rolland, N. C., Markus, C. R., Post, M. J. (2020). The effect of information content on acceptance of cultured meat in a tasting context. *PLoS One.* 15(4), e0231176.

Rubio, N. R., Xiang, N., Kaplan, D. L. (2020). Production. *Nat. Commun.* 1–11.

Schönfeldt, H. C., Gibson Hall, N. (2012). Dietary protein quality and malnutrition in *Africa. Br. J. Nutr*. 2, S69–76.

Schouteten, J. J. (2016). Emotional and sensory profiling of insect-, plant- and meat-based burgers under blind, expected and informed conditions. *Food Qual. Prefer*. 52, 27–31.

Schreuders, F. K. G., Schlangen, M., Kyriakopoulou, K., Boom, R. M., Goot, A. J. Van Der. (2021). Texture methods for evaluating meat and meat analogue structure: A review. *Food Control*. 127, 108103.

Siegrist, M., Sütterlin, B. (2017). Importance of perceived naturalness for acceptance of food additives and cultured meat. *Appetite*. 113, 320–326.

Siegrist, M., Sütterlin, B., Hartmann, C. (2018). Perceived naturalness and evoked disgust influence acceptance of cultured meat. *Meat Sci*. 139, 213–219.

Skadden, A., Slate, M., Flom, L. L. P. (2021). *Food and beverage labeling litigation: Recent trends*. [Internet]. New York (NY): Skadden.

Slade, P. (2018). If you build it, will they eat it? Consumer preferences for plant-based and cultured meat burgers. *Appetite*. 125, 428–437. PMID: 29501683.

Smetana, S., Mathys, A., Knoch, A., Heinz, V. (2015). Meat alternatives: Life cycle assessment of most known meat substitutes. *Int. J. Life Cycle Assess*. 20, 1254–1267.

Sotos-Prieto, M. (2022). Association between the quality of plant-based diets and risk of frailty. *J. Cachexia Sarcopenia Muscle*. 13(6), 2854–2862.

Stone, A. K., Teymurova, A., Dang, Q., Abeysekara, S., Karalash, A., Nickerson, M. T. (2014). Formation and functional attributes of electrostatic complexes involving napin protein isolate and anionic polysaccharides. *Eur. Food Res. Technol*. 238, 773–780.

Taneja, D. K., Rai, S. K., Yadav, K. (2020). Evaluation of promotion of iron-rich foods for the prevention of nutritional anemia in India. *Indian J. Public Health*. 64(3), 236–241.

Tenorio, A. T., Kyriakopoulou, K. E., Suarez-Garcia, E., van den Berg, C., van der Goot, A. J. (2018). Understanding differences in protein fractionation from conventional crops, and herbaceous and aquatic biomass-consequences for industrial use. *Trends Food Sci. Technol*. 71, 235–245.

Toh, D. W. K., Srv, A., Henry, C. J. (2022). Unknown impacts of plant-based meat alternatives on long-term health. *Nat Food*. 3(2), 90–91.

Tong, T. Y. N., Paul, N. A., Miranda, E. G. A., Georgina, K. F., Anika, K., Keren, P., Aurora, P. C., Ruth, C. T., Timothy, J. K. (2020). Cancer epidemiology unit, Nuffield Department of population health, University of Oxford. *BMC Med*. 18, 353.

Troitino, C. (2021). *Missouri becomes first state to start regulating meat alternative labels*. [Internet]. Jersey City (NJ): Forbes.

Tso, R., Forde, C. G. (2021). Unintended consequences: Nutritional impact and potential pitfalls of switching from animal- to plant-based Foods. *Nutrients*. 13(8), 2527.

Tuccillo, F., Lampi, A., Coda, R., Edelmann, M., Katina, K., Piironen, V. (2022). Flavor challenges in extruded plant-based meat alternatives: A review. *Compr. Rev. Food Sci. Food Saf*. 2898–2929.

Tuomisto, H. L., Teixeira de Mattos, M. J. (2011). Environmental impacts of cultured meat production. *Environ. Sci. Technol*. 45, 6117–6123.

Tyndall, S. M., Maloney, G. R., Cole, M. B., Hazell, N. G., Tyndall, S. M., Maloney, G. R., Cole, M. B., Nicholas, G. (2022). Critical food and nutrition science challenges for plant-based meat alternative products. *C. Rev. Food Sci. Nutri*. 1–16.

United Nations. (2019). World population prospects 2019: Ten key findings.

USDA, MyPlate tools. (2021). https://www.dietaryguidelines.gov/sites/default/files/2020-12/Dietary_Guidelines_for_Americans_2020-2025.pdf

Van Vliet, S., Burd, N. A., Van Loon, L. J C. (2015). The skeletal muscle anabolic response to plant-versus animal-based protein consumption. *J. Nutr*. 145, 1981–1991.

Watson, E. (2021). *Plant-based labeling battle heats up as more states challenge use of meat, dairy terms*. [Internet]. Crawley (United Kingdom): Foodnavigator-usa.com.

Whitnall, T., Pitts, N. (2019). Global trends in meat consumption. *Agric Commod*. 9, 96–99.

Wilks, M., Phillips, C. J. C. (2017). Attitudes to in vitro meat: A survey of potential consumers in the United States. *PLOS ONE*. 12, e0171904.

WRI. (2019). Creating a sustainable food future: A menu of solutions to feed nearly 10 billion people by 2050. https://research.wri.org/wrr-food. World Resources Institute.

Xiong, R., Cavitt, L. C., Meullenet, J. F., Owens, C. M. (2006). Comparison of AlloKramer, Warner-Bratzler and razor blade shears for predicting sensory tenderness of broiler breast meat. *J. Texture Stud*. 37(2), 179–199.

Yong, S., Sim, J., Srv, A., Chiang, J. H. (2021). Plant proteins for future foods: A roadmap. *Foods*. 1–31.

Zhang, C., Guan, X., Yu, S., Zhou, J., Chen, J. (2022). ScienceDirect production of meat alternatives using live cells cultures and plant proteins. *Curr. Opin. Food Sci.* 43, 43–52. https://doi.org/10.1016/j.cofs.2021.11.002.

Zhu, D., Damodaran, S., Lucey, J. A. (2008). Formation of whey protein isolate (WPI)-dextran conjugates in aqueous solutions. *J. Agric. Food Chem.* 56, 7113–7118.

Zhu, D., Damodaran, S., Lucey, J. A. (2010). Physicochemical and emulsifying properties of whey protein isolate (WPI)-dextran conjugates produced in aqueous solution. *J. Agric. Food Chem.* 58, 2988–2994.

3 Modification Methods of Plant-Based Proteins
An Overview

Priti Sharad Mali and Pradyuman Kumar

3.1 INTRODUCTION

Plant-derived proteins have recently attracted growing more attention in the food and pharmaceutical industries due to their health benefits over animal proteins in areas such as availability, contamination, infection, less religious, cultural food habit limitations, and consumer lifestyle. Proteins serve as a crucial micronutrient in the human diet and welfare. The nutritional quality substantially varies with their amino acid profile, biodigestibility, bioavailability, processing techniques, solubility, and purity. From a nutritional point of view, the therapeutic protein is now extensively used in medical field practices to treat various diseases (Park et al., 2011). In addition, protein ingredients are widely used in the food industry because of their health benefits and functionality in food formulations.

Plant protein-based food as an alternative to meat in many countries appears to overcome several barriers, including environmental issues. The development of a protein-rich plant-based product with the potential to replace meat analogs as a nutritional source is previously explored conventionally with the production of tempeh, tofu, seitan, etc. (Kyriakopoulou et al., 2021). Recently, there have been research focuses on the development and production of sustainable food products that recreate meat substitutes, not only nutritionally, but also in all aspects of their physical sensations including appearance, texture, taste, and smell. Respective products that are available on the market are chunks, strips, burgers, patties, chicken-like blocks, ground beef-like products, sausages, steaks, etc.

Concerning nutritional quality and health benefits, food protein directly impacts food product quality, which is associated with its specific functional properties. However, the functional properties of food protein are affected by various factors such as pH, temperature, applied pressure, time, and ionic strength. Furthermore, the novel extraction method considerably impacts on functionality and structural properties of the isolated protein (Ochoa-Rivas et al., 2017). The stability of protein molecules plays a key role in the structural components of the food matrices. The broad-spectrum advances for the modification of plant protein functionality involve the appliance of physical, chemical, and biological processing or a combination of these technologies at the minute or large scale of protein ingredient development (Akharume et al., 2021). Commonly used protein modification methods are salt precipitation, solvent precipitation, and alkaline extraction, but novel plant protein-processing techniques such as heat treatment, high-pressure processing, ultrasound treatment, microwave heating, radio frequency treatment, extrusion, supercritical carbon dioxide, and plasma technology are being used to induce significant changes in protein modification without alteration in its indigenous properties. Novel technique consumes less time and less energy, and the inexpensive, eco-friendly technology has been considered an alternative to conventional processing (Jha & Sit, 2021). However, the alteration of protein functionality during extraction due to modification in protein structure, protein denaturation, is considered a drawback of the conventional extraction method. Therefore, it is imperative to develop novel modification methods to improve the functionality of the plant-based proteins.

DOI: 10.1201/9781003369790-3

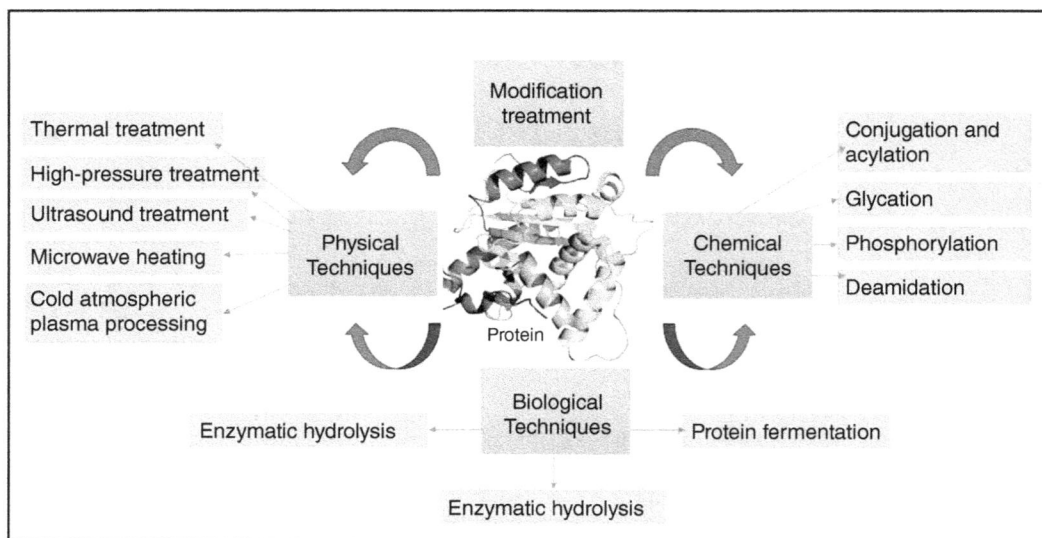

FIGURE 3.1 Modification approaches of plant-based proteins using various processing techniques.

The main purpose of this study is to introduce protein modification techniques such as physical, chemical, and biological, representing a substantial challenge in the changes in protein functionality and utilization in future foods.

3.2 MODIFICATION APPROACHES OF PLANT-BASED PROTEINS USING VARIOUS PROCESSING TECHNIQUES

Several physical, chemical, and biological approaches to plant protein modifications have been altering the molecular structure and chemical properties of proteins, providing the possibility to improve their bioactivity and techno-functionality. Modification of plant-based proteins provides the opportunity to make them multi-functional ingredients for food systems by changing their improved functional attributes, digestibility, bioactivity, nutritional content, physicochemical properties, and reduction in antinutritional factors. Generally, the protein modification methods can be classified into physical, chemical, and biological fields as schematically shown in Figure 3.1.

3.3 MECHANISM OF PROTEIN MODIFICATION

The mechanism of protein modifications of plant-based proteins is schematically shown in Figure 3.2. Protein modifications are very comprehensive due to their complex structure and properties. So, it is important to know the nature and functionality of protein as well as the mechanism. It is required to achieve a customized target and better functional property.

Proteins are complex biomolecules, and their structure modification is not a "one fits all" approach. It is important to know the nature of the polymers and structural properties of the protein as well as the mechanism required to achieve a target (Phillips, 2013). Thus, it requires developing an approach adapted to modifying such functional properties and structural properties of the protein and the mechanism needed to achieve such modification. Many researchers demonstrated that the mechanism to cause an improvement in the solubility of protein ingredients may require electrostatic repulsion, size reduction, and an increase in hydrophilicity by conjugation with a more hydrophilic polymer, manipulating the protein isoelectric point and net charge (Akharume et al.,

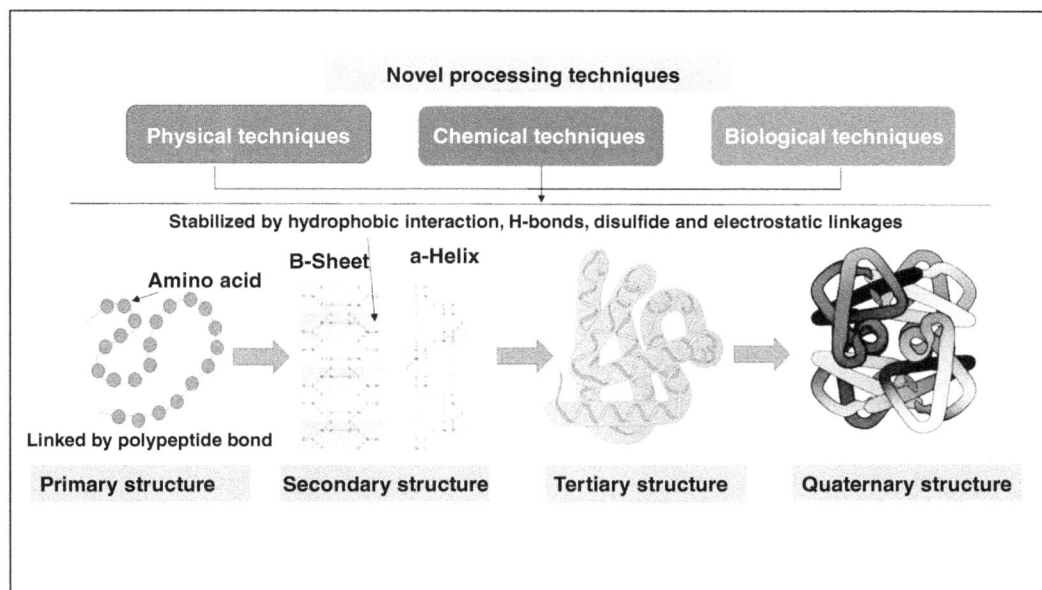

FIGURE 3.2 Mechanism of protein modifications of plant-based proteins.

2021). Briefly, any approach such as pH shifting, high pressure, ultrasonication, hydrolysis, proteolysis, glycosylation, conjugation, esterification, and acylation that could cause these changes to the protein molecules leads to increased solubility. Similarly, increasing the foaming and emulsifying activities of a protein may require mechanisms that promote a balance of its hydrophilic–hydrophobic property in addition to its solubility (Wouters et al., 2016). Other mechanisms include cavitation, cause disruption, aggregation using ultrasonication, and modifications that reveal the hydrophobic core of the protein.

3.4 PHYSICAL TECHNIQUES

Physical methods to expand the functionality of proteins are simple approaches that are not based on chemicals or enzymes. Since there are no chemicals used in the processing, these methods of protein modification have gained significant interest, avoiding harmful consequences.

3.4.1 THERMAL TREATMENT

Thermal treatment is the initial processing step in roasting, cooking, high-temperature extrusion, and drying. In addition, heat treatment can cause permanent unfolding and thermal mobility of the amino acid chain which is responsible for the inter- and intramolecular hydrophobic interaction between molecules. This permanent denaturation is accompanied by the loss of the secondary and tertiary structures of the protein (Akharume et al., 2021).

3.4.1.1 Effect of Thermal Treatment on the Functionality of Plant Protein Ingredients

Thermal treatment is commonly applied to various protein structure modifications and functionality such as mung bean, red bean, and kidney bean protein isolates (Tang et al., 2009). The researchers observed that heat-induced protein denaturation was followed by the modification in the secondary and tertiary structures of the protein. Thermal treatment was done to increase the surface hydrophobicity of proteins due to the separation of previously covered hydrophobic constituents. Further, the

TABLE 3.1

A Summary of Plant-based Protein, Thermal Treatments, and Target Functionality Impacted

Protein Type	Treatment Condition	The Effect Observed on Protein Functionality	Reference
Faba bean protein	75–175°C, 60 min	Increased water-holding capacity, decreased solubility	Bühler et al., 2020
Pea protein	50–100°C, 30 min, pH 3.0–7.0	Decreased foaming properties regardless of pH, Improved emulsifying activity at pH 7.0.	Chao & Aluko, 2018
Mung bean, red kidney bean, and red bean proteins	95°C, 30 min	Improved emulsifying activity and solubility	Tang et al., 2009
Pea protein	50–100°C, 30 min, pH 3.0–7.0	Improved emulsifying activity at pH 7.0, decreased foaming properties regardless of pH	Chao & Aluko, 2018
Red kidney bean protein	95°C, 15–120 min	Improved solubility, foaming, and emulsifying activities with moderate heating (15–30 min)	Tang & Ma, 2009

findings attributed protein denaturation and a significant increase in solubility, as well as hydrophobicity to heat treatment. Possible changes such as denaturation and partial unfolding were observed due to the increased surface charge on protein molecules. Peng et al. (2016) reported that the emulsifying properties, interfacial properties, and stability of pea protein were increased compared to the control protein sample due to increased solubility and hydrophobicity after heat treatment (95°C, 30 min). Emulsifying properties of pea protein were able to improve after heat treatment with higher stability over creaming due to higher adsorption of protein at the oil–water interface. The effect of thermal treatment observed on the functionality of plant protein is shown in Table 3.1.

3.4.2 HIGH-PRESSURE TREATMENT

High hydrostatic pressure (HHP) is applied in the range of 100–800 MPa for a few minutes to modify the structure of the plant protein. Several factors affect protein functionality such as applied pressure, ionic strength, duration of treatment, pH, and temperature of the solution, subsequently impacting the protein structure and functionality during the high-pressure processing. The mechanism of high-pressure treatment for protein modification is based on the ability to change the structure of protein molecules in the medium leading to denaturation, aggregation, and rupturing of the cell of the plant protein molecule.

3.4.2.1 Effect of HPP Treatment on the Functionality of Plant Protein Ingredients

The high-pressure treatment induces protein denaturation and aggregation with elevated textural properties. High-pressure processing does not affect the covalent bond but does affect the secondary, tertiary, and quaternary structures of the protein. Higher pressure in the range of 300–700 MPa is associated with non-reversible denaturation, and a moderate pressure range of 150–200 MPa showed dissociation of quaternary structure and protein (Mulla et al., 2022). The effect of high-pressure treatment on protein functionality varies with the type of protein due to its intrinsic factors. Some authors confirmed that the effect of high-pressure treatment reduced the solubility of approximately 2.5% (at 600 MPa) of soybean protein isolate compared to the native protein. The surface hydrophobicity of soybean protein isolates increased by about 32% at pH 6–8 under room

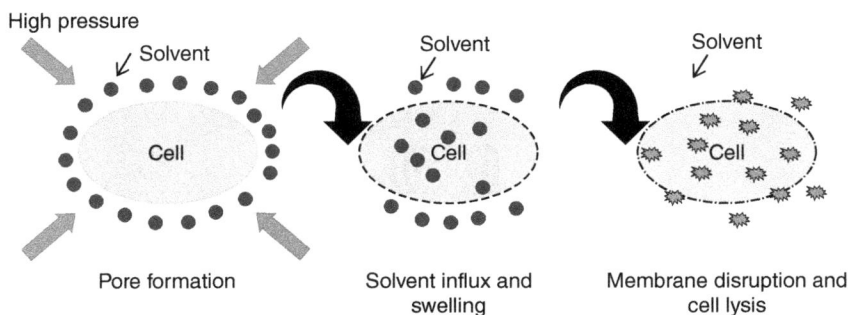

FIGURE 3.3 Schematic representation of cell lysis by high-pressure processing.

TABLE 3.2

A Summary of Plant-based Protein, HPP Treatments, and Target Functionality Impacted

Type of Protein	Treatment Condition	The Effect Observed on Protein Functionality	Reference
Kidney bean protein	200–600 MPa, 15 min	Improved water- holding capacity, foaming, and emulsifying properties, increased viscosity	Ahmed et al., 2018
Lentil protein isolate	300–600 MPa, 15 min	Improved water-holding capacity, foaming, emulsifying properties, and rheological properties	Ahmad et al., 2019
Soybean	200–600 MPa, 10 min	Improved rheological properties	Puppo et al., 2008
Lentils *(Lens culinaris Medik.)*	100–300 MPa, 15 min	Highest inhibitory and antioxidant activities	Garcia Mora et al., 2015
Soy protein isolate	600 MPa, 5 min	WHC, Solubility, gelation,	Manassero et al. 2018
Lentil protein	25–150 MPa	Improved solubility, emulsifying and foaming properties with increasing pressure up to 100 MPa, improved gelling properties at 50–150 MPa	Saricaoglu, 2020
Yellow field pea protein	200–600 MPa, 5 min	Improved emulsifying and foaming properties	Chao et al., 2018

temperature (Floury et al., 2002). Saricaoglu (2020) demonstrated improved solubility, emulsifying, and foaming properties of lentil protein with increasing pressure up to 100 MPa and improved gelling properties at 50–150 MPa. Although some studies reported the decreased solubility of plant-based protein after high-pressure treatment, especially at a higher applied pressure of approximately 400 MPa due to aggregation. Schematic representation of protein cell lysis by high-pressure processing is schematically shown in Figure 3.3.

High-pressure treatment was also evaluated as an approach for increasing the functional properties of plant-based protein. Lee et al. (2016) demonstrated HPP as the most successful approach in reducing the allergenicity of soy protein isolate compared to other methods such as microwave heating, high-intensity ultrasound, and high-pressure homogenization. In addition, they reported significantly decreased allergenicity of protein from ginkgo seeds after HPP at a pressure ranging from 300 to 600 MPa (Lee et al., 2016). HPP treatment impact on the functionality of plant protein is shown in Table 3.2.

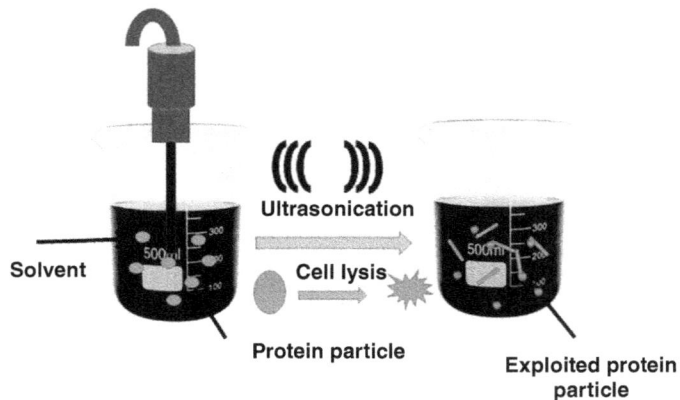

FIGURE 3.4 Ultrasonic disruption of protein particle: underlying mechanism and effects of processing parameters.

3.4.3 Ultrasound Treatment

Ultrasonic refers to high-power sound waves having frequency ranges from a lower range of 20 Hz to an upper range of 20 kHz and are mostly used in food processing. The power density generally used in liquid medium ranges from 0.01 to 1 mW/mL. Ultrasound shows cavitational effects when applied to a liquid medium. The physical, mechanical, and chemical effects of longitudinal waves transmit in the liquid medium that oscillates alternate compression and rarefaction pattern in which they propagate. The pattern creates negative pressure in the liquid medium that breaks the liquid into cavities or small bubbles. The continuous mechanical effect causes compression and rarefaction patterns making the bubble contract and expand and attain the critical size and often create micro turbulence. This dramatic formation and implosion of bubbles, referred to as cavitations, are responsible for the formation temperature and shear force in the medium. In addition, the microturbulence enhances the heat and shear forces that are required for a bubble to grow in critical size and cause disruption of cell and mass transfer in the medium. Thus, ultrasound-induced cavitations are used to enhance the extraction kinetics of the medium and increase the extraction yields by diffusing through plant cell walls and then disrupting the wall to wash out cellular content. The underlying mechanism of ultrasonic disruption of plant protein particles is shown in Figure 3.4.

3.4.3.1 Effect of Ultrasound Treatment on the Functionality of Plant Protein Ingredients

Ultrasound-assisted extraction is effective for the extraction of various plant constituents, such as protein, bioactive compounds, oil, and pigments. It has been used extensively in the extraction of protein from different components, for example, legumes, plant by-products, soybean flour, seed meal, and rice bran. However, there is much research reported on the effect of ultrasonication on the structural, physicochemical, and functional characteristics of the protein. Thus, ultrasound treatment leads to (i) the disintegration of the cell to increase hydration and reduce the particle size of the cell, (ii) an increase in the velocity and depth of the solvent penetration into the cell, (iii) an increased hydration and swelling of the cell, (iv) pitting and erosion on the cell surface, (v) freeing of molecules attached to/in the surface in the solvent, and (vi) the disruption of molecular aggregates into smaller particles and even breaking macromolecular chains impacting functionalities.

Ultrasonication has been used to modify protein structure and for the isolation of protein. Several studies reported ultrasonication treatment on the functionality of the various plant proteins, which showed partial unfolding of the protein and well-dispersed particles into nano-sized fillers as well as heating causing thermal degradation of the protein molecule. This process is not only used for the isolation of protein but is also effective in determining the functionality and structure of protein molecules depending on the molecular weight size, applied pressure, and pH. Eckert et al. (2019) observed further improvement in the foaming, oil-holding capacity, and emulsifying capacity of faba bean

TABLE 3.3

A Summary of Plant-based Protein, Ultrasound Treatments, and Target Functionality Impacted

Protein Type	Treatment Conditions	Effect Observed on Protein Functionality	Reference
Pea protein	20 kHz, 30–90% amplitude, 30 min	Improved foaming properties	Xiong et al., 2018
Chickpea protein	20 kHz, 300 W, 5–20 min	Improved solubility, emulsifying, foaming, water-holding, and gelling properties	Wang et al., 2020b
Black bean protein	20 kHz, 150–450 W, 12–24 min	Improved solubility	Jiang et al., 2014
Soybean β-conglycinin and glycinin	20 kHz, 400 W, 5–40 min	Improved solubility, emulsifying activity, and stability	Hu et al., 2015
Faba bean protein	20 kHz, 50–75% amplitude, 15–30 min	Higher adsorption at the oil–water interface, improved foaming properties	Martinez-Velasco et al., 2018

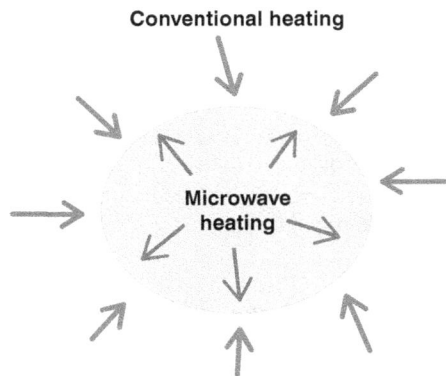

FIGURE 3.5 Mechanism of microwave heating.

protein after ultrasonication treatment of protein. However, some other authors showed better functional properties of black bean protein and faba bean protein isolates extracted by ultrasonication methods. For instance, Wang et al. (2020b) showed higher emulsifying, water-holding, gelling, and foaming ability of chickpea protein isolate extracted by ultrasound-assisted extraction method. The effect of ultrasonication observed on the functionality of plant protein is shown in Table 3.3.

3.4.4 MICROWAVE HEATING

The microwave heating technique refers to the conversion of electromagnetic energy to thermal energy through increased agitation of water molecules and charged ions when exposed to microwaves. Direct penetration of microwaves into food materials enables us to heat foods much faster than conventional heating methods that rely on surface heating such as countertop stoves or baking ovens. Mechanism of microwave heating is based on the following: (i) Dipole rotation: Flip flop rotation of dipole molecules like water—repeated changes in the polarity of the field cause rapid reorientation of the water molecules, resulting in friction and hence generation of heat. (ii) Ionic polarization: Back and forth vibration of ionic salts like sodium chloride—ions move at an accelerated pace due to their inherent charge; collisions between the ions cause the generation of thermal energy as shown in Figure 3.5.

3.4.4.1 Effect of Microwave Heating on the Functionality of Plant Protein Ingredients

Microwave heating is generally used as a modification method in food processing to change the internal structure of proteins and the degree of protein aggregation. Microwave heating can affect functional properties such as emulsification, solubility, stretching, and gel properties of the protein (Jiao et al., 2022). Polar molecules with a dipole moment in proteins undergo a torque in the MW field and produce a turning "variable pole" motion (Huang et al., 2022). The process also increases the interaction of charges on the protein surface. MW treatment results in less water being involved in the solvation layer around the protein, thus exposing the hydrophobic groups (Wang et al., 2020a). The response patterns of different protein molecules to the resonance of electromagnetic fields are related to their structural domains. The alternating electromagnetic field of microwave heating has a tearing effect on the protein, and MW easily interacts with the protein after the protein molecules are unfolded. With further increase in temperature, the protein backbone directly absorbs MW energy. MW induces the formation of more intermolecular forces among protein molecules, with stronger mechanical and heat stability. MW heating can facilitate the conversion of protein α-helix to other secondary structures, while β-sheet or random coil can promote the formation of aggregation (Zheng et al., 2020). MW can contribute to the generation of hydrophobic interactions of proteins, which leads to the formation of dense and ordered gel structures and enhances the stability of gel structures.

3.4.5 COLD ATMOSPHERIC PLASMA PROCESSING

Atmospheric pressure cold plasma (ACP) is a nonhazardous processing method that has been used to modulate the physicochemical properties of plant-based proteins. Dielectric barrier discharge (DBD) is the most popular technique used for the generation of atmospheric cold plasma in the food industries. ACP consists of free ions, electrons, and neutral particles as well as reactive oxygen species (e.g., O, OH, O_3) and reactive nitrogen species (e.g., NO, NO_2, HNO_2, ONOOH) (Mehr & Koocheki, 2020).

3.4.5.1 Effect of Cold Plasma Treatment on the Functionality of Plant Protein Ingredients

Protein oxidation is a major advantage of the power of plasma over the traditional method, which can lead to the degradation or dissociation and aggregation of the protein and increase the contact surface area of the protein molecules, thus increasing the protein oxidation rate. The significance of cold plasma modification of plant protein such as wheat gluten functionality can be implicit as researched by Nucia et al. (2021) due to getting free from surface etching and resulting in a smoother surface. This smoother gluten can be used to manufacture biodegradable film and pots because of the smooth surface area and lesser adhesion of the food product was observed. Zhang et al. (2021) reported that the solubility of soy protein isolate increased after cold plasma treatment compared to the untreated protein. The solubility of pea protein increased up to 282% after ACP treatment of 8.8 kV for 10 min. Yu et al. (2020) investigated the effect of cold plasma treatment on the solubility of flaxseed protein. The increase in protein solubility with increasing voltage and time was attributed to the increase in the surface charge of the protein and the reduction in the particle size of the protein. The effect of cold plasma treatment observed on the functionality of plant protein is shown in Table 3.4.

3.5 CHEMICAL TECHNIQUES

The functionality and stability of protein molecules are also modified by the application of a variety of chemical methods including chemical glycosylation, acylation, succinylation, deamidation, and phosphorylation via different pathways. Chemical modifications refer to the form of protein reaction with a chemical agent in which there is destruction and reformation of new bonds that alter the structural integrity of the protein molecule.

TABLE 3.4

A Summary of Plant-based Protein, Cold Plasma Treatments, and Target Functionality Impacted

Type of Protein	Treatment Conditions	Effects Observed on Protein Functionality	Reference
Grass pea protein	9.4–18.6 kVpp, 25°C, 30–60 s	Higher adsorption at the oil–water interface, improved emulsifying properties	Mehr & Koocheki, 2020
Pea protein	8.8 kVpp, 60°C, 3 kHz, 10 min	Increased water and fat-binding capacities, improved solubility	Bußler et al., 2015
Corn protein (Zein)	0–70 V, 60°C, 10 min	Increase in the content of disulfide bonds via changes in the secondary structure and rheological properties	Dong et al., 2017
Flaxseed	85 kV, 180 s	Emulsifying, foaming, and antioxidant properties	Hu et al., 2022
Whey protein isolate	70 kV, 1–70 min	Improved emulsifying and foaming properties	Segat et al., 2015
Soybean protein isolate	16–20 kV, 15 min	Improved solubility, water-holding capacity, emulsifying activity, foaming capacity	Sharafodin & Soltanizadeh, 2022
Gluten protein	Indirect plasma treatment by plasma-activated glass	Hydrophobicity, wettability, surface roughness, and gluten structure properties	Nucia et al., 2021
Flaxseed protein	5 kV, 40 kHz, 0–240 s	Increase turbidity, solubility, carbonyl content, foaming capacity, and stability	Yu et al., 2020

3.5.1 Chemical Conjugation and Acylation

Chemical acylation has been widely used to modify the functional properties of plant proteins and involves plant protein amino group conversion into a hydroxyl group of amino acid residues attached to an acylating agent. Acetic anhydride and succinic anhydride are commonly used in the acylation modification of proteins to improve functional properties. Chemical acylation is a nucleophilic substitution reaction of acetylation agents (e.g., succinic, acetic anhydride) with protein amino acid residues (e.g., lysine). A previous report demonstrated that succinylation and acetylation of pea protein improved water-holding capacity, foaming, and emulsifying properties (Johnson & Brekke, 1983). Conjugation between protein and polysaccharide is a popular approach for modification which can enable the protein molecule to be covalently linked with the polysaccharide group. Additionally, the conjugation reaction enhances the solubility and emulsifying activity of pea protein conjugated with gum arabic. Other studies also reported that the pea protein conjugated with pectin (Tamnak et al., 2016) and propylene glycol alginate (Guo et al., 2019) had significantly improved functional properties. The effect of chemical conjugation and acylation observed on the functionality of plant protein is shown in Table 3.5.

3.5.2 Glycation: The Maillard Reaction

Protein glycation is the nonenzymatic reaction between an acyclic group of reducing sugar and a derivative of protein in a complex series and parallel pathway called the Maillard reaction. Glycation of protein by reducing the sugar through the Maillard reaction has been a promising approach, which can improve the functional properties of the protein, such as solubility, emulsifying, thermal stability, and gelation properties. The aldehyde or ketone group of reducing sugar reacts with protein substrate, usually on the lysine chain amino group and N-terminal amino group. In various plants, simple reducing sugars involved in protein glycation are fructose, glucose, and galactose. The electrophoretic profile of mung bean protein and dextran showed that both legume subunits and vicilin participated in the glycation or Maillard reaction. Conjugation of protein is indicated to

TABLE 3.5

A Summary of Plant-based Protein, Chemical Conjugation Acylation, and Target Functionality Impacted

Type of Protein	Technique Used	Effect Observed on Functional Properties	Reference
Pea protein	1.0–5.0 mmol anhydride	Improved emulsifying, foaming, and water-holding capacity	Johnson & Brekke, 1983
Soybean protein	Dry sodium bisulfate 1 mM	Improved emulsifying properties	Matemu et al., 2011
Pea protein	1–5 mmol anhydride	Improved emulsifying, foaming, and water-holding capacity	Johnson & Brekke, 1983
Mung bean protein	Succinic anhydride (0–10 wt % of protein)	Surface hydrophobicity, solubility, emulsifying properties	Charoensuk et al., 2018
Oat protein	Acetic, succinic anhydride	Solubility, foaming capacity	Ma, 1984
Pea protein isolate	Succinic anhydride	Increased solubility, foaming capacity, emulsifying solubility index	Shah et al., 2019
Pea protein and pectin conjugate	Pea protein and pectin conjugate—4:4	Improved solubility and emulsifying properties	Tamnak et al., 2016
Pea protein and propylene glycol conjugate	Pea protein and propylene glycol conjugate—5:1	Improved functional properties	Guo et al., 2019
Whey protein isolate and carbonyl group conjugate	Whey protein isolate and carbonyl group conjugate—:9	Emulsifying properties and rheological properties	Ai et al., 2022

alter the secondary and tertiary structure attachment of dextran, minimizing alpha-helix content and unfolding protein structure due to heat-induced properties. The secondary and tertiary structure changes of protein were monitored by using fluorescence spectra to identify its properties such as solubility and flexibility of protein structure. The solubility of conjugates was increased compared to mung bean protein due to the increased number of hydrophilic groups. Additionally, emulsifying activity and stability of protein conjugates followed a similar trend. Several proteins, including egg white protein, whey protein, soy protein, and milk protein, have been used to achieve improved water-holding, emulsifying, and gelation properties via glycation with saccharide.

3.5.3 Phosphorylation

Protein phosphorylation is a reversible post-translation modification approach for plant-based protein in which an amino acid residue is phosphorylated by a protein kinase to regulate protein function. The type of protein, reaction conditions, and a phosphorylating agent are the main parameters that affect the degree of phosphorylation. Protein kinase catalyzes the addition of a phosphate group to three amino acids serine, threonine, and tyrosine, and the remaining phosphate group can be removed by protein phosphatases. The dynamic phosphorylation of protein plays an important role in transduction, endogenous hormone perception, and environmental stress sensing and response (Xu et al., 2019). Three commonly used phosphorylating agents include sodium trimetaphosphate (STMP), sodium tripolyphosphate (STP), and phosphorous oxychloride ($POCl_3$). According to the United States Food and Drug Administration (FDA), STMP, STP, and $POCl_3$ are generally regarded as safe (GRAS) additives when used as per good manufacturing practices (Xu et al., 2019). Most studies demonstrated that $POCl_3$ can be used in mild conditions, whereas STMP and STP are best used in alkaline conditions (>pH 9.0) and with 35–70°C. The author observed that the phosphate linkages withstand high temperatures (120°C) with pH 2.0–10.0; therefore, phosphorylated proteins

TABLE 3.6

A Summary of Plant-based Protein, Phosphorylation, and Enhancement of Phosphate Content

Type of Protein	Type of Phosphate	Incubation Time (Days)	pH	Introduced Phosphate Content (g/100 g)
Whey protein isolate	Orthophosphate	1	5.5	0.01
Casein	Orthophosphate	1–5	4.0	0.56
Egg white protein	Ovalbumin	1	7.0	0.02
Whey protein phosphate	Pyrophosphate	1	4.0	0.58
Bovine serum albumin	Pyrophosphate	5	4.0	0.45
Whey soy protein	Pyrophosphate	5	4.0	0.85

are very useable in food applications (Li et al., 2010a). The enhancement of phosphate content by phosphorylation of plant protein is shown in Table 3.6.

3.5.4 DEAMIDATION

Chemical deamidation is different from enzymatic deamidation and involves protein reaction with either mild or strongly acidic conditions at elevated temperatures along with the loss of ammonia molecules. Usually, deamidation refers to the conversion of δ-asparagines or γ- glutamine amide groups to carboxylic groups α- and γ-aspartic acid and glutamic acids with the release of ammonia (Li et al., 2010b). The rate of deamidation is influenced with so many factors including pH, temperature, water activity, nonionic catalyst, amino acid sequence, etc. The deamidation of glutamine and asparagine residues in protein results in the liberation of the carboxyl group and thereby an increase of hydration and negative charge. This is favorable, especially in the case of wheat gluten, which has a high content of glutamine consisting (of 30% amino acid), so even a low percentage of deamidation has a large effect on its charge and functionality. The deamidated proteins showed significantly improved solubility, emulsion viscosity, foam expansion, and water-binding capacity, as compared to the original proteins.

3.6 BIOLOGICAL TECHNIQUES

Functional properties of plant-based proteins can be improved by various biological techniques such as enzymatic modifications, fermentation, and crosslinking. The biological modification of plant protein was shown in Table 3.7.

3.6.1 ENZYMATIC HYDROLYSIS

Modification of the molecular structure of a plant-based protein with enzymatic hydrolysis is an attractive approach to improving the nutritional, functional properties as well as sensory properties of the protein. Enzymatic hydrolysis was reported to show positive effects on mild conditions, fast reaction rate, and high specificity. Proteases and transglutaminase are the most frequently used enzymes for modifying the polypeptide backbone of food proteins. Functional properties of proteins such as solubility, emulsifying stability, foaming, and gelation are closely related to their structure and size confirmation and ionic charge distribution. Proteolytic enzymes can enhance their functional properties over a wide range of pH and processing conditions, irrespective of the proteins substrate and protease. Enzymatic hydrolysis with trypsin enzyme (up to 4% degree of

TABLE 3.7

A Summary of Plant-based Proteins, Enzymatic Hydrolysis, and Target Functionality Impacted

Type of Proteins	Treatment Condition	Effects Observed on Protein Functionality	Reference
Chickpea protein	Hydrolysis with alcalase (4–15% DH)	Improved solubility, decreased interfacial tension, decreased emulsifying activity and stability	Ghribi et al., 2015
Lentil protein	Hydrolysis with trypsin (4–20% DH1)	Decreased interfacial tension; decreased emulsifying activity and stability	Avramenko et al., 2012
Black bean protein	Hydrolysis with pepsin and alcalase (24–28% DH)	Alcalase hydrolysates showed higher emulsion stability	do Evangelho et al., 2017
Faba bean protein	Hydrolysis with different proteases (2–16% DH)	Improved solubility, foaming capacity, oil-holding capacity	Eckert et al., 2019
Kidney bean protein	5 U enzyme/g protein, 37°C, 0–240 min	Decreased solubility, emulsifying activity, and stability	Eckert et al., 2019
Rapeseed protein concentrate	Hydrolysis with alcalase (50°C, 1 h, 1–10% DH)	Whippability, emulsion activity index, and foaming capacity increased with an increase in DH	Vioque et al., 2000

hydrolysis) increased solubility from 30% to 60% at pH 4.0–6.0 due to an increased amount of terminal COO- and NH_3^+ group. Arteaga et al. (2020) applied 11 different proteolytic enzymes used to hydrolyze pea protein isolate for the modification of functional and sensory properties. Additionally, the foaming capacity and emulsion activity index tend to increase due to the hydrolysis of protein whereas foaming stability and emulsion stability decrease. The functional properties such as solubility, foaming capacity, and emulsion stability increase due to increased protein molecule solvation, size reduction, and increased surface net charge during emulsion (Akharume et al., 2021). A list of enzymatic hydrolysis and their effects observed on the functionality of plant protein is shown in Table 3.7.

3.6.2 Enzymatic Crosslinking

Enzymatic crosslinking with transglutaminase and other oxidative enzymes is used for improving crosslinking and enhancing protein functionality. Moreover, crosslinking improves the textural properties of the protein by attributing the polypeptide to stronger structures. Transglutaminase is the only food-grade cross-linking enzyme that improves viscosity, texture, emulsion stability, foaming capacity, and gelation and increases the hydrophobicity of plant-based protein. Tang et al. (2008) observed the unfolding of the vicilin units and the formation of higher molecular weight oligomers due to the effect of enzyme crosslinking with kidney bean isolate. Additionally, protein solubility and emulsifying properties were progressively increased with increasing incubation time. Sun and Arntfield (2011) reported that transglutaminase increased the gel strength of soy protein isolate by two times and pea protein isolate by eight times compared to the gel made from untreated counterparts. Transglutaminase enzyme works by catalyzing acyl group reactions and polymerization between protein intra- or inter-chain acyl donor glutamine and acyl acceptor lysine amino acid chain residues in the protein (Gaspar & de Góes-Favoni, 2015). A list of other crosslinking enzymes, their optimal conditions, substrates, and target functionality is presented in Table 3.8.

TABLE 3.8

A Summary of Plant-based Protein, Enzymatic Crosslinking, and Target Functionality Impacted

Protein Type	Enzymes Used	Treatment Condition	Effects Observed on Protein Functionality	Reference
Potato protein	Tyrosinase	Enzyme:substrate—1:30, pH 7, 30 min	Improved emulsion stability and higher storage modulus	Glusac et al., 2018
Soy protein isolate	Transglutaminase	10 U enzyme, 10.5% substrate, pH 7.0, 24 h	Increased elasticity and gel strength	Sun & Arntfield, 2011
Faba bean protein isolate	Transglutaminase and tyrosinase	10–1,000 nkat/g of protein enzyme, 10 mg/mL substrate, pH 7, 20 h	Decreased solubility and foam height	Nivala et al., 2017
Soybean protein isolate	Transglutaminase	0.5 U/g proteins	Water-holding capacity, gelling properties	Chen et al., 2016
Chickpea protein	Transglutaminase	5 U/g protein	Emulsifying stability and gelling properties	Zhu et al., 2022
Whey protein isolate	Transglutaminase	1 U/g protein	Gelling properties	Faergemand et al., 1997

3.6.3 PROTEIN FERMENTATION

Fermentation has been used as a traditional process, a cost-effective biological method for plant-based protein modifications. Various types of starter cultures have been used for the fermentation of plant proteins such as yeast, mold, lactic acid bacteria, and bacillus strains. Among the elevated starter culture, lactobacillus was mostly used to increase the nutritional profile, sensory profile, and shelf life of the product. Fermentation was reported to improve nutritional; functional including solubility, water, and oil-holding capacity, foaming properties; and sensory properties. Liu et al. (2022) reported that fermentation can enhance biological activities. Fermentation has been applied to reduce the bitter and beany flavor of pea protein isolates (El Youssef et al., 2020). In addition, the emulsifying activity of fermented pea protein was observed to decrease at pH 4.0, whereas emulsion stability increased after 5 h. Schindler et al. (2012) also displayed the reduced bitter flavor of lupin protein isolate by lactic acid fermentation. Liquid state fermentation of soy protein isolate also decreased its bitterness and beany flavor (Meinlschmidt et al., 2016). The authors observed a decrease in surface hydrophobicity with increasing fermentation time whereas an increase in surface charge. Sometimes, there are negative effects on solubility, emulsifying properties, and foaming properties. The protein–protein interaction and aggregation were increased due to a decrease in protein solubility. The effect of fermentation, in various cultures, observed on the functionality of plant protein is shown in Table 3.9.

3.7 IMPROVE THEIR TECHNO-FUNCTIONALITY FOR USE AS FUTURE FOODS

Protein techno-functionality is critical in determining the applicability of plant-based protein; on the other hand, techno-functionality is related to the impact on the physicochemical properties of food products, influencing appearance, texture, emulsifying, foaming, stability, gelling, water- and oil-holding capacity, viscosity, and elasticity. Food constituents (e.g., protein, polysaccharides, and lipids) are imperatively functional by their molecular structure and stability of their formed complexes. Extrinsic and intrinsic factors determine the functionality, stability, and shelf life of food containing functional properties. Intrinsic factors include the protein structure, net surface charge, amino acid composition, confirmation, and surface functional group. Extrinsic factors are the pH,

TABLE 3.9

A Summary of Plant-based Protein, Fermentation, and Target Functionality Impacted

Protein Type	Culture Used	Treatment Condition	Effects Observed on Protein Functionality
Lupin protein	*Leuconostoc mesenteroides, Lactobacillus plantarum, and Lactobacillus brevis*	Emulsifying activity index, emulsifying stability index, water solubility index	Lampart-Szczapa et al., 2006
Pea protein	*Streptococcus thermophilus, Lactobacillus acidophilus, and Bifidobacterium lactis*	Functional and thermal properties	Emkani et al., 2021
Pea protein	*Pediococcus pentosaceus, Lactococcus raffinolactis, and Lactobacillus plantarum*	Inhibitor activity	Barkholt et al., 1998
Soy protein	*Lactobacillus plantarum*	Improved protein structure and increased peptide content	Rui et al., 2019
Soy protein	*Lactobacillus casei spp. Pseudoplantarum*	Inhibitor activity	Vallabha & Tiku, 2014

salts, solvents, temperature, and pressure. Protein functionality and molecular structure may change due to intrinsic and extrinsic factors. Thus, it is important to study the response setup's impact on the diversity of functional and physicochemical properties of food products.

Some studies assessed the techno-functional properties of plant proteins, such as soybean (Matemu et al., 2011), chickpea (Ghribi et al., 2015), kidney bean (Eckert et al., 2019), pea (Emkani et al., 2021), and lentils (Avramenko et al., 2012). Few studies investigated these properties such as soybean protein inhibitory activity (Vallabha & Tiku, 2014). The utilization of plant proteins is limited due to their extremely low solubility at neutral pH, except for the soybean and pea.

Therefore, this chapter discusses the techno-functionality of plant-based protein and its advantages and disadvantages. Moreover, given that there is such limited research reported on the solubility, water-holding capacity, foaming capacity, emulsifying stability, and gelling properties of plant-based protein. Potentially, plant-based proteins may be used for preparing meat analogs, protein supplements, snacks, whipped cream, desserts, bakery, sauces, soup, and salad dressing. Therefore, one can consider that exploring plant-based proteins, while aiming to develop technological alternatives for food formulation, is an open field, considering and evaluating required processing technologies for the extraction and modification of techno-functionalities to fulfill all requirements.

3.8   EFFECTS OF DIFFERENT PROCESSING TECHNIQUES ON STRUCTURE

Retaining the native protein structure as much as possible is difficult to obtain functional protein, concerning interface, foam, and emulsion stabilization. Alkaline pH step and process-related protein aggregation can be excluded to prevent phenols from oxidizing. Obvious, specific aggregation of protein-inducing processing steps might be predictable, such as microbial-related heat treatments. The exact impact of these potentially aggregate-inducing steps should be carefully examined for plant-based proteins. Modification of native proteins could be an essential step in changing the functional properties of plant-based proteins. However, several plant proteins, such as globulins, seem to lack the molecular properties to stabilize interfaces effectively. As a result, there has been a substantial effort to improve their interface-stabilizing properties by protein modification. Protein modification techniques impact on the protein structure: A schematic overview of plant protein extraction processes is shown in Figure 3.6.

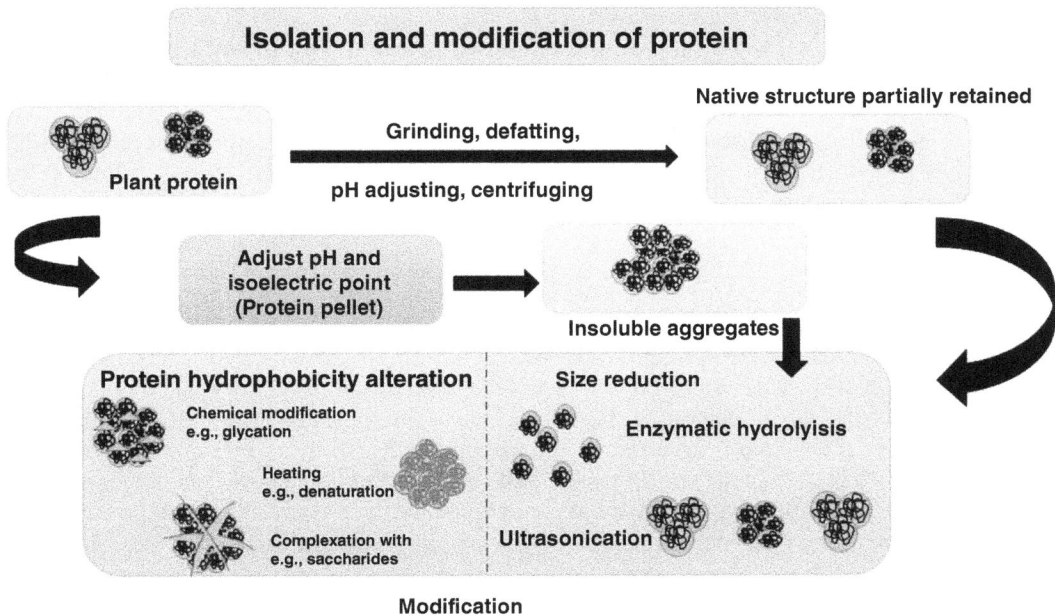

FIGURE 3.6 Schematic overview of plant protein extraction processes and their impact on the protein structure: an overview of commonly applied protein modification techniques.

Many types of novel modification techniques exist, such as physical, chemical, or biological treatments as shown in Figure 3.6. The treatments focus on improving the structure and molecular properties (surface hydrophobicity), which can be modified by chemical or controlled heat treatments. Heating might also induce protein structural alterations and denaturation and the formation of heat-induced peptides, which might dominate the interfacial properties. Protein surface properties can also be improved by complexation; an example is the complexation of insoluble gliadin protein with polysaccharides, leading to surface-active particles and effective foam stabilizers. Other treatments focus on size reduction by breaking down aggregates using ultrasound or high-pressure homogenization. Proteins can also be hydrolyzed chemically or by enzymes to obtain peptides, which can be more surface-active due to their smaller size and higher surface hydrophobicity after exposing previously buried hydrophobic regions. The formation of stiffer interfacial layers is not always guaranteed by this approach, as the peptides might be too small for effective interactions at the interface or perhaps be lacking a specific secondary structure (Sagis & Yang, 2022).

3.9 FUNCTIONAL ASPECTS AND STABILITY

In various food products, based on their emulsifying and foaming stability, surface rheological properties often play an important key role in the stability of food products. Protein globular structure tends to form stiff interfacial bonding in the oil–water and air–water interface. This consideration on protein-stabilized multiphase medium requires quantifying and understanding these rheological properties.

Protein molecules such as whey protein isolate having beta-lactoglobulin can be considered excellent stabilizers for foam and emulsion formation. Its secondary structure has been well characterized; consists of alpha helix and beta sheet structure; and forms film and viscoelastic structure after water absorption. The softening in expansion points to a gradual disruption of the solid microstructure by any given treatment.

On a macroscopic level, the modification process and consequent formation of the interfacial structure are typically studied by monitoring the surface pressure as a function of time. In general, protein diffuses toward the interface and rearranges before adsorption, an adsorption phase in which the surface pressure changes rapidly, followed by a final phase where the continuous denaturation of intramolecular rearrangements of protein takes place. But they also occur in interfaces stabilized by protein aggregates, i.e., proteins denatured before adsorption (Sagis & Yang, 2022). Moreover, proteins also show this type of slow aging in gels at room temperature.

3.10 CONCLUSION

Plant-based proteins are growing fast as an innovative ingredient in the food industry due to their advantages over animal-derived protein, especially concerning sustainability aspects and ethical implications. This chapter describes protein modification based on novel physical, chemical, and biological techniques, which determine the surface active properties of plant-based food protein ingredients. Novel processing is admired by the authors for its ability to increase the extraction yield of plant-based proteins and their function modifications. In addition, novel techniques not only increase isolation yield but also modify the secondary and tertiary structures of the protein. This is a promising technology for improving protein functionalities and nutritional qualities. However, there is very limited research on the effect of novel techniques on modified plant-based protein ingredients. High-pressure processing and ultrasonication and microwave-assisted extraction are being widely used in processing to stabilize plant protein-based emulsion in various food industries. Furthermore, the inferior quality and functionality of plant-based protein are required to be changed to be compared to animal protein to determine their quality and technological use for food applications. In this chapter, we have discussed the current opportunities and challenges of novel processing to improve and move the field forward. Novel processing including the impacts of physical, chemical, and biological methods on the plant-based protein was discussed as modification due to their ability to change its structure and functionality.

REFERENCES

Ahmed, J., Al-Ruwaih, N., Mulla, M., & Rahman, M. H. (2018). Effect of high pressure treatment on functional, rheological and structural properties of kidney bean protein isolate. *LWT-Food Science and Technology*, *91*, 191–197.

Ahmed, J., Mulla, M., Al-Ruwaih, N., & Arfat, Y. A. (2019). Effect of high-pressure treatment prior to enzymatic hydrolysis on rheological, thermal, and antioxidant properties of lentil protein isolate. *Legume Science*, *1*(1), e10.

Ai, C., Zhao, C., Guo, X., Chen, L., & Yu, S. (2022). Physicochemical properties of whey protein isolate and alkaline soluble polysaccharide from sugar beet pulp conjugates formed by Maillard reaction and genipin crosslinking reaction: A comparison study. *Food Chemistry: X*, *14*, 100358.

Akharume, F. U., Aluko, R. E., & Adedeji, A. A. (2021). Modification of plant proteins for improved functionality: A review. *Comprehensive Reviews in Food Science and Food Safety*, *20*(1), 198–224.

Arteaga, V. G., Guardia, M. A., Muranyi, I., Eisner, P., & Schweiggert-Weisz, U. (2020). Effect of enzymatic hydrolysis on molecular weight distribution, techno-functional properties and sensory perception of pea protein isolates. *Innovative Food Science & Emerging Technologies*, *65*, 102449.

Avramenko, N. A., Low, N. H., & Nickerson, M. T. (2012). The effects of limited enzymatic hydrolysis on the physicochemical and emulsifying properties of a lentil protein isolate. *Food Research International*, *51*(1), 162–169.

Barkholt, V., Jergensen, P. B., Sørensen, D., Bahrenscheer, J., Haikara, A., Lemola, E., &Frøkiær, H. (1998). Protein modification by fermentation: Effect of fermentation on the potential allergenicity of pea. *Allergy*, *53*, 106–108.

Bußler, S., Steins, V., Ehlbeck, J., & Schlüter, O. (2015). Impact of thermal treatment versus cold atmospheric plasma processing on the techno-functional protein properties from Pisum sativum 'Salamanca'. *Journal of Food Engineering*, *167*, 166–174.

Bühler, J. M., Dekkers, B. L., Bruins, M. E., & Van Der Goot, A. J. (2020). Modifying faba bean protein concentrate using dry heat to increase water holding capacity. *Foods*, *9*(8), 1077.

Chao, D., & Aluko, R. E. (2018). Modification of the structural, emulsifying, and foaming properties of an isolated pea protein by thermal pretreatment. *CyTA-Journal of Food*, *16*(1), 357–366.

Chao, D., Jung, S., & Aluko, R. E. (2018). Physicochemical and functional properties of high pressure-treated isolated pea protein. *Innovative Food Science & Emerging Technologies*, *45*, 179–185.

Charoensuk, D., Brannan, R. G., Chanasattru, W., & Chaiyasit, W. (2018). Physicochemical and emulsifying properties of mung bean protein isolate as influenced by succinylation. *International Journal of Food Properties*, *21*(1), 1633–1645.

Chen, Z., Shi, X., Xu, J., Du, Y., Yao, M., & Guo, S. (2016). Gel properties of SPI modified by enzymatic cross-linking during frozen storage. *Food Hydrocolloids*, *56*, 445–452.

do Evangelho, J. A., Vanier, N. L., Pinto, V. Z., De Berrios, J. J., Guerra Dias, A. R., & da Rosa Zavareze, E. (2017) Black bean (*Phaseolus vulgaris L.*) protein hydrolysates: Physicochemical and functional properties. *Food Chemistry*, *214*, 460–467.

Dong, S., Wang, J. M., Cheng, L. M., Lu, Y. L., Li, S. H., & Chen, Y. (2017). Behavior of zein in aqueous ethanol under atmospheric pressure cold plasma treatment. *Journal of Agricultural and Food Chemistry*, *65*(34), 7352–7360.

Eckert, E., Han, J., Swallow, K., Tian, Z., Jarpa-Parra, M., & Chen, L. (2019). Effects of enzymatic hydrolysis and ultrafiltration on physicochemical and functional properties of faba bean protein. *Cereal Chemistry*, *96*(4), 725–741.

El Youssef, C., Bonnarme, P., Fraud, S., Péron, A. C., Helinck, S., & Landaud, S. (2020). Sensory improvement of a pea protein-based product using microbial co-cultures of lactic acid bacteria and yeasts. *Foods*, *9*(3), 349.

Emkani, M., Oliete, B., & Saurel, R. (2021). Pea protein extraction assisted by lactic fermentation: Impact on protein profile and thermal properties. *Foods*, *10*(3), 549.

Faergemand, M., Otte, J., & Qvist, K. B. (1997). Enzymatic cross-linking of whey proteins by a Ca^{2+}-independent microbial transglutaminase from Streptomyces lydicus. *Food Hydrocolloids*, *11*(1), 19–25.

Floury, J., Desrumaux, A., & Legrand, J. (2002). Effect of ultra-high pressure homogenization on structure and on rheological properties of soy protein-stabilized emulsions. *Journal of Food Science*, *67*(9), 3388–3395.

Garcia-Mora, P., Peñas, E., Frías, J., Gomez, R., & Martinez-Villaluenga, C. (2015). High-pressure improves enzymatic proteolysis and the release of peptides with angiotensin I converting enzyme inhibitory and antioxidant activities from lentil proteins. *Food Chemistry*, *171*, 224–232.

Gaspar, A. L. C., & de Góes-Favoni, S. P. (2015). Action of microbial transglutaminase (MTGase) in the modification of food proteins: A review. *Food Chemistry*, *171*, 315–322.

Ghribi, A. M., Gafsi, I. M., Sila, A., Blecker, C, Danthine, S., Attia, H., Bougatef, A., Besbes, S. (2015). Effects of enzymatic hydrolysis on conformational and functional properties of chickpea protein isolate. *Food Chemistry*, *187*, 322–330.

Glusac, J., Davidesko-Vardi, I., Isaschar-Ovdat, S., Kukavica, B., & Fishman, A. (2018). Gel-like emulsions stabilized by tyrosinasecrosslinked potato and zein proteins. *Food Hydrocolloids*, *82*, 53–63.

Guo, Q., Su, J., Yuan, F., Mao, L., & Gao, Y. (2019). Preparation, characterization and stability of pea protein isolate and propylene glycol alginate soluble complexes. *LWT-Food Science and Technology*, *101*, 476–482.

Hu, H., Cheung, I. W., Pan, S., & Li-Chan, E. C. (2015). Effect of high intensity ultrasound on physicochemical and functional properties of aggregated soybean β-conglycinin and glycinin. *Food Hydrocolloids*, *45*, 102–110.

Hu, Y., Tian, H., Hu, S., Dong, L., Zhang, J., Yu, X., . . . Xu, X. (2022). The effect of in-package cold plasma on the formation of polycyclic aromatic hydrocarbons in charcoal-grilled beef steak with different oils or fats. *Food Chemistry*, *371*, 131384.

Huang, K., Shi, J., Li, M., Sun, R., Guan, W., Cao, H., . . . Zhang, Y. (2022). Intervention of microwave irradiation on structure and quality characteristics of quinoa protein aggregates. *Food Hydrocolloids*, *130*, 107677.

Jha, A. K., & Sit, N. (2021). Extraction of bioactive compounds from plant materials using combination of various novel methods: A review. *Trends in Food Science & Technology*, *119*, 579–591.

Jiang, L., Wang, J., Li, Y., Wang, Z., Liang, J., Wang, R., . . . Zhang, M. (2014). Effects of ultrasound on the structure and physical properties of black bean protein isolates. *Food Research International*, *62*, 595–601.

Jiao, X., Chen, W., & Fan, D. (2022). Behind the Veil: A multidisciplinary discussion on protein–microwave interactions. *Current Opinion in Food Science*, 100–936.

Johnson, E. A., & Brekke, C. J. (1983). Functional properties of acylated pea protein isolates. *Journal of Food Science*, *48*(3), 722–725.

Kyriakopoulou, K., Keppler, J. K., & van der Goot, A. J. (2021). Functionality of ingredients and additives in plant-based meat analogues. *Foods*, *10*(3), 600.

Lampart-Szczapa, E., Konieczny, P., Nogala-Kałucka, M., Walczak, S., Kossowska, I., & Malinowska, M. (2006). Some functional properties of lupin proteins modified by lactic fermentation and extrusion. *Food Chemistry*, *96*(2), 290–296.

Lee, C., In, S., Han, Y., & Oh, S. (2016). Reactivity change of IgE to buckwheat protein treated with high-pressure and enzymatic hydrolysis. *Journal of the Science of Food and Agriculture*, *96*, 2073–2079.

Li, C.-P., Enomoto, H., Hayashi, Y., Zhao, H., & Aoki, T. (2010a). Recent advances in phosphorylation of food proteins: A review. *LWT Food Science and Technology*, *43*(9), 1295–1300.

Li, X., Lin, C., & O'Connor, P. B. (2010b). Glutamine deamidation: Differentiation of glutamic acid and gamma-glutamic acid in peptides by electron capture dissociation. *Analytical Chemistry*, *82*(9), 3606–3615.

Liu, L., Chen, X., Hao, L., Zhang, G., Jin, Z., Li, C., . . . Chen, B. (2022). Traditional fermented soybean products: Processing, flavor formation, nutritional and biological activities. *Critical Reviews in Food Science and Nutrition*, *62*(7), 1971–1989.

Ma, C. Y. (1984). Functional properties of acylated oat protein. *Journal of Food Science*, *49*(4), 1128–1131.

Manassero, C. A., Beaumal, V., Vaudagna, S. R., Speroni, F., & Anton, M. (2018). Calcium addition, pH and high hydrostatic pressure effects on soybean protein isolates—Part 2: Emulsifying properties. *Food and Bioprocess Technology*, *11*, 2079–2093.

Martínez-Velasco, A., Lobato-Calleros, C., Hernández-Rodríguez, B. E., Román-Guerrero, A., Alvarez-Ramirez, J., & Vernon-Carter, E. J. (2018). High intensity ultrasound treatment of faba bean (*Vicia faba L.*) protein: Effect on surface properties, foaming ability and structural changes. *Ultrasonics Sonochemistry*, *44*, 97–105.

Matemu, A. O., Kayahara, H., Murasawa, H., Katayama, S., & Nakamura, S. (2011). Improved emulsifying properties of soy proteins by acylation with saturated fatty acids. *Food Chemistry*, *124*(2), 596–602.

Mehr, H. M., & Koocheki, A. (2020). Effect of atmospheric cold plasma on structure, interfacial and emulsifying properties of Grass pea (*Lathyrus sativus L.*) protein isolate. *Food Hydrocolloids*, *106*, 105899.

Meinlschmidt, P., Ueberham, E., Lehmann, J., Reineke, K., Schlüter, O., Schweiggert-Weisz, U., & Eisner, P. (2016). The effects of pulsed ultraviolet light, cold atmospheric pressure plasma, and gamma-irradiation on the immunoreactivity of soy protein isolate. *Innovative Food Science & Emerging Technologies*, *38*, 374–383.

Mulla, M. Z., Subramanian, P., & Dar, B. N. (2022). Functionalization of legume proteins using high pressure processing: Effect on techno-functional properties and digestibility of legume proteins. *LWT*, *158*, 113106.

Nivala, O., Mäkinen, O. E., Kruus, K., Nordlund, E., & Ercili-Cura, D. (2017). Structuring colloidal oat and faba bean protein particles via enzymatic modification. *Food Chemistry*, *231*, 87–95.

Nucia, A., Tomczyńska-Mleko, M., Okoń, S., Kowalczyk, K., Terpiłowski, K., Pérez-Huertas, S., . . . Mleko, S. (2021). Surface properties of gluten deposited on cold plasma-activated glass. *Food Hydrocolloids*, *118*, 106778.

Ochoa-Rivas, A., Nava-Valdez, Y., Serna-Saldívar, S. O., & Chuck-Hernández, C. (2017). Microwave and ultrasound to enhance protein extraction from peanut flour under alkaline conditions: Effects in yield and functional properties of protein isolates. *Food and Bioprocess Technology*, *10*, 543–555.

Park, K., Kwon, I. C., & Park, K. (2011). Oral protein delivery: Current status and future prospect. *Reactive and Functional Polymers*, *71*(3), 280–287.

Peng, W., Kong, X., Chen, Y., Zhang, C., Yang, Y., & Hua, Y. (2016). Effects of heat treatment on the emulsifying properties of pea proteins. *Food Hydrocolloids*, *52*, 301–310.

Phillips, L. G. (2013). *Structure-function properties of food proteins*. Cambridge, MA: Academic Press.

Puppo, M. C., Beaumal, V., Chapleau, N., Speroni, F., de Lamballerie, M., Añón, M. C., & Anton, M. (2008). Physicochemical and rheological properties of soybean protein emulsions processed with a combined temperature/high-pressure treatment. *Food Hydrocolloids*, *22*(6), 1079–1089.

Rui, X., Huang, J., Xing, G., Zhang, Q., Li, W., & Dong, M. (2019). Changes in soy protein immunoglobulin E reactivity, protein degradation, and conformation through fermentation with Lactobacillus plantarum strains. *LWT*, *99*, 156–165.

Sagis, L. M., & Yang, J. (2022). Protein-stabilized interfaces in multiphase food: Comparing structure-function relations of plant-based and animal-based proteins. *Current Opinion in Food Science*, *43*, 53–60.

Saricaoglu, F. T. (2020). Application of high-pressure homogenization (HPH) to modify functional, structural and rheological properties of lentil (*Lens culinaris*) proteins. *International Journal of Biological Macromolecules*, *144*, 760–769.

Schindler, S., Zelena, K., Krings, U., Bez, J., Eisner, P., & Berger, R. G. (2012). Improvement of the aroma of pea (*Pisum sativum*) protein extracts by lactic acid fermentation. *Food Biotechnology*, *26*(1), 58–74.

Segat, A., Misra, N. N., Cullen, P. J., & Innocente, N. (2015). Atmospheric pressure cold plasma (ACP) treatment of whey protein isolate model solution. *Innovative Food Science & Emerging Technologies*, *29*, 247–254.

Shah, N. N., Umesh, K. V., & Singhal, R. S. (2019). Hydrophobically modified pea proteins: Synthesis, characterization and evaluation as emulsifiers in eggless cake. *Journal of Food Engineering*, *255*, 15–23.

Sharafodin, H., & Soltanizadeh, N. (2022). Potential application of DBD plasma technique for modifying structural and physicochemical properties of soy protein isolate. *Food Hydrocolloids*, *122*, 107077.

Sun, X. D., & Arntfield, S. D. (2011). Gelation properties of salt-extracted pea protein isolate catalyzed by microbial transglutaminase cross-linking. *Food Hydrocolloids*, *25*(1), 25–31.

Tamnak, S., Mirhosseini, H., Tan, C. P., Amid, B. T., Kazemi, M., & Hedayatnia, S. (2016). Encapsulation properties, release behavior and physicochemical characteristics of water-in-oil-in-water (W/O/W) emulsion stabilized with pectin–pea protein isolate conjugate and Tween 80. *Food Hydrocolloids*, *61*, 599–608.

Tang, C. H., & Ma, C. Y. (2009). Heat-induced modifications in the functional and structural properties of vicilin-rich protein isolate from kidney (*Phaseolus vulgaris* L.) bean. *Food Chemistry*, *115*(3), 859–866.

Tang, C. H., Sun, X., & Yin, S. W. (2009). Physicochemical, functional and structural properties of vicilin-rich protein isolates from three Phaseolus legumes: Effect of heat treatment. *Food Hydrocolloids*, *23*(7), 1771–1778.

Tang, C. H., Sun, X., Yin, S. W., & Ma, C. Y. (2008). Transglutaminase-induced cross-linking of vicilin-rich kidney protein isolate: Influence on the functional properties and in vitro digestibility. *Food Research International*, *41*(10), 941–947.

Vallabha, V., & Tiku, P. K. (2014). Antihypertensive peptides derived from soy protein by fermentation. *International Journal of Peptide Research and Therapeutics*, *20*, 161–168.

Vioque, J., Sánchez-Vioque, R., Clemente, A., Pedroche, J., & Millán, F. (2000). Partially hydrolyzed rapeseed protein isolates with improved functional properties. *Journal of the American Oil Chemists' Society*, *77*(4), 447–450.

Wang, X., Gu, L., Su, Y., Li, J., Yang, Y., & Chang, C. (2020a). Microwave technology as a new strategy to induce structural transition and foaming properties improvement of egg white powder. *Food Hydrocolloids*, *101*, 105–530.

Wang, Y., Wang, Y., Li, K., Bai, Y., Li, B., & Xu, W. (2020b). Effect of high intensity ultrasound on physicochemical, interfacial and gel properties of chickpea protein isolate. *LWT*, *129*, 109563.

Wouters, A. G., Rombouts, I., Fierens, E., Brijs, K., & Delcour, J. A. (2016). Relevance of the functional properties of enzymatic plant protein hydrolysates in food systems. *Comprehensive Reviews in Food Science and Food Safety*, *15*(4), 786–800.

Xiong, T., Xiong, W., Ge, M., Xia, J., Li, B., & Chen, Y. (2018). Effect of high intensity ultrasound on structure and foaming properties of pea protein isolate. *Food Research International*, *109*, 260–267.

Xu, S., Xiao, J., Yin, F., Guo, X., Xing, L., Xu, Y., & Chong, K. (2019). The protein modifications of O-GlcNAcylation and phosphorylation mediate vernalization response for flowering in winter wheat. *Plant Physiology*, *180*(3), 1436–1449.

Yu, X., Huang, S., Nie, C., Deng, Q., Zhai, Y., & Shen, R. (2020). Effects of atmospheric pressure plasma jet on the physicochemical, functional, and antioxidant properties of flaxseed protein. *Journal of Food Science*, *85*(7), 2010–2019.

Zhang, Q., Cheng, Z., Zhang, J., Nasiru, M. M., Wang, Y., & Fu, L. (2021). Atmospheric cold plasma treatment of soybean protein isolate: Insights into the structural, physicochemical, and allergenic characteristics. *Journal of Food Science*, *86*(1), 68–77.

Zheng, Y., Li, Z., Zhang, C., Zheng, B., & Tian, Y. (2020). Effects of microwave-vacuum pre-treatment with different power levels on the structural and emulsifying properties of lotus seed protein isolates. *Food Chemistry*, *311*, 125932.

Zhu, G., Li, Y., Xie, L., Sun, H., Zheng, Z., & Liu, F. (2022). Effects of enzymatic cross-linking combined with ultrasound on the oil adsorption capacity of chickpea protein. *Food Chemistry*, *383*, 132641.

4 Biological Modification of Plant-Based Proteins

Subhasri Dharmaraj and Pintu Choudhary

4.1 INTRODUCTION

A sustainable food system, with optimal utilisation of natural resources, to provide nutritious whole-some food for fellow human beings is the need of the hour. Also there is an environmental, sustainable choice of plant-based diet which has many driving factors. This rising current trends of plant-based diets, particularly those that replace animal-based products, have created more awareness among the citizens. Plant-based foods, especially those that substitute animal-based products, are currently gaining momentum. However, there are still many challenges to be overcome, notably when developing plant-based products as compared to conventional animal origin products. Poor dietary flavour and technically functional qualities are still playing a major role in determining its utilisation.

Commercially, protein extracts, isolates, and concentrates are obtainable, and there is an expanding interest in additional pulse protein and legume plant protein sources such as lentils, faba beans, peas, mung, and chickpea. Due to their excellent technological, functional, and nutritional qualities, pulse protein components are garnering more interest as possible substitutes for soy and animal proteins. However, additional processing development is still required to maximise the technological performance of such plant ingredients so that they are suitable for a manufacture of diverse products.

Due to their ethical production methods, affordable prices, and superior health advantages over their animal-based equivalents, extensive research on plant-based proteins has recently received considerable interest. The prime concern is their poor aqueous solubility, formation of complexes, and also susceptibility to environmental stress conditions like pH, salt, and temperature, which have restricted their applicability. Furthermore, hemicellulose, lignin, and other indigestible polysaccharides frequently encase plant proteins, reducing their bioavailability. Therefore, it is highly desired to modify plant proteins in order to enhance their functionality and industrial production scalability to make them a potential candidate in novel plant-based products. By modifying the physicochemical characteristics of plant proteins, it is possible to enhance and diversify their technological utility while also resolving their drawbacks.

A significant amount of effort has indeed been devoted into creating multiple unconventional protein sources over the past few decades. Studies have primarily focused on producing economical protein sources rather than on making use of the resources. As a result, even while many non-traditional protein sources can be synthesised at a reasonable cost, they frequently lack important functional characteristics.

Proteins are essential to various stages of food processing and serve a variety of roles in the quality and stability of food. They are able to collaborate to generate the networks that are necessary for the creation of gel and edible films, as well as interface films that stabilise emulsification and foams. Native proteins are quite modified by variations in moisture content, screw speed, pressure, and other processing factors. When an amino acid is accessible, it has the potential to interact with other dietary components. Extrusion reduces the solubility of proteins and exposes hydrophobic regions on the surface of the protein molecule. By creating disulphide and hydrophobic linkages between protein molecules, extrusion significantly reduces the solubility of proteins.

Since these characteristics of proteins may be altered throughout various processing processes, they are referred to as "protein functionality." The three main processes used to modify proteins—chemical, biological, and physical—have a profound effect on the structure and functionality

    DOI: 10.1201/9781003369790-4

of proteins present in food. The nutritional value and functional qualities of food items may be enhanced by the physical and biological alteration of proteins.

In addition to size, modifications to the emulsion activity index (EAI) depend heavily on variations in protein solubility and hydrophobicity. The impact of protein concentration, pH, and high pressure on the rheological and functional characteristics of isolated proteins can be optimised. It is discovered that pressure treatments, pH ranges, and isolate concentrations have an impact on emulsifying action. Ionic connections and hydrophobic interactions are broken, leading in more flexible protein molecules which might absorb more efficiently. Furthermore, protein-foaming behaviour is significantly influenced by pH. It has been discovered that proteins lose some of their capacity to foam when subjected to pressure at pH 5.0, which is quite close to the pIs of many proteins. Lack of repulsive forces close to the protein pI causes protein aggregation, which lowers solubility.

Non-thermal techniques have a wide range of applications in the food and biotechnology industries, particularly because they have the capacity to change the physicochemical characteristics of food ingredients (Mirmoghtadaie et al., 2016). The development of modified proteins now benefits greatly from chemical protein modification. A vast arsenal has been created through the combined application of genetic and chemical techniques, enabling the synthesis of almost unlimited protein structures with either naturally occurring or synthetically changed residues. A number of these changes, including acylation, methylation, phosphorylation, sulfation, farnesylation, ubiquitination, and glycosylation, are crucial for key cellular functions like signalling, migration, differentiation, and trafficking. Therefore, recreating such natural protein alterations in a highly effective and controlled manner (by adding natural post translational modifications) would offer a priceless tool to investigate their specific function. Site-selective modification of proteins is a crucial method for probing the possibility of the introduction and (bio)orthogonal modification of unnatural moieties/amino acids (Boutureira & Bernardes, 2015).

This chapter concentrates on the enzymatic hydrolysis, cross-linking, and fermentation type of biological modification that has distinct advantages over physically altered protein since they are energy efficient and ecologically sound. The modification of protein results in modified size, shape, structure of the protein, along with modified techno-functionality and biological property. It is also capable of imparting new flavour and changes along with the modified textural, rheological, and sensory attributes (Figure 4.1).

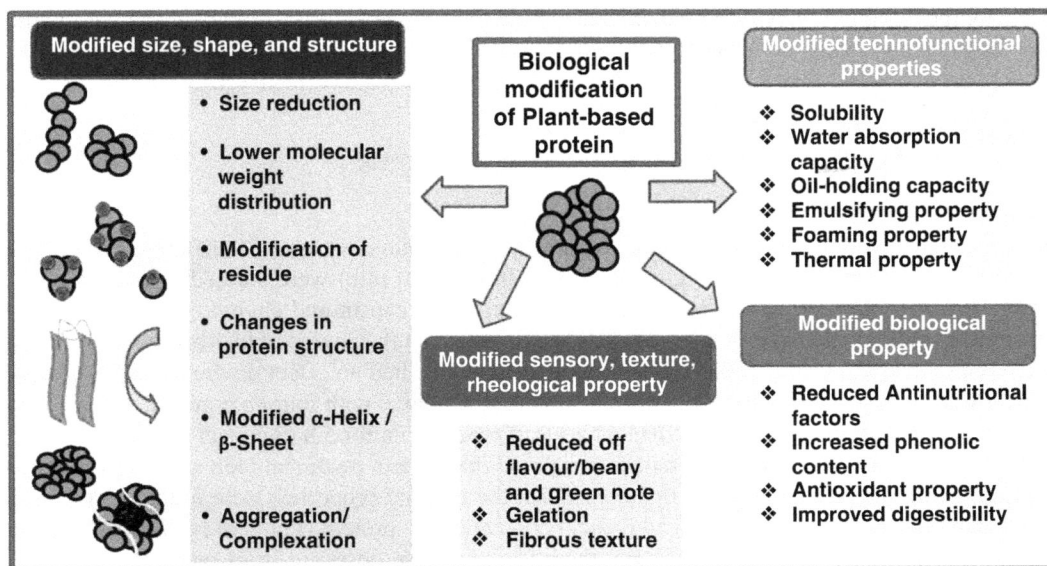

FIGURE 4.1 Biological modification of plant-based protein.

4.2 ENZYMATIC HYDROLYSIS

Enzymatic modification is one of the widely researched, used process for biological modification of plant-based proteins, especially for their absorption, availability, and non-toxic nature (Felix U. Akharume et al., 2021). Additionally, this type of change can be carried out under favourable environments with little by-products. It is reported that the chemical nature of the original protein will be preserved post enzymatic modification. And also it is found to render some benefits of quick response time and specificity over chemical methods. It is regarded as an efficient method that transforms the frameworks of proteins and changes the peptide length by rupturing peptide bonds. Because it keeps the natural composition of proteins derived from food, enzyme hydrolysis is preferred over chemical treatments. Enzymatic hydrolysis involves cleavage of the peptide linkages, resulting in a number of smaller peptides, along with free amino acids. During this process, the hydrophobic groups of the protein are exposed, resulting in certain modified techno-functionalities.

4.2.1 Enzymes

Enzymes like proteases and amylases have been used since traditional times for soy-derived foods. There are several types of enzymes used in the enzymatic hydrolysis of proteins: pepsin, pancreatin, ficin, bromelain, papain, and even microbial source of enzymes like alcalase (Tapal & Tiku, 2019). Based on the enzyme, nature of the substrate, various processing stages, number of steps, time taken for hydrolysis, degree of hydrolysis, and pH condition, the overall effect on the structure and function of plant proteins varies substantially. Diverse peptide combinations with various hydrophobic and electrostatic attributes can be produced because of differences in protease selectivity and protein structure. Usually, when the native protein constituent has poor baseline activity, most profound changes can be seen. The specificity of the enzymes can be clearly observed in the post hydrolysed electrophoretic profile of the proteins.

In general, the proteolytic enzymes participate at optimum conditions of temperature and pH, specific to the substrate, and cleave the proteins at specific peptide cleavage bonds. In case of pepsin, the enzyme is cleaved at phenylalanine or the leucine bond, whereas for the papain enzyme, the cleavage takes place at the phenylalanine, arginine, and lysine bonds, exhibiting a broad specificity. Furthermore, it hydrolyzes amides and esters. It showed that the type of enzyme used had a different effect on the protein. For example papain had a preference towards the hydrophobic residues, while bromelain had a preference towards polar amino acids. Since pea protein isolate was found to have more hydrophobic residues, it was found to cleave the peptide bonds effectively.

4.2.2 Effect of Enzymatic Modification on Plant Protein

4.2.2.1 Degree of Hydrolysis

In a research involving enzymatic modification of pea proteins isolate, 11 different proteolytic enzymes and different hydrolysis times (15, 30, 60, and 120 min) were studied. The proteolytic action can be influenced by the presence of inhibitors like trypsin and chymotrypsin. Also, with substrate specificity, the type of enzyme used had an effect on the degree of hydrolysis and the time taken (García Arteaga et al., 2020). The treatment time also had an effect on the degree of hydrolysis; it was analysed that when lupin protein isolate was treated with four enzymes in a single step process, it could attain higher value than a two-step process run for 5 h (Schlegel et al., 2020).

However, in case of rice bran protein, it was found that papain treatment increased the structural flexibility with increased degree of hydrolysis. But the thermal properties were found to decrease, which can be explained by the reduced β-sheet content of the protein (Pal et al., 2021). Also, upon the enzymatic treatment of peanut isolate, there was a change observed in its protein structure. It had increased random coil content; thus, it had formed an open structure than the native state. Thus, the decreased α-helix content and increased β-sheet resulted in more open structure, which further explained the increased surface hydrophobicity. Under natural conditions, peanut protein isolate cannot form gel under low temperatures, but when treated with alcalase, there were changes in the tertiary structure of peanut protein isolate, and it was deduced that hydrophobic interactions were the primary factor for gel formation (C. Zhang et al., 2021).

4.2.2.2 Functionality

Proteins used in food processing have certain favourable effects on their nutritive value and functional characteristics. Enzymatic hydrolysis is found to improve their techno-functionality, enhance the digestibility, absorptive capacity, flavour and texture of the proteins as well as their biological property like antioxidant activity (Wu et al., 2020).

The protein solubility was found to be improved by the action of proteolytic enzymes for pea protein isolate (Dong et al., 2011; García Arteaga et al., 2020), lupin protein isolate (Schlegel et al., 2020), lentil protein concentrate (Vogelsang-O'Dwyer et al., 2022), walnut protein (Sun et al., 2019), and rice bran protein (Pal et al., 2021).

The solubility can be explained by the size reduction of larger proteins into smaller fragment peptides, more short soluble peptides, and thus increased hydrophilicity. This is a very important property as it has major impact on the gelling, foaming, and emulsifying behaviour. It is also correlated with degree of hydrolysis, because higher degree of hydrolysis exhibited better number of smaller peptides, exposure of new ionisable groups, and thus increased solubility of the protein. Also, in the slightly acidic range, solubility is typically improved, which may also be associated with improved foaming and emulsifying capabilities.

The solubility was also examined as a function of pH. But in this study of pea protein modification, it was observed that the modification using papain yielded lower solubility, which can be explained by the protein–protein interaction which resulted in aggregation, followed by the precipitation of the protein and increased hydrophobicity (Konieczny et al., 2020). Low solubility of hydrolysed proteins had an adverse effect on solubility-dependent functional features such as foaming and emulsifying abilities. Due to the enhanced oil and water holding capacity, the resulting hydrolysates offer potential for use in baked foods and processed meat applications.

Surface hydrophobicity and charge of the protein can be found increased. It was thought that this rise in charge brought on by hydrolysis was caused by the partial unravelling of the proteins, the exposing of submerged hydrophilic amino acids, the release of peptides, and the hydrolysis process itself. Also, surface hydrophobicity can be attributed to the hydrophobic patches exposed, which in turn influences the emulsifying property. In a study of lentil protein concentrate, alcalase and novozyme exhibited higher solubility under acidic environment, while flavourzyme treated exhibited a moderate solubility (Vogelsang-O'Dwyer et al., 2022).

In case of foaming, the enzymatic hydrolysis tends to increase the foaming activity. The modification yields changes in the structure of the protein, i.e. exposed polar groups and hydrophilic groups to the liquid media, size, increased protein surface coverage, and stabilised foam bubble is thus formed (Damodaran, 2005). This stabilisation of the foam can also be attributed to the large peptides with modified structures. In case of walnut protein (present in dreg) modified by papain and protease, the change in microstructure exhibited increased water-holding capacity, emulsifying property, and not much change in foaming property (Sun et al., 2019). The enzyme combination (Alcalase 2.4 L, Papain, Corolase 7089, and Neutrase 0.8 L) was negatively correlated to enhanced protein functionalities of the lupin protein isolate. But in case of sensory evaluation, it was explained that the bitterness significantly increased when papain and alcalase combination were used. Also, alcalase and papain showed effective depletion in the IgE-reactive polypeptides (Schlegel et al., 2020).

4.2.2.3 Biological Activity and Digestibility

During enzymatic hydrolysis, while lowering the ability to chelate metals, hydrolysis had a profoundly favourable impact on antioxidant characteristics. In some studies involving rice bran protein

(Pal et al., 2021) and quinoa protein (Mudgil et al., 2019), the biological activity was studied in relation to digestibility of the protein.

In a research on proteins from quinoa and amaranth, hydrolysis using chymotrypsin, bromelain, and protease showed improved bioactive activities. Overall, compared to hydrolysates produced by bromelain and protease, chymotrypsin-treated quinoa and amaranth proteins showed much enhanced antioxidative and antihaemolytic properties (Mudgil et al., 2019). Thus, this can be based on substrate specificity and the type of enzymatic hydrolysis carried out.

In order to determine the reduction of the allergenic potential of the plant-based proteins, the degree of hydrolysis was examined together with electrophoretic profile to provide a preliminary indication of the molecular size distribution of the proteins (Schlegel et al., 2020). It was observed that trypsin and chymotrypsin inhibitory activity gradually decreased with increased degree of hydrolysis. Following enzymatic hydrolysis, the amount of total phenolic content did not undergo major change (Goertzen et al., 2021).

4.3   ASSISTED ENZYMATIC MODIFICATION

For achieving better results, a higher degree of hydrolysis, and better functional properties, enzymatic method can be used along with other physical methods like high-pressure processing (Dong et al., 2011) and ultrafiltration as utilized for faba bean protein (Eckert et al., 2019). Plant-based proteins must be fully hydrolysed in order to increase functioning. The main problems with proteolytic hydrolysis of plant proteins include the development of a bitter taste, aggregation, and the high cost of enzymes. The type of enzyme, hydrolysis conditions, and degree of hydrolysis control must all be taken into account and extensively researched. Overall, it is relatively effective to deamidate a plant protein using an enzymatic approach, and it is also fairly simple to generate the modified protein with desired functional characteristics. The functional features of the changed protein can be altered directly or indirectly by playing around the degree of hydrolysis, enzyme, substrate, catalytic sites, cleavage sites, and assisted processing methods.

4.3.1   CROSSLINKING WITH TRANSGLUTAMINASE

4.3.1.1   Transglutaminase

Transglutaminases are protein–glutamine γ glutamyl transferase enzymes which are calcium dependent. They were discovered on the basis of their transamidating property in a guinea pig liver in 1959. It has been demonstrated that transglutaminases (TGs), a large family of enzymes, are produced by mammals, animals, plants, and microorganisms. A variety of unbranched primary amines, such as the amino group of lysine and the carboxyamide group of a protein-bound glutamine residue, are converted into other amines through the acyl transfer reaction that is catalysed by TGase (acyl donor) (Parrotta et al., 2022). Hence, it is projected that the TGase-mediated approach will be more site-specific, reliable, reproducible, and adaptable than chemical modification or PEGylation (Fontana et al., 2008).

4.3.1.2   Protein-Glutaminase

Protein-glutaminase (PG) is an enzyme that selectively catalyses the deamidation of glutamine residues in proteins, which can be utilised to increase water retention and decrease allergens or offensive smells. Since it was found in *Chryseobacterium proteolyticum*, PG has received a lot of interest from the food business, because it enhances the functioning of dietary proteins. Nonetheless, due to its poor enzymatic efficacy and yield, the use of PG is still restricted (G. Zhang et al., 2021). Thus, a protein substrate's lysine and glutamine residues are cross-linked by TG, while glutamine is converted to glutamic acid by PG deamidation process. Prior to enzymatic treatment, preheating of isolate should be examined since it can increase the effectiveness of the enzymes by partially denaturing the protein structure.

a

$$\text{Gln}\!-\!\overset{\overset{\displaystyle CH_3}{|}}{\underset{\underset{\displaystyle CH_3}{|}}{C}}\!-\!NH_2 \quad + \quad RNH_2 \quad \longrightarrow \quad \text{Gln}\!-\!\overset{\overset{\displaystyle CH_3}{|}}{\underset{\underset{\displaystyle CH_3}{|}}{\overset{\parallel}{C}}}\!-\!NHR \quad + \quad NH_3$$

b

$$\text{Gln}\!-\!\overset{\overset{\displaystyle CH_3}{|}}{\underset{\underset{\displaystyle CH_3}{|}}{\overset{\parallel}{C}}}\!-\!NH_2 \quad + \quad \text{Lys}\!-\!\overset{\overset{\displaystyle CH_3}{|}}{\underset{\underset{\displaystyle CH_3}{|}}{}}NH_2 \quad \longrightarrow \quad \text{Gln}\!-\!\overset{\overset{\displaystyle CH_3}{|}}{\underset{\underset{\displaystyle CH_3}{|}}{\overset{\parallel}{C}}}\!-\!NH\!-\!\text{Lys}\overset{\overset{\displaystyle CH_3}{|}}{\underset{\underset{\displaystyle CH_3}{|}}{}} \quad + \quad NH_3$$

c

$$\text{Gln}\!-\!\overset{\overset{\displaystyle CH_3}{|}}{\underset{\underset{\displaystyle CH_3}{|}}{\overset{\parallel}{C}}}\!-\!NH_2 \quad + \quad H_2O \quad \longrightarrow \quad \text{Gln}\!-\!\overset{\overset{\displaystyle CH_3}{|}}{\underset{\underset{\displaystyle CH_3}{|}}{\overset{\parallel}{C}}}\!-\!OH \quad + \quad NH_3$$

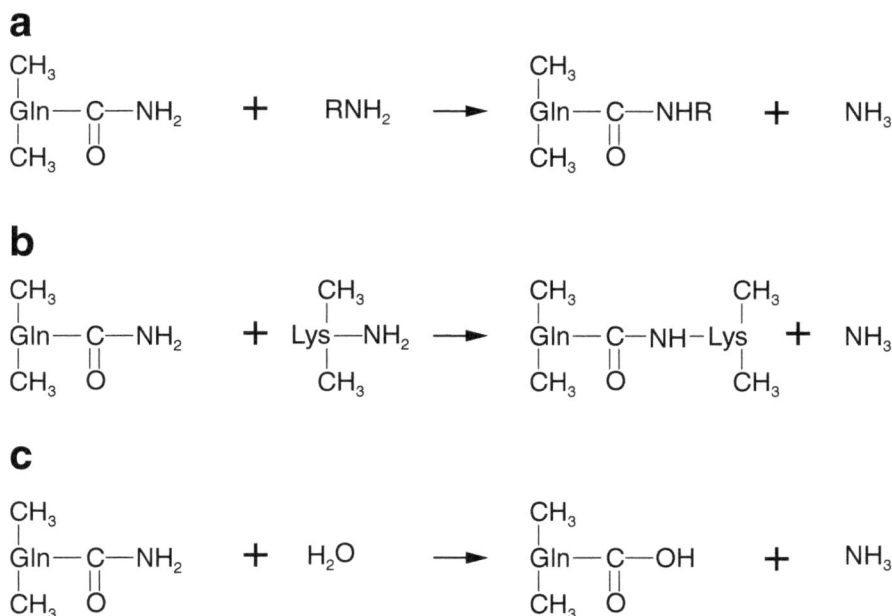

FIGURE 4.2 Transglutaminase-catalysed reaction: (a) acyl-transfer reaction, (b) cross-linking reaction between Gln and Lys residues of proteins or peptides, and (c) deamidation.

Source: Adapted from Kieliszek and Misiewicz 2014.

Many studies have looked into how TG-catalyzed cross-linking alters the functioning of proteins. Nonetheless, it appears that this kind of response is a very viable alternative for altering protein functionality. Acyl transfer and deamidation of protein, along with protein crosslinking, is found to enhance the techno-functional properties of the protein. The protein can be altered far more in an acyl-transfer reaction than it can be in a cross-linking process, depending on type of amine which is associated (Jong & Koppelman, 2002). The transglutaminase-catalysed reactions are depicted in Figure 4.2.

4.3.1.3 Microbial Transglutaminase

Microbial transglutaminase (mTG ase) catalyses acyl transfer processes, deamidation, and crosslink between glutamine and lysine residues in both intra- and extramolecular structures. Several proteins with plant origin serve as the substrate for the cross-linking enzyme mTG, which can introduce iso-peptide bonds between glutamine and lysine (Yokoyama et al., 2004). mTGase has a low molecular weight and is calcium-independent. These two characteristics both have advantages in industrial settings (Kieliszek & Misiewicz, 2014). Its use in food modifies the technical characteristics of food proteins, such as their ability to emulsify gel, become viscous, retain water, and produce and sustain foam (Gaspar & De Góes-Favoni, 2015). It is established that the majority of proteins from both animal and plant sources, including milk proteins, quinoa proteins, fish proteins, legume globulins, and many other globulins and albumins, may be effectively modified by mTGase to generate different products with enhanced technical features (Giosafatto et al., 2020).

4.3.2 FUNCTIONALITY AND SCOPE OF IMPROVEMENT

The TG treatment not only produced intermolecular covalent bonding but also changed the zeta-potential of the surface charge of the oat protein particles, evaluating possible deamidation. The

TG easily crosslinked oat globulins. As seen by SDS-PAGE under reducing conditions, the acidic and basic polypeptides of oat globulin were both crosslinked to the polymers. Several hydrophobic patches of the protein β-legumin are buried deep inside the core while hydrophilic α-legumins are exposed at the surface. Faba protein isolate demonstrated good colloidal stability (Nivala et al., 2017). With TG treatment, oat protein isolate's colloidal stability was markedly increased. When creating foams that require a lengthy shelf life in cold storage before use, this could be taken advantage of.

The PG deamidation process has altered some functional qualities. After the treatment of protein with glutaminase, the solubility under acidic environment and functional property like emulsifying stability were also found to be enhanced. Deamidation by PG, however, may result in a decline of some functional properties, including emulsifying activity and foaming stability. This decline in foam stability can be explained by the modification of protein's structure during heating for enzyme inactivation. Also, the protein's net charge rose, and resulting protein–protein interactions decreased and exhibited excessive net charge (Kunarayakul et al., 2018). Deamidation reaction of the protein during PG treatment can cause increased solubility, net surface charge, and improved structural flexibility. Also there was a decrease in solubility with the preheating step and TG treatment including oat protein concentrate. It was also noted that an additional heating step was able to enhance the cross-linkage along with improved fibrous structure (Pöri et al., 2022). Also, this study formed the proof of concept on the enzymatic crosslinking and deamidation effect on oat protein concentrate during enhanced fibrous structure formation during high moisture extrusion.

In a study on the structural modification of black soy bean protein isolate through the enzymatic biological modification, enzymatic crossing-linking, wet heating treatment glycosylation of isolate, and also transglutaminase catalysed enzymatic glycosylation. According to the findings, the TG-modified soy bean protein isolate had a secondary structure that is more open with less α-helix and β-sheet structures than the untreated protein, while the cross-linked protein isolate had more of a reduced irregular structure. Also, all the techno-functional properties of TG catalysed and cross-linked protein isolates also exhibited a variation (Y. Zhang et al., 2018). Furthermore, the glutamine level and the environment in which this amino acid is bonded are regarded to be the limiting factors of the cross-linking reaction because they culminate in the formation of a sterically defined binary complex between enzyme and substrate.

4.3.3 ROLE IN SENSORY, TEXTURAL ATTRIBUTES

The crosslinking of protein with TG can also result in the formation of gel. The formed gel is irreversible as the isopeptide bonds are very strong. Also, for the preparation of protein-based ingredients, an optimum level of crosslinking is desired, as it is not easy to dry the gel or convert it into spray powder. Hence, there are some techniques like modification of pH, or any extra addition of TG inhibitor, to inactivate the enzyme and stop the crosslinking. Canola protein isolate (CPI) was improved through enzymatic treatment using TG, which increased its gelation and raised its potential as a food additive. The amount of protein and TG can be increased to enhance gelation. The texture of commercial food protein gels and that of TG treated-CPI based gels were compared, and the results demonstrated that biologically modified CPI can be used as a food ingredient (Pinterits & Ā, 2008).

The concentration of TGase used can also have an effect on the structuring of the protein-based product. The inclusion of TGase made the pea protein gels' macrostructure stretchier but had a minimal impact on the visual macrostructure of mung bean protein isolate gels. Upon TGase addition to pea protein isolate–wheat gluten mix, it was demonstrated that TGase strengthened the gel's elastic properties as depicted in Figure 4.3. While in case of mung bean protein isolate–wheat gluten gel structures, its macrostructure was unaffected, TGase caused a decrease in the young's modulus and linear viscoelastic regime, from which it can be inferred that TGase modifies the microstructure of mung bean protein isolate gels. This difference can be attributed to the nature of crosslinking

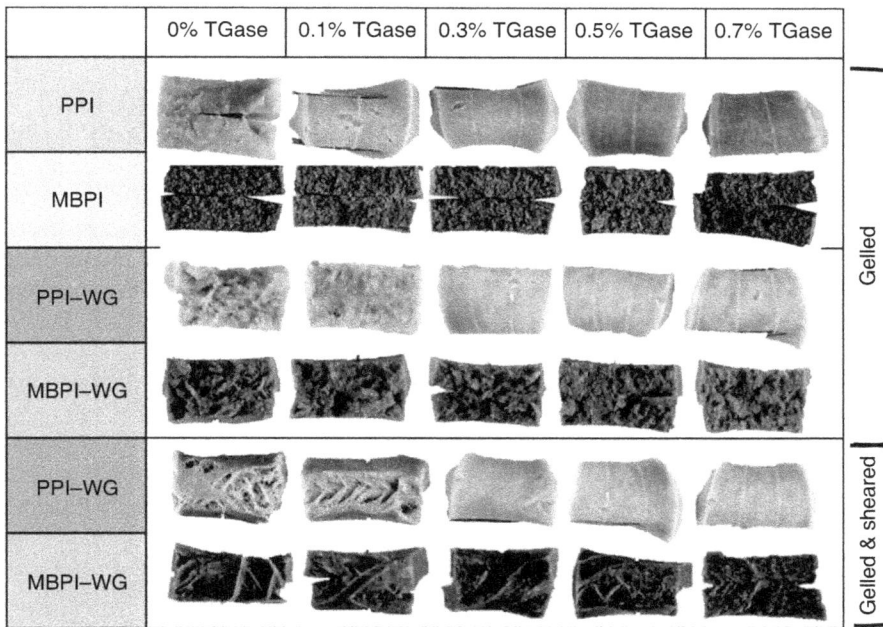

FIGURE 4.3 The depiction of the visual observation of gelled and sheared (39 s⁻¹) samples of pea protein isolate (PPI), mung bean protein isolate (MBPI), PPI–wheat gluten (WG) blends, and MBPI–WG blends with different concentrations of TGase (0.0–0.7%) incubated at 50°C. Samples were deformed manually in the direction of the shear flow.

Source: Adapted from Schlangen et al. 2023.

and the disparities between the glutamine and lysine amino acid groups' positions and access points (Schlangen et al., 2023).

4.3.4 ROLE IN NEW FOOD PRODUCT FORMULATION

Regarding the characteristics and outcomes of TG supplementation in food products, many studies have been conducted. Additionally, the inclusion of this enzyme enhances not only the rheological, flavour, and texture of food but also, in some instances, the health of the consumer. In addition to these characteristics, numerous studies have been conducted on the negative consequences of the addition of TG on the processing of food consumed by celiac and neurodegenerative disease patients, due to the formation of increased TG catalyst reaction products and its adverse effects (Amirdivani et al., 2018).

According to some research, the key possibilities for the IgE-binding epitope include the glutamine-rich tandem sequence QQQPP (Gln-Gln-Gln-Pro-Pro) found in wheat gliadin and glutenin (Tanabe et al., 1996). Glutamine can be converted into glutamic acid residues, which is an efficient method for reducing protein allergenicity (Horstmann et al., 2020). In the future, prospective uses of PG include lowering allergenicity for the production of gluten products with minimal allergenic potential (G. Zhang et al., 2021). There is potential for using coconut protein as food protein in the industry since PG deamidation can alter the distinctive properties of the protein, particularly in emulsified high-protein aqueous formulations. The full protein and low allergenicity of coconut protein make it an ideal substitute for other food ingredients (Kunarayakul et al., 2018).

mTGase modification of composite films resulted in enhanced barrier qualities of the films. mTGase was efficiently used to make a composite film made of guar gum as well as coconut protein.

The oxygen transfer rate, water vapour permeability, and tensile strength were all significantly impacted by higher mTGase concentration. Additionally, the addition of mTGase had an impact on the thermal characteristics (Sorde & Ananthanarayan, 2019).

A few years back, mTGase isolated from *Streptomyces mobaraensis* had been marketed commercially. The use of mTGase for the manufacture of various plant protein-based food products, such as tofu, noodles, bread, and bakery goods, is still constrained to raw soya bean and wheat ingredients (Dube et al., 2007). However, given the increasing demand for vegetarian foods, it holds promise to use novel modified plant proteins as functional components, such as those derived from peas, lupins, sesame, and sunflower. Resources are provided by enzymatic and physical treatments used before or after crosslinking. In a study involving brown rice, protein isolate and pea protein isolate blends resulted in the formation of stronger, more elastic gels with lower gel point temperatures as a result of mTG-induced crosslinks. The number of β-sheet structures and α-helix increased while the number of random coils decreased. As a result of the protein–protein interactions present in mixed systems, mTG's involvement in structural modification may be enhanced, allowing for the network creation of more ordered structures with improved thermal stability. Also the findings of *in-vitro* digestion revealed that the digests treated with mTG produced less hydrophobic amino acids and more essential amino acids such as Lys (Zhao et al., 2023). Overall, the incorporation of mTG into plant-based proteins has the potential to support the creation of novel food items with altered functions and distinctive digestive traits.

4.4   FERMENTATION

Proteins are some of the fundamental food biomolecules which are nutritionally rich with essential amino acids, necessary for growth and maintenance. They have wide range of functional technical attributes like solubility, dispersibility, water absorption capacity, oil absorption capacity, the efficacy of foaming and its stability, isoelectric point, and the way of precipitation around that particular point, the biological properties of the peptides which can be intended end products. All these attributes are utilised heavily with various protein-rich ingredients used in the food processing industry in various functions from the bread making, to producing the modern plant-based protein meat alternatives and new vegan products. Proteins which are made up of amino acids are unique and exhibit different surface properties based on the amino acid residue, thus any structural modification of the protein can alter its functionality. Some of the recent modifications of plant-based protein through fermentation and its changes in quality attributes are listed in Table 4.1.

Fermentation is a biochemical process through which a substrate is converted into some beneficial products or it is the synthesis of new compounds by the action of microorganisms in a controlled environmental condition. The types of fermentation might be based on the nature of environmental condition: solid state fermentation, liquid state fermentation, and solid state submerged fermentation. It is also varied with the type of starter culture used, which includes bacteria and fungal mould mat. Fermentation is used in the traditional folk cuisine for processing of legume seeds. It is used to enhance the flavour of the sauces, and other organoleptic properties (Yakubu et al., 2022). The effect of fermentation on the biological modification of protein is depicted in Figure 4.4.

4.4.1   EFFECT OF FERMENTATION ON ANTINUTRITIONAL FACTOR IN PLANT-BASED PROTEIN

In general, there are several processes which are researched to reduce the antinutritional factor (ANF) of the plant-based protein. There are several anti nutritional factors like tannins, condensed tannins, saponins, isoflavones, trypsin inhibitors, oligosaccharides, phytic acids, hemagglutinins, cyanogenic and pyrimidine glycosides and other diverse and species-dependent antinutritional substances found in legumes (Enneking & Wink, 2000). They can cause pathological conditions, or interfere with the absorption of minerals or proteins, necessitating their reduction. They can be reduced using

TABLE 4.1

List of Recent Modification of Plant-based Proteins through Fermentation

Plant-based Protein Ingredients	Microorganism	Fermentation Condition	Protein Modification	Reference
Pea Pea (*Pisum sativum* L.)	*Lactobacillus rhamnosus* L08	LAF	↑ Amino acid content from 222.18 mg/g at 0 h to 262.35 mg/g at 6 h ↑ Foaming property ↑ Emulsifying property ↓ Unpleasant flavour	Pei et al., 2022
Pea protein isolates	*Lactobacillus plantarum*	Anaerobic	↓ Protein solubility ↓ WHC ↑ OHC ↑ Foaming property	Shi et al., 2021
Pea protein extract	*Streptococcus thermophilus, Lactobacillus acidophilus, Bifidobacterium lactis* (in different combinations)	LAF	↑ Amino acid content ↑ Legumin	Emkani et al., 2021
Pea protein concentrate	*Lactobacillus plantarum*	LAF	↑ Phenolic content ↓ Trypsin inhibitor activity ↓ Chymotrypsin inhibitor activity ↓ Sulphur amino acid ↑ IVPD	Çabuk et al., 2018
Pea protein	Lactic acid bacteria + *Bifidobacterium* + (*Kluyveromyces lactis*/*Kluyveromyces marxianus*/*Torulaspora delbrueckii*)	AF	↓ Off-flavour compounds ↑ esters	Youssef et al., 2020
Lupin	*Aspergillus sojae, Aspergillus ficuum*	Solid state condition	↓ Phytic acid content, ↓water absorption capacity ↑ Swelling capacity	Olukomaiya et al., 2020
Lupin	*Leuconostoc mesenteroides, Lactobacillus Plantarum, and Lactobacillus brevis*		↓ Surface hydrophobicity, emulsifying property ↑ WAC	Lampart-Szczapa et al., 2006

(Continued)

TABLE 4.1 (Continued)
List of Recent Modification of Plant-based Proteins through Fermentation

Plant-based Protein Ingredients	Microorganism	Fermentation Condition	Protein Modification	Reference
Faba bean	*Lactobacillus delbrueckii subsp. bulgaricus and Streptococcus thermophilus.*	Anaerobic environment	↓ Phytic acid content, ↓ Trypsin inhibitor activity ↑ IVPD ↑ WAC, OAC ↓ FC, FS	Chandra-Hioe et al., 2016
Chickpea flour (*Cicer arietinum* L)	Lactic acid bacteria (LAB)	Solid state condition	↓ Phytic acid content, ↓ Foaming capacity ↑ Phenolic content ↑ Water-holding capacity	Xing et al., 2020
Chickpea flour (*Cicer arietinum* L)	*Cordyceps militaris* SN-18	Solid state condition	↑ Protein ↑ EAA ↑ WAC, OAC ↑ IVPD	Xiao et al., 2015
Chickpea protein flour	*Lactobacillus plantarum* HLJ29L	*Lactobacillus* fermentation	↑ Solubility ↓ β=sheet ↑ α-helix ↓ Disulphide bridge ↑ IVPD	Liu et al., 2023
Chickpea flour Dough–pasta fortification	*Lactiplantibacillus plantarum* T0A10		↑ Phenolic content ↓ Trypsin inhibitor activity ANF ↑ IVPD	De Pasquale et al., 2021
Chickpea fraction	*Pediococcus pentosaceus and Pediococcus acidilactici*	Solid state condition	↑ Phenolic content ↓ Phytic acid content	Xing et al., 2020
Locust bean (*Parkia biglobosa*) flour		Alkaline fermentation	↓ Gelling potential ↓WAC, ↑ OAC ↑ Phenolic content ↓ Tannin, saponin	Yakubu et al., 2022

* LAF—Lactic acid fermentation

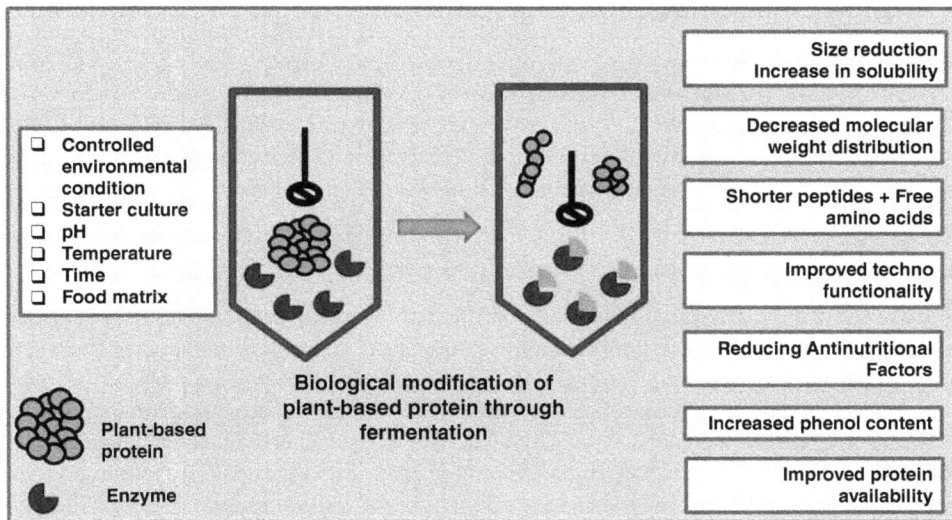

FIGURE 4.4 Biological modification of plant-based protein through fermentation.

a variety of physical processes, from soaking, boiling, germination, cooking, extrusion, to heat treatment, and also non-thermal treatments can be utilised for reducing the anti nutritional factor. Fermentation can also be carried out for plant-based proteins to reduce flatulence, for the removal of any undesirable compounds, and further to enhance the nutritional availability.

Phytic acid, an anti-nutrient, is found to have propensity to form complexes with proteins and minerals, thus reducing their bioavailability. In case of fermentation, the phytic acid inositol is found to be broken down by the substrate modulation, as the bioavailability of proteins and cations is increased to release phosphorus for fermentation process. Also, during solid state fermentation fermentation, there is an increased endogenous activity of phytase enzyme, which can result in the breakdown of non-soluble organic complexes with the phytic-acid-degrading ability and phosphatase action of microbes (Olukomaiya et al., 2020). The fermentation of legumes was found to reduce the antinutritional factors like phytate and tannins in lupin flour (Olukomaiya et al., 2020),

4.4.2 Effect of Fermentation on IVPD

In turn, the hydrolysis of proteins leads to low molecular weight protein peptides. Cleavage of large molecular weight proteins results in smaller peptides and amino acids. This causes an improved *in-vitro* protein digestibility (IVPD). Reduced antinutritional factors, such as phytic acid, may also contribute to increased IVPD because these components alter the interactions between proteins and antinutritional factors, render proteins more vulnerable to proteolytic attack, and enhance their digestibility (Chandra-Hioe et al., 2016).

Filamentous fungi have been used in solid state fermentation. It has been noted that the type of starter culture plays an important role in determining the IVPD. It was found that the proteins which were restricted within the fibre matrix reduced the enzymatic hydrolysis, thus altering the overall protein solubility, dispersibility, and thus overall reduced IVPD (Olukomaiya et al., 2020). But species like *Lactobacillus sakei, Pediococcus acidilactici, and Pediococcus pentosaceus* exhibited higher IVPD.

Also, it was found that fermentation can cause solubilisation and breakdown of proteins into smaller polypeptides. The hydrophilic groups in a protein residue bind the water molecule during fermentation; thus, there might be a decrease in water absorption capacity.

4.4.3 EFFECT OF FERMENTATION IN SOURDOUGH QUALITY

In a study, sourdoughs were prepared from unfermented and fermented whole flour, and even protein chickpea flour. It was found that the unfermented ones produced a bean raw flavour, while fermented ones had mild acidic odour. Also it was found that protein-enriched flour-based sourdough was found to have a sweet and nutty-based aroma. Thus, solid state fermentation has the specific capacity to reduce the off-flavour characteristics of the plant protein-based flour (Xing et al., 2020).

4.4.4 EFFECT OF FERMENTATION ON PHENOLIC CONTENT

The reason behind the increased level of phenolics may be because of their release due to the structural degradation of the cell during fermentation (Xing et al., 2020). It is also possible that fermentation, by loosening the lignocellulosic matrix, releases phenolic chemicals from an immobilised condition. Protein structural changes brought on by bacterial fermentation may make the substrate more accessible to the enzymes that break it down. Moreover, bacteria have the capacity to partly degrade intact proteins, which increases the amount of free amino groups. The increase in amount of essential amino acid content can also be attributed to the transaminase activity of the microbe during the fermentation process (Xiao et al., 2015).

It has been proven that proteins may break their peptide bonds as a result of proteolytic activity during the fermentation process, increasing their hydrophilicity by producing more polar groups. The improvement in fat-absorption capacity in fermented samples might be explained by a rise in the availability of these amino acids by disguising the non-polar residues from the internal protein molecules, since the binding of the fat depends on the surface availability of amino acids that are hydrophobic (Elkhalifa et al., 2010). Thus, an increase in oil absorption capacity of the flour sample can be due to enhanced insoluble hydrophobic type of protein.

Based on the proteolytic activity of bacteria, there might be an increase or decrease in the protein content of the leguminous flour post fermentation. This can be explained by the reduction in size of the protein during fermentation rather than changing its contents (Chandra-Hioe et al., 2016). The increased oil absorption capacity in fermented samples can be explained due to the increased insoluble hydrophobic protein and thus better retention of flavour and mouthfeel (Chandra-Hioe et al., 2016). The decrease in protein solubility or the lowered amount of water soluble protein after the fermentation process could be explained by the fact that larger peptides of protein were precipitated and the smaller ones remained. This is also strengthened by the decrease of pH, which is almost closer to the isoelectric point of the protein dispersion in some cases (Shi et al., 2021).

4.4.5 EFFECT OF FERMENTATION ON VOLATILE COMPOUNDS' AROMA

In general, the beany flavours of the leguminous proteins are the main reasons for their acceptability in various food-based applications. Hence, fermentation of the protein-based flours with various microbial consortia is one of the recent researches carried out in this domain. During LAF, there was an effective reduction in the off flavour of pea protein (Shi et al., 2021). Also, it was found that *Saccharomyces cerevisiae* and *Lactobacillus plantarum* could remove 79.65% and 78.94%, respectively, of the major pea protein aldehydes which are found to be responsible for the pea flavour (Xiang et al., 2023). The beany and green notes of legumes were also found to be reduced by the generation of esters during yeast-based fermentations. For a wide range of legumes and cereals, or a combination of them, the solid state fermentation method provides a technical option to enhance their nutritional value and produce edible products with palatable sensory properties. Thus, fermentation was found to improvise the overall appearance, texture, aroma, and flavour of the plant-based protein and thus help in mimicking animal-protein-based products without compromising the textural and flavour attributes.

Because of the various metabolic pathways and competences of the strains, microbial fermentation strategies will result in distinct final flavour characteristics. So, just how much the fermentation changes the aroma profile of the plant proteins depends greatly on the strains used (Tao et al., 2022). It should be remembered that no approach is ideal, and every methodology has inherent benefits and disadvantages. Certain strains may create high-acid products during fermentation, which consumers might not like. To choose the best strains and modify the fermentation process to fit consumer preferences, more study is required. By successfully promoting the Maillard process, it will improve the flavour and taste of plant-based meat substitutes and get rid of the beany flavour. The best choice for improving rheology and sensory qualities while reducing or removing antinutritional factors and fully using the nutritional and functional potential of vegetable matrices and the meals made from them is tailored fermentation.

4.5 APPLICATION OF BIOLOGICALLY MODIFIED PLANT-BASED PROTEIN IN THE PRODUCTION OF PLANT-BASED FUTURE FOODS

Protein is a major imperative biomolecule. To produce and deliver ample protein to nourish over eight billion people in a manner that is both ecologically responsible and cost-effective, we currently face a monumental challenge. Plant-based proteins are the most sought-after solution. Due to their extensive history of use and cultivation, cheaper production costs, and accessibility in many regions of the world, plant-based proteins offer a promising solution to our nutritional demands (Yong et al., 2021). There are current state-of-art and extensive researches in mimicking the animal protein products and also the development of other plant-based alternatives like plant-based milk, cheese, beverage, and other products.

Plant-based proteins differ from animal proteins. They have different techno-functionalities, like low protein solubility, foaming, gelling properties, which limit their use in various food domains. Animal proteins are thought to be of greater protein quality when compared to plant sources and are typically advocated to satisfy dietary protein demands at a low calorie load. Plant proteins are underwhelming in one or more necessary nutrients, whereas animal proteins often include a healthy mix of both essential and non-essential amino acids. Added upon, a plant protein does not give same anabolic response as an animal protein.

Thus, several modifications of the plant-based protein are carried out to meet this as a reality. Physical, chemical, and biological techniques are largely utilised in order to produce functional ingredients including hydrolysed vegetable proteins, texturised vegetable proteins, protein flour, de-flavoured protein concentrates, and isolates, along with clean label protein concentrates regardless of the fact that many food industries are gradually shifting from chemical modification of proteins. Furthermore, due to this rising demand for clean label products, an important challenge of toxicity, allergenicity reduction, and safety evaluation is needed to be addressed (Akharume et al., 2021).

Plant proteins are structured and modified for the development of new products, along with new generated flavours, emulsifiers, and varied products dependent on their modified protein functionality and changed structural arrangement to ensure that the particular note of flavour, or the enhanced texture, or the rheological property has been pertained. Active researches in cold plasma, electromagnetic waves, microfluidisation, and thermal and non-thermal treatments have led to the development of various plant-based protein products.

The several plant-based products formed can be categorised as (based on Yong et al. (2021)) follows:

a) Plant-based dairy alternatives: plant-based milk, cheese, ice cream, pudding
b) Structured plant proteins: Fibrous structure for mimicked meat analogues
c) Plant-based beverages
d) Plant-based gels: Homogeneous gels like cheese and yogurt
e) Plant protein-generated flavours: melanoidins and Strecker degradation products

Enhanced solubility and digestibility are some of the benefits of enzyme-modified plant-based proteins, which also serve as flavour enhancers and protein substitutes. The enhanced protein digestibility of the protein hydrolysates makes them a significant component of both paediatric and geriatric diets as well as sports and energy drinks (Tapal & Tiku, 2019).

Gels have recently been effectively formed with a limited amount of enzymatic hydrolysis. The earlier research mostly concentrated on whey protein, soy, and wheat gluten. In order to properly disintegrate the protein and expose its partially hydrophobic amino acid residues, limited enzymatic hydrolysis controls the protein's degree of hydrolysis to some extent. The crosslinking of molecules and the addition of polysaccharides improves the protein's gel-forming properties (C. Zhang et al., 2021).

Plant-based block-style cheese analogues have been experimented and are available commercially. It was found that though the protein quantity was lower compared to the animal origin, they exhibited almost similar textural attributes in terms of springiness, fractural strain, and also a viscoelastic rheological behaviour (Grasso et al., 2021). Plant proteins like zein are found to exhibit a self-assembly with structural modification in β-sheet and more elastic behaviour, thus can be used in the manufacture of cheddar cheese alternatives (Mattice & Marangoni, 2020).

Legume proteins have been employed successfully as nutraceutical components and also offer exceptional nutritional, functional, and anti-oxidative qualities. Legume proteins have poor solubility and lower surface activity at neutral pH, and they also cause an increase in emulsion viscosity when the solid content is large (Wen et al., 2022). Proteins' physicochemical and emulsifying capabilities are severely hampered by these restrictions, which also restrict the usage of proteins as a wall material during the encapsulation process. Thus, physical, enzymatic, and chemical modifications of legume proteins structure as well as combinations with polysaccharides can be used in encapsulation. Pickering particles with polysaccharide, protein, or polyphenol content are used to stabilise water-in-oil and oil-in-water emulsions and are researched using colloidal plant-based particles. They commonly act as structuring agents in the bulk continuous phase (Sarkar & Dickinson, 2020).

If meat substitutes that are equivalent to meat in terms of nutrients and sensory experiences are created, people may be more ready to replace (part of) their meat intake. The most current meat replacement research and development have shown that plant-based proteins and techniques like extrusion, shearing, and mixing may be utilised to produce textures that resemble meat (Kyriakopoulou et al., 2021).

Food proteins' allergenicity is assumed to be caused by the gastrointestinal system's resistance to inducing an immune response. Ninety per cent of all food allergies are related to proteins, and they affect between 1% and 2% of adults and 6% to 8% of children (Tapal & Tiku, 2019). Legumes are a common mediator of IgE-mediated food allergies. The only way to prevent a symptomatic reaction is to avoid eating foods that are allergenic. Next, to determine if the hydrolysates were allergenic, ELISA, immunoblot, stripped basophil histamine release, and skin prick tests were employed (Kasera et al., 2015).

4.6   FUTURE OUTLOOK

Numerous studies have looked into enzymatic hydrolysis as a way to enhance pulse proteins' functionality, with differing results depending on various factors as previously discussed. However, additional studies linking structural and surface properties to hydrolysate functionality might be helpful. Further investigation should be done on the effects of pre- and post-hydrolysis processes as well as ingredient pre-processing on the overall enhancement in the functionality of hydrolysates. The literature is largely devoid of research on the optimisation of enzyme inactivation settings, which could be highly helpful in upcoming investigations on pulse protein hydrolysates. Concerns about the effects of food choices on climate change and personal health were the main reasons people bought these goods.

Some of the future considerations which must be taken into account by the researcher include studies that broaden our understanding of the physical and chemical properties of protein molecules, along with investigation on the separation and restructuring of plant proteins, with further utilisation of novel technologies to produce protein materials with a wide range of adaptability in formulated foods. Apart from these concerns, research about the physical, chemical, and enzymatic methods can be used to modify the functioning of oil seed and legume proteins. Extensive research and evaluation are suggested for plant proteins, such as soya beans, to address issues with their flavour, functioning, colour, antinutritional elements, and processing methods.

4.7 CONCLUSION

In the past, research has concentrated on the detrimental impacts of interactions between protein and plant components, while more recent research has focused on the beneficial effects (e.g. enhanced emulsifying capacity, reduced allergy, and targeted production of protein pigments). The aforementioned effects can be produced by manipulating the structure of protein and by modifying the structure and overall parameters such as protein nativity under the action of proteolytic enzymes and transglutaminase. This provides an overall idea on the importance of utilisation of novel plant-based protein ingredients in the development of food products.

REFERENCES

Akharume, F. U., Aluko, R. E., & Adedeji, A. A. (2021). Modification of plant proteins for improved functionality: A review. *Comprehensive Reviews in Food Science and Food Safety*, *20*(1), 198–224. https://doi.org/10.1111/1541-4337.12688.

Amirdivani, S., Khorshidian, N., Fidelis, M., Granato, D., Koushki, M. R., Mohammadi, M., Khoshtinat, K., & Mortazavian, A. M. (2018). Effects of transglutaminase on health properties of food products. In *Current Opinion in Food Science* (Vol. 22, pp. 74–80). Elsevier Ltd. https://doi.org/10.1016/j.cofs.2018.01.008.

Boutureira, O., & Bernardes, G. J. L. (2015). Advances in chemical protein modification. *Chemical Reviews*, *115*(5), 2174–2195. https://doi.org/10.1021/cr500399p.

Çabuk, B., Nosworthy, M. G., Stone, A. K., Korber, D. R., Tanaka, T., House, J. D., & Nickerson, M. T. (2018). Effect of fermentation on the protein digestibility and levels of non-nutritive compounds of pea protein concentrate. *Food Technology and Biotechnology*, *56*(2), 257–264. https://doi.org/10.17113/ftb.56.02.18.5450.

Chandra-Hioe, M. V., Wong, C. H. M., & Arcot, J. (2016). The potential use of fermented chickpea and faba bean flour as food ingredients. *Plant Foods for Human Nutrition*, *71*(1), 90–95. https://doi.org/10.1007/s11130-016-0532-y.

Damodaran, S. (2005). Protein stabilization of emulsions and foams. *Journal of Food Science*, *70*(3), R54—R66.

De Pasquale, I., Verni, M., Verardo, V., Gómez-Caravaca, A. M., & Rizzello, C. G. (2021). Nutritional and functional advantages of the use of fermented black chickpea flour for semolina-pasta fortification. *Foods*, *10*(1), 1–21. https://doi.org/10.3390/foods10010182.

Dong, X., Zhao, M., Shi, J., Yang, B., Li, J., Luo, D., Jiang, G., & Jiang, Y. (2011). Effects of combined high-pressure homogenization and enzymatic treatment on extraction yield, hydrolysis and function properties of peanut proteins. *Innovative Food Science and Emerging Technologies*, *12*(4), 478–483. https://doi.org/10.1016/j.ifset.2011.07.002.

Dube, M., Sch, C., Neidhart, S., & Carle, R. (2007). *Texturisation and modification of vegetable proteins for food applications using microbial transglutaminase* (pp. 287–299). https://doi.org/10.1007/s00217-006-0401-2.

Eckert, E., Han, J., Swallow, K., Tian, Z., Jarpa-Parra, M., & Chen, L. (2019). Effects of enzymatic hydrolysis and ultrafiltration on physicochemical and functional properties of faba bean protein. *Cereal Chemistry*, *96*(4), 725–741. https://doi.org/10.1002/cche.10169.

Elkhalifa, A. E. O., Bernhardt, R., et al. (2010). Influence of grain germination on functional properties of sorghum flour. *Food Chemistry*, *121*(2), 387–392.

Emkani, M., Oliete, B., & Saurel, R. (2021). Pea protein extraction assisted by lactic fermentation: Impact on protein profile and thermal properties. *Foods*, *10*(3). https://doi.org/10.3390/foods10030549.

Enneking, D., & Wink, M. (2000). Towards the elimination of anti-nutritional factors in grain legumes. In *Linking research and marketing opportunities for pulses in the 21st century: Proceedings of the third international food legumes research conference* (pp. 671–683). https://doi.org/10.1007/978-94-011-4385-1_65.

Fontana, A., Spolaore, B., Mero, A., & Veronese, F. M. (2008). Site-specific modification and PEGylation of pharmaceutical proteins mediated by transglutaminase. In *Advanced drug delivery reviews* (Vol. 60, Issue 1, pp. 13–28). https://doi.org/10.1016/j.addr.2007.06.015.

García Arteaga, V., Apéstegui Guardia, M., Muranyi, I., Eisner, P., & Schweiggert-Weisz, U. (2020). Effect of enzymatic hydrolysis on molecular weight distribution, techno-functional properties and sensory perception of pea protein isolates. *Innovative Food Science and Emerging Technologies*, 65, 102449. https://doi.org/10.1016/j.ifset.2020.102449.

Gaspar, A. L. C., & De Góes-Favoni, S. P. (2015). Action of microbial transglutaminase (MTGase) in the modification of food proteins: A review. *Food Chemistry*, *171*, 315–322. https://doi.org/10.1016/j.foodchem.2014.09.019.

Giosafatto, C. V. L., Fusco, A., Al-Asmar, A., & Mariniello, L. (2020). Microbial transglutaminase as a tool to improve the features of hydrocolloid-based bioplastics. *International Journal of Molecular Sciences*, *21*(10). https://doi.org/10.3390/ijms21103656.

Goertzen, A. D., House, J. D., Nickerson, M. T., & Tanaka, T. (2021). The impact of enzymatic hydrolysis using three enzymes on the nutritional properties of a chickpea protein isolate. *Cereal Chemistry*, *98*(2), 275–284. https://doi.org/10.1002/cche.10361.

Grasso, N., Roos, Y. H., Crowley, S. V., Arendt, E. K., & O'Mahony, J. A. (2021). Composition and physicochemical properties of commercial plant-based block-style products as alternatives to cheese. *Future Foods*, *4*, 100048. https://doi.org/10.1016/j.fufo.2021.100048.

Horstmann, G., Ewert, J., Stressler, T., & Fischer, L. (2020). A novel protein glutaminase from Bacteroides helcogenes—characterization and comparison. *Applied Microbiology and Biotechnology*, *104*(1), 187–199. https://doi.org/10.1007/s00253-019-10225-2.

Jong, G. H. A. De, & Koppelman, S. J. (2002). Transglutaminase catalyzed reactions: Impact on food applications. *Journal of Food Science*, *67*(8), 2798–2806.

Kasera, R., Singh, A. B., Lavasa, S., Nagendra, K., & Arora, N. (2015). Enzymatic hydrolysis : A method in alleviating legume allergenicity. *Food and Chemical Toxicology*, *76*, 54–60. https://doi.org/10.1016/j.fct.2014.11.023.

Kieliszek, M., & Misiewicz, A. (2014). Microbial transglutaminase and its application in the food industry. A review. *Folia Microbiologica*, *59*(3), 241. https://doi.org/10.1007/s12223-013-0287-x.

Konieczny, D., Stone, A. K., Korber, D. R., Nickerson, M. T., & Tanaka, T. (2020). Physicochemical properties of enzymatically modified pea protein-enriched flour treated by different enzymes to varying levels of hydrolysis. *Cereal Chemistry*, *97*(2), 326–338. https://doi.org/10.1002/cche.10248.

Kunarayakul, S., Thaiphanit, S., & Anprung, P. (2018). Food hydrocolloids optimization of coconut protein deamidation using protein- glutaminase and its effect on solubility, emulsi fi cation, and foaming properties of the proteins. *Food Hydrocolloids*, *79*, 197–207. https://doi.org/10.1016/j.foodhyd.2017.12.031.

Kyriakopoulou, K., Keppler, J. K., & van der Goot, A. J. (2021). Functionality of ingredients and additives in plant-based meat analogues. *Foods*, *10*(3). https://doi.org/10.3390/foods10030600.

Lampart-Szczapa, E., Konieczny, P., Nogala-Kałucka, M., Walczak, S., Kossowska, I., & Malinowska, M. (2006). Some functional properties of lupin proteins modified by lactic fermentation and extrusion. *Food Chemistry*, *96*(2), 290–296. https://doi.org/10.1016/j.foodchem.2005.02.031.

Liu, Y., Zhu, S., Li, Y., Sun, F., Huang, D., & Chen, X. (2023). Alternations in the multilevel structures of chickpea protein during fermentation and their relationship with digestibility. *Food Research International*, 112453. https://doi.org/10.1016/j.foodres.2022.112453.

Mattice, K. D., & Marangoni, A. G. (2020). Evaluating the use of zein in structuring plant-based products. *Current Research in Food Science*, *3*, 59–66. https://doi.org/10.1016/j.crfs.2020.03.004.

Mirmoghtadaie, L., Shojaee Aliabadi, S., & Hosseini, S. M. (2016). Recent approaches in physical modification of protein functionality. *Food Chemistry*, *199*, 619–627. https://doi.org/10.1016/j.foodchem.2015.12.067.

Mudgil, P., Omar, L. S., Kamal, H., Kilari, B. P., & Maqsood, S. (2019). Multi-functional bioactive properties of intact and enzymatically hydrolysed quinoa and amaranth proteins. *LWT*, *110*, 207–213. https://doi.org/10.1016/j.lwt.2019.04.084.

Nivala, O., Mäkinen, O. E., Kruus, K., Nordlund, E., & Ercili-Cura, D. (2017). Structuring colloidal oat and faba bean protein particles via enzymatic modification. *Food Chemistry*, *231*, 87–95. https://doi.org/10.1016/j.foodchem.2017.03.114.

Olukomaiya, O. O., Adiamo, O. Q., Fernando, W. C., Mereddy, R., Li, X., & Sultanbawa, Y. (2020). Effect of solid-state fermentation on proximate composition, anti-nutritional factor, microbiological and functional properties of lupin flour. *Food Chemistry*, *315*(June 2019), 126238. https://doi.org/10.1016/j.foodchem.2020.126238.

Pal, T., Ahmad, R., & Singh, D. (2021). Enzymatic modification of rice bran protein : Impact on structural, antioxidant and functional properties. *LWT*, *138*(November 2020), 110648. https://doi.org/10.1016/j.lwt.2020.110648.

Parrotta, L., Tanwar, U. K., Aloisi, I., Sobieszczuk-Nowicka, E., Arasimowicz-Jelonek, M., & Duca, S. Del. (2022). *Plant transglutaminases: New insights in biochemistry, genetics, and physiology* (pp. 1–17). https://doi.org/10.3390/cells11091529.

Pei, M., Zhao, Z., Chen, S., Reshetnik, E. I., Gribanova, S. L., Li, C., Zhang, G., Liu, L., & Zhao, L. (2022). Physicochemical properties and volatile components of pea flour fermented by Lactobacillus rhamnosus L08. *Food Bioscience*, *46*, 101590. https://doi.org/10.1016/j.fbio.2022.101590.

Pinterits, A., & Ã, S. D. A. (2008). *Improvement of canola protein gelation properties through enzymatic modification with transglutaminase* (Vol. 41, pp. 128–138). https://doi.org/10.1016/j.lwt.2007.01.011.

Pöri, P., Nisov, A., & Nordlund, E. (2022). Enzymatic modification of oat protein concentrate with trans- and protein-glutaminase for increased fibrous structure formation during high-moisture extrusion processing. *LWT*, *156*(October 2021). https://doi.org/10.1016/j.lwt.2021.113035.

Sarkar, A., & Dickinson, E. (2020). Sustainable food-grade Pickering emulsions stabilized by plant-based particles. *Current Opinion in Colloid & Interface Science*, *49*, 69–81.

Schlangen, M., Ribberink, M. A., Dinani, S. T., Sagis, L. M. C., & van der Goot, A. J. (2023). Mechanical and rheological effects of transglutaminase treatment on dense plant protein blends. *Food Hydrocolloids*, *136*, 108261. https://doi.org/10.1016/j.foodhyd.2022.108261.

Schlegel, K., Sontheimer, K., Eisner, P., & Schweiggert-Weisz, U. (2020). Effect of enzyme-assisted hydrolysis on protein pattern, technofunctional, and sensory properties of lupin protein isolates using enzyme combinations. *Food Science and Nutrition*, *8*(7), 3041–3051. https://doi.org/10.1002/fsn3.1286.

Shi, Y., Singh, A., Kitts, D. D., & Pratap-Singh, A. (2021). Lactic acid fermentation: A novel approach to eliminate unpleasant aroma in pea protein isolates. *Lwt*, *150*, 111927. https://doi.org/10.1016/j.lwt.2021.111927.

Sorde, K. L., & Ananthanarayan, L. (2019). Effect of transglutaminase treatment on properties of coconut protein-guar gum composite film. *LWT—Food Science and Technology*, 108422. https://doi.org/10.1016/j.lwt.2019.108422.

Sun, Q., Ma, Z. F., Zhang, H., Ma, S., & Kong, L. (2019). Structural characteristics and functional properties of walnut glutelin as hydrolyzed: Effect of enzymatic modification. *International Journal of Food Properties*, *22*(1), 265–279. https://doi.org/10.1080/10942912.2019.1579738.

Tanabe, S., Arai, S., Yanagihara, Y., Mita, H., Takahashi, K., & Watanabe, M. (1996). A major wheat allergen has a Gln-Gln-Gln-Pro-Pro motif identified as an IgE-binding epitope. *Biochemical and Biophysical Research Communications*, *219*(2), 290–293.

Tao, A., Zhang, H., Duan, J., Xiao, Y., Liu, Y., Li, J., Huang, J., Zhong, T., & Yu, X. (2022). Mechanism and application of fermentation to remove beany flavor from plant-based meat analogs: A mini review. *Frontiers in Microbiology*, *13*. https://doi.org/10.3389/fmicb.2022.1070773.

Tapal, A., & Tiku, P. K. (2019). Nutritional and nutraceutical improvement by enzymatic modification of food proteins. In *Enzymes in food biotechnology*. Elsevier Inc. https://doi.org/10.1016/B978-0-12-813280-7.00027-X.

Vogelsang-O'Dwyer, M., Sahin, A. W., Bot, F., O'Mahony, J. A., Bez, J., Arendt, E. K., & Zannini, E. (2022). Enzymatic hydrolysis of lentil protein concentrate for modification of physicochemical and technofunctional properties. *European Food Research and Technology*, *0123456789*. https://doi.org/10.1007/s00217-022-04152-2.

Wen, C., Liu, G., Ren, J., Deng, Q., Xu, X., & Zhang, J. (2022). Current progress in the extraction, functional properties, interaction with polyphenols, and application of legume protein. *Journal of Agricultural and Food Chemistry*, *70*(4), 992–1002.

Wu, D., Tu, M., Wang, Z., Wu, C., Yu, C., Battino, M., El-Seedi, H. R., & Du, M. (2020). Biological and conventional food processing modifications on food proteins: Structure, functionality, and bioactivity. In *Biotechnology advances* (Vol. 40). https://doi.org/10.1016/j.biotechadv.2019.107491.

Xiang, L., Zhu, W., Jiang, B., Chen, J., Zhou, L., & Zhong, F. (2023). Volatile compounds analysis and biodegradation strategy of beany flavor in pea protein. *Food Chemistry*, *402*, 134275. https://doi.org/10.1016/j.foodchem.2022.134275.

Xiao, Y., Xing, G., Rui, X., Li, W., Chen, X., Jiang, M., & Dong, M. (2015). Effect of solid-state fermentation with Cordyceps militaris SN-18 on physicochemical and functional properties of chickpea (*Cicer arietinum* L.) flour. *LWT*, *63*(2), 1317–1324. https://doi.org/10.1016/j.lwt.2015.04.046.

Xing, Q., Dekker, S., Kyriakopoulou, K., Boom, R. M., Smid, E. J., & Schutyser, M. A. I. (2020). Enhanced nutritional value of chickpea protein concentrate by dry separation and solid state fermentation. *Innovative Food Science and Emerging Technologies*, *59*, 102269. https://doi.org/10.1016/j.ifset.2019.102269.

Yakubu, C. M., Sharma, R., Sharma, S., & Singh, B. (2022). Influence of alkaline fermentation time on in vitro nutrient digestibility, bio- & techno-functionality, secondary protein structure and macromolecular morphology of locust bean (Parkia biglobosa) flour. *LWT*, *161*(February), 113295. https://doi.org/10.1016/j.lwt.2022.113295.

Yokoyama, K., Nio, N., & Kikuchi, Y. (2004). Properties and applications of microbial transglutaminase. *Applied Microbiology and Biotechnology*, *64*(4), 447–454. https://doi.org/10.1007/s00253-003-1539-5.

Yong, S., Sim, J., Srv, A., & Chiang, J. H. (2021). Plant proteins for future foods : A roadmap. *Foods*, *10*(1967), 1–31.

Youssef, C. E., Bonnarme, P., Fraud, S., Péron, A. C., Helinck, S., & Landaud, S. (2020). Sensory improvement of a pea protein-based product using microbial co-cultures of lactic acid bacteria and yeasts. *Foods*, *9*(3). https://doi.org/10.3390/foods9030349.

Zhang, C., Jiang, S., Zhang, J., & Li, S. (2021). Effect of enzymatic hydrolysis on the formation and structural properties of peanut protein gels. *International Journal of Food Engineering*, *17*(3), 167–176. https://doi.org/10.1515/ijfe-2018-0356.

Zhang, G., Ma, S., Liu, X., Yin, X., Liu, S., Zhou, J., & Du, G. (2021). Protein-glutaminase: Research progress and prospect in food manufacturing. *Food Bioscience*, *43*(May), 101314. https://doi.org/10.1016/j.fbio.2021.101314.

Zhang, Y., Yin, Y., Lu, S., Yao, X., Zheng, X., Zhao, R., Li, Z., Shen, H., & Zhang, S. (2018). Effects of modified processing methods on structural changes of black soybean protein isolate. *Molecules*, *23*(9). https://doi.org/10.3390/molecules23092127.

Zhao, L., Chen, M.-H., Bi, X., & Du, J. (2023). Physicochemical properties, structural characteristics and in vitro digestion of brown rice—pea protein isolate blend treated by microbial transglutaminase. *Food Hydrocolloids*, 108673.

5 Thermal Modification of Plant-Based Proteins

Kanchan Suri, Seerat Bhinder, Seeratpreet Kaur and Mehak Katyal

5.1 INTRODUCTION

Proteins are the second-most important macronutrient needed for meeting basic calorie requirements and human survival. Besides this, proteins also act as a key ingredient in maintaining food structures through foaming, emulsification, gelation, and dough and batter formation (Ozturk & McClements, 2016). It has been estimated that in 2016, the global market for plant-based proteins was evaluated around $8 billion and is expanding fast and is anticipated to reach $9.5 billion US dollars by the year 2024. The compound annual growth rate (CAGR) for the period 2019–2024 has been estimated to about 7% (Campbell et al., 2016). Furthermore, consumer trends have evolved immensely over the last decade towards foods labelled as sustainably sourced, healthy, clean, natural, and organic. As such plant-based or vegan foods have come to be associated with labels like 'ethically sourced', 'clean', and 'environment friendly.' Reducing the environmental footprint, food-processing-associated wastes, and high demands in water and soil have become an important criteria for ethically processed and produced foods (Aschemann-Witzel & Peschel, 2019). Such a change in perspective has brought focus to vegan food products, processed without any animal-based ingredient. The rising global trends in consumer preferences for lean protein have created a market of proteins from plant origin which, unlike proteins of animal origin, are more clean, ethical, sustainable, and low in cost (Reipurth et al., 2018; Aschemann-Witzel & Peschel, 2019). The crucial aspect of basing human diet on plant sources is the lack of complete protein. Global population is highly reliant on cereals to meet 30–70% of the daily energy as well as dietary protein requirement (Poutanen et al., 2014). However, cereal proteins lack a balanced amino acid composition and are deficient in lysine, while protein-rich leguminous plant are high in lysine but limiting in sulphur-containing amino acids (Vasconcelos et al., 2010). Therefore, from the perspective of nutrition, proteins from different plant sources have the potential to meet the daily dietary requirements.

Besides the lower protein content of plant-based proteins, some particular plant residuals are believed to act as antinutritional compounds that add to another limiting factor which poses challenges to new product development and high protein claims, hence restricting their applications in food industry (Aviles-Gaxiola et al., 2018; Sim et al., 2021). The functionality of proteins can be altered and enhanced using various physical modification methods, such as thermal processing, high-pressure processing (HPP), and ultrasound treatment (Mir et al., 2020; Zheng et al., 2019). Furthermore, some of the plant-based proteins also have restricted utilization in food formulations owing to their undesirable bitter taste which can be altered by different modification methods (Zeeb et al., 2018). Among these methods, thermal modification of proteins is the simplest and the most economic method of physical modification of proteins that does not require any sophisticated instrumentation (Mir et al., 2020). Controlled heat treatments result in the denaturation of proteins, which causes changes in the secondary, tertiary, and quaternary protein structures by altering the electrostatic, ionic, hydrogen, and hydrophobic bonds responsible for the characteristic structure and functionality of proteins. This denaturation thus causes unfolding of the native protein structure, which can expose various buried reactive sites (particularly hydrophobic sites) and improve protein functionality (Campbell et al., 2016). Denaturation of whey proteins

DOI: 10.1201/9781003369790-5

and cowpea proteins has been found to significantly improve the loaf volume, texture, and overall sensory acceptability of baked foods like wheat bread and sponge cake (Erdogdu-Arnoczky et al., 1996; Campbell et al., 2016). Controlled heat treatment or controlled denaturation of proteins can alter and enhance the functionality (foaming, emulsification, and water retention capacity) of plant-based proteins which can have a positive impact on the quality parameters of final product (Campbell et al., 2016). In addition to improving functional properties of proteins, thermal treatment can inactivate antinutrients associated with plant-based proteins (Tang et al., 2009). Moreover, thermal treatment can reduce the particle size of proteins whereby the physical stability of the protein improves during storage (Li et al., 2020). However, extensive heat processing can cause protein coagulation which involves the complete loss of native protein structure, solubility, and functionality (Campbell et al., 2016). Therefore, the selection of modification methods should be done carefully especially when utilized for food and pharmaceutical purpose due to their significant influence on the functional, nutritional value, and organoleptic characteristics of plant-based proteins (Nasrabadi et al., 2021).

5.2 DIFFERENT THERMAL MODIFICATION METHODS OF PLANT-BASED PROTEINS

Plant-based proteins have poor nutritional and functional properties in comparison to animal proteins due to their harsh protein isolation methods (Sim et al., 2021). Therefore, modification of plant-based proteins by altering their physicochemical properties provides the possibility to improve and diversify their techno-functionality and biological activities as well as addressing their limitations (Nasrabadi et al., 2021). In general, various protein modification methods can be categorized into physical, chemical, biological, and other methods. However, the modification of plant-based proteins by utilizing different physical methods has gained a lot of interest these days due to several advantages like environment friendliness, easy operation conditions, less time, lower energy consumption, low cost, and no waste production in comparison to other chemical and enzymatic methods (Abbasi et al., 2019; Subasi et al., 2021). While thermal modification falls under the category of physical modification method and is one of the extensively used processing methods, which utilizes heat energy as the primary processing stress for proteins or protein-containing food materials. Mir et al. (2020) indicated that thermal modification can be effectively utilized for the modification of various plant sources including whey, zein, livetins, soy protein isolates, peanut protein isolates, cowpea protein isolates, amaranth protein isolates, and oat protein isolates. In conventional heating (CH), various processes such as drying, roasting, boiling, and steaming are among the most common methods (Sharif et al., 2018). However, CH method generates heavy stress conditions and could induce denaturation and damage to the secondary and tertiary structures of proteins, thus causing changes in their nutritional and sensory characteristics (Menta et al., 2022). Therefore, various novel thermal modification methods utilized these days are ohmic, microwave, radiofrequency, dielectric, and infrared heating (Nasrabadi et al., 2021). Besides this, thermal modification of proteins by high temperature extrusion is also in great demand these days. Various thermal modification methods are discussed in detail in Table 5.1.

5.2.1 HEATING

Heating is one of the extensively employed processing treatments for the thermal modification of proteins and utilized in various processes including cooking, roasting, drying, and high-temperature extrusion. Besides these, plant-based proteins are also heat treated to achieve sterilization, pasteurization, and anti-nutrient reduction or as part of new product development (Sharif et al., 2018; Akharume et al., 2021). In this context, controlled application of heat could be utilized for the structural modification of proteins and consequently to improve their techno-functional properties (Sun Waterhouse et al., 2014; Peng et al., 2016).

TABLE 5.1

Effects of Various Thermal Modification Methods of Plant-based Proteins on their Characteristics and Structural Changes

Sr no.	Protein Modification Method	Protein Type	Improved Protein Characteristics	Structural Changes	Reference
1.	Conventional heating	Oat protein isolate	Increased surface hydrophobicity and thermal stability of protein	Structural modification in proteins as depicted by their unfolding, increased content of α-helix, turns, and random coil along with decreased content of β-strand	Wang et al., 2022
		Lentil, pea, and faba bean protein	Increased surface hydrophobicity, stronger gels, improved water-holding capacity (WHC), emulsion stability (ES)	Protein particles' aggregation, formation of disulphide bonds between reactive -SH groups	Hall & Moraru, 2021
		Pea proteins	Enhanced emulsion properties	Protein unfolding and increase in free SH groups	Peng et al., 2016; Chao and Aluko (2018)
		Moringa olifera seed protein isolates	Increased protein solubility (PS)	Modification in secondary structure of protein as indicated by variation in content of α-helix, β-sheet, β-turn, and irregular structure	Bakwo Bassogog et al., 2022
		Album (*Chenopodium album*) protein isolates	Increased surface hydrophobicity, improved thermal stability, gelling properties, and in-vitro protein digestibility (IVPD)	Unfolding of protein structure and structural variations were observed as indicated by results of FTIR, SDS-PAGE, and fluorescence intensity	Mir et al., 2020
		Cowpea proteins	Improved gel-forming characteristics	Formation of soluble protein aggregates which are stabilized by disulphide bonds	Peyrano et al. (2016)
2.	Microwave heating	Soy protein isolate	Increased the gel strength, water-holding capacity (WHC), and storage modulus	Secondary structure gets altered as indicated by the reduction in free SH groups and variation in disulphide linkages	Mu et al., 2020
		Lotus seed protein isolates	Increased the PS and emulsion properties	Secondary structures (α-helix, β-sheet) gradually converted into β-turns and random coils	Zheng et al., 2020
		Lentil flour proteins	Increased the WHC, oil absorption capacity (OAC) and emulsion properties	Changes in secondary structure of protein were observed as indicated by variation in amide groups	Mahalaxmi et al., 2022
		Soy protein isolate	Decrease in soy protein allergenicity and gluten toxicity	α-Helix, β-turn, β-sheet, and random structure comprising the secondary structure of protein isolate were altered	Lee et al., 2016

(Continued)

TABLE 5.1 *(Continued)*

Effects of Various Thermal Modification Methods of Plant-based Proteins on their Characteristics and Structural Changes

Sr no.	Protein Modification Method	Protein Type	Improved Protein Characteristics	Structural Changes	Reference
		Soymilk proteins	Increased the digestibility of soymilk proteins and reduced the trypsin inhibitors' activity (TIA)	Secondary structure of protein gets modified as indicated by loss in α-helix and an increase of β-sheet	Vanga et al. (2020) and Kubo et al. (2021)
		Rice proteins	Increased PS, EA, ES, and foaming capacity (FC)	Secondary and tertiary structures of protein get distorted, resulting in expanded protein as indicated by reduced β-sheet content and increased content of α-helix and random coil content	Cheng et al., 2021
		Peanut proteins	Improved protein digestibility, WHC, and emulsion activity (EA)	Changes in secondary structure of protein were observed as indicated by an increase in β-sheet form and decrease in α-helix form	Ochoa-Rivas et al., 2017
3.	Ohmic heating	Soya bean protein isolates	Modulating the immuno-reactivity and reducing the allergenicity of soya bean protein	NA	Pereria et al., 2021
		Soya bean milk proteins	Modifications in surface hydrophobicity and functional properties such as FC and emulsion capacity (EC)	Increase in free amino groups, decrease in content of total sulfhydryl (SH) groups, and no significant changes in structure of protein were observed	Li et al., 2018
		Soya bean milk	Increase in protein precipitation and surface hydrophobicity resulting in firmer tofu curd	NA	Shimoyamada et al., 2015
4.	Infrared heating	Lentil flours	Improvement in the protein digestibility and protein quality	NA	Liu et al., 2022
		Cowpea flours	Increase in the protein content and improvement in the WAC, water solubility index, and IVPD	NA	Vilakati et al., 2015
		Desi chickpea and hull-less barley	Increased crude protein, amino acid content, and improved in-vitro protein digestibility (IVPD), in vitro corrected amino acid score (IVCAAS), and protein efficiency ratio (PER)	NA	Bai et al., 2018

S. No.	Method	Substrate	Functional properties/effects	Structural changes	Reference
		Green lentil and yellow pea	Increase in the protein content, WHC, and IVPD	Increase in SH-groups and the secondary protein structure of both pulse types transitioned from β-sheet and α-helix structures, to state with a higher relative percentage of random coil structure	Laing, 2022
		Proteins derived from black gram by products	Improved functional properties such as WHC, zeta potential, OAC, emulsion activity index (EAI), and FC of proteins	Secondary structure of protein remain unaltered	Kamani et al., 2021
5.	Radiofrequency heating	Rice bran protein isolates	Increased the WAC, OAC, EC, ES, and surface hydrophobicity	Primary structure of protein remained unaltered while secondary and tertiary structures of protein isolates changed with the decrease in β-sheet, β-turn, and α-helix contents and increase in random coil content	Ling et al., 2019
		Maize flour	Enhanced the OAC, ES, and EC	No significant changes in the primary structure of proteins in maize flour were observed	Hassan et al. 2019
		Soya bean seeds	Effective inactivation of TIA and improved digestibility of proteins	NA	Jiang et al., 2020; Takacs et al., 2022
		Soy protein isolate	Increased surface hydrophobicity	Modification in secondary and tertiary structures of soy protein, breakdown of disulphide bonds, increase in the content of free sulfhydryl (S–H) groups, partial unfolding of proteins, and increase in β-sheet structure on the expense of random coil structure	Guo et al., 2017
		Soya bean protein	Increased net protein utilization and net protein ratio	NA	Petres et al., 1990
6.	Extrusion	Peanut isolates	Increased accessibility for protease hydrolysis and enhanced emulsion properties	NA	Chen et al., 2018
		Red and green lentils	Enhanced crude protein, amino acid content, and amino acid score	NA	Nosworthy et al., 2018

(Continued)

TABLE 5.1 *(Continued)*

Effects of Various Thermal Modification Methods of Plant-based Proteins on their Characteristics and Structural Changes

Sr no.	Protein Modification Method	Protein Type	Improved Protein Characteristics	Structural Changes	Reference
		Chickpea	Increased the protein digestibility, protein efficiency ratio (PER), protein-digestibility-corrected amino acid score (PDCAAS), digestible indispensable amino acid score (DIASS)	NA	Nosworthy et al., 2020
		Black, faba, navy, pinto and red kidney beans	Enhanced crude protein content, protein digestibility, and PDCAAS	NA	Nosworthy et al., 2018
		Maize and soya bean protein concentrate, chickpea, and yellow pea protein	Increased the digestibility of proteins and decreased the anti-nutrients content	NA	Omosebi et al. 2018; Devi et al., 2020
		Rice protein	Increased PS, EHC, and ES	The content of ionic bond, sulfhydryl, and disulphide decreased; the hydrogen bond content increased. Formation of protein aggregates with a tight structure and the transformation of β-sheet and β-turn angles to an α-helix structure were observed	Gao et al., 2022
		Wheat gluten	Affect the celiac gluten toxicity	Protein crosslinking through intermolecular disulphide bonds to gluten proteins, increased number of free sulfhydryl groups (SH-groups), and reduced intensity of low-MW proteins in SDS-PAGE were observed	Wu et al., 2021
		Defatted wheat germ protein	Reduced surface hydrophobicity, increased PS, emulsification, and IVPD	Crosslinking of proteins, reduction in free SH groups, folding and spatial structural changes of protein conjugates, and transformation of α-helix to β-sheet were observed	Gao et al., 2023

5.2.1.1 Conventional Heating

CH is the common physical modification method which is utilized for the modification of plant-based proteins. Various CH treatments include processes like cooking, roasting, boiling, steaming, and drying which generally improve the techno-functional properties of plant-based proteins with controlled heating conditions and decrease the protein allergenicity (Venkateswara Rao et al., 2021).

5.2.1.2 Microwave Heating (MH)

Microwaves are electromagnetic waves having wavelengths ranging from 1 mm to 1 m, wherein the commonly used frequencies for heating are 0.915 GHz and 2.45 GHz (Das et al., 2009). MH generates heat within the food products and throughout their mass in response to applied electromagnetic energy by frictional interactions of ionic movement and rotation of polar dielectric molecules (Ramaswamy & Tang, 2008; Sipahioglu & Barringer, 2003). As the microwave energy is lower than the energy of chemical bonds, therefore it can modify the protein without destroying its primary structure (Han et al., 2018).

5.2.1.3 Ohmic Heating (OH)

Ohmic heating also known as resistive heating is a recent technology which offers an attractive heating option because it heats the food material through internal heat generation (Sakr & Liu, 2014). OH utilizes the electrical conductivity of food materials and depends on the directional movement of conductive ions such as electrolyte solution. In this case, when electric current flows through the food, then electric energy gets transformed into heat and this results in increased temperature of food which further influences its functional characteristics (Li et al., 2018).

5.2.1.4 Infrared Heating (IH)

Infrared heating or micronization is another processing method which utilizes electromagnetic infrared radiations to quickly heat the food materials. This heating method produces high temperature in lesser time and causes significant changes to functional properties by modifying the structure of biomolecules (Deepa & Hebbar, 2016). Moreover, IH has also received consideration as an additional heating method due to its compact equipment size, high energy efficiency, high diffusion coefficient, and low cost of equipment (Suri et al., 2019).

5.2.1.5 Radiofrequency Heating (RH)

Radiofrequency heating is a dielectric heating method which utilizes electromagnetic waves in the range of 1–300 MHz. Advantage of RH is that it can penetrate much deeper into the samples in comparison to MW, OH, or IR heating without any contact with the product and is more suitable for processing large quantities (Ling, Ouyang et al., 2019; Ling, Lyng et al., 2018). RH can significantly modify the secondary structure of protein and its functional characteristics without impacting its primary structure (Hassan et al., 2019).

5.2.2 EXTRUSION

Besides heat treatments, other thermal modification treatment is extrusion wherein a material is forced to flow through a machine via a screw system under optimum conditions of temperature, moisture, and pressure after mixing, shearing, and heating (Nosworthy et al., 2018). This treatment is quite effective in destroying the antinutritional compounds and improving the digestibility of plant proteins by increasing their amino acids availability (Sim et al., 2021). The high intense extrusion conditions for short time duration influence various functional properties such as texture, solubility, and emulsion and gelation properties of proteins (Akharume et al., 2021).

5.3 EFFECTS OF THERMAL MODIFICATION ON THE STRUCTURE OF PLANT PROTEINS

Heat is among the most common methods for significantly modifying the structure of plant-based proteins. During thermal processing, unfolding of the protein structure attributed to the exposure of the hydrophobic groups which are present in between the protein molecules leads to enhancing the surface hydrophobicity (Li et al., 2018). Protein unwinding is aided by a low-temperature environment, resulting in an intermediary molten globule state with improved functioning. Extreme heat stability, on the other hand, causes permanent changes in protein structures, resulting in hydrolysis and aggregating via various bonds such as disulphide, hydrophobic, and electrostatic, resulting in a loss of functional characteristics (Aryee et al., 2018). The heating of proteins causes a rearrangement of their secondary structure by impacting SS and -SH linkages. On the other hand, controlled heat treatments result in the denaturation of proteins, which causes changes in the secondary, tertiary, and quaternary protein structures by altering the electrostatic, ionic, hydrogen, and hydrophobic bonds responsible for the characteristic structure and functionality of proteins. This denaturation thus causes unfolding of the native protein structure, which can expose various buried reactive sites (particularly hydrophobic sites) which internally can improve functionality (Campbell et al., 2016).

Different thermal modification methods bring about different alterations in the structure of protein. OH can cause unfolding, denaturation, and the creation of uniform-sized protein aggregates with different techno-functional characteristics by delivering rapid and uniform heating as well as electrical effects (Mesias et al., 2016). On the other hand, radiofrequency like the microwave is premised on heat production and impacts protein function *via* the action of free radicals generated by dipolar and ionic flexibility in the presence of an radiofrequency field. Both RH and MH can solve the issues of low-heating rate impacts that are popular heating techniques due to their heat production (Ji et al., 2018). RH was discovered to have significant effects on the structure of soy protein by breaking disulphide linkages and increasing surface hydrophobicity (Li et al., 2019). Guo et al. (2017) reported that RH is effective in modifying the secondary and tertiary structure of soy protein isolates by breaking down the disulphide bonds and partial unfolding of proteins as indicated through the increase in β-sheet structure on the expense of random coil structure, enhancement of free sulfhydryl (-SH) contents, and surface hydrophobicity. Meanwhile, Hassan et al. (2019) reported that there were no significant changes in the primary structure of proteins in maize flour treated by RH under different temperatures even up to 60°C. Furthermore, extrusion could enable the molecules of vegetable protein to unfurl, denaturant, and realign, enhancing their techno-functionality while also giving them a meat-like texture (Zahari et al., 2020). Beck et al. (2017) observed the changed molecular weight of pea protein after low-moisture extrusion, and they confirmed the changes in the secondary structure by the identification of formed α-helices, β-sheet, non-covalently bonded β-turn, or anti-parallel β-sheet structures, as revealed by FTIR and SDS-PAGE. High temperatures during extrusion can cause the unfolding of proteins due to the breakage of hydrogen bonds. By further enhancement of the temperature, the breakdown of the intramolecular disulphide bonds, and the creation of new intermolecular ones, the formation of protein aggregates with subsequent increase in their collective molecular weight can be achieved (Beck et al., 2017). Under extrusion, the content of sulfhydryl and disulphide bonds of rice protein decreased significantly; the hydrogen bond content increased, and the ionic bond content decreased; the hydrophobic effect decreased, except at 200 rpm, 130°C, and 40%. The microstructure changed significantly after extrusion, producing protein aggregates with a tight structure. No new characteristic peaks appeared after extrusion, but transformation occurred between the components of the secondary structure: β-sheet and β-turn angles were transformed to an α-helix structure, but β-sheet was still the main component (Gao et al., 2022). For the structural elucidations of proteins, especially their secondary structure, FTIR spectroscopy can be effectively utilized (Carbonaro et al., 2008). In FTIR region, each compound has a specific set of absorption bands, and the characteristic bands found in the infrared spectra of proteins and polypeptides are the amide I and amide II (Murayama & Tomida 2004). Mir et al. (2020) reported the impact of controlled heat treatment of *Chenopodium album* PIs on its structural and conformational changes and reported the absorption peaks in amide-I, amide-II, and amide-III zones for the heat-treated PIs, which confirm the presence of structural modifications. Ellepola et al. (2005) also conducted a conformational analysis on rice globulins under different conditions using FTIR, and high α-helical, β-sheet, and β-turn contents in rice globulins were reported. Processing techniques have been reported to bring variations in secondary structures of legume seed flour as ß-sheets of legumes form intermolecular ß-sheets aggregates (Carbonaro et al., 2012). Several researchers reported structural modifications in various plant-based proteins as discussed in Table 5.1.

5.4 EFFECTS OF THERMAL MODIFICATION ON THE RHEOLOGY OF PLANT-BASED PROTEINS

The ability of a protein to entrap molecules of water to form a continuous gel structure is a particularly important quality attribute that can impact the sensory tactile response and textural attributes of a food matrix, thus establishing the end use purpose of a protein. During gelation, the interaction between aggregation arrangement and protein polypeptide chains has significant impact on gelation properties of proteins (Yang et al., 2014). Campbell et al. (2009) have stated that protein gelation depends upon the aggregation, denaturation, and dissociation of protein molecule. Rheological studies are an excellent means to ascertain the gelling properties of proteins. Rheology or flow behaviour of food material is an important criterion for optimum food product development. The rheological properties of protein gels have been previously studied in terms of storage and viscous modulus (G′ and G″, respectively) indicating the gelling ability, gel strength, stability, and viscosity (Zhao et al., 2020; Shevkani et al., 2019; Mir et al., 2020). G′ (elastic component) indicates the gel strength and stability associated with the molecular interactions that contribute to the complex structure of the gel, whereas G″ (viscous component) relates to the molecular interactions that do not contribute to the complex structure of the protein gel. The impact of the changing temperature (heating and cooling cycles) on rheology of proteins provides information that can help to control the mouthfeel and textural attributes of food products like puddings, sausages, texturized meat products, and yogurts. Dynamic rheological studies with temperature sweep tests of protein gels have been conducted to record the gelation temperature of proteins, which represents the thermal stability and denaturation characteristics of plant proteins (Shevkani et al., 2019). Studies have observed that controlled thermal treatment brings about conformational and structural changes in protein, which result in the formation of soluble globular protein complexes that possess enhanced heat stability (Taktak et al., 2018; Mir et al., 2020).

Generally, in temperature sweep rheological testing, protein solutions are first heated to 25–95°C, during which G′ remains lower than G″, until the gelation temperature reaches at which a gel starts to form and the values of G′ overtake that of G″. To test for gelling strength and thermal stability, the protein gel is held at 95 °C for a few minutes, following which the gel is cooled to 25°C (Mir et al., 2020; Shevkani et al., 2019). Mir et al. (2020) studied the impact of controlled heat treatment on rheological properties of *Chenopodium album* protein isolates (PIs) and observed that during heating phase of rheological testing, G′ and G″ of thermally modified PIs were higher than native PIs indicating an improvement in gelling capacity of PI upon thermal modification. The study also stated that the gelling capacity of PIs improved greatly at higher intensity of thermal modification (100°C for 15 and 30 min). During this heating phase, proteins denature and uncoil to expose buried reactive hydrophobic and hydrogen bonds that have the ability to bind water molecules. The study associated higher protein modification temperatures with the maximum exposure of these hydrogen and hydrophobic bonds, thus increasing the intermolecular interactions between protein chains and water molecules and producing stronger protein gels. Hall and Moraru (2021) observed a similar strengthening of pulse protein gels with heat treatment. Similarly, Peyrano et al. (2016) also observed an improvement in gelling properties, especially least gelation concentration upon thermal treatment of cowpea protein isolates. Studies have attributed the enhancement of protein gelling properties to the ability of thermal treatment to readily disrupt non-covalent bonding between polar side chains, causing the proteins to unfold and lead to sulfhydryl bond formation (Peyrano et al., 2016). During cooling phase, Mir et al. (2020) observed that G′ and G″ of the protein gels increased, wherein thermally treated PIs yielded firmer gels upon cooling. This increase in G′ and G″ has been mainly attributed to hydrogen bonding between denatured peptide chains.

Mir et al. (2020) and Hall and Moraru (2021) performed frequency sweep rheological test of thermally modified protein gels. Mir et al. (2020) observed that PIs (*Chenopodium album*) thermally modified at 100°C exhibited firmer gels than their native counterpart. Further, gel yielded from PIs modified at 100°C for 30 min had the highest G′ and G″ and the lowest tan δ (loss tangent; the ratio of G″ to G′), indicating enhanced elasticity or firmness of resulting gels. In comparison to native proteins, lower values of tan δ, in case of thermally treated proteins, have been associated with greater intermolecular bonding between protein chains through increased disulphide bridges and hydrophobic interactions. Hall and Moraru (2021) compared the rheological properties of pulse proteins (lentil, faba bean, and pea) after thermal modifications and HPP treatment. The study observed that thermally treated pulse proteins had a superior gel network with higher G′ and G″ than native and HPP-treated proteins. Furthermore, in comparison to native and HPP-treated proteins, higher disulphide bond formation was observed in thermally treated protein molecules due to which greater number of -SH groups were exposed, hence contributed to a higher increment in G′.

5.5 EFFECTS OF THERMAL MODIFICATION ON THE FUNCTIONALITY OF PLANT PROTEINS

Owing to the unique macromolecular structure, proteins are surface-active biomolecules with hydrophilic and hydrophobic regions that confer unique functional properties to them. Protein functionality mainly depends upon the primary sequence of amino acid and the ability of the protein to undergo diverse interfacial conformational changes depending upon their molecular structure and intermolecular and intramolecular interactions (Ozturk & McClements, 2016). Various properties of proteins which affect their functionality include surface hydrophobicity, protein solubility, water absorption and oil absorption capacity, foaming properties, emulsifying properties, in-vitro protein digestibility, and thermal properties, which are discussed later in the chapter.

5.5.1 Surface Hydrophobicity

Surface hydrophobicity of PIs influences protein solubility, emulsification and foaming properties, water- and oil-binding capacities, etc. Heat treatment has been found to cause significant increase in surface hydrophobicity of proteins (Mir et al., 2020; Peyrano et al., 2016; Hall & Moraru, 2021) Mir et al. (2020) conducted studies on the impact of heat treatment (time and temperature) on the physicochemical, rheological, and functional attributes of album PIs and observed that surface hydrophobicity increased with the increase in intensity of thermal modification. The study stated that PIs heated at 100°C (30 min) had the highest surface hydrophobicity. Similarly, Peyrano et al. (2016) and Wang et al. (2014) also observed an increase in surface hydrophobicity of cowpea PIs and soy proteins, respectively, after thermal treatment. Thermal treatment of PIs under controlled conditions causes ideal conformational and structural changes in protein, which expose new hydrophobic interaction sites that are otherwise buried in the protein globular molecule and result in an increase in surface hydrophobicity (Sorgentini et al., 1995).

5.5.2 Protein Solubility (PS)

Heat treatment has been found to cause significant changes in protein solubility. Solubility of heat-treated proteins has been established to be improved as well as deteriorated with thermal treatment (Mir et al., 2021; Hall & Moraru, 2021). Chao and Aluko (2018) observed that in comparison to native pea protein isolates, pea PIs thermally treated within the temperature range of 50–90°C did not exhibit much change in PS, indicating that controlled thermal treatments did not hamper protein–water interactions through protein coagulation or aggregation. However, noticeable decline in PS was observed at 100°C. In regards to quinoa PIs, Mir et al. (2021) observed a significant increase in PS upon thermal treatment, wherein quinoa PIs treated at 80°C for 30 min had the highest PS. The increase in PS has been attributed to the unfolding of protein globular structures under controlled heating, revealing hydrophobic as well as hydrophilic reaction sites, thus resulting in increased protein–water interactions. The study also observed that PS declined with an increase in intensity of thermal treatment. Similarly, Hall and Moraru (2021) observed a decline in PS of faba bean, pea, and lentil PIs after thermal treatment (95°C, 15 min) and attributed this decline to protein unfolding or denaturation which internally augmented the surface hydrophobicity of thermally treated proteins. Apart from this, the increased reaction sites for hydrophobic interactions could have caused the formation of larger insoluble protein aggregates consequently resulting in decreased PS (Sorgentini et al., 1995; Mir et al., 2021).

5.5.3 Water and Oil Absorption Capacity

Water absorption capacity (WAC) and oil absorption capacity (OAC) of a protein are important parameters for effective utilization of proteins, where the former property influences the texture and keeping quality of a food product, while, the latter improves mouthfeel and encourages the retention of flavour in food products (Bhinder et al., 2020; Das et al., 2021). Mir et al. (2021) observed that thermally modified quinoa PIs had higher WAC and OAC than native quinoa PIs; however, with the increase in temperature and time of thermal protein modification, WAC of PIs declined, while OAC increased. The heat-induced unfolding of protein structures occurs above 70°C, wherein some hydrophilic and hydrophobic protein groups buried within the globular protein core get exposed that internally enhance the polarity and solubility of proteins and consequently augment intermolecular interactions between water molecules and proteins (Morr & Ha, 1993). However, excessive protein aggregation at higher temperatures reduces the hydrophilic nature of proteins, because of which a decline in WAC was noticed at higher protein thermal modification temperatures (Mir et al., 2020, 2021). In another study, microwave heating (560 W for 3 min) of quinoa PIs was observed to enhance PS and WAC. These enhancements were attributed to a decline in -SH groups and surface hydrophobicity caused as a result of crosslinking and aggregation of quinoa PIs.

In case of OAC of proteins, Mir et al. (2021) suggested that higher heat treatment time and temperatures (100°C, 30 min) caused significant denaturation of proteins resulting in greater exposure of hydrophobic amino acid groups which have the ability to interact and bind oil readily. Similar observations in regards to WAC and OAC were made in case of sunflower PIs (Malik & Saini, 2018) and pulse PIs (Hall & Moraru, 2021).

5.5.4 Foaming Properties

Foaming properties, namely, foaming capacity (FC) and foaming stability (FS), are important functionality features of a protein, which help to identify the potential utilization of a protein in whipped or aerated foods like sponge/chiffon cakes, soufflé, mousses, and ice creams. Stable protein foams have been known to positively influence the final texture of aerated foods (Bhinder et al., 2020). Mir et al. (2021) found that in comparison to native PIs, the heat treatment of quinoa PIs caused an increase in both FC as well as FS. In case of thermally treated PIs, some losses in FS were observed with an increase in foam standing time. Similar results have been reported previously in case of thermally modified rice glutelins (Chen et al., 2020). On the other hand, Hall and Moraru (2021) observed a significant increase in FC as well as FS (30-min standing time) in case of heat-treated faba bean, pea, and lentil PIs. The studies suggested that a partial unfolding of proteins caused by controlled denaturation of proteins during thermal modification may have augmented protein adsorption at air–water interface to readily form a foam upon whipping (Mir et al., 2021; Santiago et al., 2020). However, Mir et al. (2021) proposed that with the increase in intensity of thermal treatment, proteins denatured fully to form larger aggregates which impaired the quick adsorption of proteins at the water and air interface causing a decline in foaming properties. Thus, the increase in foaming properties was attributed to the conformational changes induced in thermally treated PIs and a subsequent increase in surface hydrophobicity that leads to enhanced protein interfacial properties (Mir et al., 2021; Santiago et al., 2020).

5.5.5 EMULSIFYING PROPERTIES

Proteins are amphiphilic in nature which enables them to readily adsorb at oil and water interfaces to form emulsions. Emulsifying properties of a protein can be measured in terms of emulsifying activity index (EAI), defined as the ability of a protein to readily form an emulsion and emulsifying stability index corresponding to the formation of a stable emulsion that does not separate upon keeping (Bhinder et al., 2020). Controlled thermal modification of proteins has been found to significantly enhance both emulsion activity as well as stability index of the proteins (Mir et al., 2021; Hall & Moraru, 2021). Protein denaturation and partial unfolding induced through thermal modification may result in improved diffusion and subsequent adsorption of proteins at water and oil interface. Controlled thermal modification of globular proteins has been thus suggested to cause protein dissociation leading to the partially unfolding of proteins that exposes buried and unavailable hydrophobic protein residues and that improves surface activity and adsorption of the protein at the oil and water interface (Mir et al., 2020). Mir et al. (2021) observed that thermally modified quinoa PIs had higher emulsion activity and stability than native quinoa PIs; however, with the increase in temperature and time of thermal protein modification, both emulsion activity and stability declined. A plethora of protein characteristics and properties like surface hydrophobicity, charge, molecular mass, conformational stability, and solubility have a considerable impact on emulsifying properties of a protein (Moure et al., 2006). Since controlled thermal modification of proteins at lower intensity of heat treatment has been found to enhance surface hydrophobicity and solubility, the modified proteins were observed to possess improved emulsifying properties (Mir et al., 2020; Hall & Moraru, 2021; Mir et al., 2021). On the other hand, an increased intensity of thermal treatment results in the formation of larger protein aggregates which are highly inflexible and possess poor solubility. Such large aggregates of protein are thus unable to adsorb fat and water droplets effectively, hence causing a decline in emulsification ability and stability of the protein (Raikos, 2010).

5.5.6 IN-VITRO PROTEIN DIGESTIBILITY (IVPD)

Protein digestibility is an important criterion that helps to ascertain the bioavailability and absorption of proteins during digestion. Good-quality proteins have a high IVPD and tend to undergo complete proteolysis to yield the highest amount of amino acids. Mir et al. (2020) studied the effect of thermal modification on the IVPD of *Chenopodium album* PIs and observed a significant increase in IVPD of PIs after heat treatment. The study also stated that Album proteins heated at 100°C (30 min) and casein had comparable IVPDs. Further, thermal treatment carried out at high temperature and long duration yielded PIs with the highest IVPD—thus indicating that higher intensities of heat treatment caused greater protein unfolding and denaturation, which lead to an exposure of a large number of buried hydrophobic reaction sites that can readily interact with digestive enzymes and yield maximum number of amino acids. Wang et al. (2018) reported an increase in digestibility of egg white albumin after exposure to dry heat. It has been quantified that the heat treatment at 75°C (15 min) can improve the digestibility of egg whites by a factor of 4.8 in comparison to raw egg whites (Van et al., 2003). Ren et al. (2018) also observed an increment in IVPD of soya PIs after thermal sterilization.

5.5.7 THERMAL PROPERTIES

Thermal modification has been known to improve thermal stability of PIs from various sources like *Chenopodium album*, oats, and peanuts. Studies have attributed an increase in temperature for denaturation (T_d) of these PIs to thermal modification or pretreatment (Mir et al., 2020; Zhong et al., 2019; Hu et al., 2019). In case of Album PIs, the highest T_d measured using differential scanning calorimetry (DSC) was 96.77°C, observed for PIs thermally modified at 100°C (30 min) (Mir et al., 2020). T_d indicates the compactness of a protein molecular structure (Tang & Sun, 2011). It was also noted that with the increase in time and temperature of thermal treatment of PIs, T_d and end set temperature also increased thus indicating the formation of thermally stable protein aggregates at higher intensities of heat treatment. These aggregates possess a structure that is more compact and have higher surface hydrophobicity. On the other hand, enthalpy of denaturation, that is the energy needed to denature a protein, declined with an increase in the intensity of thermal treatment (Mir et al., 2020). Meng and Ma (2001) suggested that the decline in denaturation enthalpy was as a result of partial denaturation or unfolding of protein structure which would therefore require little energy to achieve denaturation.

5.6 APPLICATIONS OF THERMALLY MODIFIED PLANT PROTEIN IN DIFFERENT FOOD INDUSTRIES

Thermal processing has both beneficial and detrimental effects on food systems; therefore, the food scientist attempts to optimize the beneficial effects and to minimize the negative effects of heat processing. After obtaining highly functional plant proteins, the challenge is to transform these ingredients into delicious and nutritious foods and further increase their applications in various food industries. Plant-based proteins are considered as functional ingredients with various roles in food formulations, including thickening and gelling agents, stabilizers of emulsions and foams, and binding agents for fat and water. Moreover, some proteins have biological activities such as anti-oxidant or antimicrobial characteristics (Jafari et al., 2020; Doost et al., 2019; Warnakulasuriya & Nickerson, 2018). Various food industries in which plant-based proteins are utilized or researched for potential applications are as given in Table 5.2.

Proteins and polysaccharides are extensively employed to make complex coacervates with magnificent physicochemical properties. These coacervates had high encapsulation efficiency, thermal stability, and a complex structure that permits sustained release of core materials (Kayitmazer, 2017). The protein–polysaccharide conjugates had better emulsification properties, solubility (particularly below or near isoelectric pH), and thermal stability compared to their native proteins. Mustafa et al. (2018) obtained aquafaba (the viscous liquid derived from cooked pulses) from chickpea-affected functional properties, which makes it suitable for substituting egg white in different food formulations (Stantiall et al., 2018). Some of the thermally modified plant-based proteins are also used as a carrier for bioactive compounds. Yang et al. (2014) utilized thermally modified lactoferrin protein for the carrier of epigallactocatechin-3-gallate. Vnuc̆ec D et al. (2015) reported that the commercial soy protein isolate powder was thermally modified in a vacuum chamber at different temperatures (50, 100, 150, and 200 °C) and can be significantly utilized to improve the adhesive bonding strength of natural soy-based adhesives. Plant protein-based meat analogues produced using pulse proteins, which resemble the sensory properties of meat may help in the reduction of the consumption of animal-based products (Elzerman et al., 2011; Hoek et al., 2011).

TABLE 5.2

Applications of Thermally Modified Plant-based Proteins in Different Food Industries

Plant-based Protein	Food Industry	Properties	Potential Application	Reference
Pea protein	Bakery industry	Good water absorption Good gelling ability	Additive and supplement in gluten-free products	Shivkani & Singh, 2014
	Meat industry	Tenderness, less fat retention	Additive/supplement	Baugreet et al., 2016
	Sports	Muscle thickness	Nutritional supplement	Babault et al., 2015
	Liquid emulsion	Emulsification	Emulsifier	Liang & Tang, 2014
	Oil microencapsulation	Emulsification	Emulsifier	Bajaj et al., 2017
	Bakery products	Foaming stability	Foam stabilizer	Lu et al., 2020
	Fortified beverages	Solubility, rheology, and thermal stability	Nutritional supplement	Lu et al., 2020
	Bakery products	Good water absorption	Binder	Boukid et al., 2022
	Extruded products	Good water absorption Good gelling ability	Texturizer	Boukid et al., 2022
Pulse proteins (isolates and concentrates)	Extruded products	Supplementation and fortification	Dough stability, viscoelasticity	Lu et al., 2020
Cowpea protein isolates	Bakery industry (bread and sponge cakes)	Supplementation and fortification	Increased WAC and texture improver	Campbell et al., 2016
Cereal proteins	Extruded products	Supplementation and fortification	Concentrates and isolates	Lu et al., 2020
Oat protein concentrates	Dairy (yoghurt)	Replacer of SMP—manages syneresis	Concentrates and isolates	Brückner-Gühmann, 2019
Quinoa protein	Gluten-free products	Foaming ability	Bakery products	Dakhili et al., 2019
	Meat	High water absorption capacity	Texturizer	Dakhili et al., 2019
Hemp protein isolate	Dairy	Prevents syneresis	Texturizer	Schweiggert-Weisz et al., 2020
	Fruit juice industry	Source of antioxidant	Enrichment	Schweiggert-Weisz et al., 2020
Soya protein	Bakery and confectionary	Enrichment	Texturizer	Boukid et al., 2022
	Bakery products	Dough stability	Texturizer	Boukid et al., 2022
Wheat protein	Meat industry	Water absorption, structure building	Texturizer	Boukid et al., 2022

5.7 CONCLUSION

The market for sustainable and cruelty-free plant-based proteins has expanded globally. The use of these proteins is, however, limited in food applications due to their inadequate functionality, antinutritional components, presence of allergens, amino acid bioavailability, and digestibility. Thermal treatment is a simple physical processing technique that can be employed to bring about suitable desirable manipulations in protein structure, which can consequently improve the overall functional properties of proteins. Thermal modification has been found to enhance the quality of proteins, in terms of in-vitro digestibility, functionality, thermal stability, gelling ability, and rheological properties, through controlled denaturation and unfolding of protein tertiary structure that alters the ability of proteins to interact with the surrounding food matrix. Research has particularly emphasized on the unfolding of proteins, increase in surface hydrophobicity, and the creation of new aggregates of protein molecules after thermal modification. Conventional heating (cooking, roasting, boiling, steaming, and drying) and novel heating treatments (MH, OH, IR, and RH) along with extrusion have been studied for thermal modification of plant proteins. However, limited and incomplete information is available on the impact of these heating methods on the modified protein application in foods, influence of heating conditions on extent of protein modification, and heat treatment optimization correlation. Also, an information gap exists on the impact of heat treatment on proteins from different plant sources. Thus, a deeper focus is required to gain an insight into conformational and structural alterations that are induced as a result of different methods of thermal modification techniques used for processing plant proteins. Further, extensive research is required to create a plant-based protein with enhanced functionality and nutritional properties that can be used to design and build superior novel food matrices without compromising the taste or nutritional value of the end-food product. Proper and extensive research on these knowledge gaps would help in implementing the processing of modified proteins or modified protein-based food products at a commercial level.

REFERENCES

Abbasi, F., Samadi, F., Jafari, S. M., Ramezanpour, S., & Shargh, M. S. (2019). Ultrasound-assisted preparation of flaxseed oil nanoemulsions coated with alginate-whey protein for targeted delivery of omega-3 fatty acids into the lower sections of gastrointestinal tract to enrich broiler meat. *Ultrasonics Sonochemistry*, *50*, 208–217. https://doi.org/10.1016/j.ultsonch.2018.09.014.

Akharume, F. U., Aluko, R. E., & Adedeji, A. A. (2021). Modification of plant proteins for improved functionality: A review. *Comprehensive Reviews in Food Science and Food Safety*, *20*(1), 198–224. https://doi.org/10.1111/1541-4337.12688.

Aryee, A. N. A., Agyei, D., & Udenigwe, C. C. (2018). Impact of processing on the chemistry and functionality of food proteins. In *Proteins in food processing* (pp. 27–45). Woodhead Publishing. https://doi.org/10.1016/B978-0-08-100722-8.00003-6.

Aschemann-Witzel, J., &Peschel, A. O. (2019). Consumer perception of plant-based proteins: The value of source transparency for alternative protein ingredients. *Food Hydrocolloids*, *96*, 20–28. https://doi.org/10.1016/j.foodhyd.2019.05.006.

Avilés-Gaxiola, S., Chuck-Hernández, C., del Refugio Rocha-Pizaña, M., García-Lara, S., López-Castillo, L. M., & Serna-Saldívar, S. O. (2018). Effect of thermal processing and reducing agents on trypsin inhibitor activity and functional properties of soybean and chickpea protein concentrates. *LWT-Food Science and Technology*, *98*, 629–634. https://doi.org/10.1016/j.lwt.2018.09.023.

Babault, N., Païzis, C., Deley, G., Guérin-Deremaux, L., Saniez, M. H., Lefranc-Millot, C., & Allaert, F. A. (2015). Pea proteins oral supplementation promotes muscle thickness gains during resistance training: A double-blind, randomized, Placebo-controlled clinical trial vs. Whey protein. *Journal of the International Society of Sports Nutrition*, *12*(1), 3. https://doi.org/10.1186/s12970-014-0064-5.

Bai, T., Nosworthy, M. G., House, J. D., & Nickerson, M. T. (2018). Effect of tempering moisture and infrared heating temperature on the nutritional properties of desi chickpea and hull-less barley flours, and their blends. *Food Research International*, *108*, 430–439. https://doi.org/10.1016/j.foodres.2018.02.061.

Bajaj, P. R., Bhunia, K., Kleiner, L., Joyner, H. S., Smith, D., Ganjyal, G., & Sablani, S. S. (2017). Improving functional properties of pea protein isolate for microencapsulation of flaxseed oil. *Journal of Microencapsulation*, *34*(2), 218–230. https://doi.org/10.1080/02652048.2017.1317045.

Bassogog, C. B. B., Nyobe, C. E., Ngui, S. P., Minka, S. R., & Mune, M. A. M. (2022). Effect of heat treatment on the structure, functional properties and composition of Moringa oleifera seed proteins. *Food Chemistry*, *384*, 132546. https://doi.org/10.1016/j.foodchem.2022.132546.

Baugreet, S., Kerry, J. P., Botineştean, C., Allen, P., & Hamill, R. M. (2016). Development of novel fortified beef patties with added functional protein ingredients for the elderly. *Meat Science*, *122*, 40–47. https://doi.org/10.1016/j.meatsci.2016.07.004.

Beck, S. M., Knoerzer, K., & Arcot, J. (2017). Effect of low moisture extrusion on a pea protein isolate's expansion, solubility, molecular weight distribution and secondary structure as determined by Fourier Transform Infrared Spectroscopy (FTIR). *Journal of Food Engineering*, *214*, 166–174. https://doi.org/10.1016/j.jfoodeng.2017.06.037.

Bhinder, S., Kaur, A., Singh, B., Yadav, M. P., & Singh, N. (2020). Proximate composition, amino acid profile, pasting and process characteristics of flour from different Tartary buckwheat varieties. *Food Research International*, *130*, 108946. https://doi.org/10.1016/j.foodres.2019.108946.

Boukid, F., Rosell, C. M., Rosene, S., Bover-Cid, S., & Castellari, M. (2022). Non-animal proteins as cutting-edge ingredients to reformulate animal-free foodstuffs: Present status and future perspectives. *Critical Reviews in Food Science and Nutrition*, *62*(23), 6390–6420. https://doi.org/10.1080/10408398.2021.1901649.

Brückner-Gühmann, M., Benthin, A., & Drusch, S. (2019). Enrichment of yoghurt with oat protein fractions: Structure formation, textural properties and sensory evaluation. *Food Hydrocolloids*, *86*, 146–153. https://doi.org/10.1016/j.foodhyd.2018.03.019.

Campbell, L. J., Euston, S. R., & Ahmed, M. A. (2016). Effect of addition of thermally modified cowpea protein on sensory acceptability and textural properties of wheat bread and sponge cake. *Food Chemistry*, *194*, 1230–1237. https://doi.org/10.1016/j.foodchem.2015.09.002.

Campbell, L. J., Gu, X., Dewar, S. J., & Euston, S. R. (2009). Effects of heat treatment and glucono-δ-lactone-induced acidification on characteristics of soy protein isolate. *Food Hydrocolloids*, *23*(2), 344–351. https://doi.org/10.1016/j.foodhyd.2008.03.004.

Carbonaro, M., Maselli, P., Dore, P., & Nucara, A. (2008). Application of Fourier transform infrared spectroscopy to legume seed flour analysis. *Food Chemistry*, *108*(1), 361–368. https://doi.org/10.1016/j.foodchem.2007.10.045.

Carbonaro, M., Maselli, P., & Nucara, A. (2012). Relationship between digestibility and secondary structure of raw and thermally treated legume proteins: A Fourier transform infrared (FT-IR) spectroscopic study. *Amino Acids*, *43*, 911–921. https://doi.org/10.1007/s00726-011-1151-4.

Chao, D., & Aluko, R. E. (2018). Modification of the structural, emulsifying, and foaming properties of an isolated pea protein by thermal pretreatment. *CyTA-Journal of Food*, *16*(1), 357–366. https://doi.org/10.1080/19476337.2017.1406536.

Chen, L., Chen, J., Yu, L., Wu, K., & Zhao, M. (2018). Emulsification performance and interfacial properties of enzymically hydrolyzed peanut protein isolate pretreated by extrusion cooking. *Food Hydrocolloids*, *77*, 607–616. https://doi.org/10.1016/j.foodhyd.2017.11.002.

Cheng, Y. H., Mu, D. C., Jiao, Y., Xu, Z., & Chen, M. L. (2021). Microwave-assisted Maillard reaction between rice protein and dextran induces structural changes and functional improvements. *Journal of Cereal Science*, *97*, 103134. https://doi.org/10.1016/j.jcs.2020.103134.

Dakhili, S., Abdolalizadeh, L., Hosseini, S. M., Shojaee-Aliabadi, S., & Mirmoghtadaie, L. (2019). Quinoa protein: Composition, structure and functional properties. *Food Chemistry*, *299*, 125161. https://doi.org/10.1016/j.foodchem.2019.125161.

Das, D., Mir, N. A., Chandla, N. K., & Singh, S. (2021). Combined effect of pH treatment and the extraction pH on the physicochemical, functional and rheological characteristics of amaranth (Amaranthus hypochondriacus) seed protein isolates. *Food Chemistry*, *353*, 129466. https://doi.org/10.1016/j.foodchem.2021.129466.

Das, S., Mukhopadhyay, A. K., Datta, S., & Basu, D. (2009). Prospects of microwave processing: An overview. *Bulletin of Materials Science*, *32*, 1–13. https://doi.org/10.1007/s12034-009-0001-4.

Deepa, C., & Hebbar, H. U. (2016). Effect of high-temperature short-time 'micronization' of grains on product quality and cooking characteristics. *Food Engineering Reviews*, *8*, 201–213. https://doi.org/10.1007/s12393-015-9132-0.

Devi, S., Varkey, A., Dharmar, M., Holt, R. R., Allen, L. H., Sheshshayee, M. S., . . . Kurpad, A. V. (2020). Amino acid digestibility of extruded chickpea and yellow pea protein is high and comparable in moderately stunted South Indian children with use of a dual stable isotope tracer method. *The Journal of Nutrition*, *150*(5), 1178–1185. https://doi.org/10.1093/jn/nxaa004.

Doost, A. S., Nasrabadi, M. N., Wu, J., A'yun, Q., & Van der Meeren, P. (2019). Maillard conjugation as an approach to improve whey proteins functionality: A review of conventional and novel preparation techniques. *Trends in Food Science & Technology*, *91*, 1–11. https://doi.org/10.1016/j.tifs.2019.06.011.

Ellepola, S. W., Choi, S. M., & Ma, C. Y. (2005). Conformational study of globulin from rice (Oryza sativa) seeds by Fourier-transform infrared spectroscopy. *International Journal of Biological Macromolecules*, *37*(1–2), 12–20. https://doi.org/10.1016/j.ijbiomac.2005.07.008.

Elzerman, J. E., Hoek, A. C., Van Boekel, M. A., &Luning, P. A. (2011). Consumer acceptance and appropriateness of meat substitutes in a meal context. *Food Quality and Preference*, *22*(3), 233–240. https://doi.org/10.1016/j.foodqual.2010.10.006.

Erdogdu-Arnoczky, N., Czuchajowska, Z., & Pomeranz, Y. (1996). Functionality of whey and casein in fermentation and in breadbaking by fixed and optimized procedures. *Cereal Chemistry*, *73*(3), 309–316.

Gao, C., Jia, J., Yang, Y., Ge, S., Song, X., Yu, J., & Wu, Q. (2023). Structural change and functional improvement of wheat germ protein promoted by extrusion. *Food Hydrocolloids*, *137*, 108389. https://doi.org/10.1016/j.foodhyd.2022.108389.

Gao, Y., Sun, Y., Zhang, Y., Sun, Y., & Jin, T. (2022). Extrusion modification: Effect of extrusion on the functional properties and structure of rice protein. *Processes*, *10*(9), 1871. https://doi.org/10.3390/pr10091871.

Guo, C., Zhang, Z., Chen, J., Fu, H., Subbiah, J., Chen, X., & Wang, Y. (2017). Effects of radio frequency heating treatment on structure changes of soy protein isolate for protein modification. *Food and Bioprocess Technology*, *10*, 1574–1583. https://doi.org/10.1007/s11947-017-1923-2.

Hall, A. E., & Moraru, C. I. (2021). Structure and function of pea, lentil and faba bean proteins treated by high pressure processing and heat treatment. *LWT-Food Science and Technology*, *152*, 112349. https://doi.org/10.1016/j.lwt.2021.112349.

Han, Z., Cai, M. J., Cheng, J. H., & Sun, D. W. (2018). Effects of electric fields and electromagnetic wave on food protein structure and functionality: A review. *Trends in Food Science & Technology*, *75*, 1–9. https://doi.org/10.1016/j.tifs.2018.02.017.

Hassan, A. B., von Hoersten, D., & Ahmed, I. A. M. (2019). Effect of radio frequency heat treatment on protein profile and functional properties of maize grain. *Food Chemistry*, *271*, 142–147. https://doi.org/10.1016/j.foodchem.2018.07.190.

Hoek, A. C., Luning, P. A., Weijzen, P., Engels, W., Kok, F. J., & De Graaf, C. (2011). Replacement of meat by meat substitutes. A survey on person-and product-related factors in consumer acceptance. *Appetite*, *56*(3), 662–673. https://doi.org/10.1016/j.appet.2011.02.001.

Hu, Y., Sun-Waterhouse, D., Liu, L., He, W., Zhao, M., & Su, G. (2019). Modification of peanut protein isolate in glucose-containing solutions during simulated industrial thermal processes and gastric-duodenal sequential digestion. *Food Chemistry*, *295*, 120–128. https://doi.org/10.1016/j.foodchem.2019.04.115.

Jafari, S. M., Doost, A. S., Nasrabadi, M. N., Boostani, S., & Van der Meeren, P. (2020). Phytoparticles for the stabilization of Pickering emulsions in the formulation of novel food colloidal dispersions. *Trends in Food Science & Technology*, *98*, 117–128. https://doi.org/10.1016/j.tifs.2020.02.008.

Ji, H., Dong, S., Han, F., Li, Y., Chen, G., Li, L., & Chen, Y. (2018). Effects of dielectric barrier discharge (DBD) cold plasma treatment on physicochemical and functional properties of peanut protein. *Food and Bioprocess Technology*, *11*, 344–354. https://doi.org/10.1007/s11947-017-2015-z.

Kamani, M. H., Semwal, J., & Meera, M. S. (2021). Functional modification of protein extracted from black gram by-product: Effect of ultrasonication and micronization techniques. *LWT-Food Science and Technology*, *144*, 111193. https://doi.org/10.1016/j.lwt.2021.111193.

Kayitmazer, A. B. (2017). Thermodynamics of complex coacervation. *Advances in Colloid and Interface Science*, *239*, 169–177. https://doi.org/10.1016/j.cis.2016.07.006.

Kubo, M. T., dos Reis, B. H., Sato, L. N., & Gut, J. A. (2021). Microwave and conventional thermal processing of soymilk: Inactivation kinetics of lipoxygenase and trypsin inhibitors activity. *LWT-Food Science and Technology*, *145*, 111275. https://doi.org/10.1016/j.lwt.2021.111275.

Laing, E. (2022). *The effect of infrared heating on the functional and nutritional qualities of green lentil and yellow pea flours* (Doctoral dissertation, University of Saskatchewan).

Lee, H., Yildiz, G., Dos Santos, L. C., Jiang, S., Andrade, J. E., Engeseth, N. J., & Feng, H. (2016). Soy protein nano-aggregates with improved functional properties prepared by sequential pH treatment and ultrasonication. *Food Hydrocolloids*, *55*, 200–209. https://doi.org/10.1016/j.foodhyd.2015.11.022.

Li, J., Wang, B., Fan, J., Zhong, X., Huang, G., Yan, L., & Ren, X. (2019). Foaming, emulsifying properties and surface hydrophobicity of soy proteins isolate as affected by peracetic acid oxidation. *International Journal of Food Properties*, *22*(1), 689–703. https://doi.org/10.1080/10942912.2019.1602540.

Li, Q., Zheng, J., Ge, G., Zhao, M., & Sun, W. (2020). Impact of heating treatments on physical stability and lipid-protein co-oxidation in oil-in-water emulsion prepared with soy protein isolates. *Food Hydrocolloids*, *100*, 105167. https://doi.org/10.1016/j.foodhyd.2019.06.012.

Li, X., Ye, C., Tian, Y., Pan, S., & Wang, L. (2018). Effect of ohmic heating on fundamental properties of protein in soybean milk. *Journal of Food Process Engineering*, *41*(3), e12660. https://doi.org/10.1111/jfpe.12660.

Ling, B., Lyng, J. G., & Wang, S. (2018). Effects of hot air-assisted radio frequency heating on enzyme inactivation, lipid stability and product quality of rice bran. *LWT-Food Science and Technology*, *91*, 453–459. https://doi.org/10.1016/j.lwt.2018.01.084.

Ling, B., Ouyang, S., & Wang, S. (2019). Effect of radio frequency treatment on functional, structural and thermal behaviors of protein isolates in rice bran. *Food Chemistry*, *289*, 537–544. https://doi.org/10.1016/j.foodchem.2019.03.072.

Liu, S., Ren, Y., Yin, H., Nickerson, M., Pickard, M., & Ai, Y. (2022). Improvement of the nutritional quality of lentil flours by infrared heating of seeds varying in size. *Food Chemistry*, *396*, 133649. https://doi.org/10.1016/j.foodchem.2022.133649.

Lu, Z. X., He, J. F., Zhang, Y. C., & Bing, D. J. (2020). Composition, physicochemical properties of pea protein and its application in functional foods. *Critical Reviews in Food Science and Nutrition*, *60*(15), 2593–2605. https://doi.org/10.1080/10408398.2019.1651248.

Mahalaxmi, S., Himashree, P., Malini, B., & Sunil, C. K. (2022). Effect of microwave treatment on the structural and functional properties of proteins in lentil flour. *Food Chemistry Advances*, *1*, 100147. https://doi.org/10.1016/j.focha.2022.100147.

Malik, M. A., & Saini, C. S. (2019). Heat treatment of sunflower protein isolates near isoelectric point: Effect on rheological and structural properties. *Food Chemistry*, *276*, 554–561. https://doi.org/10.1016/j.foodchem.2018.10.060.

Meng, G. T., & Ma, C. Y. (2001). Thermal properties of Phaseolusangularis (red bean) globulin. *Food Chemistry*, *73*(4), 453–460. https://doi.org/10.1016/S0308-8146(00)00329-0.

Menta, R., Rosso, G., & Canzoneri, F. (2022). Plant-based: A perspective on nutritional and technological issues. Are we ready for "precision processing"? *Frontiers in Nutrition*, *9*.

Mesías, M., Wagner, M., George, S., & Morales, F. J. (2016). Impact of conventional sterilization and ohmic heating on the amino acid profile in vegetable baby foods. *Innovative Food Science & Emerging Technologies*, *34*, 24–28. https://doi.org/10.1016/j.ifset.2015.12.031.

Mir, N. A., Riar, C. S., & Singh, S. (2020). Structural modification in album (Chenopodium album) protein isolates due to controlled thermal modification and its relationship with protein digestibility and functionality. *Food Hydrocolloids*, *103*, 105708. https://doi.org/10.1016/j.foodhyd.2020.105708.

Mir, N. A., Riar, C. S., & Singh, S. (2021). Improvement in the functional properties of quinoa (Chenopodium quinoa) protein isolates after the application of controlled heat-treatment: Effect on structural properties. *Food Structure*, *28*, 100189. https://doi.org/10.1016/j.foostr.2021.100189.

Morr, C. V., & Ha, E. Y. W. (1993). Whey protein concentrates and isolates: Processing and functional properties. *Critical Reviews in Food Science & Nutrition*, *33*(6), 431–476. https://doi.org/10.1080/10408399309527643.

Moure, A., Sineiro, J., Domínguez, H., & Parajó, J. C. (2006). Functionality of oilseed protein products: A review. *Food Research International*, *39*(9), 945–963. https://doi.org/10.1016/j.foodres.2006.07.002.

Mu, D., Li, H., Li, X., Zhu, J., Qiao, M., Wu, X., . . . Zheng, Z. (2020). Enhancing laccase-induced soybean protein isolates gel properties by microwave pretreatment. *Journal of Food Processing and Preservation*, *44*(4), e14386. https://doi.org/10.1111/jfpp.14386.

Murayama, K., & Tomida, M. (2004). Heat-induced secondary structure and conformation change of bovine serum albumin investigated by Fourier transform infrared spectroscopy. *Biochemistry*, *43*(36), 11526–11532. https://doi.org/10.1021/bi0489154.

Mustafa, R., He, Y., Shim, Y. Y., & Reaney, M. J. (2018). Aquafaba, wastewater from chickpea canning, functions as an egg replacer in sponge cake. *International Journal of Food Science & Technology*, *53*(10), 2247–2255. https://doi.org/10.1111/ijfs.13813.

Nasrabadi, M. N., Doost, A. S., & Mezzenga, R. (2021). Modification approaches of plant-based proteins to improve their techno-functionality and use in food products. *Food Hydrocolloids*, *118*, 106789. https://doi.org/10.1016/j.foodhyd.2021.106789.

Nosworthy, M. G., Medina, G., Franczyk, A. J., Neufeld, J., Appah, P., Utioh, A., . . . House, J. D. (2018). Effect of processing on the in vitro and in vivo protein quality of red and green lentils (Lens culinaris). *Food Chemistry*, *240*, 588–593. https://doi.org/10.1016/j.foodchem.2017.07.129.

Nosworthy, M. G., Medina, G., Franczyk, A. J., Neufeld, J., Appah, P., Utioh, A., . . . House, J. D. (2020). Thermal processing methods differentially affect the protein quality of Chickpea (Cicer arietinum). *Food Science & Nutrition*, *8*(6), 2950–2958. https://doi.org/10.1002/fsn3.1597.

Ochoa-Rivas, A., Nava-Valdez, Y., Serna-Saldívar, S. O., & Chuck-Hernández, C. (2017). Microwave and ultrasound to enhance protein extraction from peanut flour under alkaline conditions: Effects in yield and functional properties of protein isolates. *Food and Bioprocess Technology*, *10*, 543–555. https://doi.org/10.1007/s11947-016-1838-3.

Omosebi, M. O., Osundahunsi, O. F., & Fagbemi, T. N. (2018). Effect of extrusion on protein quality, antinutritional factors, and digestibility of complementary diet from quality protein maize and soybean protein concentrate. *Journal of Food Biochemistry*, *42*(4), e12508. https://doi.org/10.1111/jfbc.12508.

Ozturk, B., & McClements, D. J. (2016). Progress in natural emulsifiers for utilization in food emulsions. *Current Opinion in Food Science*, *7*, 1–6. https://doi.org/10.1016/j.cofs.2015.07.008.

Peng, W., Kong, X., Chen, Y., Zhang, C., Yang, Y., & Hua, Y. (2016). Effects of heat treatment on the emulsifying properties of pea proteins. *Food Hydrocolloids*, *52*, 301–310. https://doi.org/10.1016/j.foodhyd.2015.06.025.

Pereira, R. N., Rodrigues, R. M., Machado, L., Ferreira, S., Costa, J., Villa, C., . . . Vicente, A. A. (2021). Influence of ohmic heating on the structural and immunoreactive properties of soybean proteins. *LWT-Food Science and Technology*, *148*, 111710. https://doi.org/10.1016/j.lwt.2021.111710.

Petres, J., Márkus, Z., Gelencsér, É., Bogár, Z., Gajzágó, I., & Czukor, B. (1990). Effect of dielectric heat treatment on protein nutritional values and some antinutritional factors in soya bean. *Journal of the Science of Food and Agriculture*, *53*(1), 35–41. https://doi.org/10.1002/jsfa.2740530105.

Peyrano, F., Speroni, F., & Avanza, M. V. (2016). Physicochemical and functional properties of cowpea protein isolates treated with temperature or high hydrostatic pressure. *Innovative Food Science & Emerging Technologies*, *33*, 38–46. https://doi.org/10.1016/j.ifset.2015.10.014.

Poutanen, K., Sozer, N., & Della Valle, G. (2014). How can technology help to deliver more of grain in cereal foods for a healthy diet? *Journal of Cereal Science*, *59*(3), 327–336. https://doi.org/10.1016/j.jcs.2014.01.009.

Raikos, V. (2010). Effect of heat treatment on milk protein functionality at emulsion interfaces. A review. *Food Hydrocolloids*, *24*(4), 259–265. https://doi.org/10.1016/j.foodhyd.2009.10.014.

Ramaswamy, H., & Tang, J. (2008). Microwave and radio frequency heating. *Food Science and Technology International*, *14*(5), 423–427. https://doi.org/10.1177/1082013208100534.

Reipurth, M. F., Hørby, L., Gregersen, C. G., Bonke, A., & Cueto, F. J. P. (2019). Barriers and facilitators towards adopting a more plant-based diet in a sample of Danish consumers. *Food Quality and Preference*, *73*, 288–292. https://doi.org/10.1016/j.foodqual.2018.10.012.

Ren, C., Xiong, W., Peng, D., He, Y., Zhou, P., Li, J., & Li, B. (2018). Effects of thermal sterilization on soy protein isolate/polyphenol complexes: Aspects of structure, in vitro digestibility and antioxidant activity. *Food Research International*, *112*, 284–290. https://doi.org/10.1016/j.foodres.2018.06.034.

Sakr, M., & Liu, S. (2014). A comprehensive review on applications of ohmic heating (OH). *Renewable and Sustainable Energy Reviews*, *39*, 262–269. https://doi.org/10.1016/j.rser.2014.07.061.

Santiago, L. A., Fadel, O. M., & Tavares, G. M. (2021). How does the thermal-aggregation behavior of black cricket protein isolate affect its foaming and gelling properties? *Food Hydrocolloids*, *110*, 106169. https://doi.org/10.1016/j.foodhyd.2020.106169.

Schweiggert-Weisz, U., Eisner, P., Bader-Mittermaier, S., & Osen, R. (2020). Food proteins from plants and fungi. *Current Opinion in Food Science*, *32*, 156–162. https://doi.org/10.1016/j.cofs.2020.08.003.

Sharif, H. R., Williams, P. A., Sharif, M. K., Abbas, S., Majeed, H., Masamba, K. G., . . . Zhong, F. (2018). Current progress in the utilization of native and modified legume proteins as emulsifiers and encapsulants–A review. *Food Hydrocolloids*, *76*, 2–16. https://doi.org/10.1016/j.foodhyd.2017.01.002.

Shevkani, K., & Singh, N. (2014). Influence of kidney bean, field pea and amaranth protein isolates on the characteristics of starch-based gluten-free muffins. *International Journal of Food Science & Technology*, *49*(10), 2237–2244. https://doi.org/10.1111/ijfs.12537.

Shevkani, K., Singh, N., Chen, Y., Kaur, A., & Yu, L. (2019). Pulse proteins: Secondary structure, functionality and applications. *Journal of Food Science and Technology*, *56*, 2787–2798. https://doi.org/10.1007/s13197-019-03723-8.

Shimoyamada, M., Itabashi, Y., Sugimoto, I., Kanauchi, M., Ishida, M., Tsuzuki, K., . . . Honda, Y. (2015). Characterization of soymilk prepared by ohmic heating and the effects of voltage applied. *Food Science and Technology Research*, *21*(3), 439–444. https://doi.org/10.3136/fstr.21.439.

Sim, S. Y. J., Srv, A., Chiang, J. H., & Henry, C. J. (2021). Plant proteins for future foods: A roadmap. *Foods*, *10*(8), 1967. https://doi.org/10.3390/foods10081967.

Sipahioglu, O., & Barringer, S. A. (2003). Dielectric properties of vegetables and fruits as a function of temperature, ash, and moisture content. *Journal of Food Science*, *68*(1), 234–239. https://doi.org/10.1111/j.1365-2621.2003.tb14145.x.

Sorgentini, D. A., Wagner, J. R., & Añón, M. C. (1995). Effects of thermal treatment of soy protein isolate on the characteristics and structure-function relationship of soluble and insoluble fractions. *Journal of Agricultural and Food Chemistry*, *43*(9), 2471–2479. https://doi.org/10.1021/jf00057a029.

Stantiall, S. E., Dale, K. J., Calizo, F. S., & Serventi, L. (2018). Application of pulses cooking water as functional ingredients: The foaming and gelling abilities. *European Food Research and Technology*, *244*, 97–104. https://doi.org/10.1007/s00217-017-2943-x.

Subaşı, B. G., Jahromi, M., Casanova, F., Capanoglu, E., Ajalloueian, F., & Mohammadifar, M. A. (2021). Effect of moderate electric field on structural and thermo-physical properties of sunflower protein and sodium caseinate. *Innovative Food Science & Emerging Technologies*, *67*, 102593. https://doi.org/10.1016/j.ifset.2020.102593.

Sun-Waterhouse, D., Zhao, M., & Waterhouse, G. I. (2014). Protein modification during ingredient preparation and food processing: Approaches to improve food processability and nutrition. *Food and Bioprocess Technology, 7*, 1853–1893. https://doi.org/10.1007/s11947-014-1326-6.

Suri, K., Singh, B., Kaur, A., Yadav, M. P., & Singh, N. (2019). Impact of infrared and dry air roasting on the oxidative stability, fatty acid composition, Maillard reaction products and other chemical properties of black cumin (Nigella sativa L.) seed oil. *Food Chemistry, 295*, 537–547. https://doi.org/10.1016/j.foodchem.2019.05.140.

Takács, K., Szabó, E. E., Nagy, A., Cserhalmi, Z., Falusi, J., & Gelencsér, É. (2022). The effect of radiofrequency heat treatment on trypsin inhibitor activity and in vitro digestibility of soybean varieties (Glycine max.(L.) Merr.). *Journal of Food Science and Technology, 59*(11), 4436–4445. https://doi.org/10.1007/s13197-022-05523-z.

Taktak, W., Nasri, R., Hamdi, M., Gomez-Mascaraque, L. G., Lopez-Rubio, A., Li, S., . . . Karra-Chaâbouni, M. (2018). Physicochemical, textural, rheological and microstructural properties of protein isolate gels produced from European eel (*Anguilla anguilla*) by heat-induced gelation process. *Food Hydrocolloids, 82*, 278–287. https://doi.org/10.1016/j.foodhyd.2018.04.008.

Tang, C. H., & Sun, X. (2011). A comparative study of physicochemical and conformational properties in three vicilins from Phaseolus legumes: Implications for the structure–function relationship. *Food Hydrocolloids, 25*(3), 315–324. https://doi.org/10.1016/j.foodhyd.2010.06.009.

Tang, C. H., Sun, X., & Yin, S. W. (2009). Physicochemical, functional and structural properties of vicilin-rich protein isolates from three Phaseolus legumes: Effect of heat treatment. *Food Hydrocolloids, 23*(7), 1771–1778. https://doi.org/10.1016/j.foodhyd.2009.03.008.

Van der Plancken, I., Van Remoortere, M., Indrawati, Van Loey, A., & Hendrickx, M. E. (2003). Heat-induced changes in the susceptibility of egg white proteins to enzymatic hydrolysis: A kinetic study. *Journal of Agricultural and Food Chemistry, 51*(13), 3819–3823. https://doi.org/10.1021/jf026019y.

Vanga, S. K., Wang, J., & Raghavan, V. (2020). Effect of ultrasound and microwave processing on the structure, in-vitro digestibility and trypsin inhibitor activity of soymilk proteins. *LWT-Food Science and Technology, 131*, 109708. https://doi.org/10.1016/j.lwt.2020.109708.

Vasconcelos, I. M., Maia, F. M. M., Farias, D. F., Campello, C. C., Carvalho, A. F. U., de Azevedo Moreira, R., & de Oliveira, J. T. A. (2010). Protein fractions, amino acid composition and antinutritional constituents of high-yielding cowpea cultivars. *Journal of Food Composition and Analysis, 23*(1), 54–60. https://doi.org/10.1016/j.jfca.2009.05.008.

Venkateswara Rao, M., CK, S., Rawson, A., & DV, C. (2021). Modifying the plant proteins techno-functionalities by novel physical processing technologies: A review. *Critical Reviews in Food Science and Nutrition*, 1–22. https://doi.org/10.1080/10408398.2021.1997907.

Vilakati, N., MacIntyre, U., Oelofse, A., & Taylor, J. R. (2015). Influence of micronization (infrared treatment) on the protein and functional quality of a ready-to-eat sorghum-cowpea African porridge for young child-feeding. *LWT-Food Science and Technology, 63*(2), 1191–1198. https://doi.org/10.1016/j.lwt.2015.04.017.

Vnučec, D., Goršek, A., Kutnar, A., & Mikuljan, M. (2015). Thermal modification of soy proteins in the vacuum chamber and wood adhesion. *Wood Science and Technology, 49*, 225–239. https://doi.org/10.1007/s00226-014-0685-5.

Wang, H., Xiang, L., Rao, P., Ke, L., Wu, B., Chen, S., . . . Su, P. (2022). Effects of pretreatments on structural and functional changes of oat protein isolate. *Cereal Chemistry, 99*(1), 90–99. https://doi.org/10.1002/cche.10480.

Wang, J., Chi, Y., Cheng, Y., & Zhao, Y. (2018). Physicochemical properties, in vitro digestibility and antioxidant activity of dry-heated egg white protein. *Food Chemistry, 246*, 18–25. https://doi.org/10.1016/j.foodchem.2017.10.128.

Wang, Z., Li, Y., Jiang, L., Qi, B., & Zhou, L. (2014). Relationship between secondary structure and surface hydrophobicity of soybean protein isolate subjected to heat treatment. *Journal of Chemistry, 2014*. https://doi.org/10.1155/2014/475389.

Warnakulasuriya, S. N., & Nickerson, M. T. (2018). Review on plant protein–polysaccharide complex coacervation, and the functionality and applicability of formed complexes. *Journal of the Science of Food and Agriculture, 98*(15), 5559–5571. https://doi.org/10.1002/jsfa.9228.

Wu, Y., Xiao, Z., Jiang, X., Lv, C., Gao, J., Yuan, J., . . . Chen, H. (2021). Effect of extrusion on the modification of wheat flour proteins related to celiac disease. *Journal of Food Science and Technology*, 1–11. https://doi.org/10.1007/s13197-021-05285-0.

Yang, W., Liu, F., Xu, C., Yuan, F., & Gao, Y. (2014). Molecular interaction between (−)-epigallocatechin-3-gallate and bovine lactoferrin using multi-spectroscopic method and isothermal titration calorimetry. *Food Research International, 64*, 141–149. https://doi.org/10.1016/j.foodres.2014.06.001.

Zahari, I., Ferawati, F., Helstad, A., Ahlström, C., Östbring, K., Rayner, M., & Purhagen, J. K. (2020). Development of high-moisture meat analogues with hemp and soy protein using extrusion cooking. *Foods*, *9*(6), 772. https://doi.org/10.3390/foods9060772.

Zeeb, B., Yavuz-Düzgun, M., Dreher, J., Evert, J., Stressler, T., Fischer, L., . . . Weiss, J. (2018). Modulation of the bitterness of pea and potato proteins by a complex coacervation method. *Food & Function*, *9*(4), 2261–2269. https://doi.org/10.1039/C7FO01849E.

Zhao, M., Xiong, W., Chen, B., Zhu, J., & Wang, L. (2020). Enhancing the solubility and foam ability of rice glutelin by heat treatment at pH12: Insight into protein structure. *Food Hydrocolloids*, *103*, 105626. https://doi.org/10.1016/j.foodhyd.2019.105626.

Zheng, T., Li, X., Taha, A., Wei, Y., Hu, T., Fatamorgana, P. B., . . . Hu, H. (2019). Effect of high intensity ultrasound on the structure and physicochemical properties of soy protein isolates produced by different denaturation methods. *Food Hydrocolloids*, *97*, 105216. https://doi.org/10.1016/j.foodhyd.2019.105216.

Zheng, Y., Li, Z., Zhang, C., Zheng, B., & Tian, Y. (2020). Effects of microwave-vacuum pre-treatment with different power levels on the structural and emulsifying properties of lotus seed protein isolates. *Food Chemistry*, *311*, 125932. https://doi.org/10.1016/j.foodchem.2019.125932.

Zhong, L., Ma, N., Wu, Y., Zhao, L., Ma, G., Pei, F., & Hu, Q. (2019). Characterization and functional evaluation of oat protein isolate-Pleurotusostreatus β-glucan conjugates formed via Maillard reaction. *Food Hydrocolloids*, *87*, 459–469. https://doi.org/10.1016/j.foodhyd.2018.08.034.

6 Ultrasonication-Assisted Modification of Plant-Based Proteins

*Gulsah Karabulut, Ragya Kapoor, Gulcin Yildiz,
Ozan Kahraman and Hao Feng*

6.1 INTRODUCTION

Protein ingredients with their well-balanced nutritional benefits and superior functionalities are extensively utilized in the food industry. As dynamic surfactants for interfacial interactions between proteins and carbohydrates, lipids, air, and water, proteins are versatile structural components of food matrices due to their amphipathic nature (Phillips 2013; O'Sullivan et al. 2016; Karabulut and Cagri-Mehmetoglu 2018). This trend is being driven by consumers' increasing knowledge and awareness of food ingredients and their growing interest in natural, sustainable, and eco-friendly food sources.

From a nutritional point of view, numerous studies have emphasized the benefits of protein for weight management, satiety, lowering the glycemic index, heart health, muscle maintenance, and even athletic performance (Livesey et al. 2008; Paul 2009; Karabulut et al. 2018). Many plant-based proteins are inherently overperformed by their animal-based counterparts in food formulations, even though they are getting more attention. Due to their low solubility at neutral pH, most legume proteins form weak gels (Kyriakopoulou, Dekkers, and van der Goot 2019; Wouters and Delcour 2019). Understanding the interplay between protein structure and function is necessary for comprehending the mechanisms by which proteins function (Fukuda et al. 2008). There is a lack of information about changes in proteins' structural conformations in their native state and unfolding conditions during denaturation and under various processing conditions. For instance, while legume proteins lack methionine, cystine, and tryptophan, most grain proteins lack lysine (Sun-Waterhouse, Zhao, and Waterhouse 2014; Singh 2017). In addition, vegetable proteins are less bioavailable and digestible in the gastrointestinal tract than animal proteins (Joshi, Shah, and Kalantar-Zadeh 2018).

Amino acids are the building blocks of proteins, which are complex biopolymers crucial to nutrition, taste, and function. Proteins have a variety of technical and functional properties, including solubility, thickening, water/oil retention, emulsifying, foaming, and gelling properties. Ismail et al. (2020) claimed that unique functional properties of proteins are influenced by their origin, amino acid composition and sequence, molecular weight, structure, conformation, and surface hydrophobicity. Due to its rich properties, protein is a vital component in food formulations, either as a component (protein concentrates or isolates) or as a component (raw materials like soy protein in soy milk). On the other hand, the properties of vegetable proteins may have some limitations that prevent them from being used in certain food industry formulations and applications. Exploiting and perfecting novel plant-based protein extraction and modification methods, such as ultrasonication, are necessary to make the plant-based protein industry more competitive.

In recent years, the food industry has extensively explored the use of power ultrasound, or high-intensity ultrasound, which operates at a low frequency of 16–100 kHz and power intensity of 10–1,000 W/cm^3, in different process intensification applications. When ultrasound is applied for modification of protein functional properties, most of the treatments are performed in liquid

DOI: 10.1201/9781003369790-6

solutions, and thus it is often referred to as sonication or ultrasonication. This is because the mode of action of an ultrasonic treatment is acoustic cavitation, which only takes place when a longitudinal sound wave travels through a liquid medium. The increasing awareness and need for sustainable production also drive different industrial sectors to seek for low-carbon footprint processing methods. Ultrasonication, as an electrification-driven process, has thus gained new momentum in industrial applications. This chapter presents the basics of sonication, a critical evaluation of the application of sonication to modify plant protein functions, challenges in ultrasonic treatments, and future trends.

6.2 PRINCIPLE OF ULTRASOUND TECHNOLOGY

Ultrasound is constituted of mechanical waves that are generated by molecular motions that oscillate in a medium. Ultrasound, when applied to a continuous fluid, has the primary effect of adding an acoustic pressure (P_a) to the preexisting hydrostatic pressure. The acoustic pressure is a sinusoidal wave that varies with respect to three independent variables: the wave's maximum pressure amplitude P_{amax}, frequency f, and time t, as seen in Eq. (1.6) (Knorr et al. 2004; Muthukumaran et al. 2006; Patist and Bates 2008; Bhargava et al. 2021).

$$P_a = P_{amax} \sin(2\pi f t) \tag{6.1}$$

Power ultrasound is a type of ultrasound with a frequency between 20 kHz and 1 MHz. It has a wide range of potential applications in the food processing industry, including homogenization, cutting, de-forming, inactivation, protein modifications, extraction, etc. (Kentish and Feng 2014). Another type of ultrasound, named as "diagnostic ultrasound," has a frequency over 1 MHz and is often used in medical imaging and as an analytical tool to examine physicochemical attributes of foods such as their structures, compositions, and particle sizes (Kentish and Ashokkumar 2011; Tao and Sun 2015; Gallo, Ferrara, and Naviglio 2018; Kahraman 2019; Bhargava et al. 2021; Korzendorfer 2022).

The fundamental component of ultrasound is a high-frequency pressure wave, which, when it moves through a medium in the form of a longitudinal wave, generates areas of high (compression) and low (rarefaction) pressure. The changes in the magnitude of sound pressure, also known as the amplitude of the pressure wave or the acoustic pressure, are influenced by the amount of ultrasonic energy that the system is subjected to. When the pressure fluctuations approach a specific threshold, the ultrasonic energy causes the liquid medium to disintegrate (Feng and Lee 2011). Because of this, the tensile stresses that were caused by the pressure waves are relieved, which in turn results in the creation of microbubbles. The primary impacts of ultrasound were caused by acoustic cavitation, which includes the creation of these tiny air bubbles in the treated system and their subsequent collapse (Coussios et al. 2007; Yi et al. 2018; Gevari et al. 2020). As shown in Figure 6.1, the ultrasound-generated bubbles expand over the course of many oscillations before they either stabilize or collapse violently, yielding microstreaming, microjets, and shockwaves. On a structural level, both types of cavitating bubbles are created during the sonication treatment process: (i) Nonlinear: Forming enormous globule clusters with a balance scale throughout pressure cycles known as stable cavitation bubbles and (ii) nonconstant: Rapidly dropping out and breaking down into smaller transient cavitation bubbles (Majid, Nayik, and Nanda 2015).

Acoustic cavitation results in the induction of sonomechanical and sonochemical effects, both of which are utilized often in sonoprocessing (Chia et al. 2019). Sonic energy in the medium is the root cause of sonomechanical events such as shock waves, localized high temperature and pressure spots, high-speed liquid jets, and macro- and micro-streaming. Sonochemical effects, on the other hand, generally make use of cavitation phenomena to speed up or catalyze a certain chemical process (Rubio, Blandford, and Bond 2016; Taha et al. 2023).

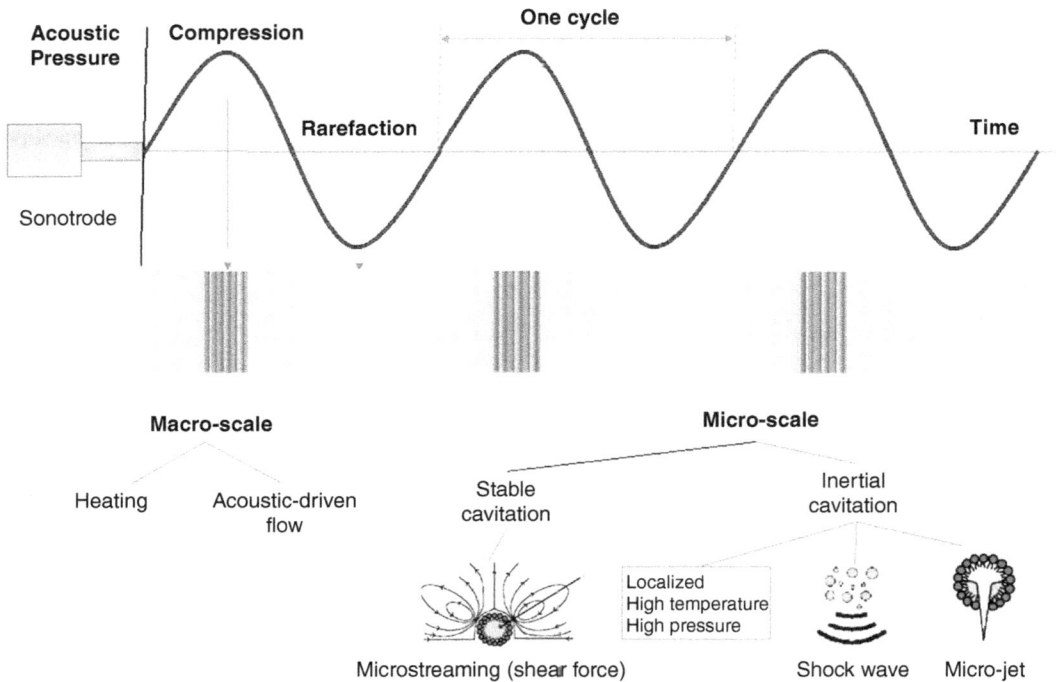

FIGURE 6.1 Ultrasound treatment-related phenomenon.

Source: Partially adopted from Nande, Howard, and Claudio 2015.

6.2.1 Influential Factors of Ultrasound Treatment

6.2.1.1 Ultrasound Source-Related Factors

6.2.1.1.1 Acoustic Power Density (CPD) and Sound Intensity (I)

Power (W = J/s) generally refers to the strength of a treatment or the energy input into a medium over a period of time. The power in an ultrasonic treatment is proportional to the amplitude of the ultrasonic wave. In ultrasonic treatments, the volumetric energy input into the food system per time is one critical parameter in process design, process control, and optimization. It is often defined by the volumetric power density in the entire treatment medium, e.g., W/L, W/cm^3, or W/m^3. In acoustic power density calculations (APD), the power of the treatment (W) can be determined directly by a calorimetric method by measuring the temperature increase in the system in a short period at the beginning of the treatment (Baumann, Martin, and Feng 2005). In other cases, the rating power of the ultrasound generator or power measured with a power meter is also used to estimate the APD. Sound intensity (*I*) is defined as the actual power output divided by the sonotrode's surface area (W/cm^2) or the sound emitting surface area of a tank unit. *I* was first proposed and used by sono-chemists and is widely used in sono-chemistry studies. Since most ultrasonic processes are a volumetric treatment, the concept of intensity (*I*), as a surface power density measurement, is less relevant for describing biochemical, biological, and mechanical events in a treatment chamber or vessel. For instance, a sonotrode with a given sound intensity (*I*) can be used to treat samples in numerous treatment chambers with different volumes. Therefore, for food processing purposes, APD is a more appropriate parameter for process design and for defining a process. It also enables comparisons of the work by different research groups.

6.2.1.1.2 *Frequency and Temperature*

The frequency (Hz) of an ultrasound generator is determined by the mechanical resonant frequency of the piezoelectric transducer. It is well known that ultrasound frequency has a significant effect on the average sizes of the cavitating bubbles in a treatment. Generally, a higher frequency ultrasonic treatment will generate more cavitating bubbles but with smaller bubble sizes. At the low-frequency end of power ultrasound, i.e., 20 kHz, less cavitating bubbles will be produced. But the bubble sizes are larger and therefore can produce more powerful cavitation activities (higher cavitation intensity). This is also the reason why most power ultrasonic treatments are performed at 20 kHz or at other low frequencies. The relation of frequency to bubble size is a key factor in determining the level of cavitation that takes place in each liquid.

Temperature is another crucial process parameter to consider in an ultrasonic process. Even if no additional heat is applied, ultrasonic treatments can cause the sample temperatures to rise. When a transducer transmits ultrasonic vibrations into a medium, the particles in that medium oscillate in response, generating heat. As a result, the properties of a liquid, such as its vapor pressure, surface tension, and viscosity, can all be impacted (Muthukumaran et al. 2006; Patist and Bates 2008).

6.2.1.2 Treatment Medium-Related Factors

Ultrasonic sound waves may travel through a variety of media, including air, water, and solids. As they do so, they create pressure fluctuations, which in turn stimulate the vibrations of particles inside the medium. Because of this, the effects of using power ultrasound on food products are dependent on the state of the medium through which the sound waves are transmitted (Li and Sun 2002; Adekunte et al. 2010).

The utilization of ultrasound in a liquid medium is the most basic and frequent technique in the food industry. When ultrasound is applied to a liquid, the primary process that causes its effects is acoustic cavitation. Acoustic cavitation refers to the generation, growth, and collapse of tiny gas- or vapor-filled bubbles or cavities in a liquid when ultrasound travels through it. The acoustic energy is released into the food system upon the collapse, which results in a variety of effects (Gallego-Juárez 2017). These effects can be put into two groups: physical and chemical. The phenomena of microjets and macro- or micro-streaming are the most important ones to consider regarding the physical effects. When microjets, which are streams of high-pressure water, are projected against the surface of solids, they create holes and erode the surface, resulting in the release of material into the medium. Microstreaming, on the other hand, takes place in the middle of the surrounding liquid and, at a high enough speed, can disrupt cell membranes and unleash intracellular enzymes and other biological effects (Figure 6.2) (Kentish and Ashokkumar 2011). When the number of cavitation spots is low, but they carry large energy, these physical consequences have a greater chance of occurring at low frequencies. The number of spots increases as the frequency increases (80–100 kHz), but the size of the bubbles reduces, thus the energy produced is less and the dominating effects are chemical (Zupanc et al. 2019). Rojas et al. (2016) reported an intriguing mechanism for how ultrasound operates in this sort of liquid medium using peach juice and an ultrasonic probe at 20 kHz, 0–15 min, and 22°C. At first, it was seen that low-intensity ultrasonic impact caused the migration and partial destruction of intracellular molecules. While the cellular skeleton is maintained, localized regions of the cell wall are disrupted when ultrasound duration and power are increased. These disruptions in cell walls may lead to swelling, due to uptake of water and posterior outflow of intracellular substances. Finally, with high ultrasonic impact, all damaged structures and internal chemical compounds combine.

When an ultrasonic wave, on the other hand, transmits through a solid medium, it causes a sequence of simultaneous contractions and expansions, known as the "sponge effect," which allows the movement of substance with the media around the solid, as illustrated in Figure 6.2. Mechanical stress can also cause the development of microchannels inside the solid, which facilitate mass transfer processes (Floros and Liang 1994; Mulet et al. 2003; Sabarez, Gallego-Juarez, and Riera 2012;

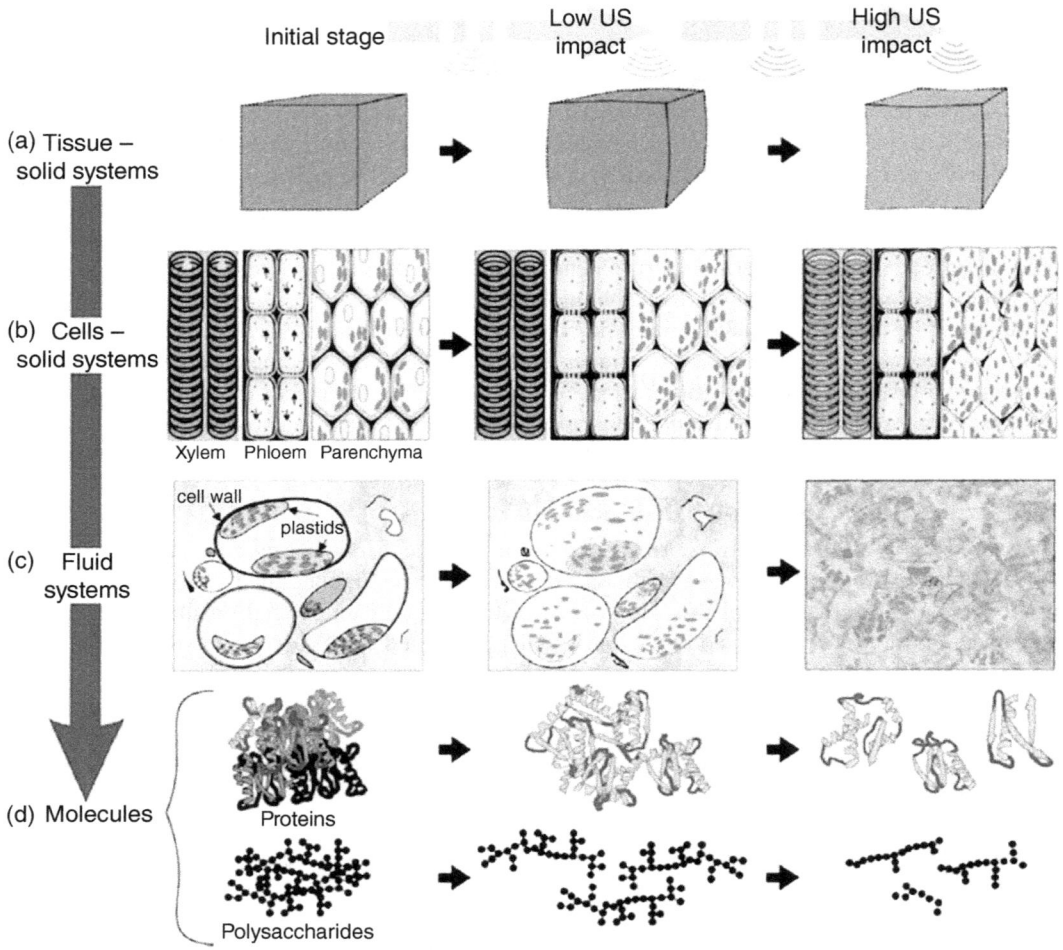

FIGURE 6.2 Effects of ultrasound on the structure of food at different stages of processing and at different levels of structure (from tissue to molecules) in both solid and liquid systems.

Source: Adapted from Rojas et al. 2021.

Astráin-Redín et al. 2020). For instance, it has been shown that drying of solid foods (vegetables, fruits, protein powders, etc.) with the use of ultrasound may increase both the drying process and product quality (Carcel et al. 2011; Ozuna et al. 2014; Rodríguez, Mulet, and Bon 2014; Kahraman et al. 2021; Kahraman and Feng 2021; Zhu et al. 2022b; Kutlu et al. 2022; Yang et al. 2022; Malvandi et al. 2022a; Waghmare et al. 2023).

Although high-intensity ultrasound can provoke localized temperature and pressure changes, oscillating flows, and formation of micro streams at the solid/liquid interface, it is more difficult to apply in gas medium. The power loss that happens when sound waves travel through air and the difference between the acoustic impedances of gases and solids or liquids make it challenging to make ultrasonic systems that work well in a gas medium (Mulet et al. 2003; Astráin-Redín et al. 2020). In addition to the solid, liquid, or gas state of treatment media, cavitation can be influenced by many factors, including solid particles, gas bubbles, and viscosity. The solid particles in a fluid act as nucleation sites, which promote the formation of bubbles and reduce the consequences of cavitation. When it comes to the viscosity of the medium, although bubble production becomes more challenging as the viscosity of the media increases, the implosion itself becomes more potent (Feng and Lee 2011; Astráin-Redín et al. 2020; Farzad et al. 2020; Malvandi et al. 2022b).

6.3   EFFECT OF HIGH-INTENSITY ULTRASOUND ON PLANT-BASED PROTEIN MODIFICATION

Ultrasound technology has attracted substantial attention recently and is increasingly being used in the food industry, especially for modifying plant proteins. Ultrasound-induced plant-protein modification is often attributed to acoustic cavitation. Additionally, phenomena such as high shear by micro- and macro-streaming, shock waves, and water jets caused by cavitation can alter the molecular structure of a protein (Kentish and Ashokkumar 2011). In the following sections, we will discuss how ultrasound modifies various plant-protein properties in detail.

6.3.1   PHYSICOCHEMICAL MODIFICATIONS OF PLANT-BASED PROTEINS BY HIGH-INTENSITY ULTRASOUND

The modification of protein functional properties is achieved by the ultrasonication of a protein solution. In this section, we will review how ultrasound can affect the primary, secondary, tertiary, and quaternary structures of proteins in such a treatment.

6.3.1.1   Modification of Primary Structure

The primary structure of a protein, which is the simple level of protein structure, can be defined as the linear sequence of amino acids linked together via peptide bonds (formed when amino group of one amino acid bonds with the carboxyl group of an adjacent amino acid) to form a polypeptide chain (Figure 6.3). Polyacrylamide gel electrophoresis (PAGE) is the most used method to study

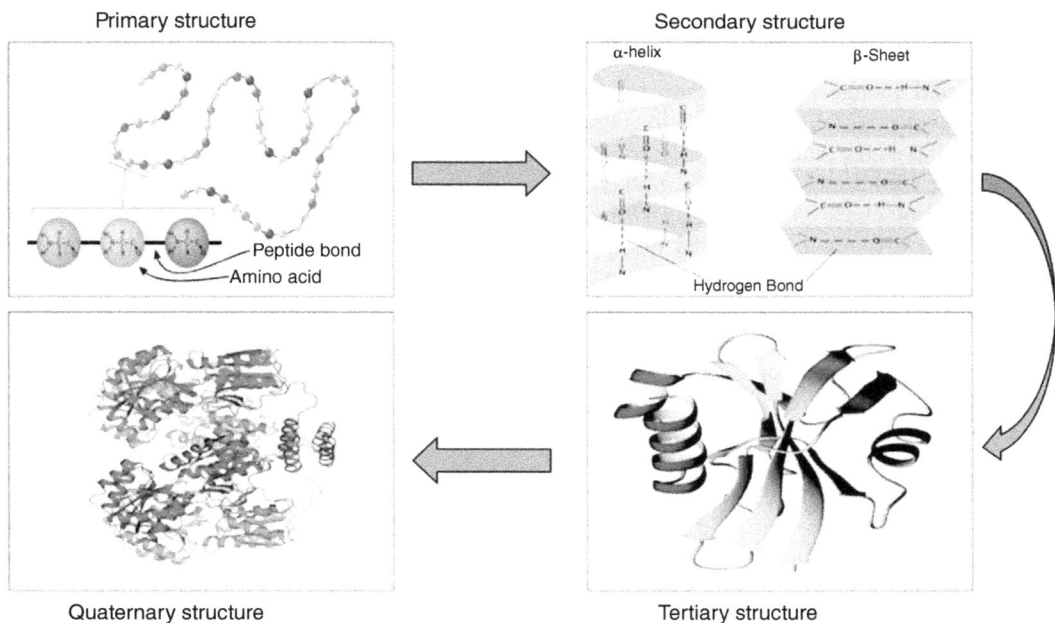

FIGURE 6.3 Schematic illustration of protein conformational levels.

Source: Adapted from Ampofo and Ngadi 2022.

proteins' primary structure. Ultrasonication can or cannot affect the primary structure of plant proteins depending on the protein type and treatment conditions. For example, Zhang et al. (2011) reported no significant changes in the protein's electrophoretic patterns after treating wheat glutens with ultrasound. The reason behind this observation was suggested to be the high viscoelasticity and strong mechanical tolerance of wheat glutens. Rahman et al. (2020) also reported no changes in the electrophoretic bands of soy protein isolates and flakes due to sonication treatment. They suggested that either the energy levels during selected ultrasound processing parameters were not high enough to modify the primary structure of the protein or radical-induced re-aggregation occurred during or right after the sonication.

However, the primary structure of proteins from other plant sources has been reported to reduce the molecular weight in the PAGE profile after an ultrasonication treatment. For example, Mir, Riar, and Singh (2019) reported that sonication induced significant changes in the SDS (Sodium Dodecyl Sulfate) pattern of album protein isolates due to the cavitational forces, micro-streaming, and turbulent forces produced during the process. Resendiz-Vazquez et al. (2017) reported similar findings on protein isolates from jack seed fruits. Lee et al. (2016), who worked on modifying pea protein by pH-shifting and sonication, reported an increase in the acidic polypeptide bands of pH 2- and pH 12-shifting combined sonicated samples. In contrast, sonication-only samples showed protein bands similar to native pea proteins. Therefore, these inconsistencies in results can arise due to variances in the source and type of protein and the acoustic powder density and treatment duration used in the sonication experiment.

6.3.1.2 Modification of Secondary Structure

The secondary structure of the protein refers to the regular and recurring arrangements arising from the hydrogen bonds formed between atoms of the polypeptide chain (as seen in Figure 6.3). The major secondary structural elements are alpha helices, beta sheets, beta turns, and random coils. Circular dichroism (CD) and Fourier transform infrared spectroscopy (FTIR) can be complementary methods for analyzing protein secondary structures based on their characteristic spectra.

Based on a recent study by Liu et al. (2022), α-helix decreased, and the proportion of β-sheet and γ-random increased, indicating the alteration of the secondary structure of hemp protein isolate (HPI) after ultrasound treatment (Figure 6.4). Similar observations were made by Hu et al. (2013a) who reported changes in secondary structures due to ultrasonic treatment of soy protein isolate (SPI). Liu et al. (2022) also reported that ultrasonication for an extended time could lead to an excessive alteration of the secondary structure of HPI due to the formation of the higher portion of the random coil. In this case, ultrasonication is likely to destabilize the native structure of proteins due to the cavitation phenomenon leading to secondary structure changes. Byanju et al. (2020) observed no significant difference in the contents of secondary structural elements in the case of soy flakes protein, soy flour protein, and pea protein; however, in the case of kidney bean protein, the β-strands and unordered structures (random coil) were reported to decrease and increase respectively when sonication was applied for extraction. Overall, ultrasound-induced changes in the secondary structure of a protein can depend on several factors, including but not limited to sources of proteins, methods of extraction, and ultrasonic treatment time and power.

6.3.1.3 Modification of Tertiary Structure

The tertiary structure of a protein refers to the overall 3D folding of proteins arising majorly from the interaction among amino acid R-groups and the external environment. Changes in 3D protein conformation and molecular structure can be determined by the intrinsic fluorescence of tryptophan (Trp), tyrosine (Try), and phenylalanine (Phe) amino acid residues and particularly Trp residue. Liu et al. (2022) reported that the fluorescence intensity of HPI decreased significantly after an ultrasound treatment (Figure 6.4), which was consistent with another study on hemp protein isolate (Wang et al. 2019). Zhou et al. (2013) reported that ultrasound treatment of wheat germ protein using a horn reactor decreased the fluorescence intensity (FI) compared with the control at low

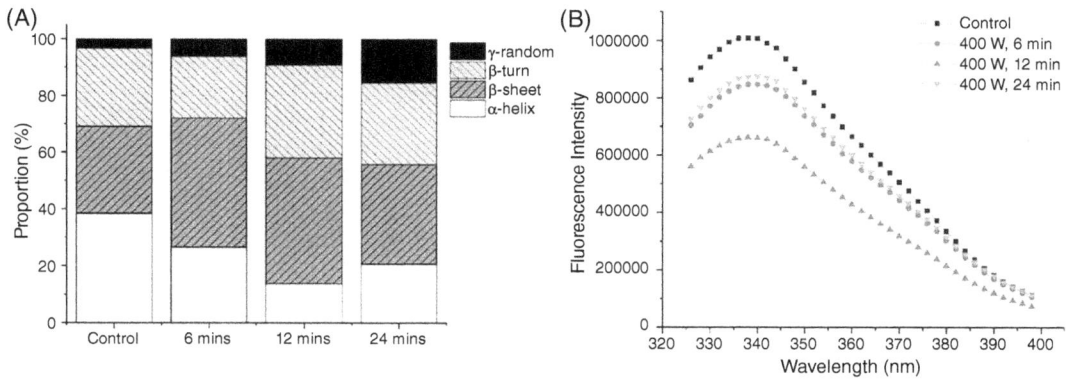

FIGURE 6.4 The secondary and tertiary structure results of hemp protein isolate (HPI) without or with HIU treatment for different times (6/12/24 min at 400 W). (a) Secondary structure and (b) tertiary structure.

Source: Adapted from Liu et al. 2022.

power. However, they also indicated that ultrasound treatment significantly increased the FI as the ultrasonic power intensity was increased. Similar observations were made by Jiang et al. (2014a), who found that black bean protein isolates undergoing ultrasonic treatments at low and medium power reduced the FI, and ultrasonication at high power increased the FI. They found that ultrasonication destroyed internal hydrophobic interactions and exposed them to the surface of the molecules dictated by the bathochromic shift. One study also found that single sonication step had no effect on the FI of treated soy protein isolate (SPI). However, acid treatment before sonication resulted in an increase in FI (Huang et al. 2017). This is due to molecular unfolding under acidic conditions (Huang et al. 2017; Keerati-u-rai et al. 2012). Since the disulfide bonds are the primary bonds stabilizing the tertiary structure in proteins' conformation, the modification in the tertiary structure after ultrasonication has also been associated with the change in free sulfhydryl (S–H) group content (Karabulut, Feng, and Yemiş 2022).

6.3.1.4 Modification of Quaternary Structure

The quaternary structure is formed by the interaction of different polypeptide chains, often called subunits, and ultimately forms functional protein complexes (Figure 6.3). The state of aggregation or dissociation of protein molecules usually reflects changes in the quaternary structure of a protein. They can be determined by measuring their particle size (laser diffraction particle size analyzer) or charge (Zeta potential (ζ)) in solution. Due to its cavitation and microstreaming, ultrasound can affect protein quaternary structure and produce smaller molecular subunits. Jiang et al. (2014a) reported that when the power of the ultrasonic treatment is increased, the particles become smaller, and the particle size distribution broadens. The underlying phenomenon behind this observation was assumed to the cavitational forces which caused these unstable aggregates to break up into smaller soluble protein aggregates.

Zeta potential (ζ) is the electrical potential difference between the stationary layer of the particle and the dispersion medium, which plays a crucial role in stabilizing the particle in dispersion. It is measured by the presence of charged amino acids, which affect the dispersion and aggregation of protein (Rahman and Lamsal 2021). The more significant zeta potential implies better system stability due to the presence of repulsive force among the dispersed protein molecules (Kamani, Semwal, and Meera 2021). Kamani, Semwal, and Meera (2021) studied the effects of ultrasonication treatment on the zeta potential of legume (black gram) protein and found more negatively charged amino acids than positively charged amino acids, decreasing the absolute zeta potential. However, they also found that the zeta potential increased over a prolonged period when ultrasonication power

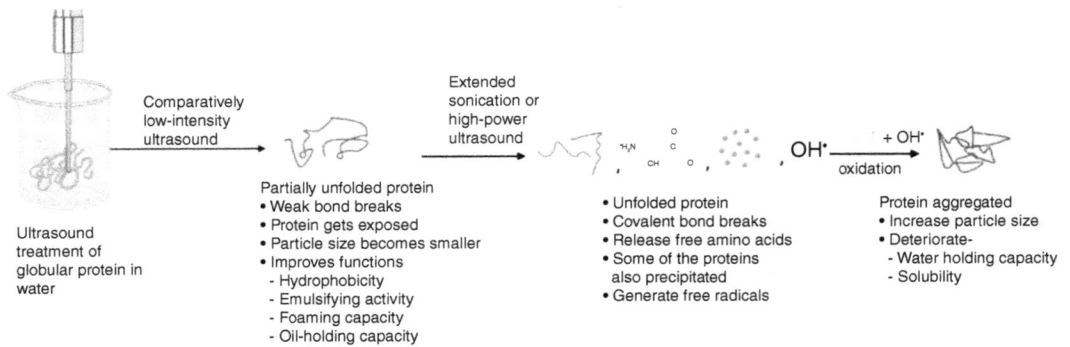

FIGURE 6.5 Schematic diagram of the effect of HIU on the structure and functions of proteins.

Source: Adapted from Rahman and Lamsal 2021.

was high. Similar observations were made by Jiang et al. (2014b) while studying the effect of ultrasonication on the zeta potential of black bean protein isolates.

6.3.2 MODIFICATION OF FUNCTIONAL PROPERTIES OF PLANT-BASED PROTEINS BY HIGH-INTENSITY ULTRASOUND

Due to the rising veganism, consumers purchase plant-based foods in increasing numbers. In such food systems, plant proteins can impact the appearance, texture, taste, color, and flavor of foods due to their physicochemical and functional properties such as solubility, emulsifying ability, foam-forming ability, gel-forming ability, and ability to hold/retain oil or water (Figure 6.5). In the following section, we will discuss some of these functional properties and how they can be modified using ultrasonication.

6.3.2.1 Solubility

Solubility determines protein's ability to be soluble or insoluble in water. This is a very fundamental protein property that has important connotations and can serve as a prerequisite for other structural and functional properties. Protein with a higher solubility can be used in applications such as beverages or soups. The solubility of a protein is usually measured as a function of pH using various methods where the amount of nitrogen in the supernatant is determined after centrifugation.

Unlike most animal proteins, plant proteins have poor solubility due to a higher globulin fraction, which can limit their application in various food and beverage systems. Physical methods such as ultrasonic treatment are effective techniques for improving solubility in the case of various plant proteins, including soy protein isolate (SPI) (Hu et al. 2013b; Jambrak et al. 2009), pea protein isolate (PPI) (Jiang et al. 2017), faba bean protein isolate (FBPI) (Martínez-Velasco et al. 2018), black bean protein isolate (BBPI) (Jiang et al. 2014a), and hemp seed protein isolate (HPI) (Karabulut, Feng, and Yemiş 2022; Karabulut and Yemiş 2022). Ultrasonic cavitation activity that results in favorable conformational/structural changes and exposure of hydrophilic amino acid residues is often ascribed for enhancing the solubility of proteins (Tang et al. 2009). Additionally, ultrasonic treatment can reduce the particle size of treated protein, which increases protein–water interactions, thus resulting in increased protein solubility (Arzeni et al. 2012). Furthermore, Hu et al. (2015) observed that an increase in sonication time and intensity or power (W) up to an optimal value could improve the solubility of the 7S fraction of SPI. Some studies also reported that combining ultrasound treatment with other chemical treatments like pH shifting could improve plant protein solubility (Jiang et al. 2014b, 2017; Kahraman, Petersen, and Fields 2022).

6.3.2.2   Water/Oil Absorption Capacity

The water absorption capacity (WAC) of a protein (water-binding capacity or water-holding capacity) measures the total amount of water absorbed by the three-dimensional structure of a protein. For example, plant protein with a high water-binding capacity could be used in a cake or pizza dough. Similarly, oil/fat absorption capacity (OAC) (oil/fat-binding capacity or oil/fat holding capacity) measures the total amount of oil that can be absorbed into the 3D structure of a protein. This ability is essential for flavor retention and can help improve the palatability of the final product, especially in the production of alternate protein products or animal meat analogs. Both properties are critical as they directly indicate the interaction of protein molecules with water/oil in the solvent. Among various methods, the most common method to evaluate WAC/OAC is to assess the amount of water/oil expelled when protein samples are subjected to external forces (e.g., pressing or centrifugation). The WAC/OAC of 1 g of protein powder is calculated as given in Eq. (6.2):

$$WAC/OAC\,(\%) = \frac{\text{Weight of water / Oil held by sample after centrifugation}}{\text{Initial weight of sample}} \times 100 \quad (6.2)$$

Paglarini, Martini, and Pollonio (2019) studied the effect of sonication and heat treatment and reported a significant increase in the OAC of soy protein isolate. Sonication treatment results in the exposure of more hydrophobic groups that enhance oil entrapment by protein molecules. On the contrary, some studies also reported improvement in the WAC due to high-intensity ultrasonication of SPI cold set gels (Hu et al. 2013a; Zhang et al. 2016). This observation was attributed to increased protein solubility and the reduced size of protein particles. Generally, WAC and OAC are inversely correlated, i.e., when the WAC of protein decreases, OAC increases. Partial unfolding and denaturation of proteins during the sonication process promote proper network formation to bind water/oil droplets.

6.3.2.3   Gelling

A protein gel is a three-dimensional network formed by unfolding, associating, and aggregating the unfolded protein molecules via protein–protein interactions. Protein's ability to form gels makes it an excellent thickening and texturizing agent in food applications such as puddings, and jellies. The gelation properties or strength of gels formed using plant proteins can be determined by either a texture analyzer or a rheometer.

As mentioned in the sections before, the application of ultrasonic energy leads to partial unfolding and aggregation of plant proteins, which could further lead to increased gelation of the proteins. Hu et al. (2013a) in their study proposed that sonication causes a reduction in particle size and results in partial unfolding of SPII, which enhances surface hydrophobicity and increases sulfhydryl groups. The increase in surface hydrophobicity and sulfhydryl groups produces a dense and uniform gel network via intermolecular hydrophobic interactions and disulfide bonds. Similar observations were made by Xue et al. (2018) regarding gels formed by plum seed protein. They inferred that the increase in gel strength (from 5% to 28%) and consequent formation of compact and uniform gel networks were due to an increase in solubility and a decrease in particle size of plum seed protein following power ultrasound treatment.

6.3.2.4   Foaming

A foam, in its most basic definition, is a dispersion of gas in a second, continuous phase (Murray and Ettelaie 2004; Zayas 1997). The ability of protein to form a stable foam is essential in the production of a variety of foods, including ice creams, dairy creamers, whipped cream, etc., The foaming capacity of a protein is measured as the amount of interfacial area that can be created by whipping the protein. Foam stability is calculated in terms of foam half-life, i.e., the time required to lose 50% of the liquid contained in foam (Mauer 2003). A protein's ability to form stable foams depends on various functional properties, including solubility, molecular flexibility, surface charge, and hydrophobicity, which is in turn depend on proteins' intrinsic features (size, shape, amino acid composition) and extrinsic parameters (e.g., pH, ionic strength, and temperature of the medium). Higher protein solubility and molecular flexibility (i.e., the ability of a protein to rapidly change its conformation upon transfer from one environment to another) are two critical properties that are widely associated with high surface activity and thus enhance foam formation (He et al. 2021).

Physical treatments like ultrasonication have been shown to improve the foaming properties of proteins owing to their ability to increase solubility and decrease protein size, which further leads to increased protein adsorption at the air–water interface. Additionally, ultrasound-assisted partial denaturation has increased protein interaction at the interface by exposing more hydrophobic groups (Jambrak et al. 2009; Malik, Sharma, and Saini 2017). The increased surface hydrophobicity indicates that ultrasound mediates some degree of unfolding in the protein molecule's structure, which can be correlated with the improvement in foaming properties. Li et al. 2020) showed that the foaming capacity of commercial soy protein isolate could be enhanced by up to 33% via ultrasonic cavitation (20 kHz, 550 W, 35°C, 5–30 min). However, some studies also found that foaming capacity only increased up to a specific treatment time. If more power is input for an extended period, proteins tend to form larger aggregates and protect hydrophobic groups resulting in decreased foaming capacity (Malik, Sharma, and Saini 2017; Mir, Riar, and Singh 2019).

6.3.2.5 Emulsifying

Due to proteins' amphipathic nature, they can act as emulsifiers by forming a film around oil droplets dispersed in an aqueous medium to form and stabilize oil-in-water emulsions. Emulsifying activity/stability (EA/ES) is the ability of emulsifiers to form and stabilize tiny droplets. Ultrasonication affects the emulsion properties of many plant-based food proteins, as investigated by researchers. O'Sullivan, Park, and Beevers (2016) investigated the effect of ultrasound on the emulsifying performance of wheat and soy protein isolate and reported that ultrasonic-induced reduction in the protein size increases the surface-area-to-volume ratio to enhance emulsifying properties of proteins of plant origin. Similarly, Karabulut and Yemiş (2022) reported increased emulsion capacity and stability of hemp protein isolate following an ultrasound treatment. They attributed this observation to an increased exposure of hydrophobic groups of protein molecules due to sonication. Surface hydrophobicity of protein has a well-proven linear correlation with emulsifying properties as it indicates the structural orientation and the number of hydrophobic groups on the protein surface of the protein molecule.

Nazari et al. (2018) found no significant improvement in emulsion properties of millet protein concentration as sonication time increased from 5 to 12.5 min (18.4 W/cm^2 intensity). However, a further increase in time from 12.5 to 20 min led to a significant improvement due to an increase in proteins' tendency to adsorb at the oil and water interfaces aligned with higher solubility results, as well as an increase in exposure of the buried hydrophobic groups. They also found that an increase in sonication time from 12.5 to 20 min at a higher intensity (73.95 W/cm^2) caused a significant decrease in the emulsion properties related to the more extensive aggregate formation through oxidation. Sun et al. (2021) showed that ultrasound could also help improve the extraction properties of plant proteins. They studied the ultrasound-assisted extraction versus traditional alkaline extraction method on the emulsifying properties of peanut protein isolate. They concluded that ultrasound treatment, due to its ability to induce conformational changes to modify the interfacial association between protein–oil phases, can help improve the emulsifying properties of peanut protein.

6.3.2.6 Viscosity

Viscosity represents the resistance of a fluid to flow and depends on the interactions between components. Physical properties like the viscosity of proteins are critical in the food industry as it helps predict the sensory properties of food or beverage formulations. The intrinsic viscosity of a protein solution also indicates the degree of hydration of proteins. It provides information about the associated hydrodynamic volume related to the structural molecular conformation of proteins in solution

(Behrouzian, Razavi, and Karazhiyan 2014). O'Sullivan et al. (2016) studied the effect of ultrasonication on the intrinsic viscosity of vegetable proteins and demonstrated that ultrasound treatment induced a significant reduction in the intrinsic viscosity of pea and soy protein isolate in solution and, consequently, a significant reduction in the hydrodynamic volume occupied by the proteins and the solvents entrained within them. Similar observations were made by other researchers where sonication of a plant protein solution resulted in a substantial reduction in the viscosity (Martínez-Velasco et al. 2018). Ashokkumar (2015) suggested that the physical shear forces generated during acoustic cavitation are responsible for viscosity reduction. Moreover, Ashokkumar (2015) also demonstrated that low-frequency ultrasound could be more effective in lowering the viscosity than high-frequency ultrasound because the strong cavitation activity associated with low-frequency ultrasonication generates strong shear forces compared to that from a high-frequency ultrasound treatment, which generates a significant number of radicals.

6.3.3 Modifications of Nutritional Properties of Plant-Based Proteins by High-Intensity Ultrasound

6.3.3.1 Digestibility

Despite their rich protein contents, most plant proteins have a low proteolytic digestibility due to low solubility, rigid disulfide-bonded structures, protein–phenolic and protein–carbohydrates interactions, high fiber content, and, in some cases, antinutritional components (such as trypsin inhibitors). Low digestibility limits the utilization of plant protein sources in industrial applications (Karabulut and Yemiş 2022; Pan et al. 2020). Ultrasonication could be an emerging method for increasing the digestibility of plant proteins under mild conditions and more sustainable ways, compared to chemical and thermal treatments. A number of studies have reported the enhancement of digestibility of plant proteins after treatment with ultrasound. In a recent study, rapeseed protein treated by HIU (10–70% amplitude, 1,200 W, 20 kHz) showed an increase in its digestibility when the sound amplitude was between 10% and 40% (Pan et al. 2020). The maximum digestibility obtained at an amplitude of 40% was attributed to the high shear, micro streaming, and pressure effects, resulting in unfolding protein interactions. However, when the ultrasound amplitude was greater than 40%, a reduced protein digestibility was observed, presumably caused by disulfide-bonded macro aggregates. In another study, the in-vitro digestibility of soy protein increased by 12% and reached 97% after a sonication treatment (30% amplitude for 10 min) (Khatkar, Kaur, and Khatkar 2020). The authors emphasized that ultrasound treatment induced the unfolding of protein by cavitation-related forces and produced more accessible sites for hydrolysis with digestive enzymes. In contrast, Vanga et al. (2020) reported no notable changes in the digestibility of ultrasound-treated almond milk proteins. The authors associated these findings with an increase in the ordered structures of α-helix and β-sheet after ultrasound treatment of almond milk protein. This may indicate that the use of ultrasound for improving the digestibility of plant proteins should be performed under suitable process conditions.

6.3.3.2 Bioactivity

The bioactivity of plant protein sources, such as bioactive peptides, has been a topic for a number of recent studies. Bioactive peptides, known as digested protein forms, have shown great potential in pharmaceutical applications for treating chronic diseases, hormones, and neurotransmitters to cancer. The significant challenges of producing bioactive peptides arise from their low solubility, low yield, excessive costs due to the enzymes, prolonged process times, and immunogenicity. Research efforts have been made to attack the drawbacks of bioactive peptides using ultrasound-assisted procedures. For example, the yield of ACE (angiotensin-I-converting enzyme) inhibitor peptides from wheat germ proteins were enhanced by 22% using a high intensity ultrasound (HIU) pretreatment (Jia et al. 2010). Ultrasound pretreatment also promoted the ACE inhibitor activity of wheat germ proteins compared to the conventional procedure. Fadimu et al. (2022) reported that the HIU pre-treatment increased the antidiabetic and ACE inhibitory activity of lupin protein hydrolysates prepared by Protamex enzyme. Low molecular weight peptides showed a higher concentration ratio in the peptide mixture after the HIU treatment. The higher bioactive potential of lupin protein hydrolysates was attributed to the transformations of amino acid composition after sonication. Similar results were found in rapeseed protein hydrolysates treated by HIU before enzymatic hydrolysis. Ultrasonication increased the degree of hydrolysis and ACE inhibitory activity over the non-HIU samples. The HIU at 600 W for 12 min recorded the highest ACE inhibitory activity of 72%. The authors suggested that the heat, cavitation, pressure, and shear effects of sonication make proteins accessible to enzymatic attacks and reduce the particle sizes more, leading to the production of bioactive peptides. The increase in ACE inhibitory activity was also attributed to hydrophobicity promoted by HIU pretreatment. In a similar study, the content of antioxidant peptides of peanut and wheat proteins was increased by ultrasound-pretreated alcalase hydrolysis (Zhu et al. 2011; Yu et al. 2012). HIU treatments increased the radical scavenging capacity of corn protein hydrolysates due to the formation of low-molecular-weight peptides containing aromatic amino acids (Liang et al. 2017). HIU treatment promotes the antioxidant activity of proteins by exposing the active groups inside the compact protein aggregates (Karabulut, Feng, and Yemis 2022; Karabulut and Yemiş 2022). The hemp seed protein showed a higher antioxidant potential after HIU treatment which could be associated with the higher free SH group and aromatic amino acid contents.

6.3.3.3 Allergen Reduction

Modification of protein structures can influence the natural allergens of proteins. Recent studies showed that ultrasonication and its combination with other methods could reduce allergen values. In the study of Li et al. (2013), the allergenicity of peanut proteins, such as Ara h 1 and Ara h 2, was reduced using HIU-enzyme-assisted treatments. The combined HIU + enzyme method was more efficient in lowering the allergen levels than ultrasonic treatments alone. The reduction of the allergens was attributed to the molecular destruction by the combined treatments. The microjets and shear forces due to acoustic cavitation led to alteration in the secondary, tertiary, and quaternary protein structures, causing the reduction of allergens. The other well-known allergen, i.e., soy protein allergens, was reduced by HIU pretreatments before germination. Ultrasonic treatment at 300 W before germination reduced the allergenicity of protein by 51% (Yang et al. 2015). In another study, the allergens in kiwi fruit protein were decreased by 50% after HIU treatment at 20 kHz, 400 W for 16 min. Additionally, ultrasonication was reported to decrease the level of antinutritional factors in proteins in oilseeds, legumes, etc. Treatment of soy protein with HIU at 20 kHz and 355 kHz reduced the trypsin-inhibitory activities (Wu et al. 2021).

6.4 HIGH-INTENSITY ULTRASOUND COMBINED TECHNIQUES

6.4.1 PROTEIN MODIFICATION BY COMBINATIONS OF SONICATION AND CHEMICAL METHODS

6.4.1.1 High-Intensity Ultrasound-Promoted Glycosylation

Glycosylation (Maillard reaction) is a non-enzymatic covalent reaction between the carbonyl group of polysaccharides and amino groups of proteins, which was first described by Louis-Camille Maillard in 1921. In the last decades, several authors have reported how Maillard reaction was used to modify plant protein properties (Ma et al. 2020). Two common methods are employed to obtain Maillard conjugates, including dry-base and wet-base procedures (Jiang et al. 2022). From an industrial view, both approaches are time consuming (up to days or weeks) and have several adverse effects on the conjugates, such as protein aggregation, low yield, browning, and melanoidin formation (Chen et al. 2022a). Therefore, ultrasound-assisted Maillard procedures, as a novel, emerging, and green approach, have been explored for improving conjugation and modification efficiency in recent years, including sonication as a pretreatment or sonication-assisted procedures (Zhao et al. 2021). In a study by Chen et al. (2022b), rice protein–dextran conjugates were prepared using HIU-assisted glycation reaction at optimum conditions (82°C, 600 W, 22 min) and compared with the classical wet base procedure. Ultrasound-treated samples had a grafting degree 2.08–fold higher than the classical one, which was attributed to cavitation-related forces that unfold protein structures, expose active sites, and accelerate glycation reaction. Ultrasound-assisted glycation might lead to a more flexible protein structure than the classical procedure by orienting hydrophobic groups and producing smaller particle sizes. The solubility of conjugates reached 90.6%, 20.63-fold higher than that of rice protein produced from classical wet base procedure. In a recent study by Chen et al. (2022a), pea protein–arabinose Maillard conjugates were prepared with different power–time combinations of sonication. The degree of glycation was higher when treated with 150 W for 30 min compared with that from the classical procedure. The emulsion activity and thermal stability of pea protein were improved with the sonication-assisted glycation. They associated the improved functional properties with the more abundant free SH group, smaller particle sizes, and surface hydrophobicity of sonication-assisted conjugates. As Jiang et al. (2022) reported, the combination of pH-shifting and sonication-assisted treatments on pea protein–inulin complexes (80°C, 400 W, 25 min) had a glycation rate 2.3 times of that in the classical wet base method. Similarly, Ma et al. (2020) reported that the glycation times of soy protein–pectin conjugates prepared using an ultrasound-assisted procedure (450 W, 70°C, 45–60 min) were reduced to 15 min, while the classical wet base procedure took 20 h to achieve the same degree of grafting. An increase in the proportion of random coil forms and formation of more flexible and looser protein structures after sonication-assisted glycation were believed to be resulting in improved emulsion properties.

HIU-assisted glycation also reduced the undesired browning of soy protein isolate–maltodextrin conjugates, while ultrasonic forces accelerated the glycation rate and the production of Maillard intermediates (Zhao et al. 2021). Therefore, sonication-assisted glycation under optimized conditions has become more attractive for industrial applications as it provides short processing time, high yield, and advanced techno-functional properties compared to conventional procedures.

6.4.1.2   High-Intensity Ultrasound-Promoted Coacervation

Plant proteins and polysaccharides can form complexes through electrostatic interactions apart from glycation to enhance the techno-functional properties of plant proteins. In the process, the positively charged protein below the isoelectric point interacts with negatively charged polysaccharides (Wang et al. 2021a). However, the protein–polysaccharide complexes produced by traditional method are unstable because of the aggregation tendency, low protein solubility, and large particle sizes.

Recently, Yildiz et al. (2018) reported the complexation of positively charged pea protein and negatively charged three types of polysaccharides (modified starch, gum arabic, and pectin) using ultrasonication. They found that sonication improved the solubility near the isoelectric point and produced more stable emulsions by reducing the particle sizes and breakdown of the protein interactions. Pea protein-modified starch complexes presented better functional properties than other complexes. In addition, the researchers reported more efficient microencapsulation of docosahexaenoic acid using the pea protein-modified starch complexes prepared by sonication-assisted procedure. The pea protein-modified starch capsules reduced the peroxide values of docosahexaenoic acid. Similarly, the soy protein–pectin complexes were produced by a different power of sonication (150–600 W for 15 min) (Wang et al. 2021b). The treatment at 450 W showed smaller sizes and better solubility than other power levels and traditional procedures, yielding smaller and soluble complexes. Another study on soy protein–pectin complexes by sonication reported increases in zeta potential and surface hydrophobicity of the complexes but reductions in particle size and turbidity (Figure 6.6, Ma et al. 2019). As a result, sonication-assisted coacervation of proteins and polysaccharides promotes electrostatic interaction through increased collisions between macromolecules, reduces particle sizes, and enhances functionality, showing more stable and soluble complexes than conventional methods.

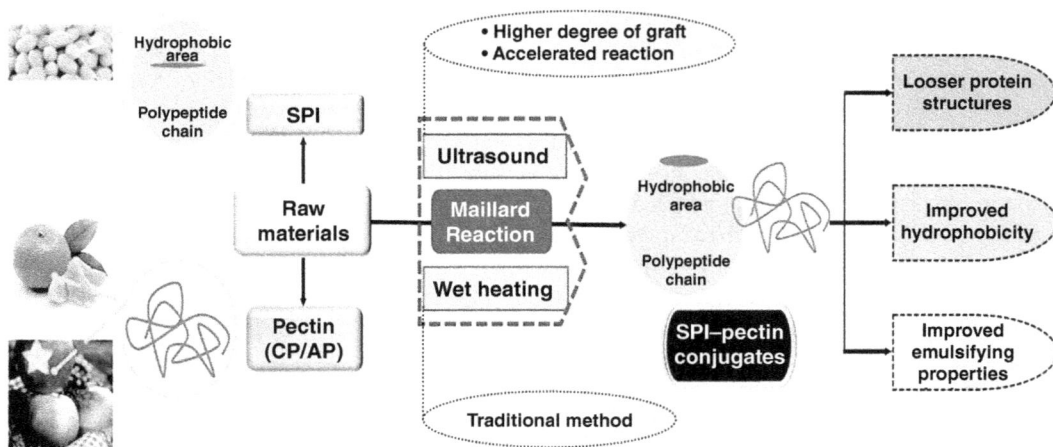

FIGURE 6.6 Schematic diagram of the effect of sonication-combined coacervation on the structure and functions of soy protein–pectin conjugates.

Source: Adapted from Ma et al. 2019.

6.4.1.3   High-Intensity Ultrasound-Promoted pH-Shifting

pH-shifting is a simple chemical modification method laid on the unfolding–refolding process under extreme pH environments. The reorganized protein aggregates with a partial loss of tertiary structures are named as "molten globule forms" (Lee et al. 2016). Recent studies have showed that the HIU and pH-shifting combined processes improved the modulation effects synergistically on protein structures. For instance, Lee et al. (2016) reported the first attempt on pH-shifting-assisted HIU treatments on soy protein. They found that alkaline pH-shifting-sonication combinations promoted functional properties, such as solubility, by making molten globule protein forms more susceptible to sonication. The combined pH-shifting and sonication treatment increased the free SH and hydrophobic group contents by disrupting intermolecular interactions while decreasing the particle sizes of pea protein (Jiang et al. 2017). In another study, the combined manothermo-sonication (MTS) and pH-shifting treatments boosted emulsion properties, bioactive encapsulation capabilities, and rheology more than other treatments (Yildiz et al. 2017). The pH-shifting and sonication-combined treatments resulted in a greater gastric enzyme sensitivity attributable to increased disordered random coils forms of partially denatured protein relative to treatments alone and native forms (Figueroa-González et al. 2022). Wang et al. (2022a) reported the effect of soni-cation and pH 2- and 12-shifting on chickpea protein isolate. The pH-shifting and HIU-combined treatments enhanced the foaming properties more than pH-shifting-alone treatment. At the same time, pH-12 shifting + sonication treatment recorded the highest solubility and smallest particle sizes. The authors attributed the enhanced foam properties to reduced interfacial tension by HIU treatment. The secondary structural changes are associated with the formation of stable interfacial areas. Silventoinen and Sozer (2020) applied sonication and pH-shifting + sonication treatments to barley protein isolate to improve its techno-functional properties. Especially, alkaline pH-shifting + sonication treatment promoted a more stable suspension with higher solubility and smaller particle sizes than pH-shifting-alone and acidic pH-shifting + sonication treatments. The researchers high-lighted that the mechanical effect of long-time sonication resulted in the breakage of non-covalent bonds and electrostatic interactions between protein molecules. Similar findings were reported by Alavi, Chen, and Emam-Djomeh (2021) in the study of pH-shifting + sonication-treated faba bean protein. The emulsions of pH-shifting + sonication-treated faba bean protein showed smaller sizes and higher stability than the pH-shifting-alone treatment. The primary structures of Faba bean pro-tein in the SDS-PAGE patterns remained unchanged after pH-shifting and pH-shifting + sonication treatments. The solubility of Faba bean protein after pH-shifting + sonication treatment increased from 12% to 40% at pH 3 and 7. The pH-shifting-sonication-combined treatment offers conveniences for plant protein modifications; however, some drawbacks related to chemical side effects need further evaluation.

6.4.2 Protein Modification by Combinations of Sonication and Enzymatic Methods

The mechanism of enzymatic hydrolysis, a method of biological modification, is based on the cleavage of inner and intermolecular bonds and reduction of molecular weights, thus enhancing functional properties of proteins. The combined effects of enzyme and sonication treatments could promote mass transfer, increase the contact sites between protein and enzyme, and change protein hydrolysates' techno and biological activities by unfolding compact forms (Wen et al. 2018). For example, enhanced protein extraction yields using combined enzymatic and HIU treatments from industrial by-products such as brewer's spent grain and sesame bran were reported in previous reports (Yu et al. 2020; Görgüç, Bircan, and Yılmaz 2019; Zhao et al. 2019). The ultrasonic waves produce mechanical vibration and localized spots with high temperatures and pressures in the aqueous environment, thus breaking the cell walls of fibrous material and releasing bound proteins. The enzymes could be proteolytic (alcalase, trypsin, pepsin, etc.) or carbohydrase (Viscozyme, peptidase, cellulose, etc.) (Görgüç, Bircan, and Yılmaz 2019). Besides the enhanced protein extraction yield, the obtained protein hydrolyzed from enzymatic-sonication pretreatments presents better solubility values than the treatments applied alone (Yu et al. 2020). In another study, the ultrasonic treatment was applied after soy protein hydrolysis (Tian et al. 2020). The proportion of molecular weights less than 10 kDa significantly increased, thus showing more antioxidant activities. The disruption of peptide aggregates resulted in a better solubility after the sonication and enzyme combinations. The changes in secondary conformation of enzymatic-sonication-treated soy proteins were attributed to an increase in the β-sheet and random coil forms relative to the enzyme-alone treatment. Similar effects were reported in the enzymatic hydrolysis of treated soy proteins by Zhao et al. (2019). The modifications in secondary and tertiary protein forms by cavitational forces exposed the sensitive and large contact sites for enzymatic attacks. The smaller peptide showed more robust antioxidant potentials after sequential treatments of sonication and enzyme.

Enzymatic crosslinking represents another way of the enzymatic modification of plant proteins. Transglutaminase and laccase enzymes are the most common crosslinking agents for improving protein functionality, including gelling, emulsifying, etc. Since the compact forms of plant proteins do not allow the crosslinkage of these types of enzymes, ultrasonication could promote the enzymatic activity by modifying protein forms. For instance, the cross-linked chickpea protein by transglutaminase enzyme assisted with HIU further improved the emulsion properties, oil absorption capacity, and surface hydrophobicity by reducing particle sizes (Zhu et al. 2022a). In addition, sonication catalyzes the transglutaminase crosslinking efficiency and thus promotes the formation of self-supporting hydrogels (Wang et al. 2022b). The gels of sonication-assisted cross-linked protein change the secondary structures and demonstrate more uniform networks by forming ε-(γ-glutamyl) lysine bonds (Cui et al. 2019). The sonication-assisted enzymatic process can be used successfully to overcome the disadvantages of enzymatic hydrolysis, including excessive cost, sensitivity, low yield, and long treatment time.

6.4.3 Protein Modification by Combinations of Sonication and Physical Methods

HIU-alone treatments are often considered insufficient for modifying the functional properties of plant proteins. In many studies, thermosonication has been used as a solution to overcome these limitations by combining heat and sonication. In a previous study, mung bean protein was modified using thermosonication (20 kHz; 30% amplitude; 5, 10, 20, and 30 min; 30, 50, and 70°C),

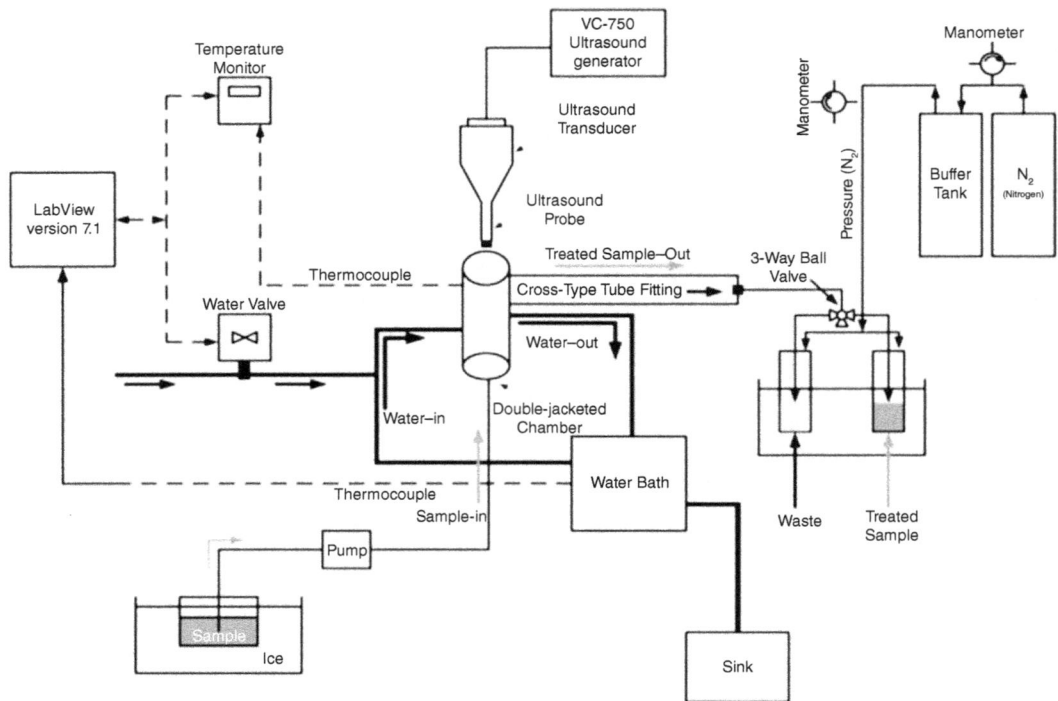

FIGURE 6.7 The manothermosonication system.

Source: Adapted from Yildiz et al. 2017.

which induced an increase in soluble protein content and a decrease in particle sizes more than a heating-alone treatment (Zhong and Xiong 2020). It is held that the active groups in the protein core are exposed to the surface by the combined effects of sonication and temperature, resulting in higher surface hydrophobicity and sulfhydryl groups. Thermosonication at 70°C significantly improved the turbidity and stability of protein solutions. The authors concluded that thermosonication promoted the temperature, cavitation, shear, and bubble implosion effects. At the same time, thermosonication pretreatments boosted the enzymatic accessibility of peanut proteins, as Chen et al. (2022a) reported. The exposure of more unfolded and flexible forms enhanced the proteolysis of peanut proteins, compared to sonication-alone treatments. The bioactive properties of mung and kidney bean protein hydrolyze, such as cholesterol-lowering and antioxidants, were improved using thermosonication by producing more bioactive peptides with higher degrees of hydrolysis than sonication alone and heat-alone treatments (Ashraf et al. 2020). The maximum changes were observed in the secondary and tertiary structures in the thermosonicated protein hydrolyzes.

The combination of low hydrostatic pressure with thermosonication treatments, known as mano-thermosonication (MTS), represents the most effective ultrasonic treatment in liquid media (Feng and Yang 2011). A typical MTS is shown in Figure 6.7 (Yildiz et al. 2017). The MTS system significantly enhanced the cavitation effect of the ultrasonic treatment and decreased the processing time (from minutes to seconds) for the modification of soy protein functional properties (Yildiz et al. 2017). More soluble and smaller soy protein nanoaggregates were produced by MTS treatments (optimum 50°C, 60 s, 200 kPa) compared to the pH-shifting alone and high-pressure homogenization treatments. The emulsion activity and stability of MTS-treated protein were increased due to smaller sizes and higher surface hydrophobicity. The MTS-modified proteins exhibited excellent stability of capsulated canola oil over 21 days at 4°C. In particular, the combined pH-shifting and MTS-treated soy protein nano aggregates showed better functional and capsulation properties than alone treatments. The combination of sonication with mild temperature and/or pressure facilitated by MTS might present an effective way to modify plant protein functional properties.

6.5 CHALLENGES OF ULTRASOUND TREATMENTS

In food applications, ultrasound can be used at low, medium, high, and extreme power/intensity levels. Even though the food industry uses high-intensity ultrasonication the most, it has been linked to protein denaturation and the formation of reactive oxygen species at certain levels of extended treatment times. During a prolonged period of sonication, hydrophobic bonds break, resulting in the unfolding, denaturation, and subsequent modifications to the protein's structure and function. It is essential to keep in mind that the kind of final food system and preferred characteristics that are being targeted for application may or may not benefit from these modifications. In aqueous systems, water molecules can be broken down into their radicals (hydroxyl radicals and hydrogen atoms) by the generated cavitational forces (Weiss, Kristbergsson, and Kjartansson 2011; Rahman et al. 2020). Free radicals produced by ultrasound can:

- alter the structures of the secondary and tertiary proteins
- convert free SH groups into SS bonds
- cause aromatic hydroxylation and the formation of carbonyl groups (Zhu et al. 2018)

Because of this, protein oxidation caused by cavitational stress-generated free radicals can alter the isolated protein's nutritional and technological properties.

The exposure time of the protein substrate to sonication is critical for plant protein applications. The acoustic power intensity and frequency used by the ultrasonic system determine the treatment time (Bhargava et al. 2021). Both low-intensity ultrasound with a long duration and high intensity short time (HIST) ultrasonic treatments may find its own applications in the food processing industries. The vegetable protein business makes use of high intensity or HIST because it has less of an impact on protein denaturation and loss of functional characteristics than low-intensity or low-power/long duration ultrasound does. According to Rahman and Lamsal (2021), ultrasonication increases temperature, which results in protein denaturation if no cooling is used. Because they have a negative impact on protein structure, treatment temperatures are normally controlled at less than 60°C for ultrasound-assisted protein extraction (Rickert et al. 2004). It is well known that an increase in temperature during sonication reduces viscosity, surface tension, and vapor pressure, thereby limiting cavitation and sonochemical effects (Chemat et al. 2017). Some studies used pulses or ice baths to control the temperature during the ultrasound-assisted extraction. To make it easy to use ultrasound for vegetable protein modification, a HIST ultrasound system should be developed that can operate quickly at low temperatures to reduce production costs.

Foaming and uneven acoustic energy distribution can occur when the sonotrode tip is not properly immersed in the medium. Some studies suggested for a more effective treatment that the sonotrode tip should be submerged in a sample to a depth of 2–3 cm (Annandarajah et al. 2018; Chittapalo and Noomhorm 2009). The tip of the sonotrode to the bottom of the sonoreactor is also critical to the distribution of acoustic field and thus the distribution of cavitation activity in the treatment chamber. The geometry of the sonoreactor, which houses the plant protein substrate, is another crucial factor. However, the effects of the sonotrode tip and sonoreactor geometry on the ultrasound-assisted processes applying to plant macromolecules have been the subject of very few studies. To improve the design and optimize the operation of an ultrasound-assisted plant protein modification process, in-depth research in this area is required. With this information, manufacturers can fabricate ultrasonic devices with specific dimensions that are made for plant protein modification without sacrificing the protein's technical and functional properties.

Since protein modification with power ultrasound is mostly performed at high acoustic powder densities, the accompanied high cavitation activities will result in pitting of the titanium sonotrode, normally starting from the tip of the probe. Therefore, titanium metal powders can be detected in the liquid samples after a treatment, and replacement of the titanium ultrasonic probe becomes necessary. This will increase the production cost and will also introduce small amount of metal powders into the food system. To minimize this problem, improved ultrasonic probe geometry design to allow a relatively uniform sound intensity distribution over the sound emitting surface and novel design concepts to contain the cavitation activities in a region away from sound emitting surface have been explored (Dion and Burns 2011).

Because most of published studies on ultrasound application are performed at the laboratory scale, it is difficult for the plant protein industry to optimize key ultrasound equipment parameters at production scale, which affect their design and applications. The scale-up of an ultrasonic process from lab scale to pilot scale and then to commercial production scale requires a good understanding of effect of the sample-to-solvent ratio, temperature, time, the characteristics of the sonotrode, and the frequency/energy intensity on the outcome of the operation. These kinds of knowledge necessitate immediate attention because they are so essential to industrial optimization. To fill in these gaps, researchers and the plant protein industry should collaborate to better comprehend the relationship between structural changes in plant proteins and their subsequent functional properties in relation to ultrasound parameters at the micro/macroscopic level.

6.6 FUTURE TRENDS

To give consumers an impression that they are consuming animal products, alternative animal protein production seeks to imitate their structure, composition, appearance, and flavor. It is difficult to replicate the meat's intricate structure and properties using plant-based ingredients. As a result, the search for plant proteins equivalent to animal proteins in nutrition and function is getting faster. In addition, food technologists working on protein products are looking for processed or structured products made with plant proteins that have the same texture and appearance as meat and desirable organoleptic properties. The production of conventional plant-based alternative protein products uses straightforward processing techniques like fermentation process, chemically protein coagulation, pressing, heating, steaming, cooling, and washing (Malav et al. 2015). The most recent processing techniques include extrusion, shear cell technology, and 3D printing. The pursuit of new protein processing methods and of those for enhancing these processes is still the primary focus.

6.6.1 EXTRUSION

To make textiles from 50% to 70% proteinaceous plant material, extrusion is a common method. Pressure, heat, and mechanical shear are used in this thermomechanical process. Defatted soybean meal, soy protein concentrates and isolates, wheat gluten, pea protein concentrates and isolates, and peanut protein are among the vegetable protein raw materials presently utilized as extrusion raw materials (Kyriakopoulou, Dekkers, and van der Goot 2019). For decades, the extrusion method has been the subject of extensive research. However, one of the most significant obstacles is process control (Zhang et al. 2019), and the extruded product design is so far in its infancy.

Based on how much water is added during the process, there are two types of extrusion methods. High moisture extrusion process (40–80% addition of moisture) and low moisture extrusion process (20–40% addition of moisture). Organized proteins with low moisture content frequently require rehydration before use and are frequently used with additional components. They might not require any additional processing. Water and oil absorption, density, and size/shape are important functional properties of extruded products for formats with low moisture content. The primary raw material, extrusion settings, tool choice, and secondary cutting are all influenced by these properties. Partial flakes that are less dense, like a product that expands too much, will have trouble rehydrating and may even collapse when processed or eaten. They might come across as a hard lump. Preconditioning is an essential first phase in the protein extrusion process to ensure that moisture can evenly penetrate the protein particles in advance of their introduction into the extruder. Proteins melt and denature when exposed to elevated temperatures and pressures by extruders (Zhang et al. 2019), thereby shedding their secondary or tertiary structure. The binding sites that enable the proteins to recombine are revealed when denatured proteins realign in the flow route. The globular plant protein is transformed by this crosslinking into a structure that is more like the fibrous and layered structure of the meat. The mixture quickly loses water due to the elevated temperature and pressure release, which causes the material to expand and form the final puffed shape. The material can also be cut further to get the size and shape of the piece you want. Extrusion can adjust the color and flavor of protein components and result in a structure that resembles meat. Proteins can also gain nutritional value through extrusion.

6.6.2 SHEAR CELL TECHNOLOGY

Recently, researchers at Wageningen University in the Netherlands introduced shear cell technology (Manski et al. 2007). Another way to create a plant-based meat substitute with the mouthfeel and texture of authentic steak is to use heat and shear to create a layered fibrous structure. A shear cell is a shearing device used in this method and can apply substantial shear. Shear cells come in two varieties: A cylindrical Form-Couette cell and a conical cell based on a cone-plate rheometer developed for scale-up processes (Manksi, van der Goot, and Boom 2007). With this technology, the ingredients and processing parameters determine the structure of the finished product. During structuring, the shear cell's well-defined and constant protein deformation results in a low mechanical energy input. As a result, product quality has been less variable with shear cell technology than with extrusion technology (Manksi, van der Goot, and Boom 2007; Krintiras et al. 2016). The device's capacity and throughput are enhanced by increasing the quette cell's length and size. The ability of several vegetable protein combinations to form fibrous structures (soy protein concentrate, soy protein isolate, wheat gluten, or soy protein isolate and pectin) was tested using the shear cell technique (Manksi, van der Goot, and Boom 2007; Dekkers, Nikiforidis, and van der Goot 2016). On the other hand, plant-based shear cells are not available in stores, and the throughput of the device has been a hurdle for its large-scale applications.

6.6.3 3D PRINTING

Three-dimensional (3D) printing is an innovative and adaptable digital technology for rapid prototyping and additive manufacturing. The vegetable paste is used to micro-extrude filaments in the 3D printing to create a muscle-like matrix. The pastes are incorporated into a 3D printer matrix using modeling software called Auto Computer-Aid Design (AutoCAD) (Carrington 2020). NOVAMEAT, one of the food technology enterprises that makes 3D-printed plant-based meat products, has said that it can make steaks that look and feel like meat. These steaks are made from pea protein, rice protein, seaweed, canola fat, and beetroot (Carrington 2020). Food development applications stand to benefit from the speed and adaptability of 3D-printed substrates. Plant-based manufacturers now have more tools to recreate and enhance their products' taste, texture, and dining experience thanks to these evolving technologies. They are laying the groundwork for a new generation of alternative protein nutrition products that will be more adaptable.

6.7 CONCLUSION

The quest for novel and sustainable methods to modify plant-based proteins has stimulated and will continue to drive the development of ultrasound-cased technologies. In the wave of promoting green and low-carbon footprint production, ultrasonication, as an electrification technology, can be foreseen to experience continued development. The commercialization of ultrasound-based technology for plant protein modification repiques a joint effort among the food industry, academia, and ultrasound equipment manufacturers to fully understand the relationship between acoustic cavitation/sonication parameters and protein functional properties, as well as operation and production costs.

REFERENCES

Adekunte, A. O., B. K. Tiwari, P. J. Cullen, A. G. M. Scannell, and C. P. O'Donnell. 2010. "Effect of sonication on colour, ascorbic acid and yeast inactivation in tomato juice." *Food Chemistry* 122(3): 500–507. https://doi.org/10.1016/j.foodchem.2010.01.026.

Alavi, F., L. Chen, and Z. Emam-Djomeh. 2021. "Effect of ultrasound-assisted alkaline treatment on functional property modifications of faba bean protein." *Food Chemistry* 354: 129494. https://doi.org/10.1016/j.foodchem.2021.129494.

Ampofo, J., and M. Ngadi. 2022. "Ultrasound-assisted processing: Science, technology and challenges for the plant-based protein industry." *Ultrasonics Sonochemistry* 84: 105955. https://doi.org/10.1016/j.ultsonch.2022.105955.

Annandarajah, C., D. Grewell, J. N. Talbert, D. R. Raman, and S. Clark. 2018. "Batch thermosonication for the reduction of plasmin activity in skim milk." *Journal of Food Processing and Preservation* 42(5): 13616. https://doi.org/10.1111/jfpp.2018.42.issue-5 10.1111/jfpp.13616.

Arzeni, C., K. Martínez, P. Zema, A. Arias, O. E. Pérez, and A. M. R. Pilosof. 2012. "Comparative study of high intensity ultrasound effects on food proteins functionality." *Journal of Food Engineering* 108(3): 463–472. https://doi.org/10.1016/j.jfoodeng.2011.08.018.

Ashokkumar, M. 2015. "Applications of ultrasound in food and bioprocessing." *Ultrasonics Sonochemistry* 25(1): 17–23. https://doi.org/10.1016/j.ultsonch.2014.08.012.

Ashraf, J., L. Liu, M. Awais, T. Xiao, L. Wang, X. Zhou, . . . S. Zhou. 2020. "Effect of thermosonication pre-treatment on mung bean (*Vigna radiata*) and white kidney bean (*Phaseolus vulgaris*) proteins: Enzymatic hydrolysis, cholesterol lowering activity and structural characterization." *Ultrasonics Sonochemistry* 66: 105121. https://doi.org/10.1016/j.ultsonch.2020.105121.

Astráin-Redín, L., S. Ciudad-Hidalgo, J. Raso, S. Condón, G. Cebrián, and I. Álvarez. 2020. "Application of high-power ultrasound in the food industry." *Sonochemical Reactions*, 90444. https://doi.org/10.5772/intechopen.90444.

Baumann, A., S. E. Martin, and H. Feng. 2005. "Power ultrasound treatment of Listeria monocytogenes in apple cider." *Journal of Food Protection* 68(11): 2333–2340. https://doi.org/10.4315/0362-028X-68.11.2333.

Behrouzian, F., S. M. Razavi, and H. Karazhiyan. 2014. "Intrinsic viscosity of cress (*Lepidium sativum*) seed gum: Effect of salts and sugars." *Food Hydrocolloids* 35: 100–105. https://doi.org/10.1016/j.foodhyd.2013.04.019.

Bhargava, N., R. S. Mor, K. Kumar, and V. S. Sharanagat. 2021. "Advances in application of ultrasound in food processing: A review." *Ultrasonics Sonochemistry* 70: 1–12. https://doi.org/10.1016/j.ultsonch.2020.105293.

Byanju, B., M. M. Rahman, M. P. Hojilla-Evangelista, and B. P. Lamsal. 2020. "Effect of high-power sonication pretreatment on extraction and some physicochemical properties of proteins from chickpea, kidney bean, and soybean." *International Journal of Biological Macromolecules* 145: 712–721. https://doi.org/10.1016/j.ijbiomac.2019.12.118.

Carcel, J. A., J. V. García-Perez, E. Riera, and A. Mulet. 2011. "Improvement of convective drying of carrot by applying power ultrasound influence of mass load density." *Drying Technology* 29(2): 174–182. https://doi.org/10.1080/07373937.2010.483032.

Carrington, D. 2020. "Most realistic plant-based steak revealed." [accessed January 08, 2023]. www.theguardian.com/food/2020/jan/10/most-realistic-plant-based-steak-revealed.

Chemat, F., N. Rombaut, A. G. Sicaire, A. Meullemiestre, A. S. Fabiano-Tixier, and M. Abert-Vian. 2017. "Ultrasound assisted extraction of food and natural products. Mechanisms, techniques, combinations, protocols and applications. A review." *Ultrasonics Sonochemistry* 34: 540–560. https://doi.org/10.1016/j.ultsonch.2016.06.035.

Chen, X., Y. Dai, Z. Huang, L. Zhao, J. Du, W. Li, and D. Yu. 2022a. "Effect of ultrasound on the glycosylation reaction of pea protein isolate–arabinose: Structure and emulsifying properties." *Ultrasonics Sonochemistry* 89: 106157. https://doi.org/10.1016/j.ultsonch.2022.106157.

Chen, X., H. Zhao, H. Wang, P. Xu, M. Chen, Z. Xu, . . . Y. Cheng. 2022b. "Preparation of high-solubility rice protein using an ultrasound-assisted glycation reaction." *Food Research International* 161: 111737. https://doi.org/10.1016/j.foodres.2022.111737.

Chia, S. R., K. W. Chew, P. L. Show, M. Sivakumar, T. C. Ling, and Y. Tao. 2019. "Isolation of protein from Chlorella sorokiniana CY1 using liquid biphasic flotation assisted with sonication through sugaring-out effect." *Journal of Oceanology and Limnology* 37(3). https://doi.org/10.1007/s00343-019-8246-2.

Chittapalo, T., and A. Noomhorm. 2009. "Ultrasonic assisted alkali extraction of protein from defatted rice bran and properties of the protein concentrates." *International Journal of Food Science and Technology* 44(9): 1843–1849. https://doi.org/10.1111/j.1365-2621.2009.02009.x.

Coussios, C. C., C. H. Farny, G. ter Haar, and R. A. Roy. 2007. "Role of acoustic cavitation in the delivery and monitoring of cancer treatment by high-intensity focused ultrasound (HIFU)." *International Journal of Hyperthermia* 23(2):105–20. https://doi.org/10.1080/02656730701194131.

Cui, Q., X. Wang, G. Wang, R. Li, X. Wang, S. Chen, . . . L. Jiang. 2019. "Effects of ultrasonic treatment on the gel properties of microbial transglutaminase crosslinked soy, whey and soy–whey proteins." *Food Science and Biotechnology* 28(5): 1455–1464. https://doi.org/10.1007/s10068-019-00583-y.

Dion, J. R., and D. H. Burns. 2011. "Determination of volume fractions in multicomponent mixtures using ultrasound frequency analysis." *Applied Spectroscopy* 65(6): 648–656. https://doi.org/10.1366/10-06126.

Dekkers, B. L., C. V. Nikiforidis, and A. J. van der Goot. 2016. "Shear-induced fibrous structure formation from a pectin/SPI blend." *Innovative Food Science and Emerging Technology* 36: 193–200. https://doi.org/10.1016/j.ifset.2016.07.003.

Fadimu, G. J., A. Farahnaky, H. Gill, and T. Truong. 2022. "Influence of ultrasonic pretreatment on structural properties and biological activities of lupin protein hydrolysate." *International Journal of Food Science & Technology* 57(3): 1729–1738. https://doi.org/10.1111/ijfs.15549.

Farzad, M., H. E. Ferouali, O. Kahraman, and J. Yagoobi. 2020. "Enhancement of heat transfer and product quality using jet reattachment nozzles in drying of food products." *Drying Technology* 40: 352–370. https://doi.org/10.1080/07373937.2020.1804927.

Feng, H., and H. Lee. 2011. "Effect of power ultrasound on food quality." In *Ultrasound Technologies for Food and Bioprocessing*; Springer: New York, NY, USA.

Feng, H., and W. Yang. 2011. "Ultrasonic processing." In *Nonthermal Processing Technologies for Food*, H. Q. Zhang, G. V. Barbosa, V. M. Balasubramaniam, C. P. Dunne, D. F. Farkas, J. T. C. Yuan, editors; John Wiley & Sons: Ames, IA.

Figueroa-González, J. J., C. Lobato-Calleros, E. J. Vernon-Carter, E. Aguirre-Mandujano, J. Alvarez-Ramirez, and A. Martínez-Velasco. 2022. "Modifying the structure, physicochemical properties, and foaming ability of amaranth protein by dual pH-shifting and ultrasound treatments." *LWT* 153: 112561. https://doi.org/10.1016/j.lwt.2021.112561.

Floros, J. D. and H. Liang. 1994. "Acoustically assisted diffusion through membranes and biomaterials." *Food Technology* 48: 79–84.

Fukuda, T., N. Maruyama, M. R. M. Salleh, B. Mikami, and S. Utsumi. 2008. "Characterization and crystallography of recombinant 7s globulins of adzuki bean and structure– function relationships with 7s globulins of various crops." *Journal of Agricultural and Food Chemistry* 56(11): 4145–4153. https://doi.org/10.1021/jf072667b.

Gallego-Juárez, J. A. 2017. "Basic principles of ultrasound.' In *Ultrasound in Food Processing. Recent Advances*, M. Villamiel et al. editors; Wiley Blackwell: Chichester, pp. 4–26. https://doi.org/10.1002/9781118964156.ch1.

Gallo, M., L. Ferrara, and D. Naviglio. 2018. "Application of ultrasound in food science and technology: A perspective." *Foods* 7(10): 164. https://doi.org/10.3390/foods7100164.

Gevari, M. T., T. Abbasiasl, S. Niazi, M. Ghorbani, and A. Koşar. 2020. "Direct and indirect thermal applications of hydrodynamic and acoustic cavitation: A review." *Applied Thermal Engineering* 171: 115065. https://doi.org/10.1016/j.applthermaleng.2020.115065.

Görgüç, A., C. Bircan, and F. M. Yılmaz. 2019. "Sesame bran as an unexploited by-product: Effect of enzyme and ultrasound-assisted extraction on the recovery of protein and antioxidant compounds." *Food Chemistry* 283: 637–645. https://doi.org/10.1016/j.foodchem.2019.01.077.

He, X., J. Chen, X. He, Z. Feng, C. Li, W. Liu, . . . C. Liu. 2021. "Industry-scale microfluidization as a potential technique to improve solubility and modify structure of pea protein." *Innovative Food Science & Emerging Technologies* 67: 102582. https://doi.org/10.1016/j.ifset.2020.102582.

Hu, H., I. W. Cheung, S. Pan, and E. C. Li-Chan. 2015. "Effect of high intensity ultrasound on physicochemical and functional properties of aggregated soybean β-conglycinin and glycinin." *Food Hydrocolloids* 45: 102–110. https://doi.org/10.1016/j.foodhyd.2014.11.004.

Hu, H., E. C. Li-Chan, L. Wan, M. Tian, and S. Pan. 2013a. "The effect of high intensity ultrasonic pre-treatment on the properties of soybean protein isolate gel induced by calcium sulfate." *Food Hydrocolloids* 32(2): 303–311. https://doi.org/10.1016/j.foodhyd.2013.01.016.

Hu, H., J. Wu, E. C. Li-Chan, L. Zhu, F. Zhang, X. Xu, . . . S. Pan. 2013b. "Effects of ultrasound on structural and physical properties of soy protein isolate (SPI) dispersions." *Food Hydrocolloids* 30(2): 647–655. https://doi.org/10.1016/j.foodhyd.2012.08.001.

Huang, L., X. Ding, C. Dai, and H. Ma. 2017. "Changes in the structure and dissociation of soybean protein isolate induced by ultrasound-assisted acid pretreatment." *Food Chemistry* 232: 727–732. https://doi.org/10.1016/j.foodchem.2017.04.077.

Ismail, B. P., L. Senaratne-Lenagala, A. Stube, and A. Brackenridge. 2020. "Protein demand: Review of plant and animal proteins used in alternative protein product development and production." *Animal Frontiers* 10(4): 53–63. https://doi.org/10.1093/af/vfaa040.

Jambrak, A. R., V. Lelas, T. J. Mason, G. Krešić, and M. Badanjak. 2009. "Physical properties of ultrasound treated soy proteins." *Journal of Food Engineering* 93(4): 386–393. https://doi.org/10.1016/j.jfoodeng.2009.02.001.

Jia, J., H. Ma, W. Zhao, Z. Wang, W. Tian, L. Luo, and R. He. 2010. "The use of ultrasound for enzymatic preparation of ACE-inhibitory peptides from wheat germ protein." *Food Chemistry* 119(1): 336–342. https://doi.org/10.1016/j.foodchem.2009.06.036.

Jiang, J., B. Zhu, Y. Liu, and Y. L. Xiong. 2014b. "Interfacial structural role of pH-shifting processed pea protein in the oxidative stability of oil/water emulsions." *Journal of Agricultural and Food Chemistry* 62(7): 1683–1691. https://doi.org/10.1021/jf405190h.

Jiang, L., J. Wang, Y. Li, Z. Wang, J. Liang, R. Wang., . . . M. Zhang. 2014a. "Effects of ultrasound on the structure and physical properties of black bean protein isolates." *Food Research International* 62: 595–601. https://doi.org/10.1016/j.foodres.2014.04.022.

Jiang, S., J. Ding, J. Andrade, T. M. Rababah, A. Almajwal, M. M. Abulmeaty, and H. Feng. 2017. "Modifying the physicochemical properties of pea protein by pH-shifting and ultrasound combined treatments. *Ultrasonics sonochemistry* 38: 835–842. https://doi.org/10.1016/j.ultsonch.2017.03.046.

Jiang, W., Y. Wang, C. Ma, D. J. McClements, F. Liu, and X. Liu. 2022. "Pea protein isolate-inulin conjugates prepared by pH-shift treatment and ultrasonic-enhanced glycosylation: Structural and functional properties." *Food Chemistry* 384: 132511. https://doi.org/10.1016/j.foodchem.2022.132511.

Joshi, S., S. Shah, and K. Kalantar-Zadeh. 2018. "Adequacy of plant based proteins in chronic kidney disease." *Journal of Renal Nutrition* 29(2): 112–117. https://doi.org/10.1053/j.jrn.2018.06.006.

Kahraman, O. 2019. "Separation of solvents from selected biopolymeric matrixes and aqueous mixtures by ultrasound (PhD Dissertation)." University of Illinois at Urbana-Champaign.

Kahraman, O., and H. Feng. 2021. "Continuous-flow manothermosonication treatment of apple-carrot juice blend: Effects on juice quality during storage." *LWT-Food Science and Technology* 110360. https://doi.org/10.1016/j.lwt.2020.110360.

Kahraman, O., A. Malvandi, L. Vargas, and H. Feng. 2021. "Drying characteristics and quality attributes of apple slices dried by a non-thermal ultrasonic contact drying method." *Ultrasonics Sonochemistry* 73: 105510. https://doi.org/10.1016/j.ultsonch.2021.105510.

Kahraman, O., G. E. Petersen, and C. Fields. 2022. "Physicochemical and functional modifications of hemp protein concentrate by the application of ultrasonication and pH shifting treatments." *Foods* 11(4): 587. https://doi.org/10.3390/foods11040587.

Kamani, M. H., J. Semwal, and M. S Meera. 2021. "Functional modification of protein extracted from black gram by-product: Effect of ultrasonication and micronization techniques." *LWT* 144: 111193. https://doi.org/10.1016/j.lwt.2021.111193.

Karabulut, G., and A. Cagri-Mehmetoglu. 2018. "Antifungal, mechanical, and physical properties of edible film containing *Williopsis saturnus* var. *saturnus* antagonistic yeast." *Journal of Food Science* 83(3): 763–769. https://doi.org/10.1111/1750-3841.14062.

Karabulut, G., B. Efendioğlu, B. Kurtuluş, E. Turan, H. Kuyumcu, Ş. Esen, and A. Ç. Mehmetoğlu. 2018. "Effects of edible coating incorporated with *Bacillus subtilis* on shelf life of strawberry." *GIDA-Journal of Food* 43(1): 53–63. https://doi.org/10.15237/gida.GD17054.

Karabulut, G., H. Feng, and O. Yemiş. 2022. "Physicochemical and antioxidant properties of industrial hemp seed protein isolate treated by high-intensity ultrasound." *Plant Foods for Human Nutrition* 77(4): 577–583. https://doi.org/10.1007/s11130-022-01017-7.

Karabulut, G., and O. Yemiş. 2022. "Modification of hemp seed protein isolate (Cannabis sativa L.) by high-intensity ultrasound treatment. Part 1: Functional properties." *Food Chemistry* 375: 131843. https://doi.org/10.1016/j.foodchem.2021.131843.

Keerati-u-rai, M., M. Miriani, S. Lametti, F. Bonomi, and M. Corredig. 2012. "Structural changes of soy proteins at the oil–water interface studied by fluorescence spectroscopy." *Colloids and Surfaces B: Biointerfaces* 93: 41–48. https://doi.org/10.1016/j.colsurfb.2011.12.002.

Kentish, S., and M. Ashokkumar. 2011. "The physical and chemical effects of ultrasound." In *Ultrasound Technologies for Food and Bioprocessing*, H. Feng, G. V. Cánovas, and J. Weiss, editors; Springer: New York, NY, pp. 1–12.

Kentish, S., and H. Feng. 2014. "Applications of power ultrasound in food processing." *Annual Review of Food Science and Technology* 5: 14.1–14.22. https://doi.org/10.1146/annurev-food-030212-182537.

Khatkar, A. B., A. Kaur, and S. K. Khatkar. 2020. "Restructuring of soy protein employing ultrasound: Effect on hydration, gelation, thermal, in-vitro protein digestibility and structural attributes." *LWT* 132: 109781. https://doi.org/10.1016/j.lwt.2020.109781.

Knorr, D., M. Zenker, V. Heinz, and D. Lee. 2004. "Applications and potential of ultrasonics in food processing." *Trends in Food Science & Technology* 15: 261–266.

Korzendorfer, A. 2022. "Vibrations and ultrasound in food processing- Sources of vibrations, adverse effects, and beneficial applications- An overview." *Journal of Food Engineering* 324: 110875. https://doi.org/10.1016/j.jfoodeng.2021.110875.

Krintiras, G. A., J. G. Diaz, A. J. van der Goot, A. I. Stankiewicz, and G. D. Stefanidis. 2016. "On the use of the Couette cell technology for large scale production of textured soy-based meat replacers." *Journal of Food Engineering* 169: 205–213. https://doi.org/10.1016/j.jfoodeng.2015.08.021.

Kutlu, N., R. Pandiselvam, A. Kamiloglu, I. Saka, N. U. Sruthi, A. Kothakota, C. T. Socol, and C. M. Maerescu. 2022. "Impact of ultrasonication applications on color profile of foods." *Ultrasonics Sonochemistry* 89: 106109. https://doi.org/10.1016/j.ultsonch.2022.106109.

Kyriakopoulou, K., B. Dekkers, and A. J. van der Goot. 2019. Chapter 6—plant-based meat analogues. In *Sustainable Meat Production and Processing*, C. M. Galanakis, editor; Academic Press: San Diego, CA, pp. 103–126. https://doi.org/10.1016/B978-0-12-814874-7.00006-7.

Lee, H., G. Yildiz, L. C. Dos Santos, S. Jiang, J. E. Andrade, N. J. Engeseth, and H. Feng. 2016. "Soy protein nano-aggregates with improved functional properties prepared by sequential pH treatment and ultrasonication." *Food Hydrocolloids* 55: 200–209. https://doi.org/10.1016/j.foodhyd.2015.11.022.

Li, B., and D. W. Sun. 2002. "Effect of power ultrasound on freezing rate during immersion freezing of potatoes." *Journal of Food Engineering* 55: 277–282. https://doi.org/10.1016/S0260-8774(02)00102-4.

Li, C., F. Yang, Y. Huang, C. Huang, K. Zhang, and L. Yan. 2020. "Comparison of hydrodynamic and ultrasonic cavitation effects on soy protein isolate functionality." *Journal of Food Engineering* 265: 109697. https://doi.org/10.1016/j.jfoodeng.2019.109697.

Li, H., J. Yu, M. Ahmedna, and I. Goktepe. 2013. "Reduction of major peanut allergens Ara h 1 and Ara h 2, in roasted peanuts by ultrasound assisted enzymatic treatment." *Food Chemistry* 141(2): 762–768. https://doi.org/10.1016/j.foodchem.2013.03.049.

Liang, Q., X. Ren, H. Ma, S. Li., K. Xu, and A. O. Oladejo. 2017. "Effect of low-frequency ultrasonic-assisted enzymolysis on the physicochemical and antioxidant properties of corn protein hydrolysates." *Journal of Food Quality*. https://doi.org/10.1155/2017/2784146.

Liu, X., M. Wang, F. Xue, and B. Adhikari. 2022. "Application of ultrasound treatment to improve the technofunctional properties of hemp protein isolate." *Future Foods* 6: 100176. https://doi.org/10.1016/j.fufo.2022.100176.

Livesey, G., R. Taylor, T. Hulshof, and J. Howlett. 2008. "Glycemic response and health—a systematic review and meta-analysis: Relations between dietary glycemic properties and health outcomes." *The American Journal of Clinical Nutrition* 87(1): 258S–268S. https://doi.org/10.1093/ajcn/87.1.258S.

Ma, X., F. Hou, H. Zhao, D. Wang, W. Chen, S. Miao, and D. Liu. 2020. "Conjugation of soy protein isolate (SPI) with pectin by ultrasound treatment." *Food Hydrocolloids* 108: 106056. https://doi.org/10.1016/j.foodhyd.2020.106056.

Ma, X., T. Yan, F. Hou, W. Chen, S. Miao, and D. Liu. 2019. "Formation of soy protein isolate (SPI)-citrus pectin (CP) electrostatic complexes under a high-intensity ultrasonic field: Linking the enhanced emulsifying properties to physicochemical and structural properties." *Ultrasonics Sonochemistry* 59: 104748. https://doi.org/10.1016/j.ultsonch.2019.104748.

Majid, I., G. A. Nayik, and V. Nanda. 2015. "Ultrasonication and food technology: A review." *Cogent Food & Agriculture* 1(1): 1071022. https://doi.org/10.1080/23311932.2015.1071022.

Malav, O. P., S. Talukder, P. Gokulakrishnan, and S. Chand. 2015. "Meat analog: A review." *Critical Reviews in Food Science and Nutrition* 55: 1241–1245. https://doi.org/10.1080/10408398.2012.689381.

Malik, M. A., H. K. Sharma, and C. S. Saini. 2017. "High intensity ultrasound treatment of protein isolate extracted from dephenolized sunflower meal: Effect on physicochemical and functional properties." *Ultrasonics Sonochemistry* 39: 511–519. https://doi.org/10.1016/j.ultsonch.2017.05.026.

Malvandi, A., D. N. Coleman, J. J. Loor, and H. Feng. 2022a. "A novel sub-pilot-scale direct-contact ultrasonic dehydration technology for sustainable production of distillers dried grains (DDG)." *Ultrasonics Sonochemistry*. https://doi.org/10.1016/j.ultsonch.2022.105982.

Malvandi, A., R. Kapoor, H. Feng, and M. Kamruzzaman. 2022b. "Non-destructive measurement and real-time monitoring of apple hardness during ultrasonic contact drying via portable NIR spectroscopy and machine learning." *Infrared Physics & Technology* 122: 104077.

Manski, J. M., A. J. van der Goot, and R. M. Boom 2007. "Advances in structure formation of anisotropic protein-rich foods through novel processing concepts." *Trends in Food Science and Technology* 18: 546–557. https://doi.org/10.1016/j.tifs.2007.05.002.

Martínez-Velasco, A., C. Lobato-Calleros, B. E. Hernández-Rodríguez, A. Román-Guerrero, J. Alvarez-Ramirez, and E. J. Vernon-Carter. 2018. "High intensity ultrasound treatment of faba bean (*Vicia faba* L.) protein: Effect on surface properties, foaming ability and structural changes." *Ultrasonics Sonochemistry* 44: 97–105. https://doi.org/10.1016/j.ultsonch.2018.02.007.

Mauer, L. 2003. "Protein I Heat treatment for food proteins." *Encyclopedia of Food Sciences and Nutrition*: 4868–4872.

Mir, N. A., C. S. Riar, and S. Singh. 2019. "Physicochemical, molecular and thermal properties of high-intensity ultrasound (HIU) treated protein isolates from album (*Chenopodium album*) seed." *Food Hydrocolloids* 96: 433–441. https://doi.org/10.1016/j.foodhyd.2019.05.052.

Mulet, A., J. A. Cárcel, J. Benedito, C. Roselló, and S. Simal. 2003. "Ultrasonic mass transfer enhancement in food processing." In *Transport Phenomena in Food Processing*, J. Welti-Chanes, J. Vélez-Ruiz, G. Barbosa-Canova, editors; CRC Press: New York.

Murray, B. S., and R. Ettelaie. 2004. "Foam stability: Proteins and nanoparticles." *Current Opinion in Colloid & Interface Science* 9(5): 314–320. https://doi.org/10.1016/j.cocis.2004.09.004.

Muthukumaran, S., S. E. Kentish, G. W. Stevens, and M. Ashokkumar. 2006. "Application of ultrasound in membrane separation processes: A review." *Reviews in Chemical Engineering* 22: 155–194. https://doi.org/10.1515/REVCE.2006.22.3.155.

Nande, R., C. M. Howard, and P. P. Claudio. 2015. "Ultrasound-mediated oncolytic virus delivery and uptake for increased therapeutic efficacy: State of art." *Oncolytic Virotherapy* 193–205. https://doi.org/10.2147/OV.S66097.

Nazari, B., M. A. Mohammadifar, S. Shojaee-Aliabadi, E. Feizollahi, and L. Mirmoghtadaie. 2018. "Effect of ultrasound treatments on functional properties and structure of millet protein concentrate." *Ultrasonics Sonochemistry* 41: 382–388. https://doi.org/10.1016/j.ultsonch.2017.10.002.

O'Sullivan, J., B. Murray, C. Flynn, and I. Norton. 2016. "The effect of ultrasound treatment on the structural, physical and emulsifying properties of animal and vegetable proteins." *Food Hydrocolloids* 53: 141–154. https://doi.org/10.1016/j.foodhyd.2015.02.009.

O'Sullivan, J., M. Park, and J. Beevers. 2016. "The effect of ultrasound upon the physicochemical and emulsifying properties of wheat and soy protein isolates." *Journal of Cereal Science* 69: 77–84. https://doi.org/10.1016/j.jcs.2016.02.013.

Ozuna, C., J. A. Carcel, P. M. Walde, and J. V. García-Perez. 2014. "Low-temperature drying of salted cod (*Gadus morhua*) assisted by high power ultrasound: Kinetics and physical properties." *Innovative Food Science & Emerging Technologies* 23: 146–155. https://doi.org/10.1016/j.ifset.2014.03.008.

Paglarini, C. S., S. Martini, and M. A. Pollonio. 2019. "Physical properties of emulsion gels formulated with sonicated soy protein isolate." *International Journal of Food Science & Technology* 54(2): 451–459. https://doi.org/10.1111/ijfs.13957.

Pan, M., F. Xu, Y. Wu, M. Yao, X. Xiao, N. Zhang, . . . L. Wang 2020. "Application of ultrasound-assisted physical mixing treatment improves in vitro protein digestibility of rapeseed napin." *Ultrasonics Sonochemistry* 67: 105136. https://doi.org/10.1016/j.ultsonch.2020.105136.

Patist, A., and D. Bates. 2008. "Ultrasonic innovations in the food industry: From the laboratory to commercial production." *Innovative Food Science and Emerging Technologies* 9: 147–154. https://doi.org/10.1016/j.ifset.2007.07.004.

Paul, G. L. 2009. "The rationale for consuming protein blends in sports nutrition." *Journal of the American College of Nutrition* 28(4): 464S–472S. https://doi.org/10.1080/07315724.2009.10718113.

Phillips, L. G. 2013. *Structure-Function Properties of Food Proteins*; Academic Press: Cambridge, MA.

Rahman, M. M., B. Byanju, D. Grewell, and B. P. Lamsal. 2020. "High-power sonication of soy proteins: Hydroxyl radicals and their effects on protein structure." *Ultrasonics Sonochemistry* 64: 105019. https://doi.org/10.1016/j.ultsonch.2020.105019.

Rahman, M. M., and B. P. Lamsal. 2021. "Ultrasound-assisted extraction and modification of plant-based proteins: Impact on physicochemical, functional, and nutritional properties." *Comprehensive Reviews in Food Science and Food Safety* 20: 1457–1480. https://doi.org/10.1111/1541-4337.12709.

Resendiz-Vazquez, J. A., J. A. Ulloa, J. E. Urías-Silvas, P. U. Bautista-Rosales, J. C. Ramírez-Ramírez, P. Rosas-Ulloa, and L. J. U. S. González-Torres. 2017. "Effect of high-intensity ultrasound on the techno-functional properties and structure of jackfruit (Artocarpus heterophyllus) seed protein isolate." *Ultrasonics Sonochemistry* 37: 436–444. https://doi.org/10.1016/j.ultsonch.2017.01.042.

Rickert, D. A., M. A. Meyer, J. Hu, and P. A. Murphy. 2004. "Effect of extraction pH and temperature on isoflavone and saponin partitioning and profile during soy protein isolate production." *Journal of Food Science* 69: 623–631. https://doi.org/10.1111/j.1365-2621.2004.tb09910.x.

Rodríguez, J., A. Mulet, and J. Bon. 2014. "Influence of high-intensity ultrasound on drying kinetics in fixed beds of high porosity." *Journal of Food Engineering* 127: 93–102. https://doi.org/10.1016/j.jfoodeng.2013.12.002.

Rojas, M. L., M. T. K. Kubo, M. E. C. Silva, and P. E. D. Augusto. 2021. "Ultrasound processing of fruits and vegetables, structural modification and impact on nutrient and bioactive compounds: A review." *International Journal of Food Science and Technology* 56: 4376–4395. https://doi.org/10.1111/ijfs.15113.

Rojas, M. L., T. S. Leite, M. Cristianini, I. D. Alvim, and P. E. D. Augusto. 2016. "Peach juice processed by the ultrasound technology: Changes in its microstructure improve its physical properties and stability." *Food Research International* 82: 22–33. https://doi.org/10.1016/j.foodres.2016.01.011.

Rubio, F., E. D. Blandford, and L. J. Bond. 2016. "Survey of advanced nuclear technologies for potential applications of sonoprocessing." *Ultrasonics* 71: 211–222. https://doi.org/10.1016/j.ultras.2016.06.017.

Sabarez, H. T., J. A. Gallego-Juarez, and E. Riera. 2012. "Ultrasonic-assisted convective drying of apple slices." *Drying Technology* 30(9): 989–997. https://doi.org/10.1080/07373937.2012.677083.

Silventoinen, P., and N. Sozer. 2020. "Impact of ultrasound treatment and pH-shifting on physicochemical properties of protein-enriched barley fraction and barley protein isolate." *Foods* 9(8): 1055. https://doi.org/10.3390/foods9081055.

Singh, N. 2017. "Pulses: An overview." *Journal of Food Science and Technology* 54(4): 853–857. https://doi.org/10.1007/s13197-017-2537-4.

Sun, X., W. Zhang, L. Zhang, S. Tian, and F. Chen. 2021. "Effect of ultrasound-assisted extraction on the structure and emulsifying properties of peanut protein isolate." *Journal of the Science of Food and Agriculture* 101(3): 1150–1160. https://doi.org/10.1002/jsfa.10726.

Sun-Waterhouse, D., M. Zhao, and G. I. Waterhouse. 2014. "Protein modification during ingredient preparation and food processing: Approaches to improve food processability and nutrition." *Food and Bioprocess Technology* 7(7): 1853–1893. https://doi.org/10.1007/s11947-014-1326-6.

Taha, A., T. Mehany, R. Pandiselvam, S. Anusha Siddiqui, N. A. Mir, M. A. Malik, O. J. Sujayasree, K. C. Alamuru, A. C. Khanashyam, F. Casanova, X. Xu, S. Pan, and H. Hu. 2023. "Sonoprocessing: Mechanisms and recent applications of power ultrasound in food." *Critical Reviews in Food Science and Nutrition*: 1–39. https://doi.org/10.1080/10408398.2022.2161464.

Tang, C. H., X. Y. Wang, X. Q. Yang, and L. Li. 2009. "Formation of soluble aggregates from insoluble commercial soy protein isolate by means of ultrasonic treatment and their gelling properties." *Journal of Food Engineering* 92(4): 432–437. https://doi.org/10.1016/j.jfoodeng.2008.12.017.

Tao, Y., and D. Sun. 2015. "Enhancement of food processes by ultrasound: A Review." *Critical Reviews in Food Science and Nutrition* 55(4): 570–594. https://doi.org/10.1080/10408398.2012.667849.

Tian, R., J. Feng, G. Huang, B. Tian, Y. Zhang, L. Jiang, and X. Sui. 2020. "Ultrasound driven conformational and physicochemical changes of soy protein hydrolysates." *Ultrasonics sonochemistry* 68: 105202. https://doi.org/10.1016/j.ultsonch.2020.105202.

Vanga, S. K., J. Wang, V. Orsat, and V. Raghavan. 2020. "Effect of pulsed ultrasound, a green food processing technique, on the secondary structure and in-vitro digestibility of almond milk protein." *Food Research International* 137: 109523. https://doi.org/10.1016/j.foodres.2020.109523.

Wang, N., X. Zhou, W. Wang, L. Wang, L. Jiang, T. Liu, and D. Yu. 2021b. "Effect of high intensity ultrasound on the structure and solubility of soy protein isolate-pectin complex." *Ultrasonics Sonochemistry* 80: 105808. https://doi.org/10.1016/j.ultsonch.2021.105808.

Wang, S., J. Wang, F. Xue, and C. Li. 2019. "Effects of heating or ultrasound treatment on the enzymolysis and the structure characterization of hempseed protein isolates." *Journal of Food Science and Technology* 56: 3337–3346. https://doi.org/10.1007/s13197-019-03815-5.

Wang, T., K. Chen, X. Zhang, Y. Yu, D. Yu, L. Jiang, and L. Wang. 2021a. "Effect of ultrasound on preparing soy protein isolate-maltodextrin embedded hemp seed oil microcapsules and establishing oxidation kinetics models." *Ultrasonics Sonochemistry* 77: 105700. https://doi.org/10.1016/j.ultsonch.2021.105700.

Wang, Y., Q. Liu, Y. Yang, R. Zhang, A. Jiao, and Z. Jin. 2022a. "Construction of transglutaminase covalently cross-linked hydrogel and high internal phase emulsion gel from pea protein modified by high-intensity ultrasound." *Journal of the Science of Food and Agriculture*. https://doi.org/10.1002/jsfa.12372.

Wang, Y., S. Wang, R. Li, Y. Wang, Q. Xiang, K. Li, and Y. Bai. 2022b. "Effects of combined treatment with ultrasound and pH shifting on foaming properties of chickpea protein isolate." *Food Hydrocolloids* 124: 107351. https://doi.org/10.1016/j.foodhyd.2021.107351.

Wanghmare, R., M. Kumar, R. Yadav, P. Mhatre, S. Sonawane, S. Sharma, Y. Gat, D. Chandran, R. M. Hasan, A. Dey, T. Sarkar, K. Banwo, M. Alao, J. Balakrishnan, D. Suryawanshi, and J. M. Lorenzo 2023. "Application of ultrasonication as pre-treatment for freeze drying: An innovative approach for the retention of nutraceutical quality in foods." *Food Chemistry* 404: 134571. https://doi.org/10.1016/j.foodchem.2022.134571.

Weiss, J., K. Kristbergsson, and G. T. Kjartansson. 2011. "Engineering food ingredients with high-intensity ultrasound." In *Ultrasound Technologies for Food and Bioprocessing*, H. Feng, G. Barbosa-Canovas, J. Weiss, editors; Springer: Berlin, pp. 239–285. https://doi.org/10.1007/978-1-4419-7472-3_10.

Wen, C., J. Zhang, J. Zhou, Y. Duan, H. Zhang, and H. Ma. 2018. "Effects of slit divergent ultrasound and enzymatic treatment on the structure and antioxidant activity of arrowhead protein." *Ultrasonics Sonochemistry* 49: 294–302. https://doi.org/10.1016/j.ultsonch.2018.08.018.

Wouters, A. G., and J. A. Delcour 2019. "Cereal protein-based nanoparticles as agents stabilizing air-water and oil-water interfaces in food systems." *Current Opinion in Food Science* 25: 19–27. https://doi.org/10.1016/j.cofs.2019.02.002.

Wu, Y., W. Li, G. J. Martin, and M. Ashokkumar. 2021. "Mechanism of low-frequency and high-frequency ultrasound-induced inactivation of soy trypsin inhibitors." *Food Chemistry* 360: 130057. https://doi.org/10.1016/j.foodchem.2021.130057.

Xue, F., C. Zhu, F. Liu, S. Wang, H. Liu, and C. Li. 2018. "Effects of high-intensity ultrasound treatment on functional properties of plum (Pruni domesticae semen) seed protein isolate." *Journal of the Science of Food and Agriculture* 98(15): 5690–5699. https://doi.org/10.1002/jsfa.9116.

Yang, H., J. Gao, A. Yang, and H. Chen. 2015. "The ultrasound-treated soybean seeds improve edibility and nutritional quality of soybean sprouts." *Food Research International* 77: 704–710. https://doi.org/10.1016/j.foodres.2015.01.011.

Yang, Y., D. Liang, X. Wang, F. Li, X. Fan, and Y. Liu. 2022. "Effects of contact ultrasound & far-infrared radiation strengthening drying on water migration and quality characteristics of taro slices." *Journal of Processing and Preservation* 46: 17030. https://doi.org/10.1111/jfpp.17030.

Yi, C., Q. Lu, Y. Wang, Y. Wang, and B. Yang. 2018. "Degradation of organic wastewater by hydrodynamic cavitation combined with acoustic cavitation." *Ultrasonics Sonochemistry* 43: 156–165. https://doi.org/10.1016/j.ultsonch.2018.01.013.

Yildiz, G., J. Andrade, N. E. Engeseth, and H. Feng. 2017. "Functionalizing soy protein nano-aggregates with pH-shifting and mano-thermo-sonication." *Journal of Colloid and Interface Science* 505: 836–846. https://doi.org/10.1016/j.jcis.2017.06.088.

Yildiz, G., J. Ding, S. Gaur, J. Andrade, N. E. Engeseth, and H. Feng. 2018. "Microencapsulation of docosahexaenoic acid (DHA) with four wall materials including pea protein-modified starch complex." *International Journal of Biological Macromolecules* 114: 935–941. https://doi.org/10.1016/j.ijbiomac.2018.03.175.

Yu, D., Y. Sun, W. Wang, S. F. O'Keefe, A. P. Neilson, H. Feng, . . . H. Huang. 2020. "Recovery of protein hydrolysates from brewer's spent grain using enzyme and ultrasonication." *International Journal of Food Science & Technology* 55(1): 357–368. https://doi.org/10.1111/ijfs.14314.

Yu, L., J. Sun, S. Liu, J. Bi, C. Zhang, and Q. Yang. 2012. "Ultrasonic-assisted enzymolysis to improve the antioxidant activities of peanut (Arachin conarachin L.) antioxidant hydrolysate." *International Journal of Molecular Sciences* 13(7): 9051–9068. https://doi.org/10.3390/ijms13079051.

Zayas, J. F. 1997. "Foaming properties of proteins." In *Functionality of Proteins in Food*, pp. 260–309. https://link.springer.com/chapter/10.1007/978-3-642-59116-7_6.

Zhang, H., I. P. Claver, K. X. Zhu, and H. Zhou. 2011. "The effect of ultrasound on the functional properties of wheat gluten." *Molecules* 16(5): 4231–4240. https://doi.org/10.3390/molecules16054231.

Zhang, J., L. Liu, H. Liu, A. Yoon, S. H. Rizvi, and Q. Wang. 2019. "Changes in conformation and quality of vegetable protein during texturization process by extrusion." *Critical Reviews in Food Science and Nutrition* 59: 3267–3280. https://doi.org/10.1080/10408398.2018.1487383.

Zhang, P., T. Hu, S. Feng, Q. Xu, T. Zheng, M. Zhou, . . . H. Hu. 2016. "Effect of high intensity ultrasound on transglutaminase-catalyzed soy protein isolate cold set gel." *Ultrasonics Sonochemistry* 29: 380–387. https://doi.org/10.1016/j.ultsonch.2015.10.014.

Zhao, C., H. Yin, J. Yan, X. Niu, B. Qi, and J. Liu. 2021. "Structure and acid-induced gelation properties of soy protein isolate–maltodextrin glycation conjugates with ultrasonic pretreatment." *Food Hydrocolloids* 112: 106278. https://doi.org/10.1016/j.foodhyd.2020.106278.

Zhao, F., X. Liu, X. Ding, H. Dong, and W. Wang. 2019. "Effects of high-intensity ultrasound pretreatment on structure, properties, and enzymolysis of soy protein isolate." *Molecules* 24(20): 3637. https://doi.org/10.3390/molecules24203637.

Zhong, Z., and Y. L. Xiong. 2020. "Thermosonication-induced structural changes and solution properties of mung bean protein." *Ultrasonics Sonochemistry* 62: 104908. https://doi.org/10.1016/j.ultsonch.2019.104908.

Zhou, C., H. Ma, X. Yu, B. Liu, A. E. G. A. Yagoub, and Z. Pan. 2013. Pretreatment of defatted wheat germ proteins (by-products of flour mill industry) using ultrasonic horn and bath reactors: Effect on structure and preparation of ACE-inhibitory peptides. *Ultrasonics Sonochemistry* 20(6): 1390–1400. https://doi.org/10.1016/j.ultsonch.2013.04.005.

Zhu, G., Y. Li, L. Xie, H. Sun, Z. Zheng, and F. Liu. 2022a. "Effects of enzymatic cross-linking combined with ultrasound on the oil adsorption capacity of chickpea protein." *Food Chemistry* 383: 132641. https://doi.org/10.1016/j.foodchem.2022.132641

Zhu, K. X., C. Y. Su, X. N. Guo, W. Peng, and H. M. Zhou. 2011. "Influence of ultrasound during wheat gluten hydrolysis on the antioxidant activities of the resulting hydrolysate." *International Journal of Food Science & Technology* 46(5): 1053–1059. https://doi.org/10.1111/j.1365-2621.2011.02585.x.

Zhu, R., S. Jiang, D. Li, C. L. Law, Y. Han, Y. Tao, H. Kiani, and D. Liu. 2022b. "Dehydration of apple slices by sequential drying pretreatments and airborne ultrasound-assisted air drying: Study on mass transfer, profiles of phenolics and organic acids and PPO activity." *Innovative Food Science & Emerging Technologies* 75: 102871. https://doi.org/10.1016/j.ifset.2021.102871.

Zhu, Z., W. Zhu, J. Yi, N. Liu, Y. Cao, J. Lu, E. A. Decker, and D. J. McClements. 2018. "Effects of sonication on the physicochemical and functional properties of walnut protein isolate." *Food Research International* 106: 853–861. https://doi.org/10.1016/j.foodres.2018.01.060.

Zupanc, M., Z. Pandur, T. S. Perdih, D. Stopar, M. Petkovsek, and M. Dular. 2019. "Effects of cavitation on different microorganisms: The current understanding of the mechanisms taking place behind the phenomenon. A review and proposals for further research." *Ultrasonics Sonochemistry* 57: 147–165. https://doi.org/10.1016/j.ultsonch.2019.05.009.

7 High-Pressure Processing of Plant-Based Proteins

Priyanka Suthar, Anna Aleena Paul, Kanchan Bhatt and Riya Barthwal

7.1 INTRODUCTION

In the past, the strategy of the food industry was to develop safe food items with extended shelf life, which is less appreciated nowadays to focus on the safety alone; the customer seeks commodities with high nutritional contribution, bioactive substances, and good sensory characteristics (Hameed et al., 2018). The food industry is now interested in cutting-edge non-thermal procedures as a green option for food processing, such as high-pressure processing (HPP), ultrasonic processing, and pulsed electric fields (PEFs). Due to its simplicity of use and minimal impact on the bioactive components, flavours, and physical look of food, HPP treatment is becoming progressively more acceptable these days. The Le Chatelier's principle regulates this specific processing, which states that if a constant pressure applies to a system, pressure adjusts the system's equilibrium to the condition that occupies the least volume (Hameed et al., 2018). No covalent bonds are broken during the HPP process, and hence tiny molecular structures like vitamins, minerals, and some volatile compounds are often only marginally impacted (Muntean et al., 2016). Currently, food technologists and researchers are using HPP to inactivate microbial spores, for increasing the shelf life of milk and other food stuffs and extracting bioactive compounds from meats, fruits, and vegetables. It is also being used as a pretreatment in combination with other novel techniques of food processing, thus giving rise to effective hurdle technology (Stratakos et al., 2019; Evelyn & Silva, 2019). The process of extracting bioactive chemicals can be achieved in a smaller time period at room temperature (preventing the thermal destruction of heat-labile components) and with greater extraction yields (Huang et al., 2013).

Protein is a crucial macronutrient for maintaining the growth and development of the body. As the value of high-quality proteins in the diet is becoming more understood, researchers and nutritionists are searching for environmentally friendly and sustainable protein sources and their extraction techniques. Traditional techniques were widely used in the past to extract proteins; however, this resulted in lower extraction yields due to protein breakdown. Thus, in order to increase extraction efficiency and minimize protein breakdown during extraction, researchers are currently concentrating on non-thermal green methods (Kumar et al., 2021). The extraction of proteins with the use of high-pressure processing is a new sector that has been trending positively in customers' perceptions. Extraction of proteins with the help of high-pressure processing is affected by a number of parameters, including extraction pressure, operation duration, the composition and concentration of the extraction solvent, and the solid–liquid ratio (Mustafa & Turner, 2011). It has been observed that the final product is extracted at faster rate with minimum impurities. When proteins are subjected to high pressure, they fold and refold, disrupting intra-molecular hydrophobic and electrostatic connections but having little effect on covalent bonds, thus maintaining the quality of proteins (Bolumar et al., 2016).

Plant proteins are more desired than animal-based proteins because of a number of environmental (greenhouse gas emissions), cultural, and economic concerns. Beans, wheat, rice, oats, peanuts, maize, and other pulses are among the plant sources utilized to extract plant-based proteins. Additionally, there is an expanding amount of clinical evidence that shows the health benefits of

DOI: 10.1201/9781003369790-7

protein at or above the current recommendations for dietary protein intake, particularly in older adults. It has been recognized that protein helps with the weight control, enhancement of HDL cholesterol level, and boosting of thermogenesis and satiety and promotes bone mineralization (Pasiakos et al., 2015; Pasiakos, 2015). This chapter provides an updated information about the importance of plant-based proteins in modern day life and effect of high-pressure processing in extraction and modification of these proteins.

7.2   PLANT-BASED PROTEINS (PBPS) AND THEIR SIGNIFICANCE

According to estimates, 39 million children under the age of five are overweight, and 149 million have stunted growth and development as a result of a chronic shortage of nutrient-rich foods in their diets (UNICEF, 2022). New sources must be looked at, in order to supply the rising demand of food and proper nutrition. Looking forward to this, traditionally used animal-based proteins are now being substituted by PBPs due to a number of constraints including higher costs and narrow range of nutrients and due to a number of restrictions, including rising costs, a scarcity of nutrients, a risk to human health, the loss of freshwater supplies, and susceptibility to climate change. On the other hand, proteins made from plants are recognized as vegan diets since they include a lot of amino acids, are readily absorbed by the body, and help treat a variety of illnesses. There are various plant sources from where the proteins are derived, such as soy, canola, potato, pea, quinoa, corn, rice, kidney bean, peanut, faba bean, chick pea, flax seed, chia, buckwheat (kuttu), amaranth, and algae. Except for sulphur-containing amino acids like methionine, soy proteins (35–45%) contain a well-balanced range of essential amino acids (Qin et al., 2022). Similarly, faba beans contain around 27–34% of protein (dry weight basis) and are also regarded as sustainable crop since it provides ecosystem benefits through biological nitrogen fixing and agricultural diversification (Samaei et al., 2020). Similarly, quinoa (seeds, stem, leaves, and roots) and chickpea were reported as well.

The utilization of proteins derived from plants has recently become more popular as a result of their biodegradable composition along with extensive edible and non-edible uses. Adults and children who are allergic to protein supplement formulas made with cow's/bovine milk often use soy-based protein as a replacement. Cereal–legume mixtures have shown increased levels of essential amino acids and nutritional aspects of various food products. Chickpea protein has a neutral flavor, light color, and bland taste, which may make it highly acceptable for the development of new products such as noodles, cookies, and sausages (Boukid, 2021). Recently, a variety of proteins, including wheat gluten, corn zein, and soy protein, has been utilized for preparing edible food coatings. In order to create hydrophobic edible packaging, whey protein isolate and yellow pea protein have been used (Acquah, 2020). Antimicrobial bioactive peptides provide a shield to us from a wide range of pathogens. By using selective toxicity without damaging the host tissue, antimicrobial BAPs, which are derived from cumin, buckwheat, and soya bean, protect mammals from a wide range of pathogens, including bacteria, fungus, and protozoans (Ciociola et al., 2021). Pulse proteins, having remarkable amino acid profile and high gelling nature, are used as a basic component for yoghurt substitutes. Faba beans proteins are used as emulsifiers which have the potential of replacing synthetic emulsifiers (Liu et al., 2022). Soya bean proteins are used to create mixed hydrogels for controlled medication delivery systems because of their amphiphilic nature, superior functional qualities, and high nutritious value.

Apart from all these applications, various researches are being conducted in order to increase use of these proteins for food supplementation and fortification at global levels. PBPs are preferred over other sources including animal-based proteins, as they are advantageous in maintaining ecological and environmental integrity, ensuring greater food security, meeting increased consumer demands, and limiting energy deficiency from protein. Whole plant meals, such as pulses, legumes, and vegetables, tend to be higher in fibre and antioxidants while being lower in saturated fat, salt, and cholesterol when compared to animal proteins like meat and dairy. PBPs are studied for a

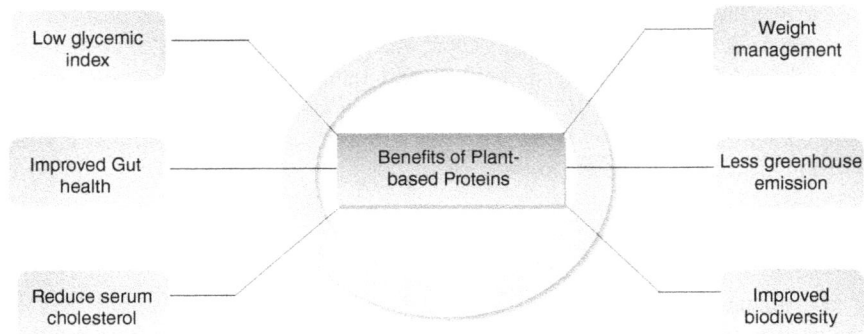

FIGURE 7.1 Benefits of plant-based proteins.

broader range of health areas including diabetes, cardiovascular health, blood pressure, cancer, metabolic syndrome, cholesterol, kidney function, muscle mass, hypertension, etc. They have been linked to improved glycemic control; decreased risk of cardiovascular disease and type 2 diabetes; decreased levels of total cholesterol, triglycerides, apolipoproteins; and increased levels of HDL-cholesterol. Researches indicate that the intake of plant protein reduces markers of cardiovascular disease, blood lipids, lower low-density lipoprotein cholesterol, and non-high-density lipoprotein cholesterol when compared to animal protein consumption (Zhao et al., 2020). Another study illustrated that 23% lower incidence of type 2 diabetes was linked to consuming 5% of one's daily calories from vegetable protein as compared to animal protein (Malik et al., 2016). PBPs are also examined for cancer risk reduction. Intake of animal proteins is linked with high risk of colorectal cancer, which can be lowered by substituting animal proteins with plant proteins (Andersen et al., 2019). Figure 7.1 shows benefits of plant-based proteins.

Antinutrients found in plant foods are one of the health issues linked with the increased dietary intake of PBPs. These antinutritional substances including phytates, phenolics compounds, saponins, glucosinolates, and enzyme inhibitors are difficult to digest, absorb, and utilize. Another health concern related to PBPs is increasing allergenicity. Common foods such as peanut, wheat, sesame, buckwheat, mustard, and soy have been identified as severe allergens in various countries. Prolamins, albumins, and trypsin inhibitors are the major allergens which make the use of limited PBPs (Hertzler et al., 2020). Despite of the fact that PBPs are of higher nutritional, biological, and ethical values, they are having difficulty in finding perfect spot in the global market. A number of methods are employed for the extraction of plant-based proteins, namely, solvent extraction, alkali-based extraction, biochemical extraction, physical extraction using ultrasounds, pulse electric field, microwave, irradiation, supercritical carbon dioxide, and high pressure. Conventional extraction techniques like solvent extraction include the use of different solvents like acetone, phenol, and ethanol. However, the use of novel technologies has made protein extraction much efficient and facile. The non-thermal nature of these technologies is beneficial in order to alter protein structure and modulation without changing its inherent properties. One of the newest technologies, high-pressure processing (HPP), is used to produce proteins with little-to-no impact on their sensory, nutritional, and physical qualities. Also, with HPP, the covalent bonds of protein are unaffected. Consequently, it can be a useful tool to customize the technological and functional characteristics of dietary proteins.

7.3   FUNDAMENTALS AND PRINCIPLE OF HIGH-PRESSURE PROCESSING

High-pressure processing (HPP), a cold pasteurization technology, is widely recognized by the food industry as it is capable of yielding food products retaining the freshness and high quality. This non-thermal technology thus plays a vital role in extending the shelf life without comparatively impacting its characteristics with that of the traditional thermal techniques. High-pressure processing is a novel preservation technique, where the food products are hermetically sealed in a flexible container. At room temperature, the high pressure of 100–600 MPa is applied. In this technique, the liquid (commonly water) acts as the pressure transfer medium, distributing the effect of pressure uniformly throughout the interior and surface of the food. Today, HPP technology is adopted by a number of nations from Europe, Asian, and South America regions. Interestingly, the highest revenue-generated food products are in the markets of Dips, Deli Salads, and Salsa (Abera, 2019; Balasubramaniam et al., 2015). A few manufacturers including Avure, Baotou, Toyo Koatsu, Stansted Fluid Power, Kobelco, etc., produce laboratory HPP equipment with varying specifications and capacities of 0.3–10 L in order to meet the varying experimental requirements. In Japan, HPP jam was initially developed, making it the primary commercial HPP food product (Huang et al., 2017). In food industry, the application of pressure is inevitable from our ancestral period onwards. According to Yordanov and Angelova (2010), pressure brings out numerous changes in food products in the perspective of physical and chemical attributes including decrease in volume and rise in temperature, energy, etc. Hence, the use of HPP is in compliance with the three elements of physical and chemical principles, which are explained in the subsequent sections.

7.3.1 LE CHATELIER'S PRINCIPLE

The application of pressure causes changes in equilibrium, which are demonstrated by Le Chatelier's principle. This concept states that when phenomena, such a chemical reaction, a phase transition, etc., are accompanied by the decrease in the characteristic volume, then the equilibrium shifts to the direction in such a way that the change is reduced—that is by an increase in pressure. The system changes to achieve the lowest volume as a result of the pressure.

7.3.2 MICROSCOPIC ORDERING

When a component is kept at a fixed temperature, its degree of molecular order grows in relation to an increase of pressure. In other words, the molecular structure and chemical reactions are affected by the opposing forces of pressure and temperature.

7.3.3 ISOSTATIC PRINCIPLE

According to the isostatic principle, the application of pressure shall be uniform—acting equally in all the directions. An ideal hydrostatic condition should be independent of time and space. In high-pressure processes, regardless of the characteristic of food component (geometry and size), the pressure and its effects are evenly dispersed within the food product. Also, this principle provides an insight on why a non-porous food material having high amount of moisture content is not macroscopically damaged by undergoing a pressure treatment. These salient features enable the development and commercialization. However, the variation in compressibility of air and water affects the food products containing the air pockets upon treatment involving pressure, in exception to the perfectly elastic foods and components having closed-cell foam where the air cannot escape (Srinivas et al., 2018). A typical high hydrostatic pressure system consists of pressure vessel and its closure, pressure generation system, temperature controller, and material handling system.

7.3.3.1 High-Pressure Vessel and Its Closure

It consists of a forged monolithic cylindrical vessel made of low-alloy steel having high tensile strength and is considered as the heart of a high-pressure processing system. The thickness of the wall is determined by (i) maximum working pressure, (ii) diameter of the vessel, and (iii) number of cycles designated for the vessel.

Generally, interrupted threaded closures are used as they are capable of automatic open and close processes. This methodology is useful for fast-cycling CIP systems; thus minimizing the down time for loading and unloading.

7.3.3.2 Pressure Generation System

After the exhaustion of high-pressure vessel, a high pressure is generated using direct compression (piston type) or indirect compression type (pump type).

7.3.3.3 Temperature Controller and a Material Handling System

In cold processing techniques, water acts as the pressure medium, where it is mixed with a small amount of lubricant (suitable soluble oil), along with the anticorrosion purpose (Srinivas et al., 2018).

HPP has gained the attention of food market in the recent years over other non-thermal food processing techniques; however, the high capital cost and technical barriers and investment in importing, etc., limit the application of food companies worldwide. For HPP, the prevailing production condition is as follows: 200–600 MPa pressure for a duration of 5 min approximately. The approval of USFDA for HPP as a non-thermal pasteurization technology helps the food sector and its allies including research laboratories, government organizations, etc., to replace traditional methods for pasteurization. Defined regulations are necessary to facilitate the use of HPP in terms of its application, retaining quality, development, and consumer trust. As a whole, HPP is considered a novel technology and subjected to the Novel Food Regulation, and the containers of HPP have to comply with the standards of the Pressure Equipment Directive (Huang et al., 2017). Protein structure can respond to pressure in two ways: general compression (elastic effects that are generally reversible) and conformation modifications (elastic effects that are primarily irreversible). The initial phenomenon is a linear pressure that occurs at reduced pressures, typically 200 MPa, and is mostly a volume reduction inside the conformer sub-ensemble i.e., volume is reduced without impacting conformation. Because the primary chemical linkages in proteins are only very slightly compressible, it means pressure has little effect on their fundamental structure. As a result, the impact on volume reduction is negligible. A shift from a large-scale structure to a negligible-volume structure as a result of an irregular action to pressure is part of the second reaction, which involves changes in protein conformation that usually take place at higher pressures. Following the hydrogen bonding, the cavities are the parts of the protein that can be compressed most significantly. As a result, the most substantial pressure implications are the hydrogen bond weakening and cavity rupture caused by intrusion of water. The tertiary arrangement is disrupted by increased intermolecular interactions, which finally cause protein unfolding. As a result, proteins can transition from their native conformation to partially unfolded intermediates and/or fully unfolded structures (Queirós et al., 2018). Figure 7.2 depicts the effects of HPP on protein structure.

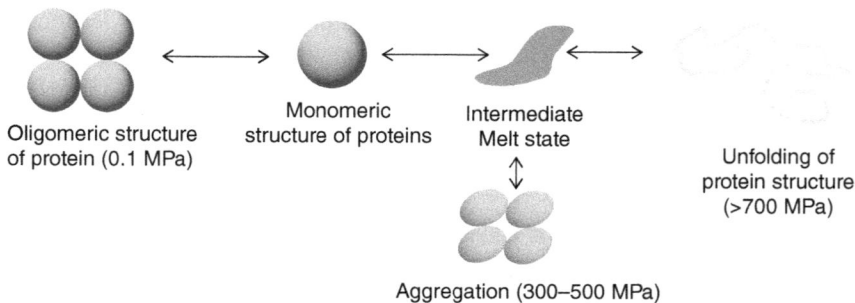

FIGURE 7.2 Depiction of the effects of HPP on protein structure.

7.4 EFFECT OF HIGH-PRESSURE PROCESSING ON PLANT PROTEINS

In the new era of food market, consumer-liking towards vegan or plant-based diets is budding. Plant products being a rich source of bioactive and other essential components, it is a true challenge for the food companies to provide food products with maximum nutritional value, sensory appeal, and high quality. Since the digestibility of plant-based bioactive components such as proteins is hard, it requires the assistance of an appropriate processing technique. Thus, the emerging cold processes like HPP piqued the interest of food sector. Gross and Jaenicke (1994) stated that the single-chain proteins tested at room temperature at neutral pH do not denature at pressures less than 400 MPa. This suggests that protein denaturation, as a stress process, cannot play a substantial role in microorganism adaptation to deep-sea environment.

Though HPP is widely used for different products, its application for developing plant-based proteins is slowly emerging. The negligible impact on the physical, chemical, nutritional and sensorial properties of high hydrostatic pressure processing on the food components upsurges its relevance. According to Mulla et al. (2022) study, the impact of HPP on legume proteins might change their rheological as well as functional characteristics (foaming ability, potential for emulsification, water holding capacity, surface hydrophobicity, etc.), opening the door to plant-based meat substitutes. Additionally, another research on plant-based yogurt products was developed, where plant-based milks are fermented using traditional methodology. Even though the fermentation process imparts probiotic culture and specific flavor, plant protein often undergoes acidification process which weakens the gel formed and phase separation. Time-consuming fermentation process and decrease in the amount of protein content decrease the acceptance of traditional plant-based yogurt production process. However, the application of HPP helps to homogenize the protein gels, customize the texture, speed up the fermentation process and improves the protein content, making it notable alternative technique (Sim et al., 2020). HHP was examined as a strategy for improving the nutritive value of proteins from plants. For instance Nasrabadi et al. (2021) claimed that, when compared to other techniques including high-intensity ultrasound, microwaving, and high-pressure homogenization, HHP was the most effective way for lowering the allergenicity of isolated soy protein for use in infant formula. Additionally, they noted a considerable reduction in the allergenicity of ginkgo seed proteins following HHP at pressures ranging from 300 to 700 MPa. These treatments mostly alter the molecular structure of allergens (through deamidation, clustering, and hydrolysis) or encourage allergen interactions with other dietary ingredients. These plant proteins can respond differentially to both thermally and nonthermally induced conformational changes since it appears that their structural patterns are distinctive from one another. HPP treatment results in greater hydrolysis rate and more release of antioxidant peptides (i.e., 51.26% @ 500–1,000 Da), which improve proteolysis accessibility. These peptides are said to have antithrombosis and angiotensin-converting enzyme (ACE) inhibition in addition to antioxidant effects (Mulla et al., 2022).

Effect of HPP on pea protein concentrates was described in the study conducted by Sim et al. (2019). Here, HPP was performed at 250–550 MPa for the time period of 15 min. It is evident that the gel was strengthened with the pressure level and protein concentration more than the heat treatment. Because high pressure disrupts intramolecular electrostatic and hydrophobic interactions while having little-to-no effect on covalent bonds, it is a useful tool for tuning non-covalent interactions and destabilizing the hydrophobic effect. As a result, while having little impact on the proteins' secondary and primary structures, HPP treatments cause changes to their quaternary and tertiary structures. Covalent bonds are typically affected by high pressure, and even α-helix or β-sheet structures appear to be nearly incompressible. HPP can also dissociate protein subunits. Due to the conformation changes of tertiary and quaternary proteins with the increase in pressure level, greater degree of protein was denatured and aggregated, in accordance with the network formation. HPP can also dissociate protein subunits. Solvent electrostriction at the level of salt linkages at the interfaces of monomer subunits, poor van der Waals contacts between monomers, and solvation of the nonpolar groups at the boundaries of contact in the oligomers are the reasons of pressure-induced separation of oligomeric proteins (Bolumar et al., 2016). Also, at 550 MPa, the starch granules in pea protein concentrates retained their structure without undergoing gelatinization process. These features help the researchers to develop novel products of pea proteins with comparatively better palatability and nutritional value. Chen et al. (2019) observed an interaction between tea polyphenol (TP) and HHP-treated soya bean protein (SP). While 0.1% (w/v) tea polyphenol appeared to preserve the helix structure, high pressure at 400 MPa significantly changed the extracellular arrangement of SP through enhancing the content of β-sheets and reducing the content of α-helix. The surface hydrophobicity reduced after HHP treatment and TP incorporation. The best dissolution of native SP appeared at 0.258 g/mL with 0.08% (w/v) TP. TP increased protein solubility to 0.50 g/mL as well as emulsifying activity by around three times, i.e. up to 43.5%, when paired with HHP treatment. The HP treatment significantly enhanced the bio-functional (such as digestibility and antioxidant in vitro) and other functional (such as gelation, emulsifying activity, water-holding capacity, emulsion physicochemical stability, and foaming capacity stability) characteristics of legume-based PIs as well as PCs. Additionally, HP has the capacity to significantly reduce the antinutritional aspects and allergenicity of foods based on pulse proteins. The integrity of protein secondary structures, constant shear viscosity, and enthalpy of legume proteins were all considerably decreased by HP-assisted enzymatic hydrolysis. According to Gharibzahedi & Smith, 2021, the HP-assisted proteolytic process boosted the production of bioactive peptides with good antihypertensive, antiradical, and antioxidant characteristics.

7.5 APPLICATIONS OF HIGH PRESSURE ON PLANT PROTEIN-BASED FOOD PRODUCTS

The inactivation of spoilage and harmful bacteria is known to be effective when food is subjected to thermal treatment (e.g. pasteurization and sterilization processes). Despite this, it is known that these treatments can cause changes to sensory attributes like color, taste, or even cause severe nutritional losses. Proteins are generally considered to unfold under pressure and then refold once the pressure is released. Depending on the particular protein and the circumstances used, this folding/unfolding process results in partial or complete denaturation and electrostatic interaction tuning. The goal of the research was to find out how barley-based non-dairy milk that has been enriched with beta-carotene would fare during pulsed high-pressure processing in both non-thermal as well as thermal conditions. It was determined how the soluble protein content was affected by the pressure, the number of pulses, and the temperature. According to the pressure and pulse settings, the samples were then subjected under pressure for about 2 min, and total treatment period was reported from 2.5 to 9 min. The homogeneity of the system was favored by higher pressure pulses at 100 MPa. The results showed that increasing the number of pulses from 1 to 3 at 100 MPa enhanced the solubility of the barley proteins (Strieder et al., 2022). Plant proteins can be altered by HPH (high pressure homogenization) to increase their functioning. Because mechanical pressures from HPH induce the reduced particle size of plant proteins, that leads to macromolecules to cavitate and fragment. Sim et al. (2021) observed that as homogenization pressure was raised, the volume-weighted average value of hazelnut protein molecules gradually decreased. Potato protein isolate (PPI), a different plant protein source, uses HPH to produce a fermented product that resembles yoghurt without the use of stabilizers. As a pre-processing phase, using HPH 200 MPa and temperature of 15°C decreased downstream PPI sedimentation. Lactic acid fermentation was made possible with less creaming because of the HPH's ability to create stable (against separation) PPI and emulsions of canola oil. Raising the homogenization pressure caused the particle size to decrease, resulting in the formation of a finer, whiter emulsion with improved physical stability against separation (Levy et al., 2021). According to Dhakal et al. (2014), almond milk's protein solubility reduced as pressure increased. Both the dot blot as well as Western blot verified the reduction of immuno-reactivity for high-pressure processing of almond milk by both antibodies. The polypeptides, 61 and 63 kDa, showed lower band intensity. Western-blot test results showed the simultaneous emergence of high molecular weight polypeptides, which suggested that amandin aggregation was a contributing factor in the observed reduced immuno-reactivity. Protein solubility was reduced by a maximum of 70% and 75% as a result of the studied HPP and TP treatments, respectively. Known to induce unfolding of proteins owing to distortion in their various forms of structure (i.e., secondary, tertiary, and quaternary), HPP and heat treatment may also promote protein precipitation and/or aggregation, which may reduce protein solubility. Protein unfolding and/or denaturation can result in the breaking of barrier over the free energy when proteins are in their lowest free energy state and then they are forced into another state by pressure. Proteins experience irreversible unfolding and/or denaturation at pressure of 300 MPa. The disulphide linkage aggregates through creation of covalent interactions which results in the oxidation of sulfhydryl groups present within the polypeptide chains and has been implicated in the irreversible denaturation. Furthermore, pressure induces the changes in volume, which is a crucial thermodynamic parameter, as well as changes in structural properties like protein's conformational transition. The creation of novel products in the dairy sector is made possible by the distinct physical and sensory qualities of foods that have undergone pressure processing. Non-animal sources of proteins are being explored as prospective alternatives for animal-originated proteins like meat products to address the increased demand for protein-enriched components for the population. Several proteins produced from plants or fungi are being used in meat products to substitute animal tissue. The most popular components are fibrous extruded or shredded vegetable proteins which are also known as texturized vegetable protein and are commonly made from slurries of wheat gluten, cooked vegetables emulsion, or pea protein.

The use of bioreactor-grown fungus, like as Quorn™, to make patties or "faux-sausages" is another prevalent component. Like many fungi, Quorn™ includes significant levels of fibre and carbohydrate in addition to protein, giving meat analogues a more palatable texture and healthier nutritional profile. Traditional high-protein foods like tempeh (which is made from fermented soya bean cake), tofu (which is prepared by agglomerated soy protein), and seitan (wheat gluten is cooked) are other alternatives for meat substitutes (Jones, 2016). In the year 2022, Janardhanan et al. (2022) studied the potential for increasing the quality of veal patties using contemporary technologies like HPP in combination with sous-vide cooking (SVCOOK). The prepared samples with various formulations were then subjected to SVCOOK at 55–65°C for 15 min after three different pressure treatments, i.e. at 350–600 MPa for 5–15 min. Plant-based and hybrid patties that had been treated with HPP tended to have less reddish color tone and more of a yellowish hue. The difference in the physicochemical characteristics of plant-based patties from those of veal as well as hybrid patties was reported, whereas both texture and color parameters of the dual technology treated hybrid patties resembled veal patties. The impact of HPP on hybrid patties was not equivalent to that on veal patties. It was concluded that the use of the both technologies i.e. HPP-SVCOOK, might be beneficial to create unique hybrid goods that are physicochemically similar to burgers made from veal.

7.6   OPPORTUNITIES AND CHALLENGES

The characteristics of plant proteins can be tailored by HPP application because varied processing conditions can result in various unfolding processes and expositions of buried sites. The degree to which these sites are exposed determines the behavior of the proteins, causing them to aggregate by encouraging protein–protein interactions or to dissolve by encouraging protein–solvent interactions. Additionally, exposing hydrophobic areas could aid to enhance the emulsifying characteristics. Therefore, HPP may make it possible to modify the solubility and surface hydrophobicity of proteins by adjusting the ratio between partial unfolding and aggregation. In order to accomplish the desired goals, it is currently challenging to understand how changes in protein functionality relate to conformational as well as structural changes, which could be related to the HPP conditions such as temperature, pressure, and time along with medium composition factors like ionic strength, pH, and co-solutes. In order to effectively use this technology to enhance food quality and customer acceptance, further study is needed to comprehend the complex processes that proteins go through

TABLE 7.1
Effect of HPP on Various Plant Proteins

Plant-based Protein Products	Parameters/ Conditions Required	Effects Of HPP on Developed Product	Reference
Soy protein isolates	400 MPa for 15 min	Low concentrated soy protein isolate solution (4.0–4.5%) has improved solubility and viscosity with increase in pressure. Depolymerization of 11S globulin protein to simpler base units.	Yan-ping et al. 2021
Almond milk	450 and 660 MPa at 30°C	Reduces the immuno-reactivity by inducing any physicochemical modification of the almond milk proteins epitopes.	Briviba et al. 2016
Phycocyanin in spirulina	450 and 660 MPa for 3.5 min	Due to less effect of HPP on chromophore in phycocyanin due to smaller size, it may encapsulate the blue chromophores by aggregation; resulting in reduced oxidation of the protein's blue color on light exposure through limited interaction with oxidizing radicals.	Zhang et al. 2021
Peanut paneer	600 MPa for 5 min	Product with higher yield was obtained as the dissociation and association of the proteins to its subunits resulting in insoluble aggregates.	Chauhan et al. 2015
Soy protein	400 MPa	The improvement in the binding property of soy protein with other smaller molecules with conformational rearrangement as a result of the dissociation of the 7S (β-conglycinin), unfolding of 11S structure, and increase in the β-sheet structure, along with the loss of α-helix secondary structure.	Chen et al. 2019
Wheat dough	500 or 600 MPa for 5 min and 33% moisture content	Formation of hydrogen bond due to Van der Waals forces and hydrophobic interactions led to molecular density and thus reduction of volume. On contrary, further increase in pressure (800 MPa for 50 min) causes the gluten to lose its elasticity resulting from the extensive cross-link and the development of insoluble protein aggregates.	McCann et al. 2013
Soy milk products	400 MPa or more	Tofu gels exhibited higher strength with an increase in pressure on treatment with $CaCl_2$ as coagulant due to the denaturation of soy protein or by the coagulation promoted by the cations. Reduced the microbial load; improvising the safety of food.	Yan-ping et al. 2021; Zhang et al. 2005

at high pressure. To translate the usage of these proteins to practical applications, future research should concentrate on understanding how other substances interact and affect the aforementioned modifications. Additionally, integrating HPP with additional variables like pH or temperature might be helpful to develop novel protein components that may have superior characteristics, form fresh aggregates or gels, or exhibit various thermal stabilities. It is also interesting to modify the textural characteristics of food by adding proteins that have undergone HPP treatment. These attributes may be further enhanced by combining the proteins with chemicals that may enhance their activity. The nature of the protein and its native properties, the makeup of the medium, and the HPP settings are the key non-controllable parameters that will determine how the changes to the protein's structure induced by HPP will alter the techno-functional qualities of the protein. In order to determine if high-pressure-processed plant proteins are suitable for human ingestion when they are included in more complex food products, it is also important to evaluate their *in vitro* and *in vivo* digestibility (Queirós et al., 2018). Bridgman (1914) observed the coagulation of albumen under pressure in 1914, and he discovered that high pressure may cause protein denaturation. Pressure generally shows two different types of effects on the structure of proteins i.e., elastic effect (reversible) and plastic effect (mostly irreversible). Less than 200 MPa pressure results in general compression with a negligible impact on proteins' primary structure. On the other hand, high pressure results in alterations in the proteins' structure as their cavities collapse due to hydration, water penetration, and shortening of the hydrogen bonds (Akasaka, 2006). Furthermore, this led to intermolecular interactions which result in weakening of tertiary structure and eventually trigger the protein to unfold. Structures of protein will vary from their natural structures once pressure is released, resulting in altered behaviours including foaming, emulsification, or gelation. Various protein types respond uniquely to increasing pressure. Peanut, cowpea and amaranth proteins are susceptible to high pressure, whereas kidney bean proteins are comparatively resistant (Yin et al., 2008). Denaturation also depends on the concentration of proteins; the higher the protein concentration, minimum will be the changes. Furthermore, compared to α helices, β sheet areas are less susceptible to deformation, and, therefore, they are less responsive to pressure. Various studies have shown that HPP results in the reduction of protein solubility mainly because of the formation of insoluble macro-aggregates (Condes et al., 2015). On the contrary, HPP raises the water-holding capacity of protein, which could potentially have a good effect on the textural qualities of many protein-rich food products. This is confirmed that HPP alters and modifies protein structure both in good and bad ways. The combination of HPP with other attributes will surely help in the creation of novel protein-based products with improved and promising properties.

7.7 CONCLUSION

This chapter provides an overview of how HHP technology has affected the safety, health, and edible qualities of plant-based commodities. By altering the technological functions of plant protein, HHP enhances the flavour of plant-based goods like vegan yoghurt and vegetarian meat. Additionally, the HHP treatment successfully lowers plant proteins' allergenicity and prevents the development of mycotoxins, assuring the safety of plant-based goods for ingestion. Despite the fact that several elements still need investigation, HHP technology has entered the industrial realm. The majority of plant-based goods now processed under HHP that are sold commercially include plant-based milks, yoghurts, meat, instant rice, plant-based dips, and ready-to-eat oatmeal. The application of HHP treatment can help formulate plant-based goods more effectively. HHP increases the quality and safety of food, increases shelf life, and enables a clearer label with fewer substances. Therefore, the use of HHP technology for the manufacturing of plant-based foods will grow over the next years. According to the current context of rising public knowledge about health and food, the need for nutritious and superior food is expanding. Thus, keeping in view the ultimate requirement of the production of plant-based foods by novel technologies, this chapter provides an overview of how HHP technology has affected the safety, health, and edible qualities of plant-based commodities. In former times, the extraction of plant proteins was done by the application of chemicals and heat treatments which reportedly destroy the structure of proteins and found undesirable. As a result, the trend is shifting towards the use of innovative or green technologies that are ideal for providing high-quality food. Consumers are now prepared to pay a premium for high-quality food goods. HHP enhances the flavour of plant-based goods like vegan yoghurt and vegetarian meat. Thus, the desire for higher-quality and more useful plant protein components can replace or compete with existing animal-based protein ingredients on the market. Furthermore, HPP has the potential to be used to modify the functional and rheological characteristics of legume protein. HPP (high-pressure processing) and UHT (ultra-high temperature treatments), in conjunction with ultrasonication and microfluidization, are commonly employed for the stabilization of protein emulsion products derived from plants. These alteration techniques are extensively used and highly commercialized in the food businesses. Additionally, the HHP treatment successfully lowers plant proteins' allergenicity and prevents the development of mycotoxins, assuring the safety of plant-based goods for ingestion. In order to develop novel proteins that may have superior properties, the combination of HPP and other parameters such as pH or temperature can be a game changer. Despite several elements still need investigation, HHP technology has entered the industrial realm. The majority of plant-based goods now processed under HHP that are sold commercially include plant-based milks, yoghurts, meat, instant rice, plant-based dips, and ready-to-eat oatmeal. The application of HHP treatment can help formulate plant-based goods more effectively by creating new structural approaches, adding plant protein-generated flavours, and enhancing plant protein nutritional values. HHP increases the quality and safety of food, increases shelf life, and enables a clearer label with fewer substances. Therefore, the use of HHP technology for the manufacturing of plant-based foods will grow over the next years.

REFERENCES

Abera, G. (2019). Review on high-pressure processing of foods. *Cogent Food & Agriculture*, 5(1), 1568725.

Acquah, C., Zhang, Y., Dubé, M. A., & Udenigwe, C. C. (2020). Formation and characterization of protein-based films from yellow pea (Pisum sativum) protein isolate and concentrate for edible applications. *Current Research in Food Science*, 2, 61–69.

Akasaka, K. (2006). Probing conformational fluctuation of proteins by pressure perturbation. *Chemical Review*, 106, 1814–1835.

Andersen, V., Halekoh, U., Tjonneland, A., Vogel, U., & Kopp, T. I. (2019). Intake of red and processed meat, use of non-steroid anti-inflammatory drugs, genetic variants and risk of colorectal cancer: A prospective study of the Danish "diet, cancer and health" cohort. *International Journal of Molecular Sciences*, 20, 1121.

Balasubramaniam, V. M., Martinez-Monteagudo, S. I., & Gupta, R. (2015). Principles and application of high pressure-based technologies in the food industry. *Annual Review of Food Science and Technology*, 6(1), 435–462.

Bolumar, T., Middendorf, D., Toepfl, S., & Heinz, V. (2016). Structural changes in foods caused by high-pressure processing. In *High pressure processing of food* (pp. 509–537). Springer, New York, NY.

Boukid, F. (2021). Chickpea (*Cicer arietinum* L.) protein as a prospective plant-based ingredient: A review. *International Journal of Food Science & Technology*, 56(11), 5435–5444.

Bridgman, P. W. (1914). The coagulation of albumen by pressure. *Journal of Biological Chemistry*, 18, 511–512.

Briviba, K., Gräf, V., Walz, E., Guamis, B., & Butz, P. (2016). Ultra high pressure homogenization of almond milk: Physico-chemical and physiological effects. *Food Chemistry*, 192, 82–89.

Chauhan, O. P., Kumar, S., Nagraj, R., Narasimhamurthy, R., & Raju, P. S. (2015). Effect of high pressure processing on yield, quality and storage stability of peanut paneer. *International Journal of Food Science & Technology*, 50(6), 1515–1521.

Chen, G., Wang, S., Feng, B., Jiang, B., & Miao, M. (2019). Interaction between soybean protein and tea polyphenols under high pressure. *Food Chemistry*, 277, 632–638.

Ciociola, T., Giovati, L., Conti, S., & Magliani, W. (2021). Anti-infective antibody-derived peptides active against endogenous and exogenous fungi. *Microorganisms*, 9(1), 143.

Condes, M. C., Anon, M. C., & Mauri, A. N. (2015). Amaranth protein films prepared with high-pressure treated proteins. *Journal of Food Engineering*, 166, 38–44.

Dhakal, S., Liu, C., Zhang, Y., Roux, K. H., Sathe, S. K., & Balasubramaniam, V. M. (2014). Effect of high pressure processing on the immunoreactivity of almond milk. *Food Research International*, 62, 215–222.

Evelyn, F., & Silva, F. V. (2019). Heat assisted HPP for the inactivation of bacteria, moulds and yeasts spores in foods: Log reductions and mathematical models. *Trends in Food Science & Technology*, 88, 143–156.

Gharibzahedi, S. M. T., & Smith, B. (2021). Effects of high hydrostatic pressure on the quality and functionality of protein isolates, concentrates, and hydrolysates derived from pulse legumes: A review. *Trends in Food Science & Technology*, 107, 466–479.

Gross, M., & Jaenicke, R. (1994). Proteins under pressure: The influence of high hydrostatic pressure on structure, function and assembly of proteins and protein complexes. *European Journal of Biochemistry*, 221(2), 617–630.

Hameed, F., Ayoub, A., & Gupta, N. (2018). Novel food processing technologies: An overview. *IJCS*, 6(6), 770–776.

Hertzler, S. R., Lieblein-Boff, J. C., Weiler, M., & Allgeier, C. (2020). Plant proteins: Assessing their nutritional quality and effects on health and physical function. *Nutrients*, 12(12), 3704. doi:10.3390/nu12123704

Huang, H. W., Hsu, C. P., Yang, B. B., & Wang, C. Y. (2013). Advances in the extraction of natural ingredients by high pressure extraction technology. *Trends in Food Science & Technology*, 33(1), 54–62.

Huang, H. W., Wu, S. J., Lu, J. K., Shyu, Y. T., & Wang, C. Y. (2017). Current status and future trends of high-pressure processing in food industry. *Food Control*, 72, 1–8.

Janardhanan, R., Huerta-Leidenz, N., Ibañez, F. C., & Beriain, M. J. (2022). High-pressure processing and sous-vide cooking effects on physicochemical properties of meat-based, plant-based and hybrid patties. *LWT*, 114273.

Jones, O. G. (2016). Recent advances in the functionality of non-animal-sourced proteins contributing to their use in meat analogs. *Current Opinion in Food Science*, 7, 7–13.

Kumar, M., Tomar, M., Potkule, J., Verma, R., Punia, S., Mahapatra, A., . . . Kennedy, J. F. (2021). Advances in the plant protein extraction: Mechanism and recommendations. *Food Hydrocolloids*, 115, 106595.

Levy, R., Okun, Z., Davidovich-Pinhas, M., & Shpigelman, A. (2021). Utilization of high-pressure homogenization of potato protein isolate for the production of dairy-free yogurt-like fermented product. *Food Hydrocolloids*, 113, 106442.

Liu, C., Pei, R., & Heinonen, M. (2022). Faba bean protein: A promising plant-based emulsifier for improving physical and oxidative stabilities of oil-in-water emulsions. *Food Chemistry*, 369, 130879.

Malik, V. S., Li, Y., Tobias, D. K., Pan, A., & Hu, F. B. (2016). Dietary protein intake and risk of type 2 diabetes in US men and women. *American Journal of Epidemiology*, 183, 715–728.

McCann, T. H., Leder, A., Buckow, R., & Day, L. (2013). Modification of structure and mixing properties of wheat flour through high-pressure processing. *Food Research International*, 53(1), 352–361.

Mulla, M. Z., Subramanian, P., & Dar, B. N. (2022). Functionalization of legume proteins using high pressure processing: Effect on technofunctional properties and digestibility of legume proteins. *LWT*, 113106.

Muntean, M. V., Marian, O., Barbieru, V., Cătunescu, G. M., Ranta, O., Drocas, I., & Terhes, S. (2016). High pressure processing in food industry–characteristics and applications. *Agriculture and Agricultural Science Procedia*, 10, 377–383.

Mustafa, A., & Turner, C. (2011). Pressurized liquid extraction as a green approach in food and herbal plants extraction: A review. *Analytica Chemical Acta*, 703(1), 8–18.

Nasrabadi, M. N., Doost, A. S., & Mezzenga, R. (2021). Modification approaches of plant-based proteins to improve their techno-functionality and use in food products. *Food Hydrocolloids*, 118, 106789.

Pasiakos, S. M. (2015). Metabolic advantages of higher protein diets and benefits of dairy foods on weight management, glycemic regulation, and bone. *Journal of Food Science*, 80(S1), A2–A7.

Pasiakos, S. M., Lieberman, H. R., & Fulgoni III, V. L. (2015). Higher-protein diets are associated with higher HDL cholesterol and lower BMI and waist circumference in US adults. *The Journal of Nutrition*, 145(3), 605–614.

Qin, P., Wang, T., & Luo, Y. (2022). A review on plant-based proteins from soybean: Health benefits and soy product development. *Journal of Agriculture and Food Research*, 7, 100265.

Queirós, R. P., Saraiva, J. A., & da Silva, J. A. L. (2018). Tailoring structure and technological properties of plant proteins using high hydrostatic pressure. *Critical Reviews in Food Science and Nutrition*, 58(9), 1538–1556.

Samaei, S. P., Ghorbani, M., Tagliazucchi, D., Martini, S., Gotti, R., Themelis, T., & Babini, E. (2020). Functional, nutritional, antioxidant, sensory properties and oil in- water emulsions. *Journal of Agricultural and Food Chemistry*, 53(26), 10248–10253.

Sim, S. Y. J., Karwe, M. V., & Moraru, C. I. (2019). High pressure structuring of pea protein concentrates. *Journal of Food Process Engineering*, 42(7), e13261.

Sim, S. Y. J., Srv, A., Chiang, J. H., & Henry, C. J. (2021). Plant proteins for future foods: A roadmap. *Foods*, 10(8), 1967.

Sim, S. Y. J., Xin, Y. H., & Christiani, J. H. (2020). A novel approach to structure plant-based yogurts using high pressure processing. *Foods*, 9(8), 1126.

Srinivas, M. S., Madhu, B., Srinivas, G., & Jain, S. K. (2018). High pressure processing of foods: A review. *Agricultural of Journal*, 65, 467–476.

Stratakos, A. C., Inguglia, E. S., Linton, M., Tollerton, J., Murphy, L., Corcionivoschi, N., . . . Tiwari, B. K. (2019). Effect of high pressure processing on the safety, shelf life and quality of raw milk. *Innovative Food Science & Emerging Technologies*, 52, 325–333.

Strieder, M. M., Silva, E. K., Mekala, S., Meireles, M. A. A., & Saldaña, M. D. (2022). Pulsed high-pressure processing of barley-based non-dairy alternative milk: β-carotene retention, protein solubility and anti-oxidant activity. *Innovative Food Science & Emerging Technologies*, 103212.

UNICEF. (2022). UNICEF data: Monitoring the situation of children and women. data.unicef.org/ Accessed on 21/12/2022.

Yan-ping, L., Sukmanov, V. O., & Hanjun, M. (2021). The effect of high pressure on soy protein functional features: A review. *Journal of Chemistry and Technologies*, 29(1), 77–91.

Yin, S.-W., Tang, C.-H., Wen, Q.-B., Yang, X.-Q., & Li, L. (2008). Functional properties and in vitro trypsin digestibility of red kidney bean (*Phaseolus vulgaris* L.) protein isolate: Effect of high-pressure treatment. *Food Chem*, 110, 938–945.

Yordanov, D. G., & Angelova, G. V. (2010). High pressure processing for foods preserving. *Biotechnology & Biotechnological Equipment*, 24(3), 1940–1945.

Zhang, H., Li, L., Tatsumi, E., & Isobe, S. (2005). High-pressure treatment effects on proteins in soy milk. *LWT-Food Science and Technology*, 38(1), 7–14.

Zhang, Z., Cho, S., Dadmohammadi, Y., Li, Y., & Abbaspourrad, A. (2021). Improvement of the storage stability of C-phycocyanin in beverages by high-pressure processing. *Food Hydrocolloids*, *110*, 106055.

Zhao, H., Song, A., Zheng, C., Wang, M., & Song, G. (2020). Effects of plant protein and animal protein on lipid profile, body weight and body mass index on patients with hypercholesterolemia: A systematic review and meta-analysis. *Acta Diabetologica*, 57, 1169–1180.

8 Pulsed Electric Field Modifications of Plant-Based Proteins

Shweta Suri and Deepika Kathuria

8.1 INTRODUCTION

Protein in food plays an important role in human nutrition. It is a macromolecule formed by a long chain of amino acid residue. It is required in the human body for the structure, function, and regulation of the body's tissues and organs. They are also key ingredients in the formulation of food products owing to their necessary functions such as thickening and gelling capacity, emulsification, water binding, foaming, and fat absorption (Cao et al., 2019). Dairy proteins are an excellent source of functional proteins with higher nutritional values, acting as a raw material in many food products. However, maintaining sustainability in food system has increased the utilization of plant proteins as alternative sources of animal proteins (FAO, 2017). In addition, the consumption of animal protein can have adverse impacts on human health, including, hypertension and obesity (Richter et al., 2015). There is also concern about the allergenic potential of animal proteins, which can be significantly lowered by the application of plant proteins (Sá et al., 2020).

In addition, plant proteins have lower production cost, more environment-friendliness as well as higher acceptability among consumers (Fasolin et al., 2019). Plant-based foods are said to be healthier and relatively "no" more worrisome. Plant-based foods are considered more environmentally friendly and better for the climate. Plant-based foods can be favored to provide a varied diet, and because they are new and trendy (Aschemann-Witzel et al., 2021). Based on nutritional characteristics, plant proteins are widely accepted for partial or complete replacement of animal proteins in order to reduce the health-related diseases like cardiovascular disease and type 2 diabetes (Tharrey et al., 2018). This has increased the demand of plant-based protein in the market which was projected as $35 billion in 2018 and might get raised to $45 billion by 2023 at a CAGR of 4.93% (Akharume et al., 2021). Moreover, plant proteins act as a versatile ingredient in different food products due to their desirable characteristics such as thickening, gelling, stabilizing, foaming, and water-holding and oil-absorption abilities. However, the plant proteins are obstinate due to high molecular weight, weak electrostatic repulsion, and lower solubility. Furthermore, these proteins have poor digestibility and possess some specific plant residue including antinutrients. The modulation in protein characteristics improves its functionality as well as eliminates the adverse effect of these antinutrients. The application of different processing methods such as physical (heat or mild alkali treatment), chemical (acylation, phosphorylation, and deamidation), enzymatic (proteases), and other novel techniques has been proposed (Nasrabadi et al., 2021) that can result in the modification of structure of food biopolymers and thereby considerably influence the technological functional properties (Knorr et al., 2011; Esteghlal et al., 2019). Among different methods, nonthermal processing methods have gained popularity in the food industry, which recognizes their potential to replace or counter traditional thermal processing methods. By combining the safety of pasteurization with higher quality and lower processing costs, pulsed electric fields (PEF) technology has ushered in a revolution of food processing. It involves the application of high voltage pulses to liquid or semi-solid foods placed between two electrodes, which has been used successfully for enhanced chemical reactions and intracellular components extraction (Lin et al., 2012). Most of the PEF studies

DOI: 10.1201/9781003369790-8

are associated with microbial and enzyme inactivation, and less information is available about the structural modification of proteins, enzymes as well as overall quality and acceptability. Physical, chemical, enzymatic, and high-pressure techniques all modify the functional properties of food proteins, but PEF offers a technique that modifies these properties in a consumer-friendly, "green" manner, resulting in the production of high-value food products. Hence, this chapter provides an insight on the changes in structural and functional properties of plant-based proteins by using PEF.

8.2   PLANT-BASED PROTEINS AND THEIR UTILIZATION CHALLENGES IN FOOD PRODUCTS

Plant-based proteins are gaining consideration in the food as well as pharmaceutical sectors owing to their benefits over animal-based materials, including lower cost. Cereals (wheat, rice, millet, sorghum, corn, and barley); legumes (peas, cowpeas, soybeans, beans, lupines, field peas, and chickpeas); pseudocereals (buckwheat, quinoa, and amaranth); nuts; almonds; and seeds (pumpkin, flax, chia, sesame, and sunflower) are among the sources of plant protein (Pojić et al., 2018). In addition, soy and pea proteins (legumes) and wheat gluten (cereals) are the most commonly used components of plant-based protein products due to their ease of use, proportion, and their processing capacity. Rice protein and mung bean protein are also often used in combination with key legume proteins to create a nutritionally balanced amino acid profile (Sha & Xiong, 2020).

Protein is a particularly important nutritional element. The proteins and amino acids that make them up contribute to the building of muscle fibers, the immune system, and the maintenance of many important functions. Human protein requirements should be met primarily through diet (Gorska-Warsewicz et al., 2018). Since protein is an important nutrient, the global demand for protein is determined by the size of the human population. Proteins of animal origin are by far the main sources of protein (Henchion et al., 2017).

Plant proteins are substitute to animal proteins and can be obtained from cheap and sustainable food sources, *viz.* agricultural plant waste and by-products from the vegetable and oil industries, which could also have an impact on reducing food waste (Sá et al., 2020). Plant-based proteins possess antioxidant, antibacterial, anti-hypertensive, anti-inflammatory, and antidiabetic biological properties (Jafari et al., 2020). Plant proteins can also be used to produce bioactive peptides. However, most plant proteins are poorly water-soluble; complex; and highly sensitive to pH, ionic strength, as well as temperature, making them difficult to handle and restraining their usage (Warnakulasuriya & Nickerson, 2018). Most plant proteins—for example, flaxseeds do not have a single protein component but a variety of proteins with variable fractions having broad isoionic points (Nasrabadi et al., 2019). Therefore, there is an urgent need to modify the properties of plant proteins to enhance their function. An additional drawback of plant proteins is the incidence of certain plant residues that are considered antinutrients. Also, some plant proteins have less applicability because of their undesirable taste. Figure 8.1 depicts certain challenges in the applicability of plant proteins.

8.3   MODIFICATION APPROACHES: PHYSICAL, CHEMICAL, AND BIOLOGICAL

The term "protein modification" states the course of altering the molecular structure or some chemical groups of a protein by specific methods with an aim of improving its technical functionality and biological activity. Modification of plant proteins offers great potential of becoming multifunctional components of food systems by altering their physicochemical action and addressing their drawbacks (Nasrabadi et al., 2019). A number of researchers suggested the different protein modification approaches, *viz.* the physical, chemical, and biological, to enhance the quality of plant protein. Figure 8.2 shows the different approaches to plant protein modification.

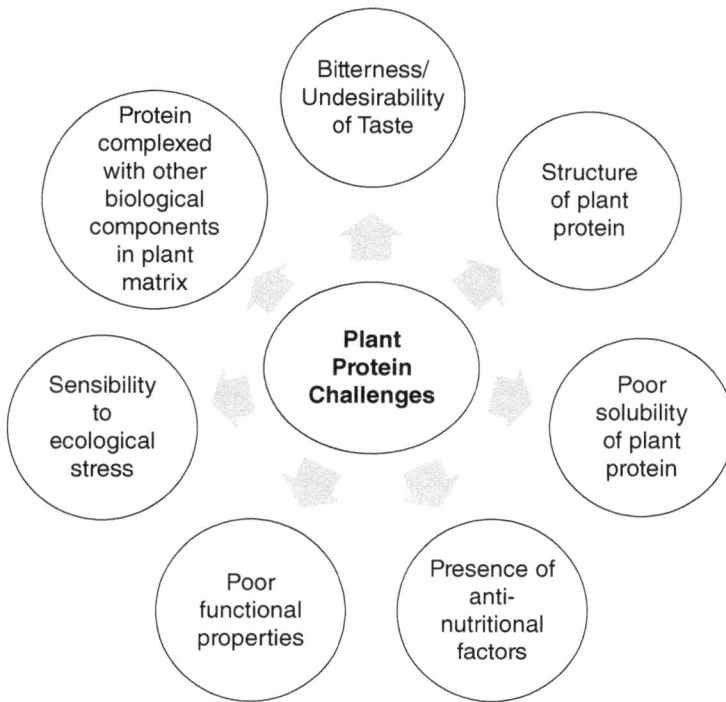

FIGURE 8.1 Challenges in the application of plant proteins.

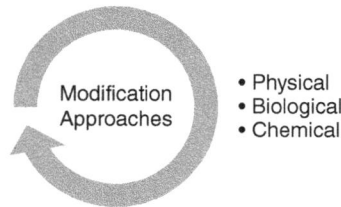

FIGURE 8.2 Modification approaches for plant proteins.

Physical methods are the modest approaches for extending protein functionality. They do not rely on chemicals or enzymes. These protein modification methods are of great interest because they do not use chemicals in their processing and avoid the detrimental consequences of potential chemical residues. Physical methods include thermal processing (conventional heating, ohmic, microwave, radio frequency, and infrared), high-pressure treatment, ultrasonication, irradiation, pulsed electric field (PEF), cold plasma treatment, etc. PEF technology is a physical way of protein modification that uses high electric field pulses with intensities ranging from 10 to 80 kV/cm and lasting for few microseconds to milliseconds (Syed et al., 2017). PEF technique is effective in enhancing the efficiency of extraction of plant proteins. It also possesses the capacity to activate and obstruct the endogenous enzymes for the preservation of foods. Furthermore, PEF technique is helpful in improving the biological properties of food-based peptides (Zhang et al., 2021b). PEF has several benefits relating to its nonthermal performance, speed, efficiency, low power consumption, and low pollution. PEF technology has been shown to be a novel way that can be utilized to enhance bioactivity, especially the antioxidant capacity of different types of peptides in plant-based foods.

PEF may therefore be a good option to be used in the processing of plant proteins (Park & Boxer, 2002), and also it leads to macroscopic changes that can affect the quality of a wide variety of foods (Zhang et al., 2021b).

Besides, the chemical ways of modification of plant-based proteins have been broadly applied owing to their effectiveness, low cost, and ease of use. The chemical ways of protein modification can be accomplished by adding novel functional fragments or removing components from the protein structure. Most chemical modification methods have regulatory issues because they apply chemicals as well as produce chemical by-products in most cases. Thus, some of these chemical plant protein ways are not promising for applications in food (Zhang et al., 2019). Chemical protein modification ways include phosphorylation, cationation, acylation, glycation, pH shifting, deamidation, etc. Apart from the physical and chemical ways of protein modification, there are certain biological ways which include the use of enzymes and the fermentation process. Biological ways of modification of plant-based proteins are environmental friendly, less energy-intensive, and do not generate toxic by-products. However, the price of enzymes and cultures must be considered for large-scale usage. Moreover, biological ways of modification also have the potential to enhance the nutritional qualities of food, such as digestibility, nutrient absorption capacity, as well as antioxidant and antimicrobial action (Nasrabadi et al., 2019).

8.4 MECHANISM/PRINCIPLE OF PULSED ELECTRIC FIELD TREATMENT

PEF processing is one of the emerging techniques in the field of electromagnetic for medicinal, environmental, and food usage (Raso et al., 2016). The technique is mostly dependent on the number of pulses applied to the product held between two electrodes. Between these electrodes, there is a precise gap known as the processing gap of the chamber. A high voltage is used during the PEF process for the inactivation of microorganisms found in the food product. Electric fields are applied in various forms, such as exponentially decaying waves, dipolar waves, and oscillating pulses. This method can also be performed at different temperature ranges, *viz.* ambient, below ambient, and above ambient. Food is packaged and refrigerated after PEF treatment. The concept involved is the transmission of electrical pulses in the food, and the food comprises certain ions delivering a defined degree of electrical conductivity (Syed et al., 2017) (Zhang et al., 1995). The mechanism/principle behind the functioning of the PEF technique is to provide pulse power to the product placed between a space-limited set of electrodes that process the PEF chamber. A typical system includes a high-voltage pulse generator, a processing/treatment chamber that manages the products to be processed, and control and monitoring devices (Figure 8.3). Food samples are placed in a treatment/processing chamber equipped with electrodes that are bonded together with a nonconductive material to prevent electrical currents from passing from one object to another. High-voltage electrical pulses are generated and transmitted to the product. The product placed between electrodes is subjected to a force per unit charge that ruptures the bacterial cell membrane (Fernandez-Diaz et al., 2000). In general, PEF technology is suggested for the pasteurization of various food products, including milk, juice, yogurt, liquid eggs, and soups (Bendicho et al., 2003). In addition, combining PEF with ultrasonic, high-pressure, and UV treatments can improve process productivity.

8.5 APPLICATION OF PULSED ELECTRIC FIELD TREATMENT FOR PROTEIN MODIFICATION

PEF is commonly used in food preservation due to its tremendous potential for preserving high-quality products at reduced temperature with shorter residence time, resulting in maximum retention of the product's freshness and nutritional value (Toepfl et al., 2006). The PEF treatment is significantly implemented in the preservation of liquid foods; hence, there is a need to study the

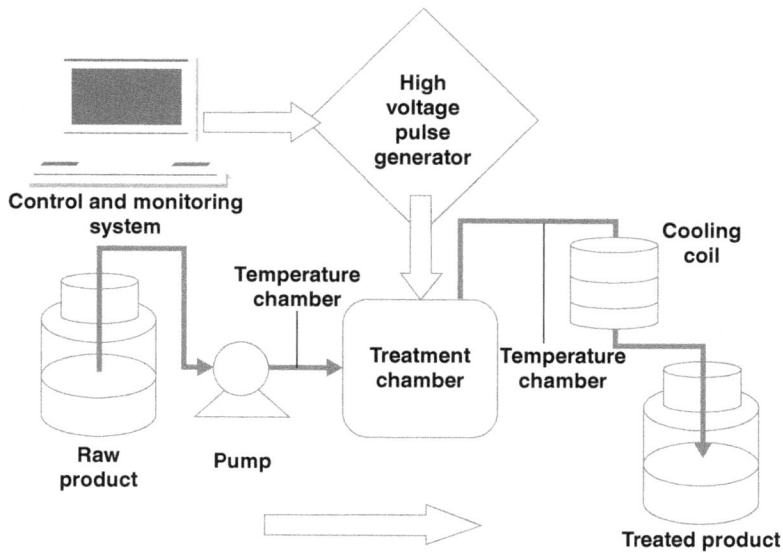

FIGURE 8.3 Schematic diagram showing PEF-food processing system with basic component.

Source: Mohamed, M. E., and A. H. A. Eissa. 2012. "Pulsed electric fields for food processing technology." *Structure and function of food engineering* 11: 275–306. https://doi.org/10.5772/48678.

possible changes taking place in food proteins on the application of PEF. This technique disrupts electrostatic interactions, resulting in the modification of secondary and tertiary structures of protein via unfolding, denaturation, followed by final aggregation. PEF treatment of protein molecules causes free electron, ion, and other charged particle movement; polarization; dipole moment orientation; and dielectric constant changes at lower electric field strength. At initial electric field strength, the hydrophobic amino acid gets exposed to the solvent side, resulting in the unfolding of protein; however, at very higher field strength, the unfolded protein gets aggregated due to the formation of weak covalent and noncovalent bonds (Han et al., 2018). These changes subsequently affect the functional properties, i.e., increase the protein solubility. Xiang et al. (2011) and Singh et al. (2013) studied the effect of PEF on soybean protein isolate and reported changes in protein conformation. The effects of PEF in food products are related to the PEF system and the liquid food properties (Barsotti et al., 2001). In the case of a PEF system, the electric field intensity, pulse waveform, number of pulses, pulse width, treatment temperature, and time are important factors that affect protein structure, including monomer unfolding, outright denaturation, aggregation, and gel formation (Barbosa-Cánovas et al., 1999; Barsotti et al., 2001). At relatively low electric field intensity, the stress involved may not be sufficient for some food products to induce permanent denaturation (Niu et al., 2020). Waveforms such as exponential decay and square pulse applied in bipolar fashion exert more stress on the cell membrane and enhance its electric breakdown due to alternating changes in the movement of charged molecules (Barbosa-Cánovas et al., 1999). PEF treatment time is calculated by multiplying the number of pulses with the pulse width (Barbosa-Cánovas et al., 2000). Further, to avoid unnecessary excessive heating and undesirable electrolytic reactions, PEF treatment is generally applied in the form of short pulses (Sepulveda et al., 2005). However, in order to structurally modify protein in foods, the system must have a wide pulse width (Perez & Pilosof, 2004). Furthermore, the temperature of PEF treatment has an impact on protein modification. PEF treatments are more effective at mild temperatures (50–60°C) (Dunn & Pearlman, 1987). According to Zhang et al. (2017), PEF treatment increases the solubility and emulsifying and foaming properties of canola seed-extracted protein; however, depending upon the electric field intensity and duration of process, it causes denaturation and aggregation which thereby reduce the solubility. On the other hand, the delicate equilibrium that maintains the protein's native structure is also affected by environmental factors such as pH, salinity, and ionic strength, and elevated pressure (Walstra et al., 2006).

8.6 CHANGES IN THE PROPERTIES OF PLANT-BASED PROTEIN UPON PULSED ELECTRIC TREATMENT

Protein is an important food ingredient used for determining the physical behavior of the food during preparation, processing, and storage. Processing conditions such as heat treatment, change in pH, and salt addition result in unfolding or denaturation of a protein. The subtle alteration in size, charge distribution, hydrophilicity, hydrophobicity, molecular flexibility, and steric properties of proteins affects their functional properties which thereby changes the textural and sensory characteristics of food. Among these, molecular flexibility and hydrophobicity significantly influence various functional properties such as solubility, water-holding capacity, foaming, emulsification, and gelling ability. The protein modification involves structural modifications by changing the balance between forces required to maintain the native structure of the protein. The initial step of structure modification includes protein unfolding followed by change in secondary structure due to covalent and noncovalent changes resulting in incorrect folding or in aggregation (Mañas & Vercet, 2006). For instance, protein solubility is determined by the hydrophilic and hydrophobic residues on the protein surface, as well as the number of hydrogen bonds. The changes in the contents of sulfhydryl (SH) groups and disulfide bonds alter the gelling property. Furthermore, the emulsifying property is more closely related to protein surface activity (Han et al., 2018). Despite the fact that the main mechanism underlying the effects of electric fields on proteins is not well understood, few authors have reported changes in the apparent charge of proteins following PEF treatment (Jaeger et al., 2014; Nunes et al., 2019). During PEF treatment, protein polar groups absorb energy and generate free radicals, impacting intramolecular interactions such as hydrophobic and electrostatic interactions, disulfide bridges, hydrogen bonds, salt bridges, and Van der Waals forces (Han et al., 2018). PEF treatment causes protein molecules to gradually stretch or deform in response to an external electric field. Furthermore, protein molecules have heterogeneous charge distributions that are pulled in opposite directions in the presence of an external electric field, hence destabilizes the secondary and tertiary structure of protein (Liu et al., 2019). As a result, PEF treatment dissociates protein electrostatic fields, impacting the electronic interactions of polypeptide chains and exposing the inner buried hydrophobic and SH groups (Zhao et al., 2014). The structural and functional properties of proteins can be altered. The detailed effect of PEF on structural and functional properties of protein is discussed in the next section.

8.6.1 EFFECT ON STRUCTURE OF PROTEIN

Protein is an important component of food that has different nutritional, functional, and textural properties depending on its molecular structure. Primary structure (straight chain of amino acids linked by peptide bond), secondary structure (repeated local structure formed by folding of peptide chain by hydrogen bond), and tertiary structure are all components of protein structure (unevenly folded irregular 3D pattern of peptide chain). In case of plant protein such as soya, protein is mainly composed of two globular proteins—i.e., 7S (β-conglycinin) and 11S (glycinin) (Riblett et al., 2001) whose hierarchical structure, tight conformation, and lower molecular flexibility result in lower solubility and limited functional properties and thereby hinder its application in food industry (Lee et al., 2016). For the modification of protein, unfolding of protein is the major step. The structural changes occurring due to covalent changes which result in chemical derivatization of functional groups of amino acid, peptides, and protein or by non-covalent modification lead to conformational changes in protein special structure (polymer-chain-acquired 3D structure) (Zhao et al., 2012). PEF treatment causes minor or insignificant changes in the protein's primary structure (peptides and amino acids). The modification in protein secondary structure caused by PEF treatment is due to the strong dipole moment created by the protein's α-helix nature. The applied external electric field affects these dipole moments, which align in the same direction to form a macrodipole, causing structural instability. As a result, an external electric field dissociates the protein's local electrostatic field and influences the electronic interaction of polypeptide chains (Liu et al., 2019). Various studies have reported the disruption of α-helix structure of protein by PEF treatment for the inactivation of enzyme activity (Zhong et al., 2007; Zhao & Yang, 2008). On the other hand, β-sheet results in random coiling on the application of PEF treatment. Furthermore, Zhang et al. (2017) observed the transformation of α-helix and β-turn structural forms into random coil and β-sheet structural forms in canola protein isolate after PEF treatment at 30 kV voltage for 180 s. Melchior et al. (2020) observed a similar transformation in pea, rice, and gluten protein concentrates with an increase in the number of pulses at 1.65 kV/cm electric field strength, 55s pulse width, and 400 Hz frequency of PEF treatment. Furthermore, Liang et al. (2018) discovered α-helix transformations in two peptides of pine nut protein (Lys-Cys-His-Gln-Pro and Gls-Cys-His-Lys-Pro) with an increase in PEF intensity (5–20 kV/cm). However, Liu et al. (2010) observed a decrease in the α-helix and β-sheet structure in soy protein isolate after a PEF treatment of 50 kV/cm for 1,600–2,400 s. In the case of tertiary structure, the amino acid residues generate asymmetric spatial distributions of charges, resulting in strongly polarized and charged regions. Surface hydrophobicity is one of the structural properties, which reflects the change in 3D conformation caused by PEF (Kato & Nakai, 1980). The surface hydrophobicity of soya protein isolate increased with the number of pulses at 25 kV/cm and 60 pulses (Xiang et al., 2011). Due to intermolecular hydrophobic interaction, Li et al. (2007) discovered hydrophobic collapse in soya protein isolate on PEF treatment above 30 kV/cm and time above 288 ms. Wang et al. (2023) observed a similar effect of PEF treatment on hydrophobicity in soya protein isolate, which increased up to 10 kV/cm, but beyond this resulted in hydrophobic collapse. Furthermore, the presence of disulfide bonds or cysteine residues results in proper protein folding, structure, and function. Disulfide bond conformational changes, cleavage, or formation of disulfide bonds and exposure or embeddedness of sulfhydryl groups under external stress all contribute to protein unfolding (Betz, 1993). Protein aggregation is caused by both hydrophobic interactions and disulfide bonds during PEF treatment. PEF treatment increased the surface sulfhydryl group in soya protein isolate, but PEF treatment above 30 kV/cm for 288 s resulted in a decrease in surface sulfhydryl groups due to protein aggregation by disulfide bonds. PEF treatment facilitates the crosslinking of amino acid side chains by breaking hydrophobic bonds (Subaşi et al., 2021). The effect of PEF treatment on the structural properties of various plant proteins is shown in Table 8.1.

8.6.2 Effect on Function of Protein

Pulsed electric field (PEF) treatment of food is nowadays of scientific interest because this technique favors nonthermal ways and helps to preserve natural flavors and nutrients intact. PEF is a commercially used technique because of its high performance, low energy usage, and light treatment technology (Rahaman et al., 2019). The application of pulsed electric field (PEF) leads to modifications of plant-based proteins. PEF treatment leads to changes in the structural properties of plant foods by supporting the release of bioactive metabolites or improving the shelf life of foods without the use of chemicals (Manzoor et al.,2019). PEF is considered a promising sublethal technique to increase the permeability of cell membranes, where exposure of food materials to short, repetitive high-voltage pulses can lead to the formation of temporary or permanent pores in cell membranes (Siddeeg et al., 2019). This technology was originally introduced in the treatment of liquid foods to inactivate microorganisms and enzymes, primarily in fruit juices, eggs, and milk and in the disruption of plant cells to improve targeted extraction of substances such as vegetable oils (Zhang et al., 2017). The processing of plant-based food using PEF technology poses significant challenges. In comparison to fruit juices and vegetable products, foods that are rich sources of

TABLE 8.1

Effect of PEF on the Structural Properties of Plant-based Protein

Sr. No.	Protein Source	PEF Condition	Changes in Structural Properties of Protein	Reference
1.	Soya protein isolate	0–40 kV/cm for 0–547 μs, 2 ms pulse width, and 500 pulse frequency	• PEF strength above 30 kV/cm and treatment time above 288 μs caused changes in the secondary structures, i.e., decrease in α-helix and increase in β-sheet along with constant random coiling. • PEF treatment at 547 μs caused higher denaturation and aggregation of soyabean protein isolate.	Li et al. 2007
2.	Soya protein isolate	0–50 kV/cm; 40 μs pulse width; treatment time 4.8 ms	• PEF treatment beyond 35 kV/cm relocates β-turns into structured α-helix • Slight increase in anti-parallel β-sheets while decrease in β-sheets content	Liu et al. 2011
3.	Soya protein isolate	0–15 kV; 1–8 μs pulse width; 1–9 ms pulse cycle	• Decrease in α-helix while increased random coils and β-sheets. • Collapse of hydrophobic core; however, strong PEF treatment causes reburial of hydrophobic residues back into the core. • Changes in disulfide bonds.	Li 2012
4.	Canola protein	10–35 kV/cm; pulse width 8 μs; residence time 180 s	• Increase in pulse intensity time reduced β-sheets and α-helix, number of total sulfhydryl group while resulting in free sulfhydryl groups • Increased surface hydrophobicity • PEF caused protein molecule aggregation • Alter secondary and tertiary structures changing a-helices and β-sheets in amide I	Zhang et al. 2017
5.	Pine nut protein	5–20 kV/cm at 1,800 and 2,400 Hz	• Alteration in tertiary structure due to higher surface hydrophobicity and free sulfhydryl group	Liang et al 2018
6.	Rice, pea, and gluten protein	1.65 kV/cm for 5 μs, 400 Hz and 20,000/60,000 pulse per second	• Significant changes only in gluten protein due to the presence of sulfhydryl group • Cleavage of disulfide bonds, reduction of α-helix and β-sheet structures with concomitant increase in random coil	Melchior et al. 2020
7.	Vital wheat gluten	2.5–12.5 kV/cm at 100–900 Hz, pulse width for 2–10 μs for 1–9 min	• Decrease in α-helix and β-turns while increase in β-sheets. • Exposure of inner molecules i.e., sulfhydryl groups. • Surface hydrophobicity was reduced,	Zhang et al. 2021a
8.	Sunflower protein	10–150 V/cm for 5 s–2 h at 25–45°C	• Change in secondary and tertiary structures at moderate electric field at 150 V for 20 s	Subasi et al. 2021
9.	Pea protein isolate	5, 10, and 20 V/cm and frequencies of 50 Hz and 20 kHz	• Unfolding of α-helix into β-sheet structure • Exposure of aromatic amino acid to solvent	Chen et al. 2022

protein usually score higher in heat sensitivity, viscosity, and electrical conductivity (Zhao et al., 2012). The effect of PEF treatment on soluble proteins in wheat seeds was studied. Soluble proteins are essential components of many plant enzymes, which play an important role in plant growth and stimulate metabolism. Unbound proteins within cellular and/or organelle membranes constitute soluble proteins (Sinha, 2004). PEF treatment led to an increase in the soluble protein content of wheat seeds from 8.94 mg/g in untreated seeds to 8.98 mg/g in PEF-treated seeds at 2 kV/cm with

25 pulses followed by 9.46 mg/g in PEF treatment at 2 kV/cm with 50 pulses, 9.51 mg/g in PEF-treated seeds with 4 kV/cm with 25 pulses, and 10.02 mg/g in PEF-treated samples with 6 kV/cm with 50 pulses (Ahmed et al., 2020). Previous studies found that the enhancement in soluble protein content by PEF treatment was owing to the voltage stress or voltage-sensitive channels. This channel can be stimulated at insufficient levels of voltage, opening the critical transmembrane potential. Due to the electrical damage, the nutrients became more readily available and therefore increased (Toepfl et al., 2006).

Further, increasing the electric field strength to 2, 4, and 6 kV/cm and pulse numbers (25 and 50) significantly increased total free amino acids in wheat seedlings; nevertheless, individually decreasing most amino acids was shown. The largest enhancement in total free amino acids (21.7%) was observed with PEF-treated samples with 6 kV/cm with 50 pulses (Toepfl et al., 2006). Studies revealed that PEF causes permeabilization, and plant matrix or cells undergo significant mutations when exposed to high-voltage electrical pulses. These electrical pulses result in structural changes and cell membrane disassembly as a function of pulse duration, EF intensity, and pulse number (Toepfl et al., 2006). PEF-treated milk showed improved rentability in comparison to heat-pasteurized milk and improved texture and organoleptic properties of cheddar cheese derived from PEF-pasteurized milk (Yu et al., 2009). Therefore, the application of PEF may be a good alternative to heat treatment of protein-based foods.

The solubility of plant protein decreases and increases with the exposure to PEF treatment. In a study, it was found that the solubility of the canola protein increased upon PEF treatment, whereas compared to the control sample, the PEF-treated sample showed 43.25% increase; however, a 50.07% increase was observed at 35 kV voltage followed by 52.74% increase when the residence time period was set at 180 s, 61.30% when pulse width was set at 8 µs, and 63.35% when the pulsed frequency was set at 1,000 Hz. However, a further increase in residence time above 180 s resulted in a significant decrease in solubility (Zhang et al., 2017). On contrary, a significant decrease in solubility was observed in most cases upon the application of a moderate-intensity pulsed electric field, regardless of protein type as well as pH conditions. The solubility of the pea protein decreases from 21.1% with no PEF treatment to 20% with the application of 60,000 pulses at 5 pH, while at 6 pH with the application of pulses. The decrease in the solubility is observed from 23.2% to 14.9%. Also, in rice at a pH of 5, when pulses were applied, the solubility increased from 13.6% in control to 8.4%. Likewise, at pH 6, there was a decrease in the solubility of rice protein from 16.4% to 10.8% (Melchior et al., 2020). Remarkably, the solubility of wheat protein gluten nearly doubled after 20,000 pulses at pH 5. Applying an electric field to a protein at a pH below its isoelectric point facilitates the exposure of the protein's charge, enhances water–protein interactions, and improves overall solubility (Melchior et al., 2020).

A number of researchers studied the effect of PEF treatment on the functional properties of plant protein like water-holding capacity, oil-holding capacity, and emulsion stability. Table 8.2 presents the information about the effect of PEF on functional attributes of plant proteins. The water-holding capacity of canola protein increased at the lower end of the PEF parameters and residence time from the control sample (2.45 g/g) and decreased at the higher end of these parameters. The maximum water-holding capacity was found to be 3.64 g/g at a voltage (25 kV), 3.82 g/g when residence time was set at 150 s, 4.12 g/g at 4 µs pulse width, and 3.92 g/g for a pulse frequency of 400 Hz (Zhang et al., 2017). Water-holding capacity is mainly due to the protein type as well as pH of the suspension. A manifest enhancement in functional activities was detected only in the case of wheat protein gluten at pH 6. Usually, near the isoelectric point, the application of moderate-intensity pulsed electric field treatments resulted in the formation of aggregates which are stabilized by disulfide bonds. The pulsed-electric-field-treated sample is perhaps better able to hold both water and oil in the resultant network. Alternatively, PEF will change the ability of wheat protein gluten to relate with solvents by a pH-influenced mechanism. Near the isoelectric point, gluten proteins will interact with water largely by interactions with contact hydrophilic groups, though at lower pH, solvents will also be trapped in the protein lattice when interacting with contact groups. Furthermore, it is not

TABLE 8.2

Effect of PEF on the Functional Properties of Plant-based Protein

Sr. No.	Protein Source	PEF Condition	Changes in Functional Properties of Protein	Reference
1.	Canola protein isolate	10–35 kV at 5 kV pulse frequency of 600, pulse width of 8 ms, and residence time of 180 s	• Maximum water-holding capacity was found to be 3.64 g/g at a voltage (25 kV), 3.82 g/g when residence time was set at 150 s, 4.12 g/g at 4 µs pulse width, and 3.92 g/g for a pulse frequency of 400 Hz.	Zhang et al., 2017
2.	Wheat (*Triticum aestivum* L.) seeds	2–6 kV/cm; 25 and 50 pulses	• An increase in the soluble protein content of wheat seeds from 8.94 mg/g in untreated seeds to 8.98 mg/g in PEF-treated seeds at 2 kV/cm with 25 pulses followed by 9.46 mg/g in PEF treatment at 2 kV/cm with 50 pulses, 9.51 mg/g in PEF-treated seeds with 4 kV/cm with 25 pulses, and 10.02 mg/g in PEF-treated samples with 6 kV/cm with 50 pulses	Ahmed et al., 2020
3.	Soy milk	12–25 kV cm^{-1}; 10–120 pulses	• Apparent viscosity of soy milk increased from 6.62 to 7.46 (10^{-3} Pa s) by increasing electric field intensity from 18 to 22 kV/cm and number of pulses increasing from 0 (the no treated control) to 100.	Xiang, 2009
4.	Rice protein	Pulse width (5 µs), frequency (400 Hz), number of pulses was set at 20,000 or 60,000	• Oil-holding capacity of rice protein was found to be 129.7% without PEF treatment, which reached to 145.4% with 20,000 pulse treatment at pH 5; however, a significant decrease (131.1%) in oil-holding capacity of protein was noticed with an increase in the number of pulses to 40,000.	Melchior et al., 2020
5.	Soy Protein Isolate (SPI)	Pulsed electric field strength (5, 10, 20 kV/cm) Pulse frequency 1,000 Hz; pulse width: 40 µs	• Moderate PEF strength (10 kV/cm) and alkaline conditions (pH 11) enhanced the solubility of SPI from 26.06% to 70.34%.	Wang et al., 2023
6.	Wheat gluten	Electric field intensity (0–12.5 kV/cm); pulse width (2–10 µs); pulse frequency (100–900 Hz); Retention time (1–9 min)	• Water-holding capacity increased with an increase in the electric field intensity from 0 to 12.5 kV/cm. • Peak value of water-holding capacity was reached at 10 kV/cm. • Pulse width did not affect the water holding capacity. • Oil-holding capacity was significantly affected (P < 0.01) by PEF (except for PF at P = 0.05). • Emulsion stability was not influenced by pulse width (0–10 µs and retention time (except for 1 min of RT). • Foaming capacity was influenced by pulse width. • PEF parameters did not show significant effect on foaming stability.	Zhang et al., 2021a

omitted that the dry protein network formed at pH 6 can physically capture the solvents by capillary and better retain them in the structure (Melchior et al., 2020).

In context to emulsibility and emulsion stability, a significant increase was observed with voltage and residence time of 30 kV and 180 s, occurring up to 61% and 64%, respectively. EC and ES remained stable after 180 and 120 s, respectively. A pulse frequency of 600 Hz and a width of 6 µs was the inflection point between increases and decreases in emulsibility (61%) and emulsion stability (60%). The control EC and ES values were 54% and 53%, respectively (Zhang et al., 2017). Further research is required to know about the impact of PEF treatment on the emulsifying and

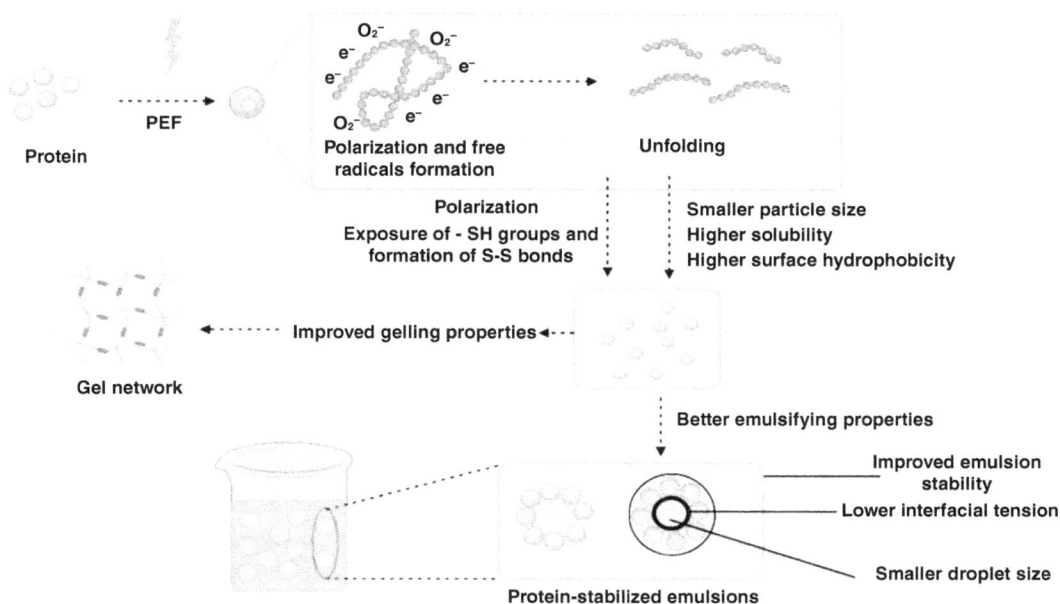

FIGURE 8.4 Illustration showing the mechanism of PEF in enhancing the functional properties of plant-based foods.

Source: Taha, A., F. Casanova, P. Šimonis, V. Stankevič, M.A.E. Gomaa, and A. Stirkė. 2022. "Pulsed Electric Field: Fundamentals and Effects on the Structural and Techno-Functional Properties of Dairy and Plant Proteins". *Foods* 11(11): 1556.

foaming properties of plant-based proteins. The modification of protein structure induced by PEF treatment can enhance the technical functional attributes of plant- based proteins. As suggested in previous work, PEF can polarize and expand protein molecules, exposing hydrophobic groups on the surface of molecules (Figure 8.4).

Oil-holding capacity of canola protein of control was 8.12 mL/g which increased to 13.19 mL/g with the rise in voltage (25 kV) and decreased to 14.15 mL/g when pulse width was below 6 μs. Further, the oil-holding capacity increased to 13.04 mL/g with a residence time of 150 s and to 13.15 mL/g when pulse frequency reached 800 Hz. However, oil-holding capacity remained stable after a further increase in these processing parameters (Zhang et al., 2017). Besides, oil-holding capacity of rice protein was found to be 129.7% without PEF treatment, which reached 145.4% with 20,000 pulse treatment at pH 5; however, a significant decrease (131.1%) in oil-holding capacity of protein was noticed with increase in number of pulses to 40,000 (Melchior et al., 2020).

8.7 CONCLUSION

The pulsed electric field is a promising green technology that is widely used in many food applications. The current need for sustainable development has increased the use of PEF treatment in the food industry. This chapter concludes that PEF treatment has a significant impact on the structure and techno-functional properties of plant proteins. PEF treatment promotes polarization, subunit dissociation, and unfolding, exposing the hydrophobic and sulfhydryl groups and thus affects the protein's functional properties. Although the effect of PEF on food protein has been studied more on animal protein, more research using different types of plant protein is needed. Furthermore, collaborations between the food industry and academic institutions are critical to increasing the use of

PEF treatment for food protein modification, particularly with plant proteins, which lag behind due to their poor functionality.

REFERENCES

Ahmed, Z., M. F. Manzoor, N. Ahmad, X. A. Zeng, Z. U. Din, U. Roobab, A. Qayum, R Siddique, A. Siddeeg, and A. Rahaman. 2020. "Impact of pulsed electric field treatments on the growth parameters of wheat seeds and nutritional properties of their wheat plantlets juice." *Food Science & Nutrition* 8 (5): 2490–2500. https://doi.org/10.1002/fsn3.1540.

Akharume, F. U., R. E. Aluko, and A. A. Adedeji. 2021. "Modification of plant proteins for improved functionality: A review." *Comprehensive Reviews in Food Science and Food Safety* 20 (1): 198–224.

Aschemann-Witzel, J., R. F. Gantriis, P. Fraga, and F. J. Perez-Cueto. 2021. "Plant-based food and protein trend from a business perspective: Markets, consumers, and the challenges and opportunities in the future." *Critical Reviews in Food Science and Nutrition* 61 (18): 3119–3128. https://doi.org/10.1080/10408398.2020.1793730.

Barbosa-Canovas, G. V., M. M. Congora-Nieto, and B. G. Swanson. 2000. "Processing fruits and vegetables by pulsed electric field technology." In: *Minimally Processed Fruits and Vegetables: Fundamental Aspects and Applications* (pp. 223–235). S. M. Alzamora, M. S. Tapia, and A. López-Malo, edited. Aspen Publishers, Inc., Gaithersburg, MD.

Barbosa-Cánovas, G. V., M. M. Gongora-Nieto, U. R. Pothakamury, and B. G. Swanson. 1999. "Chapter 2e design of PEF processing equipment." In: *Preservation of Foods with Pulsed Electric Fields* (p. 20e46). Academic Press, San Diego, CA.

Barsotti, L., E. Dumay, T. H. Mu, M. D. F. Diaz, and J. C. Cheftel. 2001. "Effects of high voltage electric pulses on protein-based food constituents and structures." *Trends in Food Science & Technology* 12 (3–4): 136–144. https://doi.org/10.1016/S0924-2244(01)00065-6.

Bendicho, S., G. V. Barbosa-Cánovas, and O. Martín. 2003. "Reduction of protease activity in simulated milk ultrafiltrate by continuous flow high intensity pulsed electric field treatments." *Journal of Food Science* 68 (3): 952–957. https://doi.org/10.3168/jds.S0022-0302(03)73649-2.

Betz, S. F. 1993. "Disulfide bonds and the stability of globular proteins." *Protein Science* 2 (10): 1551–1558. https://doi.org/10.1002/pro.5560021002.

Cao, Y., S. Bolisetty, G. Wolfisberg, J. Adamcik, and R. Mezzenga. 2019. "Amyloid fibril-directed synthesis of silica core–shell nanofilaments, gels, and aerogels." *Proceedings of the National Academy of Sciences* 116 (10): 4012–4017. https://doi.org/10.1073/pnas.1819640116.

Chen, Y., T. Wang, Y. Zhang, X. Yang, J. Du, D. Yu, and F. Xie. 2022. "Effect of moderate electric fields on the structural and gelation properties of pea protein isolate." *Innovative Food Science & Emerging Technologies* 77: 102959. https://doi.org/10.1016/j.ifset.2022.102959.

Dunn, J. E., and J. S. Pearlman. 1987. *Methods and Apparatus for Extending the Shelf-Life of Fluid Food Products.* U.S. Patent No. 4,695,472. U.S. Patent and Trademark Office, Washington, DC.

Esteghlal, S., H. H. Gahruie, M. Niakousari, F. J. Barba, A. E. D. Bekhit, K. Mallikarjunan, and S. Roohinejad. 2019. "Bridging the knowledge gap for the impact of non-thermal processing on proteins and amino acids." *Foods* 8 (7): 262. https://doi.org/10.3390/foods8070262.

FAO. 2017. *The Future of Food and Agriculture–Trends and Challenges.* FAO, Rome. ISBN 978-92-5-109551-5. https://www.fao.org/3/i6583e/i6583e.pdf

Fasolin, L. H., R. N. Pereira, A. C. Pinheiro, J. T. Martins, C. C. P. Andrade, O. L. Ramos, and A. A. Vicente. 2019. "Emergent food proteins–Towards sustainability, health and innovation." *Food Research International* 125: 108586. https://doi.org/10.1016/j.foodres.2019.108586.

Fernandez-Diaz, M. D., L. Barsotti, E. Dumay, and J. C. Cheftel. 2000. "Effects of pulsed electric fields on ovalbumin solutions and dialyzed egg white." *Journal of Agricultural and Food Chemistry* 48 (6): 2332–2339. https://doi.org/10.1021/jf9908796.

Górska-Warsewicz, H., W. Laskowski, O. Kulykovets, A. Kudlińska-Chylak, M. Czeczotko, and K. Rejman. 2018. "Food products as sources of protein and amino acids—The case of Poland." *Nutrients* 10 (12): 1977. https://doi.org/10.3390/nu10121977.

Han, Z., M. J. Cai, J. H. Cheng, and D. W. Sun. 2018. "Effects of electric fields and electromagnetic wave on food protein structure and functionality: A review." *Trends in Food Science & Technology* 75: 1–9. https://doi.org/10.1016/j.tifs.2018.02.017.

Henchion, M., M. Hayes, A. M. Mullen, M. Fenelon, and B. Tiwari. 2017. "Future protein supply and demand: Strategies and factors influencing a sustainable equilibrium." *Foods* 6 (7): 53. https://doi.org/10.3390/foods6070053.

Jaeger, H., N. Meneses, and D. Knorr. 2014. "Food technologies: Pulsed electric field technology." In: *Encyclopedia of Food Safety* (pp. 239–244). Y. Motarjemi, edited. Academic Press, Waltham, MA.

Jafari, S. M., A. S. Doost, M. N. Nasrabadi, S. Boostani, P. and Van der Meeren. 2020. "Phytoparticles for the stabilization of Pickering emulsions in the formulation of novel food colloidal dispersions." *Trends in Food Science & Technology* 98: 117–128. https://doi.org/10.1016/j.tifs.2020.02.008.

Kato, A., and S. Nakai. 1980. "Hydrophobicity determined by a fluorescence probe method and its correlation with surface properties of proteins." *Biochimica et biophysica acta (BBA)-Protein structure* 624 (1): 13–20. https://doi.org/10.1016/0005-2795(80)90220-2.

Knorr, D., A. Froehling, H. Jaeger, K. Reineke, O. Schlueter, and K. Schoessler. 2011. "Emerging technologies in food processing." *Annual Review of Food Science and Technology* 2: 203–235.

Lee, H., G. Yildiz, L. C. Dos Santos, S. Jiang, J. E. Andrade, N. J. Engeseth, and H. Feng. 2016. "Soy protein nano-aggregates with improved functional properties prepared by sequential pH treatment and ultrasonication." *Food Hydrocolloids* 55: 200–209. https://doi.org/10.1016/j.foodhyd.2015.11.022.

Li, Y. Q. 2012. "Structure changes of soybean protein isolates by pulsed electric fields." *Physics Procedia* 33: 132–137. https://doi.org/10.1016/j.phpro.2012.05.040.

Li, Y. Q., Z. Chen, and H. Mo. 2007. "Effects of pulsed electric fields on physicochemical properties of soybean protein isolates." *LWT-Food Science and Technology* 40 (7): 1167–1175. https://doi.org/10.1016/j.lwt.2006.08.015.

Liang, R., S. Cheng, and X. Wang. 2018. "Secondary structure changes induced by pulsed electric field affect antioxidant activity of pentapeptides from pine nut (*Pinus koraiensis*) protein." *Food Chemistry* 254: 170–184. https://doi.org/10.1016/j.foodchem.2018.01.090.

Lin, Z. R., X. A. Zeng, S. J. Yu, and D. W. Sun. 2012. "Enhancement of ethanol–acetic acid esterification under room temperature and non-catalytic condition via pulsed electric field application." *Food and Bioprocess Technology* 5 (7): 2637–2645. https://doi.org/10.1007/s11947-011-0678-4.

Liu, Y. F., I. Oey, P. Bremer, A. Carne, and P. Silcock. 2019. "Modifying the functional properties of egg proteins using novel processing techniques: A review." *Comprehensive Reviews in Food Science and Food Safety* 18 (4): 986–1002. https://doi.org/10.1111/1541-4337.12464.

Liu, Y. Y., X. A. Zeng, Z. Deng, S. J. Yu, and S. Yamasaki. 2011. "Effect of pulsed electric field on the secondary structure and thermal properties of soy protein isolate." *European Food Research and Technology* 233 (5): 841–850. https://doi.org/10.1007/s00217-011-1580-z.

Liu, Y. Y., X. A. Zeng, and Z. Han. 2010. "Raman spectra study of soy protein isolate structure treated with pulsed electric fields." *Spectroscopy and Spectral Analysis* 30 (12): 3236–3239. https://doi.org/10.3964/j.issn.1000-0593(2010)12-3236-04.

Mañas, P., and A. Vercet. 2006. "Effect of pulsed electric fields on enzymes and food constituents." In: *Pulsed electric fields Technology for the Food Industry, Fundamentals and Applications* (pp. 131–152). J. Raso and V. Heinz, edited. Springer, New York, NY.

Manzoor, M. F., N. Ahmad, R. M. Aadil, A. Rahaman, Z. Ahmed, A. Rehman, A., . . . A. Manzoor. 2019. "Impact of pulsed electric field on rheological, structural, and physicochemical properties of almond milk." *Journal of Food Process Engineering* 42 (8): e13299. https://doi.org/10.1111/jfpe.13299.

Melchior, S., S. Calligaris, G. Bisson, and L. Manzocco. 2020. "Understanding the impact of moderate-intensity pulsed electric fields (MIPEF) on structural and functional characteristics of pea, rice and gluten concentrates." *Food and Bioprocess Technology* 13 (12): 2145–2155. https://doi.org/10.1007/s11947-020-02554-2.

Mohamed, M. E., and A. H. A. Eissa. 2012. "Pulsed electric fields for food processing technology." *Structure and Function of Food Engineering* 11: 275–306. https://doi.org/10.5772/48678.

Nasrabadi, M. N., A. S. Doost, and R. Mezzenga. 2021. "Modification approaches of plant-based proteins to improve their techno-functionality and use in food products." *Food Hydrocolloids* 118: 106789. https://doi.org/10.1016/j.foodhyd.2021.106789.

Nasrabadi, M. N., S. A. H. Goli, A. S. Doost, K. Dewettinck, and P. Van der Meeren. 2019. "Bioparticles of flaxseed protein and mucilage enhance the physical and oxidative stability of flaxseed oil emulsions as a potential natural alternative for synthetic surfactants." *Colloids and Surfaces B: Biointerfaces* 184: 110489. https://doi.org/10.1016/j.colsurfb.2019.110489.

Niu, D., X. A. Zeng, E. F. Ren, F. Y. Xu, J. Li, M. S. Wang, and R. Wang. 2020. "Review of the application of pulsed electric fields (PEF) technology for food processing in China." *Food Research International* 137: 109715. https://doi.org/10.1016/j.foodres.2020.109715.

Nunes, L., and G. M. Tavares. 2019. "Thermal treatments and emerging technologies: Impacts on the structure and techno-functional properties of milk proteins." *Trends in Food Science & Technology* 90: 88–99. https://doi.org/10.1016/j.tifs.2019.06.004.

Park, E. S., and S. G. Boxer. 2002. Origins of the sensitivity of molecular vibrations to electric fields: Carbonyl and nitrosyl stretches in model compounds and proteins. *The Journal of Physical Chemistry B* 106 (22): 5800–5806.

Perez, O. E., and A. M. Pilosof. 2004. "Pulsed electric fields effects on the molecular structure and gelation of β-lactoglobulin concentrate and egg white." *Food Research International* 37 (1): 102–110. https://doi. org/10.1016/j.foodres.2003.09.008.

Pojić, M., A. Mišan, and B. Tiwari. 2018. "Eco-innovative technologies for extraction of proteins for human consumption from renewable protein sources of plant origin." *Trends in Food Science & Technology* 75: 93–104. https://doi.org/10.1016/j.tifs.2018.03.010.

Rahaman, A., A. Siddeeg, M. F. Manzoor, X. A. Zeng, S. Ali, Z. Baloch, . . . Q. H. Wen. 2019. "Impact of pulsed electric field treatment on drying kinetics, mass transfer, colour parameters and microstructure of plum." *Journal of Food Science and Technology* 56 (5): 2670–2678. https://doi.org/10.1007/ s13197-019-03755-0.

Raso, J., W. Frey, G. Ferrari, G. Pataro, D. Knorr, J. Teissie, and D. Miklavčič. 2016. "Recommendations guidelines on the key information to be reported in studies of application of PEF technology in food and biotechnological processes." *Innovative Food Science & Emerging Technologies* 37: 312–321. https:// doi.org/10.1016/j.ifset.2016.08.003.

Riblett, A. L., T. J. Herald, K. A. Schmidt, and K. A. Tilley. 2001. "Characterization of β-conglycinin and glycinin soy protein fractions from four selected soybean genotypes." *Journal of Agricultural and Food Chemistry* 49 (10): 4983–4989. https://doi.org/10.1021/jf0105081.

Richter, C. K., A. C. Skulas-Ray, C. M. Champagne, and P. M. Kris-Etherton. 2015. "Plant protein and animal proteins: Do they differentially affect cardiovascular disease risk?" *Advances in Nutrition* 6 (6): 712–728. https://doi.org/10.3945/an.115.009654.

Sá, A. G. A., Y. M. F. Moreno, and B. A. M. Carciofi. 2020. "Plant proteins as high-quality nutritional source for human diet." *Trends in Food Science & Technology* 97: 170–184. https://doi.org/10.1016/j.tifs.2020.01.011.

Sepulveda, D. R., M. M. Góngora-Nieto, J. A. Guerrero, and G. V. Barbosa-Cánovas. 2005. "Production of extended-shelf life milk by processing pasteurized milk with pulsed electric fields." *Journal of Food Engineering* 67 (1–2): 81–86.

Sha, L., and Y. L. Xiong. 2020. "Plant protein-based alternatives of reconstructed meat: Science, technology, and challenges." *Trends in Food Science & Technology* 102: 51–61. https://doi.org/10.1016/j. tifs.2020.05.022.

Siddeeg, A., X. A. Zeng, A. Rahaman, M. F. Manzoor, Z. Ahmed, and A. F. Ammar. 2019. "Effect of pulsed electric field pretreatment of date palm fruits on free amino acids, bioactive components, and physicochemical characteristics of the alcoholic beverage." *Journal of Food Science* 84 (11): 3156–3162. https:// doi.org/10.1111/1750-3841.14825.

Singh, A., V. Orsat, and V. Raghavan. 2013. "Soybean hydrophobic protein response to external electric field: A molecular modeling approach." *Biomolecules* 3 (1): 168–179. https://doi.org/10.3390/biom3010168.

Sinha, R. 2004. "Absorption and translocation of water." In: *Modern Plant Physiology* (pp. 64–82). R. K. Sinha, edited, 1st ed. Alpha Science International Ltd, Pangbourne.

Subaşı, B. G., M. Jahromi, F. Casanova, E. Capanoglu, F. Ajalloueian, and M. A. Mohammadifar. 2021. "Effect of moderate electric field on structural and thermo-physical properties of sunflower protein and sodium caseinate." *Innovative Food Science & Emerging Technologies* 67: 102593. https://doi.org/10.1016/j. ifset.2020.102593.

Syed, Q. A., A. Ishaq, U. U. Rahman, S. Aslam, and R. Shukat. 2017. "Pulsed electric field technology in food preservation: A review." *Journal of Nutritional Health & Food Engineering* 6 (6): 168–172.10.15406/ jnhfe.2017.06.00219.

Taha, A., F. Casanova, P. Šimonis, V. Stankevič, M. A. E. Gomaa, and A. Stirkė. 2022. "Pulsed electric field: Fundamentals and effects on the structural and techno-functional properties of dairy and plant proteins." *Foods* 11 (11): 1556. http://dx.doi.org/10.3390/foods11111556.

Tharrey, M., F. Mariotti, A. Mashchak, P. Barbillon, M. Delattre, and G. E. Fraser. 2018. "Patterns of plant and animal protein intake are strongly associated with cardiovascular mortality: The Adventist health study-2 cohort." *International journal of Epidemiology* 47 (5): 1603–1612. https://doi.org/10.1093/ije/dyy030.

Toepfl, S., V. Heinz, and D. Knorr. 2006. "Applications of pulsed electric fields technology for the food industry." In: *Pulsed Electric Fields Technology for the Food Industry* (pp. 197–221). J. Raso and V. Heinz, edited. Springer, Boston, MA.

Walstra, P., J. T. M. Wouters, and T. J. Geurts. 2006. "Proteins preparations." In: *Dairy Science and Technology* (pp. 537–573). P. Walstra, J. T. M. Wouters, and T. J. Geurts, edited. Taylor & Francis Group, Boca Raton, FL.

Wang, R., L. H. Wang, Q. H. Wen, F. He, F. Y. Xu, B. R. Chen, and X. A. Zeng. 2023. "Combination of pulsed electric field and pH shifting improves the solubility, emulsifying, foaming of commercial soy protein isolate." *Food Hydrocolloids* 134: 108049. https://doi.org/10.1016/j.foodhyd.2022.108049.

Warnakulasuriya, S. N., and M. T. Nickerson. 2018. "Review on plant protein–polysaccharide complex coacervation, and the functionality and applicability of formed complexes." *Journal of the Science of Food and Agriculture* (15): 5559–5571. https://doi.org/10.1002/jsfa.9228.

Xiang, B. Y. 2009. "Effects of pulsed electric fields on structural modification and rheological properties for selected food proteins." PhD Thesis, Department of Bioresource Engineering, Macdonald Campus, McGill University, Sainte-Anne-de-Bellevue, Québec, Canada.

Xiang, B. Y., M. O. Ngadi, B. K. Simpson, and M. V. Simpson. 2011. "Pulsed electric field induced structural modification of soy protein isolate as studied by fluorescence spectroscopy." *Journal of Food Processing and Preservation* 35 (5): 563–570. https://doi.org/10.1111/j.1745-4549.2010.00501.x.

Yu, L. J., M. Ngadi, and G. S. V. Raghavan. 2009. Effect of temperature and pulsed electric field treatment on rennet coagulation properties of milk. *Journal of Food Engineering* 95 (1): 115–118.

Zhang, C., Y. H. Yang, X. D. Zhao, L. Zhang, Q. Li, C. Wu, . . . J. Y. Qian. 2021a. "Assessment of impact of pulsed electric field on functional, rheological and structural properties of vital wheat gluten." *LWT* 147: 111536. https://doi.org/10.1016/j.lwt.2021.111536

Zhang, J., L. Liu, H. Liu, A. Yoon, S. S. Rizvi, and Q. Wang. 2019. "Changes in conformation and quality of vegetable protein during texturization process by extrusion." *Critical Reviews in Food Science and Nutrition* 59 (20): 3267–3280. https://doi.org/10.1080/10408398.2018.1487383.

Zhang, L., L. J. Wang, W. Jiang, and J. Y. Qian. 2017. "Effect of pulsed electric field on functional and structural properties of canola protein by pretreating seeds to elevate oil yield." *LWT* 84: 73–81. https://doi.org/10.1016/j.lwt.2017.05.048.

Zhang, Q., G. V. Barbosa-Cánovas, and B. G. Swanson. 1995. "Engineering aspects of pulsed electric field pasteurization." *Journal of Food Engineering* 25 (2): 261–281. https://doi.org/10.1016/0260-8774(94)00030-D.

Zhang, S., L. Sun, H. Ju, Z. Bao, X. A. Zeng, and S. Lin. 2021b. "Research advances and application of pulsed electric field on proteins and peptides in food." *Food Research International* 139: 109914. https://doi.org/10.1016/j.foodres.2020.109914

Zhao, W., Y. Tang, L. Lu, X. Chen, and C. Li. 2014. "Pulsed electric fields processing of protein-based foods." *Food and Bioprocess Technology* 7 (1): 114–125. https://doi.org/10.1007/s11947-012-1040-1.

Zhao, W., and R. Yang. 2008. "The effect of pulsed electric fields on the inactivation and structure of lysozyme." *Food Chemistry* 110 (2): 334–343. https://doi.org/10.1016/j.foodchem.2008.02.008.

Zhao, W., R. Yang, and H. Q. Zhang. 2012. "Recent advances in the action of pulsed electric fields on enzymes and food component proteins." *Trends in Food Science & Technology* 27 (2): 83–96. https://doi.org/10.1016/j.tifs.2012.05.007.

Zhong, K., J. Wu, Z. Wang, F. Chen, X. Liao, X. Hu, and Z. Zhang. 2007. "Inactivation kinetics and secondary structural change of PEF-treated POD and PPO." *Food Chemistry* 100 (1): 115–123. https://doi.org/10.1016/j.foodchem.2005.09.035.

9 Cold Atmospheric Plasma Processing of Plant-Based Proteins

Srutee Rout and Prem Prakash Srivastav

9.1 INTRODUCTION

The topic of sustainable proteins is popular. The advantages of a plant-based diet for our health and the environment are driving this change. Plant-based proteins were expected to account for 7.7% of the global protein industry by 2020 and expand from 29.4 billion dollars to 162 billion dollars by 2030 (Kim & Shin, 2022). Post-COVID developments have spurred the search for new protein sources. Plant-based proteins can improve endurance and meet the physiological needs of a rapidly rising population when food supply and biodiversity are challenges (Sá & Serpa, 2020; Galanakis, 2021). Plant proteins meet nutritional needs and are environmentally friendly, therefore consumers like them. Consumers seem to prioritise nutritional quality and healthiness over preservation and animal welfare (Profeta et al., 2021). Low solubility and emulsification prevent plant proteins from being incorporated into food matrices (Amagliani et al., 2021). Thus, these proteins must be valued sustainably. Physical, chemical, and enzymatic processes can use proteins in food matrices (Charoensuk et al., 2018). Plant proteins are being studied for meat replacements, dairy alternatives, lipophilic component encapsulation, emulsifying, and foaming (Kamath et al., 2022; Sha & Xiong, 2020). Plant proteins make beef burgers look better and are better for the environment (Smetana et al., 2021).

The agro-chemical industry seeks sustainable solutions by minimising resource use, particularly energy. Cold plasma (CP) may be a long-term, "green" option. CP is a sustainable way to improve seed germination, clean food-contact surfaces, modify food components, and inactivate enzymes (Bourke et al., 2018; Misra & Roopesh, 2019). Plasma is "clean label" since it uses less solvents and chemicals and leaves no debris. Misra et al. (2019) found that low-current (1–5 mA) and high-voltage (more than 50 kV) CP therapy required the same energy as lighting a bulb. Plasma changes proteins "greenly" and cheaply. The 12th Sustainable Development Goal, "sustainable consumption and production patterns," aims to end food loss and use by-products. CP treatment of plant proteins to boost nutritional content is a well-known example of using sustainable proteins with "green" methods that reduce waste and chemical solvents.

Non-thermal protein modification has been studied (Rahman & Lamsal, 2021; Venkateswara Rao et al., 2021). Plasma affects milk proteins, animal-based proteins, and enzymes (Nikmaram & Keener, 2022). Plant proteins and CP have not been well studied. This chapter critically discusses how plant proteins can be physically and chemically altered to improve solubility, interfacial behaviour, and structure. CP has been extensively studied for enzymatic stability in fruit- and vegetable-based products (Pohl et al., 2022; Mayookha et al., 2022). Since CP can modify proteins and inactivate enzymes, this chapter only discusses its potential uses. This chapter aims to cover protein structure physical and chemical changes.

9.2 CP SOURCES

Numerous energy sources, such as heat, electricity, and electromagnetic waves like microwaves and radio waves, can generate CP. Dielectric barrier discharges (DBDs), atmospheric glow discharges,

 DOI: 10.1201/9781003369790-9

corona glow discharges, high voltage pulsed discharges, gliding arc discharges, plasma jets, radio frequency discharges, microwave-induced plasma (MIP), and inductively coupled plasma (ICP) are a few examples of the apparatus that can be used for this (Garcia-Rey et al., 2022; Waskow et al., 2022). There are numerous choices available when selecting the carrier gas: Air, oxygen, nitrogen, helium, argon, any mix of these gases, or any one of these gases alone (Bharti et al., 2022). Although there are other advantages as well, accessibility and cost are the key advantages of using air as a carrier gas.

9.2.1 DIELECTRIC BARRIER DISCHARGE (DBD)

Atmospheric cold plasma (ACP) is often produced using DBD. This creates a plasma discharge between two parallel electrodes wrapped in dielectric (Moreau & Defoort, 2022). DBD excels in geometrical flexibility, operational parameters, miniaturisation, simple design, safety, accessibility, and power supply. DBD ACP's dielectric barrier layer limits discharge and prevents sparks and arcs. These layers include alumina, silica glass, glass, enamel, ceramics, and polymers. Process variables, device configuration, packaging material, and substrate affect this system's efficiency (Nageswaran et al., 2019). The dielectric can lose its dielectric characteristics and limit current at extremely high frequencies. DBD treatments commonly use 10^4–10^6 Pa gas pressure, 50 Hz to 10 MHz frequency band, AC or pulsed DC current, 1–100 kVrms voltage amplitude, and 0.1 mm to several centimeters of electrode gap. DBD device is simple, reliable, steady, and affordable (Li et al., 2020).

DBD is efficient for treatments under 1 min (Sima et al., 2023). By employing ambient air in DBDs instead of helium or argon, ACP can be widely adopted and operational expenses reduced (Akhtar et al., 2022). Electricity into electrons maintains ambient gas temperature in this system (Yepez et al., 2022). High-pressure non-equilibrium plasma DBDs work. Their many industrial uses make them remarkable (Kaavya et al., 2022).

9.2.2 PARTICLE JET

Atmospheric pressure plasma jets release plasma species across a large region when operated with an unsealed electrode arrangement (Domonkos et al., 2021). Plasma jets can directly treat any size object. They can treat cancer, wounds, and inactivate bacteria (Figure 9.1) (Šimončicová et al., 2019). Plasma jets can utilise microwave, AC, or pulsed DC power. Helium makes plasma jet creation easier. Noble gases are combined with a limited amount of reactive gases like air or oxygen because they have fewer reactive species. They can process large and small volumes.

9.2.3 CORONAL EJECTION

This plasma arises when high magnetic and electric fields are placed between two contacting surfaces with different radii, like a point electrode or thin wire electrode. The collector electrode is larger than the emitter electrode. Coronas' polarity depends on the emitter's voltage relative to the collection electrode. The negative corona's cathode is the high-voltage electrode, while the positive's anode is low-voltage electrode (Narimisa et al., 2022; Stryczewska & Boiko, 2022). Corona plasma is used in electrostatic precipitation, electrophotography, ionisation sources, ozone production, hazardous material dissolution, gas and liquid purification, and so on. Corona remains the favoured method for treating fruits and vegetables with ozone.

9.3 PROTEINS AND THEIR STRUCTURE

Proteins, polymers of amino acids, are formed by peptide bonds between the carboxyl groups of two amino acids. The amino acid sequence and side chains determine protein structure and polypeptide chain properties. Protein secondary structure is formed by backbone atom interactions. Interactions

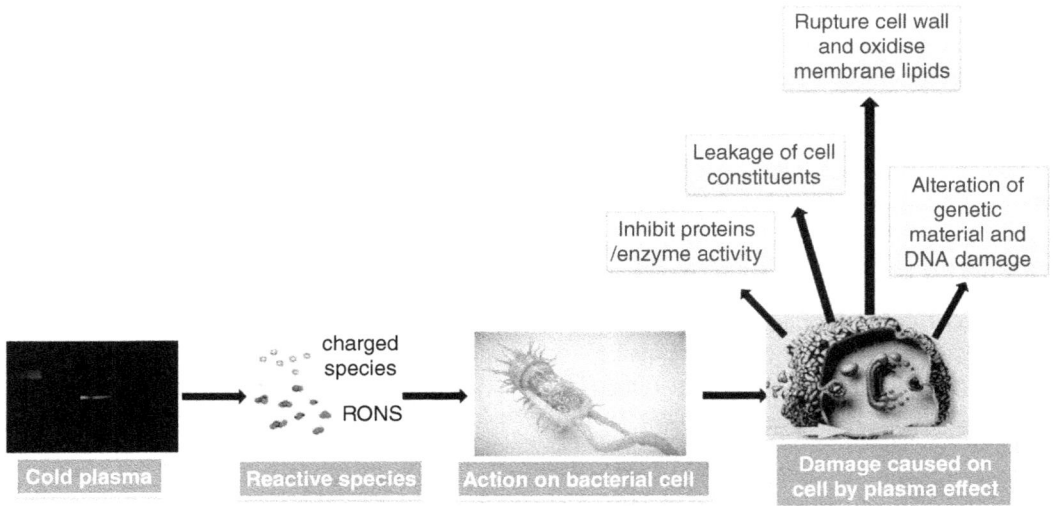

FIGURE 9.1 Inactivation of bacterial cell by CP.

affect protein conformation. The secondary structure of proteins—helixes and sheets—is more complex than the basic structure. The helix is formed by hydrogen bonds between the nth amino acid's carbonyl oxygen and the (n + 4)th amide proton. Peptide chains have pleated sheet structures supported by hydrogen bonds. Protein secondary structure depends on polypeptide chain residue size. Smaller residues form sheets, while larger residues form helices in polypeptide chains. Sheet structures have closer intrinsic amino acid side groups than helical structures. As a result, bulky or similarly charged residues often group together in helical structures, consuming more space. Other secondary structures that are likewise stabilised by hydrogen bonding include random coils and turns. The various interactions that stabilise the secondary structure can be disturbed by outside forces. Protein conformation is influenced by environmental and processing variables such as pH, ionic strength, and temperature. Even though the sequence of amino acids is the only information a gene transmits, each protein in its native form has unique biological and functional characteristics that are mostly controlled by both primary and secondary structures (Pillai et al., 2022). The polypeptides fold into a three-dimensional tertiary structure with the help of ionic and non-ionic forces. The folding of various helices, twists, sheets and other sections of the protein into its native structure is specifically referred to as tertiary structure. The tertiary structure develops on its own and is held in place by side chain interactions as well as extracellular proteins. A protein's solubility, for example, is impacted by its tertiary structure (Soni et al., 2022). Many writers have offered comprehensive explanations of the design and operation of many types of plasma devices, making their works an excellent source of information (Garcia-Rey et al., 2022; Waskow et al., 2022). Hydrophilic amino acids can freely interact with water whereas the hydrophobic side chains are buried inside soluble proteins to avoid their accessibility to the water–protein interface (Soni et al., 2022). Other researchers concentrating on protein structure have comprehensive information on the tertiary structure of proteins (Moore-Kelly et al., 2019). A protein's quaternary structure depicts how numerous protein molecules come together to form a single protein complex (Marciano et al., 2022). It is interesting that very few native proteins have had their conformation defined, despite the fact that the structure, conformation, and properties of proteins are completely governed by the individual amino acid sequence.

Amino acids are a source of nitrogen, which can be used in the synthesis of proteins and other biomolecules. Based on their rate of protein synthesis in vivo, either relative or absolute, these can be categorised as follows: Valine, tryptophan, phenylalanine, methionine, lysine, leucine, isoleucine, and histidine. Tyrosine, cysteine, and arginine are conditionally essential amino acids. Serine, proline, glycine, glutamic acid, glutamine, aspartic acid, asparagine, and alanine are dispensable amino acids (Fallah et al., 2022). The ratios and concentrations of the individual amino acids that make up the protein under examination have an impact on the protein's nutritional quality. The quality or biological value of the protein increases with the ratio of essential amino acids. The distribution of a particular amino acid inside a protein is also very significant.

9.4 DIFFERENT PLANT PROTEINS

Consumers' perceptions of protein are trending more positively, and demand for both plant-based and animal-based sources of protein is rising. Additionally, a growing amount of clinical evidence, particularly in older persons, supports the health advantages of protein at or above the current recommendations for dietary protein intake. To begin with, it's crucial to choose the right combination of plant-based ingredients to make a particular plant-based diet, such as meat, egg, milk, or fish analogue. These components could be simple entire materials or isolated elements (such as proteins, carbs, lipids, vitamins, or minerals) (such as beans, peas, rice, wheat, and mushrooms). These substances differ greatly from those present in animal products in terms of their compositions, structures, and physicochemical characteristics. Therefore, putting these components together to create animal product analogues is one of the main hurdles. Sometimes, materials derived from plants can be used directly (such as mushrooms), but other times they may need to be broken down into certain structural components and then put back together to create products that are analogous to animal products (e.g. soy proteins).

Because of their numerous functional qualities, including their capacity to thicken, gel, emulsify, froth, and retain fluids, plant proteins are frequently utilised in plant-based foods. They are also a significant source of vital amino acids. These proteins can be obtained from a variety of plant sources, each with specific properties, such as soybeans, peas, faba beans, mung beans, lentils, algae, and microalgae. The majority of plant proteins have globular shapes and are frequently found as complex multimers, which are collections of many protein types bound together by chemical and/or physical connections. The biological origin of these proteins, as well as any modifications made to their original associations and states during separation and purification, all affect how functional they are. The absence of plant proteins with consistent functional characteristics presents a significant obstacle in the production of plant-based foods. Future study will be needed to pinpoint the best botanical sources and isolation techniques for creating dependable functional components. Another difficult task is getting plant proteins to adopt the same structural arrangements as those in animal products, which would produce comparable physicochemical properties.

9.4.1 CHICKPEA

Chickpeas are the least expensive source of protein for the underprivileged, families with little resources, and vegans. Legumes are noted for having a high protein content. Chickpea protein has been reported to have increased protein bioavailability, with globulins accounting for the majority of this increase (Yust et al., 2003). Contrary to cereals, which are abundant in sulphur-containing amino acids and in which lysine is the limiting amino acid, chickpea seeds are lacking in methionine and cysteine (sulphur-containing amino acids) and are rich in arginine and lysine (Rachwa-Rosiak et al., 2015). India produces the most chickpeas in the world (with a share of roughly 66.19% and is followed by Australia and Turkey) and accounts for 86.03% of all the chickpeas produced in Asia. Chickpeas contain about 1.42% and 27.84% of soluble and insoluble dietary fibre, respectively (Khatoon & Prakash, 2004). The method of processing has a significant impact on the ratio of soluble and insoluble dietary fibre in the pulse flour.

9.4.2 PEANUT

Peanuts are one of the most important oilseed crops, with China producing the majority of the world's supply (Ye et al., 2020). Most peanuts are used to produce edible oil. Defatted peanut flour (DPF), a protein-rich by-product of oil extraction, has low levels of antinutritional chemicals and contains between 47% and 55% high-quality protein. Because it has a superb amino acid profile, a tempting fragrance, it is used as a commercial animal protein alternative. The three main components of peanut protein are arachin, arachin I, and arachin II, according to Luo et al. (2019). The most important portion of peanut protein is arachin I, the smallest of the three main parts. Trp, Tyr, and Phe made up a considerable amount of arachin II, arachin I, and arachin when the amino acid composition was examined (Sanyal et al., 2023). However, due to its weak practical qualities, it has a restricted utility (Ge et al., 2020). It may be necessary to enhance the protein from peanuts in defatted meal's functional properties to increase its utilisation in the food industry.

9.4.3 SOY PROTEIN

The most significant oilseed crop with widespread recognition is soya bean (*Glycine max* L.), which provides large volumes of affordable, functional, and nutrient-rich proteins. The FDA's advice to take 25 g of soy protein per day to significantly lower the risk of coronary heart disease in 1999 helped further for soy's reputation as a nutrient-dense food component. Soya bean protein-containing foods are consequently one of the product categories in the food sector with the quickest rate of growth. Examples include infant formula, milk-based beverages, and processed meats. Soya bean-containing foods are usually treated both thermally and nonthermally. Due to the negative impacts of conventional heat treatments, researchers have been seeking for alternative non-thermal methods to destroy viruses and unwanted microorganisms while preserving shelf life and food quality. Some of these methods include CP, ultrasonication, high hydrostatic pressure, pulsed light, and others (Mukhtar et al., 2022). It has been found that macromolecules like proteins and carbohydrates can go through dynamic structural reformation under particular circumstances, which may change their physicochemical and technical qualities to variable degrees.

9.4.4 QUINOA

Quinoa (*Chenopodium quinoa* Willd.) has recently gained popularity as a gluten-free pseudocereal with a full nutritional profile, thanks to its high levels of dietary fibre, vitamins, bioactive compounds, and minerals (Okon, 2021). It has high concentrations of vitamins (including pyridoxine, riboflavin, alpha-tocopherol, and folic acid) and minerals (such as magnesium, calcium, iron, zinc, and copper), as well as sulphur-containing amino acids like lysine (Graziano et al., 2022). The utilisation of quinoa flour has drawn increased attention for its technological properties in addition to its nutritional worth. The technological qualities of quinoa flour are any criteria that define if it is suitable for use in the creation of dishes like cookies, bakery items, pasta, and snacks (Sarwar et al., 2023; Dabija et al., 2022).

9.4.5 AMARANTH

Recently, the plant amaranth (*Amaranthus* spp.) was rediscovered. It was produced by the Aztecs as a staple diet between 6,000 and 8,000 years ago, but the invading Spanish in the New World prohibited it as food for religious reasons (Coţovanu et al., 2021). According to Martínez-López et al. (2019), it is a pseudocereal with a moderately high protein content (15–17%) and a balanced composition of amino acids, notably important ones (Meza et al., 2022). It also possesses a rare starch quality and premium oil (including squalene). Amaranth seeds are used in an increasing number of baked goods, including cookies, cakes, chocolates, pancakes, and pasta (Martinez-Lopez et al., 2020; Iftikhar & Khan, 2019). The global amaranth market was valued at $5.88 billion USD in 2017; from 2018 to 2025, it is anticipated to grow at a compound annual growth rate of 11.3% (Gupta, 2019). According to Wang et al. (2023), amaranth protein (APN) may be able to lower the rising food demand in some developing countries. 7S globulin, albumin, 11S globulin, glutelin, and P globulin are thought to make up the majority of APN. Because APN has a poor solubility around the isoelectric point and a very low solubility in the acidic zone, its use in food items is constrained (Cruz-Morán et al., 2023).

9.4.6 WHEAT PROTEINS

Wheat grain proteins have been examined for about 300 years, with a systematic review of the literature. Due to their function as storage components, gluten proteins make up to 80% of the total nitrogen in grains, while their quantity rises as total grain protein concentration does. Based on their sequential extraction in a series of solvents—gliadins being extractable in aqueous ethanol and glutenins being extractable in dilute acid or alkali—gluten proteins are divided into two groups: The gliadins (classified as prolamins) and glutenins (classified as glutelins) (Zang et al., 2022). Based on the characteristics (solubility and amino acid content) provided by their repeating domains, wheat gluten proteins are classified as prolamins. However, wheat grains also contain a variety of proteins, particularly those with conserved cysteine residues, whose sequences are connected to the prolamins' nonrepetitive domains.

9.5 NOVEL FUTURE FOODS USING PLANT PROTEINS

For many years, plant proteins have been the main source of ingredients for foods, such as tofu and fried bean curd rolls because of gelling properties. Recent developments in technology and food science have increased awareness of the use of this category of ingredient for texturizing veggie meat. Despite this development, only a small selection of plant proteins, such as those from soy, pea, and coconut, are capable of performing the promising effects. However, economically mature goods are extremely hard to come by for those that are more widely available, including rice proteins (RPs) and wheat proteins (WPs). This is explained by the inferior solubility, gelation, foaming, and emulsification functional characteristics, which are mostly the result of macroscopic aggregation brought on by hydrophobic attractions and/or disulphide bonding. The complexed proteins' capacity to permit different rheological and textural qualities also led to modern contemporary applications of these structures in 3D food printing, foam stabilisation, and fresh-keeping. For instance the sustained release of eugenol with prolonged bacteriostasis was made possible by self-emulsification of eugenol by wheat proteins mixed with soy proteins via pH-cycle technology (Wang & Luo, 2023). Zein and sodium caseinate (NaCas) loaded with curcumin were molded into edible films by the pH-cycle method, which can be utilised to package edible ingredients directly to prevent oxidation and increase shelf life (Wang et al., 2019). In order to create amphiphilic binary nanostructures, hydrophilic pea proteins were co-folded. This protein network was used for stabilising liquid foams. With the reinforcement of starch nanocrystals, the foams had a lifespan of up to 96 h, as opposed to several hours for normal protein-stabilised foams. The pH-cycle-treated ternary nanocomplexes with the same mass of zein and NaCas coupled with propylene glycol alginate (PGA) showed outstanding emulsification abilities to create Pickering that resembled a gel. The Pickering emulsions produced by preparation exhibited better qualities and delivery characteristics appropriate for bioactive substances in food. Additionally, following a pH cycle, the alternate protein–polysaccharide motifs produced beautiful beads-on-a-string nanostructures that had good biphasic wettability at a water–oil interface. When used as an ink for 3D printed foods, such distinctive structures were good stabilisers of High Internal Phase Pickering emulsions (HIPPEs). The HIPPEs were giving outstanding performance in stability against heat, long-term storage, and freeze-thaw, which was even better than their commercialised equivalents, by delicately controlling the rheological characteristics using salt as a stimulus (Yi et al., 2023). All-plant margarine, which had the potential to take the place of the current whipped creams containing excessive amounts of saturated acids and cholesterol, was developed as a result of the remarkable mechanical qualities.

9.6 EFFECT OF CP ON PROTEIN FUNCTIONALITY

Since proteins are used in food and have significant nutritional value, they are the primary dietary components. Functional properties are characteristics of proteins that influence how useful they are in diets. Functional qualities of proteins are important for studying because of their nutritional or technological components in food production (Le Thanh-Blicharz et al., 2022). Proteins' dynamic and functional properties are represented by physicochemical indices, which are intricate relationships among their structure, torsional reconfiguration, and physicochemical properties.

Proteins are divided into primary, secondary, and tertiary structures based on the spatial arrangement of their amino acid polymers. Proteins' basic structure is created by peptide bonds between the carboxylic acid groups of two adjacent amino acids. The secondary structure, also known as a helix or a sheet structure, is produced through the process of inter- or intra-hydrogen binding. The tertiary structure, which is folded through the formation of hydrophobic, hydrogen, and/or disulphide bond interactions, determines the three-dimensional shape of a polypeptide or protein. To obtain the greatest stability, the polar and non-polar groups are placed within and outside of the protein molecules, respectively. A few polypeptide chains coming into contact with one another will also help build the quaternary structure (Huang et al., 2022).

The type and order of amino acids, known as a protein's major structure, are the main factors in determining how effectively it functions as food. On the other hand, secondary and tertiary structures are as valuable since they regulate compactness, which influences biological availability (Arora et al., 2022). The kind, arrangement, and conformation of amino acids also affect the fundamental functional and structural characteristics of proteins present in food (Kang et al., 2021; Nwachukwu & Aluko, 2019). These three elements—net charge, structure, and interaction with other substances—are the primary determinants of a protein's utility (either technologically or nutritionally). Since their fundamental structures typically determine their nutritional and commercial applications, proteins are extensively impacted by extrinsic elements such as pH, temperature, and processing (Amagliani et al., 2021).

Numerous atoms in the excited state are produced by ACP, such as Reactive Oxygen Species (ROS), singlet oxygen, Reactive Nitrogen Species (RNS), and ozone, all of which are very beneficial for boosting the microbiological safety of food components. Therefore, it may be effective to use reactive species to change the chemical structure of proteins and subsequently their functionality (Figure 9.2). According to some research, radical species can lower the amount of free sulfhydryl groups while increasing the concentration of carbonyl. When whey protein concentrate solution was treated at 80 kV for 1 h, specifically with NH or NH_2, or by disulphide bond cleavages, the amino acid side chain groups changed according to Zhang et al. (2022). As a result, samples treated with plasma had a considerably greater carbonyl content. A crude enzymatic extract of squid mantle was treated using a DBD device at 70 kV for various treatment times, and the results showed a decrease in free sulfhydryl groups and an increase in carbonyl content (Nyaisaba et al., 2019). According to their findings, processing time and the DBD plasma treatment enhanced the amount of carbonyl in the beef myofibrillar proteins.

ACP may alter protein functionality directly by reactive species or indirectly by neighbouring components that have been activated. Identification of the functional changes caused by this type of ACP protein handling requires knowledge of the matrix in which proteins are handled. Additionally, the reactivity of the exposed functional groups and crosslinks have the ability to cause direct injury to proteins in an embedded matrix. However, additional research is required in this area.

Proteins are utilised in the preparation of food from a technological perspective because they have a variety of functional properties, such as the capacity to solubilise, reorganise surfaces, texturise,

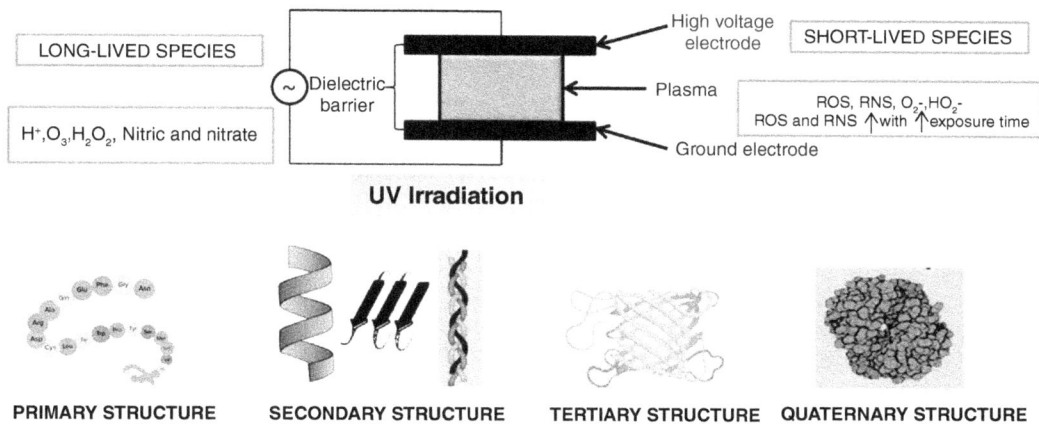

FIGURE 9.2 The effect of CP on different structures of protein.

and tolerate heat (Wen et al., 2022). Techno-functional qualities, which consider the technological aspects of proteins and their responsiveness to various phases of food processing and storage, are used to evaluate these properties (Bou et al., 2022).

In view of the ionising nature of ACP treatment, its impact on protein structure, and consequently, their scientific applicability, the primary physical and chemical properties of proteins that have received ACP treatment are discussed separately in the sections that follow.

9.6.1 MODIFICATION

9.6.1.1 Solubility

The equilibrium state between protein–protein and/or solvent interactions, which is among the fundamental properties of protein and the external effects, can be used to indicate the solubility of proteins, i.e. a critical metric in determining how functional they are. It has been shown that proteins are zwitterions at a specific pH level. Each particle's charge will vary with any pH change, which will have an impact on how soluble they are. The key determinants of the solubility of proteins are thought to be their molecular weight, amino acid ratio, aggregation or denaturation state, and charge density on the protein surface (Mollakhalili-Meybodi et al., 2021).

Protein solubility is hypothesised to affect potential applications, particularly in gels, emulsions, and foams. A protein's suitability for usage in a pharmaceutical, food, or beverage must be determined because denaturation and aggregation have a substantial impact on the protein's solubility (Gantumur et al., 2023). The physicochemical characteristics of protein surface and the structural conformation, which are altered by the solvent properties and exposure treatments, are closely related to protein solubility (Zou et al., 2022). Protein solubility in buffers with varied additions has often been investigated in order to predict how proteins will respond to treatments. Treatments that make proteins more soluble should enhance their functional characteristics, such as surface activity, hydrodynamic features, and rheological qualities (Ran et al., 2022; Wang et al., 2020). Due to the influx of extremely potent ions, polymeric proteins exposed to air CP either produce new oxygen-containing groups or polar radicals (Bahrami et al., 2022). The effects of CP on protein solubility have produced inconsistent results. It has been demonstrated that the kind of gas, pressure and flow, treatment duration, and electric current strength all have a substantial impact on the functional properties of proteins.

As far as we know, a DBD system has been used to test two voltages (9.8 and 18.8 kV) and exposure times (30 s and 1 min) to find out how well grass pea protein isolate (GPPI) dissolves in CP. Although the hydrophobicity was enhanced by the CP treatment, the solubility increased with longer treatment durations and higher treatment voltages. Applying CP reduced both the surface charge and particle size. According to Gantumur et al. (2023), protein solubility is impacted more by protein particle size than by protein hydrophobicity, and protein solubility is boosted by an increase in the electrostatic contacts between protein molecules. Therefore, it is thought that the ability of CP treatment to decrease protein particle size and increase protein surface tension contributes to the enhancement of protein solubility. However, Pérez-Andrés et al. (2019) also reported that gelatin, bovine lung protein, and pig-made haemoglobin were less soluble in CP. Protein aggregation and alteration, according to Jiang et al. (2023), increased the hydrophobic groups' exposure, which decreased the proteins' propensity to interact with water. Based on the hardness of the wheat flour, Mollakhalili-Meybodi et al. (2021) studied the reactivity of proteins to ACP. In this regard, the effects of radio frequency (RF) CP therapy have been explored in relation to the solubility behaviour of three types of flour: *Thinopyrum intermedium*, *Triticeae tribe* (soft), and *Triticum aestivum* L. (middle wheat grass). These three proteins' solubilities don't seem to vary much (Attri et al., 2021).

9.6.1.2 Interfacial Behaviour

The area between two insoluble components, such as oil–water and air–water in multistage systems, is known as the interface. It is well known that protein behaviour at interfaces has a big impact on food quality. Although not all proteins have this capability, the ability to change the shape of a protein, which is kinetically explained by the displacement of side groups from the unfavourable aqueous phase, can be utilised to determine if it can adsorb at the interface. Protein structural adaptability, interface potential and its distribution pattern, and hydrophobicity and its magnitude, among other things, can be used to define how well proteins can attach to surfaces. Rapid protein adsorption at the interface reduces the amount of hydrophobic moieties exposed to the aqueous phase. Additionally, the distribution of both hydrophilic and hydrophobic groups in proteins has a higher impact on surface activity than the total hydrophobicity of their surfaces (Sánchez-Morán et al., 2019). By generating a zwitterionic layer that is about 10 nm thick and accounting for surface charge, proteins can also stabilise an emulsion.

Proteins' structural flexibility, capacity to adsorb at the oil–air–water interface, and ability to form an interfacial, cohesive layer are what give them their foaming and emulsifying properties (Mollakhalili-Meybodi et al., 2021). The ability of a protein to form stable emulsions or foam is determined by its protein structure, which is directly related to exposure treatment. Therefore, it is crucial to investigate the impact of CP therapy on the foaming and emulsifying properties of proteins. The interfacial characteristics of proteins treated with ACP will be determined on the basis of the kind and volume of reactive species produced as well as how the conformation of the proteins has been assessed (Sánchez-Morán et al., 2019).

The impact of CP treatment on the GPPI's interfacial characteristics has been investigated by determining how well it reduces interfacial tension (IFT). After receiving a CP treatment of 9.4 kV for 60 s on GPPI, the IFT was decreased to its lowest content, demonstrating that it was more successful at reorganising near the interface and generating an interfacial layer. Despite having tight secondary and tertiary structures, globular proteins were found to be resistant to conformational change by Bergfreund et al. in 2021. This is true even though CP treatment enhanced the proteins' ability to shed the tertiary structure, which may be related to their ability to greatly reduce interfacial tension. The absence of tertiary structure facilitates the reorganisation of proteins at the surface. In the early stages of emulsification, native GPPI may more quickly lower interfacial tension. It appears that native GPPI's increased hydrophobicity and decreased surface charge will also have a significant impact on its capacity to absorb at hydrophobic surfaces and reach a thermodynamic equilibrium, given that its solubility is lower than previously believed (Mehr & Hadi, 2023; Mozafarpour et al., 2022). The high absorption rate and lack of interfacial tension reduction observed in native GPPI may be explained by the fact that too many native proteins prevented their reorganisation at the interface (Ghobadi et al., 2021).

After the first 10 min of treatment, the DBD CP's capacity to emulsify myofibrillar proteins began to decline. The first 10 min of treatment increase the exposure of hydrophobic groups on protein molecules, and a decline beyond this time could be due to chain-oxidation-induced aggregation formation between amino acid chains. A previous study found that samples treated at lower frequencies (70 Hz) were more successful at altering the functional properties of soy proteins, while samples treated at higher frequencies (120 Hz) demonstrated the greatest foaming stability. However, extensive protein-to-protein oxidation by plasma resulted in the creation of insoluble aggregates, which negatively affected the proteins' interfacial characteristics. However, soy protein isolates outperformed oxidation precipitate or aggregate molecules via ionic or non-ionic interactions, but dietary proteins performed better when their ordered structures were partly destroyed.

Hydrophobic groups are strengthened by increased sulfhydryl groups and crosslinking, which also increases the viscoelasticity of aggregated proteins at the contact (Mozafarpour et al., 2022). In general, it is possible to think of the effects of ACP treatment on the properties of the protein interface as the reactive species produced causing harm to the original protein structure. At the beginning of the CP therapy, the emulsification index rose; however, as the treatment went on, the value decreased. Initial partial unfolding appears to have boosted hydrophobic group exposure, which boosted surface rearrangement to cut down on free energy. Lengthening the ACP time may cause proteins to re-aggregate in order to hasten the synthesis of crosslinks through newly formed carbonyl and sulfhydryl groups. ACP treatment of proteins changes their interfacial properties in a variety of ways depending on the features of the plasma (gas type, time, mode of action, flow rate, and voltage), the qualities of the protein, and the characteristics of the two non-soluble liquids. The pace of protein absorption at the surface, the amount of proteins adsorbed, the level of surface reduction, and a protein's ability to form a layer of viscoelastic cohesion all have a substantial impact on a protein's interfacial properties. For an oil or air bubble in water to be stable, proteins must be able to construct a viscoelastic film continually.

9.6.1.3 Structure-Alteration

Proteins are a common component of many diets since they are organic substances with potential structural characteristics. The mouthfeel and overall texture of the dish can vary if the proteins are thickened. The evaluation of food proteins, which defines the final product's textural qualities and juiciness, places a high priority on the protein's capacity to gel. Given that gels usually behave as a solid-like substance with the majority of liquid attributes, foods containing gels vary widely in their rheological characteristics. It is believed that gel formation and the higher water retention it produces play a substantial role in the enhanced viscosity and viscoelastic properties of protein-rich meals.

The plasma treatment of pea protein isolate (PPI) is taken into consideration (voltage: 40 kV, current: 20.2 A). The water-holding capacity (WHC) of $CaCl_2$-produced PPI gels has increased at various intervals (1–4 min), and as the treatment time was extended, the WHC decreased. The WHC of every sample that was treated was higher than the WHC of the control sample. Proteins' WHC behaviour changed in a manner similar to how PPI solution solubility changed. The sulphide bond (-S–S) is reported to have a major impact on the WHC determination of proteins (Rusu et al., 2022). Incomplete degradation and the lower energy required to start protein unfolding in pork gelatin and lung protein extract (ELP) may be the cause for the lower gelation temperature of these substances, according to the effects of ACP treatment (70 kV, 20 min) on structural and functional properties like rheological and gelling of gelatin, haemoglobin, and ELP. The WHC of both pork gelatin and ELP after ACP treatment improved. Oil-holding capacity (OHC) was not shown to be affected by haemoglobin that has had ACP therapy, although ELP has. There are several proteins that make up

ELP, and it seems that each one is affected by ACP therapy in a distinct way (Saremnezhad et al., 2021). In a different study, Chaple et al. (2020) looked into the effects of plasma on whole-wheat grain flours when exposed for 10–30 min at 70 kV. The OHC and WHC of the plasma-treated flour increased over the course of the study, according to the findings. The wheat proteins did not alter following the plasma treatments, according to FTIR analyses.

Depending on how compact the protein is, the plasma treatment affects the protein's OHC/WHC characteristics. The treatment at 8.6 kV for 12 min improved the WHC/OHC of pea testa flour when compared to Tenebrio flour that had insect damage. It is important to keep in mind that fundamental characteristics of proteins, such as the types of amino acids, protein structure, and hydrophobicity, affect their ability to hold substances together. In this regard, it is taken into account that the protein's oxidation rate, state, free-SH and carbonyl group content, and hydrophobic surface are all significant influences on the protein's texturizing property after ACP treatment (Chen et al., 2020).

The most important criteria in determining a protein's function are its branching, compactness, length, hydrophobicity, and hydrophilicity; all of these aspects are significantly impacted by plasma therapy. From a technological standpoint, it has been demonstrated that the presence of specific amino acid residues (cysteine, lysine, histidine, and tryptophan) and the encouragement of amino acid cross-link formation boosted the consistency of ACP-treated proteins.

9.6.2 CHEMICAL ALTERATION OF PROTEINS

By utilizing CP, researchers have tried to modify protein molecules at the molecular level (Bao et al., 2021). Plasma engages with the bonds and aids in the dissolution of particular connections. For instance, a highly denatured peanut protein isolate was treated with plasma (70 W, 1.0–6.0 min) after being grafted with sesbania gum (Yu et al., 2021). Both the degree of browning (DB) and the degree of grafting (DG) increased with the duration of the CP treatment. An amphiphilic derivative of PPI can be made. When the treatment period lasted more than 3 min, the DG and DB values of aggregated peanut protein started to decline. Additionally, the prolonged therapy may result in over-oxidation, which would lessen the graft's efficacy (Yu et al., 2021). The enhanced peanut protein had better antioxidant and gelation properties. The grafted protein became more soluble due to the protein surface's coarse, uneven texture, and an increase in -OH radicals. Sesbania gum and denatured PPI were used to create a glycoprotein emulsion for the encapsulation of beta-carotene by Yu et al. (2021). It was discovered that the stability to heat and salt addition had improved, and the encapsulation efficiency was over 97%. In comparison to PPI-based emulsion, glycoprotein-stabilised emulsion showed greater bioaccessibility and a more controlled release of carotene.

Table 9.1 gives a list of changes induced by the effect of CP on different plant proteins.

9.7 CONCLUSION

In nature, plant proteins are extraordinarily abundant and offer numerous advantages over animal proteins, including quick production, widespread accessibility, high sustainability, environmentally friendly manufacturing, etc. In view of the severe requirements for a diet that is low in carbon emissions, favourable to the environment, and healthier, plant proteins are currently being emphasised as being prospective substitutes for animal proteins for a variety of culinary uses. Numerous steps have already been taken to turn many plant proteins into useful food ingredients to acquire structural features which can be made possible due to advancements in food production technology. However, constant efforts must be made in the highly targeted area of plant protein-based meat, such as "cereal meat," if we are to make a difference in our society's future. Because they are biocompatible, biodegradable, and bioaccessible, in addition to having a variety of tailored material qualities, plant proteins have potential applications outside of food in the areas of biomedicine, environmental protection, and biomaterials.

TABLE 9.1

Different Changes Induced by CP on Plant Proteins

Protein	Process Parameters	Salient Results	Conclusion	Reference
Whey	DBD for 5, 30, and 60 min at 70 kV	Foaming was increased with CP treatment up to 15 min, but it thereafter declined as treatment time increased	Partial unfolding of proteins during early stages of CP treatment	Song et al., 2019
Zein-chitosan	DBD for 2 min at 1.5 A and 40, 50, 60 V	Particle sizes reduced from 1191 nm to 370 nm and conductivity reached 43.4 ds/cm from 36.8 ds/cm	Interactions between zein and chitosan increased, and more receptors for carrying resveratrol are produced	Chen et al., 2020
Grass pea protein	DBD at 9.4 and 18.6 kV for 30 and 60 s	The most efficient proteins are those treated at 9.4 kV for 60 s because they have the lowest interfacial tension	Carbonyl groups, disulphide linkages, di-tyrosine crosslinks, and surface charge in grass pea protein isolate augmented with more treatment time and voltage	Mozafarpour et al., 2022
Soy protein isolate	Radio frequency At 13.56 MHz and 18 W	Soy powder has rough, highly developed micro-scale surfaces, and these surfaces significantly affect the powder's wetting regime. Following plasma treatment, the oxygen concentration rose to 22% from 19.5%	Powder becomes coarser on a micro scale after plasma treatment. The soy protein isolate is markedly hydrophilised by cold air plasma discharge, and it has been shown that there is no hydrophobic recovery for up to 1 month after treatment	Sharafodin & Soltanizadeh, 2022
Pea protein concentrate	DBD at 30 kV and 1 A for 2–10 min	The compressive strength of gel produced at 70°C was 0.53 kPa; when the temperature was elevated to 80°C and 90°C, respectively, the values were 2.70 kPa and 6.27 kPa. Following plasma treatment, the intensity of the fluorescence reduced, indicating protein unfolding	Denaturation at low temperature (70–90°C) was made possible by CP treatment, which enhanced the pea protein's gelling characteristics	Bu et al., 2022
Peanut protein isolate	DBD for 1–4 min at 35 V, 0.2 A	After receiving a 2-min CP treatment, the turbiscan stability index rose from 0.29 0.08 (0.5 h) to 2.92 0.38 (7 h)	Enhanced emulsion stability with a 2-min treatment period. The partial unfolding of the free -SH caused by the plasma treatment at 2 min increased the free -SH content of PPI	Venkataratnam et al., 2020

REFERENCES

Akhtar, J., Abrha, M. G., Teklehaimanot, K., & Gebrekirstos, G. (2022). Cold plasma technology: Fundamentals and effect on quality of meat and its products. *Food and Agricultural Immunology*, *33*(1), 451–478.

Amagliani, L., Silva, J. V., Saffon, M., & Dombrowski, J. (2021). On the foaming properties of plant proteins: Current status and future opportunities. *Trends in Food Science & Technology*, *118*, 261–272.

Arora, A., Castro-Gutierrez, R., Moffatt, C., Eletto, D., Becker, R., Brown, M., . . . Taliaferro, J. M. (2022). High-throughput identification of RNA localization elements in neuronal cells. *Nucleic Acids Research*, *50*(18), 10626–10642.

Attri, H., Dey, T., Singh, B., & Kour, A. (2021). Genetic estimation of grain yield and its attributes in three wheat (*Triticum aestivum* L.) crosses using six parameter model. *Journal of Genetics*, *100*, 1–9.

Bahrami, R., Zibaei, R., Hashami, Z., Hasanvand, S., Garavand, F., Rouhi, M., . . . Mohammadi, R. (2022). Modification and improvement of biodegradable packaging films by cold plasma; A critical review. *Critical Reviews in Food Science and Nutrition*, *62*(7), 1936–1950.

Bao, Y., Ertbjerg, P., Estévez, M., Yuan, L., & Gao, R. (2021). Freezing of meat and aquatic food: Underlying mechanisms and implications on protein oxidation. *Comprehensive Reviews in Food Science and Food Safety*, *20*(6), 5548–5569.

Bergfreund, J., Bertsch, P., & Fischer, P. (2021). Adsorption of proteins to fluid interfaces: Role of the hydrophobic subphase. *Journal of Colloid and Interface Science*, *584*, 411–417.

Bharti, B., Li, H., Ren, Z., Zhu, R., & Zhu, Z. (2022). Recent advances in sterilization and disinfection technology: A review. *Chemosphere*, 136404.

Bou, R., Navarro-Vozmediano, P., Domínguez, R., López-Gómez, M., Pinent, M., Ribas-Agustí, A., . . . Jorba-Martín, R. (2022). Application of emerging technologies to obtain legume protein isolates with improved techno-functional properties and health effects. *Comprehensive Reviews in Food Science and Food Safety*, *21*(3), 2200–2232.

Bourke, P., Ziuzina, D., Boehm, D., Cullen, P. J., & Keener, K. (2018). The potential of cold plasma for safe and sustainable food production. *Trends in Biotechnology*, *36*(6), 615–626.

Bu, F., Nayak, G., Bruggeman, P., Annor, G., & Ismail, B. P. (2022). Impact of plasma reactive species on the structure and functionality of pea protein isolate. *Food Chemistry*, *371*, 131135.

Chaple, S., Sarangapani, C., Jones, J., Carey, E., Causeret, L., Genson, A., . . . Bourke, P. (2020). Effect of atmospheric cold plasma on the functional properties of whole wheat (Triticum aestivum L.) grain and wheat flour. *Innovative Food Science & Emerging Technologies*, *66*, 102529.

Charoensuk, D., Brannan, R. G., Chanasattru, W., & Chaiyasit, W. (2018). Physicochemical and emulsifying properties of mung bean protein isolate as influenced by succinylation. *International Journal of Food Properties*, *21*(1), 1633–1645.

Chen, S., Han, Y., Jian, L., Liao, W., Zhang, Y., & Gao, Y. (2020). Fabrication, characterization, physicochemical stability of zein-chitosan nanocomplex for co-encapsulating curcumin and resveratrol. *Carbohydrate Polymers*, *236*, 116090.

Coțovanu, I., Ungureanu-Iuga, M., & Mironeasa, S. (2021). Investigation of quinoa seeds fractions and their application in wheat bread production. *Plants*, *10*(10), 2150.

Cruz-Morán, Y., Morales-Camacho, J. I., Delgado-Macuil, R., de Fátima Rosas-Cárdenas, F., & Luna-Suárez, S. (2023). Improvement of techno-functional properties of acidic subunit from amaranth 11S globulin modified by bioactive peptide insertions. *Electronic Journal of Biotechnology*, *61*, 45–53.

Dabija, A., Ciocan, M. E., Chetrariu, A., & Codină, G. G. (2022). Buckwheat and Amaranth as Raw Materials for Brewing, a Review. *Plants*, *11*(6), 756.

Domonkos, M., Tichá, P., Trejbal, J., & Demo, P. (2021). Applications of cold atmospheric pressure plasma technology in medicine, agriculture and food industry. *Applied Sciences*, *11*(11), 4809.

Fallah, M., Najafi, F., & Kavoosi, G. (2022). Proximate analysis, nutritional quality and anti-amylase activity of bee propolis, bee bread and royal jelly. *International Journal of Food Science & Technology*, *57*(5), 2944–2953.

Galanakis, C. M. (2021). Functionality of food components and emerging technologies. *Foods*, *10*(1), 128.

Gantumur, M. A., Hussain, M., Li, J., Hui, M., Bai, X., Sukhbaatar, N., . . . Jiang, Z. (2023). Modification of fermented whey protein concentrates: Impact of sequential ultrasound and TGase cross-linking. *Food Research International*, *163*, 112158.

Gantumur, M. A., Sukhbaatar, N., Shi, R., Hu, J., Bilawal, A., Qayum, A., . . . Hou, J. (2023). Structural, functional, and physicochemical characterization of fermented whey protein concentrates recovered from various fermented-distilled whey. *Food Hydrocolloids*, *135*, 108130.

Garcia-Rey, S., Nielsen, J. B., Nordin, G. P., Woolley, A. T., Basabe-Desmonts, L., & Benito-Lopez, F. (2022). High-resolution 3D printing fabrication of a microfluidic platform for blood plasma separation. *Polymers*, *14*(13), 2537.

Ge, S., Wu, Y., Peng, W., Xia, C., Mei, C., Cai, L., . . . Tsang, Y. F. (2020). High-pressure CO2 hydrothermal pretreatment of peanut shells for enzymatic hydrolysis conversion into glucose. *Chemical Engineering Journal*, *385*, 123949.

Ghobadi, M., Koocheki, A., Varidi, M. J., & Varidi, M. (2021). Encapsulation of curcumin using Grass pea (Lathyrus sativus) protein isolate/Alyssum homolocarpum seed gum complex nanoparticles. *Innovative Food Science & Emerging Technologies*, *72*, 102728.

Graziano, S., Agrimonti, C., Marmiroli, N., & Gullì, M. (2022). Utilisation and limitations of pseudocereals (quinoa, amaranth, and buckwheat) in food production: A review. *Trends in Food Science & Technology, 125*, 154–165.

Gupta, C. (2019). *Development of Gluten Free Pasta Using Amaranth Flour and Pea Protein Flour* (Doctoral dissertation). Iowa State University.

Huang, L. Y., Li, W., Du, N., Lu, H. Q., Meng, L. D., Huang, K. Y., & Li, K. (2022). Preparation of quaternary ammonium magnetic chitosan microspheres and their application for Congo red adsorption. *Carbohydrate Polymers, 297*, 119995.

Iftikhar, M., & Khan, M. (2019). Amaranth. *Bioactive Factors and Processing Technology for Cereal Foods*, 217–232.

Jiang, K., Koob, J., Chen, X. D., Krajeski, R. N., Zhang, Y., Volf, V., . . . Abudayyeh, O. O. (2023). Programmable eukaryotic protein synthesis with RNA sensors by harnessing ADAR. *Nature Biotechnology, 41*(5), 698–707.

Kaavya, R., Pandiselvam, R., Gavahian, M., Tamanna, R., Jain, S., Dakshayani, R., . . . Hemeg, H. A. (2022). Cold plasma: A promising technology for improving the rheological characteristics of food. *Critical Reviews in Food Science and Nutrition*, 1–15.

Kamath, R., Basak, S., & Gokhale, J. (2022). Recent trends in the development of healthy and functional cheese analogues-a review. *LWT, 155*, 112991.

Kang, Z. L., Zou, X. L., Meng, L., & Li, Y. P. (2021). Effects of NaCl and soy protein isolate on the physicochemical, water distribution, and mobility in frankfurters. *International Journal of Food Science & Technology, 56*(12), 6572–6579.

Khatoon, N., & Prakash, J. (2004). Nutritional quality of microwave-cooked and pressure-cooked legumes. *International Journal of Food Sciences and Nutrition, 55*(6), 441–448.

Kim, Y. H., & Shin, W. S. (2022). Evaluation of the physicochemical and functional properties of aquasoya (Glycine max Merr.) powder for vegan muffin preparation. *Foods, 11*(4), 591.

Le Thanh-Blicharz, J., Lewandowicz, J., Małyszek, Z., Baranowska, H. M., & Kowalczewski, P. Ł. (2022). Chemical modifications of normal and waxy potato starches affect functional properties of aerogels. *Gels, 8*(11), 720.

Li, S., Chen, H., Wang, X., Dong, X., Huang, Y., & Guo, D. (2020). Catalytic degradation of clothianidin with graphene/TiO$_2$ using a dielectric barrier discharge (DBD) plasma system. *Environmental Science and Pollution Research, 27*, 29599–29611.

Luo, H., Pandey, M. K., Khan, A. W., Guo, J., Wu, B., Cai, Y., . . . Jiang, H. (2019). Discovery of genomic regions and candidate genes controlling shelling percentage using QTL-seq approach in cultivated peanut (Arachis hypogaea L.). *Plant Biotechnology Journal, 17*(7), 1248–1260.

Marciano, S., Dey, D., Listov, D., Fleishman, S. J., Sonn-Segev, A., Mertens, H., . . . Schreiber, G. (2022). Protein quaternary structures in solution are a mixture of multiple forms. *Chemical Science, 13*(39), 11680–11695.

Martinez-Lopez, A., Millan-Linares, M. C., Rodriguez-Martin, N. M., Millan, F., & Montserrat-de la Paz, S. (2020). Nutraceutical value of kiwicha (Amaranthus caudatus L.). *Journal of Functional Foods, 65*, 103735.

Martínez-López, S., Sarriá, B., Mateos, R., & Bravo-Clemente, L. (2019). Moderate consumption of a soluble green/roasted coffee rich in caffeoylquinic acids reduces cardiovascular risk markers: Results from a randomized, cross-over, controlled trial in healthy and hypercholesterolemic subjects. *European Journal of Nutrition, 58*, 865–878.

Mayookha, V. P., Pandiselvam, R., Kothakota, A., Ishwarya, S. P., Khanashyam, A. C., Kutlu, N., . . . Abd El-Maksoud, A. A. (2022). Ozone and cold plasma: Emerging oxidation technologies for inactivation of enzymes in fruits, vegetables, and fruit juices. *Food Control*, 109399.

Mehr, H. M., & Hadi, N. B. A. (2023). Pickering stabilizing capacity of plasma-treated grass pea protein nanoparticles. *Journal of Food Engineering, 350*, 111458.

Meza, S. L. R., de Castro Tobaruela, E., Pascoal, G. B., Magalhães, H. C. R., Massaretto, I. L., & Purgatto, E. (2022). Induction of metabolic changes in amino acid, fatty acid, tocopherol, and phytosterol profiles by exogenous methyl jasmonate application in tomato fruits. *Plants, 11*(3), 366.

Misra, N. N., & Roopesh, M. S. (2019). Cold plasma for sustainable food production and processing. In *Green Food Processing Techniques* (pp. 431–453). Academic Press.

Misra, N. N., Yadav, B., Roopesh, M. S., & Jo, C. (2019). Cold plasma for effective fungal and mycotoxin control in foods: Mechanisms, inactivation effects, and applications. *Comprehensive Reviews in Food Science and Food Safety, 18*(1), 106–120.

Mollakhalili-Meybodi, N., Yousefi, M., Nematollahi, A., & Khorshidian, N. (2021). Effect of atmospheric cold plasma treatment on technological and nutrition functionality of protein in foods. *European Food Research and Technology, 247*, 1579–1594.

Moore-Kelly, C., Welsh, J., Rodger, A., Dafforn, T. R., & Thomas, O. R. (2019). Automated high-throughput capillary circular dichroism and intrinsic fluorescence spectroscopy for rapid determination of protein structure. *Analytical Chemistry, 91*(21), 13794–13802.

Moreau, E., & Defoort, E. (2022). Effect of the high voltage waveform on the ionic wind produced by a needle-to-plate dielectric barrier discharge. *Scientific Reports, 12*(1), 18699.

Mozafarpour, R., Koocheki, A., & Nicolai, T. (2022). Modification of grass pea protein isolate (Lathyrus sativus L.) using high intensity ultrasound treatment: Structure and functional properties. *Food Research International, 158*, 111520.

Mukhtar, K., Nabi, B. G., Arshad, R. N., Roobab, U., Yaseen, B., Ranjha, M. M. A. N., . . . Ibrahim, S. A. (2022). Potential impact of ultrasound, pulsed electric field, high-pressure processing, microfludization against thermal treatments preservation regarding sugarcane juice (saccharum officinarum). *Ultrasonics Sonochemistry*, 106194.

Nageswaran, G., Jothi, L., & Jagannathan, S. (2019). Plasma assisted polymer modifications. In *Non-thermal Plasma Technology for Polymeric Materials* (pp. 95–127). Elsevier.

Narimisa, M., Ghobeira, R., Onyshchenko, Y., De Geyter, N., Egghe, T., & Morent, R. (2022). Different techniques used for plasma modification of polyolefin surfaces. *Plasma Modification of Polyolefins: Synthesis, Characterization and Applications*, 15–56.

Nikmaram, N., & Keener, K. M. (2022). The effects of cold plasma technology on physical, nutritional, and sensory properties of milk and milk products. *LWT, 154*, 112729.

Nwachukwu, I. D., & Aluko, R. E. (2019). A systematic evaluation of various methods for quantifying food protein hydrolysate peptides. *Food Chemistry, 270*, 25–31.

Nyaisaba, B. M., Miao, W., Hatab, S., Siloam, A., Chen, M., & Deng, S. (2019). Effects of cold atmospheric plasma on squid proteases and gel properties of protein concentrate from squid (Argentinus ilex) mantle. *Food Chemistry, 291*, 68–76.

Okon, O. G. (2021). The nutritional applications of quinoa seeds. *Biology and Biotechnology of Quinoa: Super Grain for Food Security*, 35–49.

Pérez-Andrés, J. M., Álvarez, C., Cullen, P. J., & Tiwari, B. K. (2019). Effect of cold plasma on the techno-functional properties of animal protein food ingredients. *Innovative Food Science & Emerging Technologies, 58*, 102205.

Pillai, A. S., Hochberg, G. K., & Thornton, J. W. (2022). Simple mechanisms for the evolution of protein complexity. *Protein Science, 31*(11), e4449.

Pohl, P., Dzimitrowicz, A., Cyganowski, P., & Jamroz, P. (2022). Do we need cold plasma treated fruit and vegetable juices? A case study of positive and negative changes occurred in these daily beverages. *Food Chemistry, 375*, 131831.

Profeta, A., Siddiqui, S. A., Smetana, S., Hossaini, S. M., Heinz, V., & Kircher, C. (2021). The impact of Corona pandemic on consumer's food consumption: Vulnerability of households with children and income losses and change in sustainable consumption behavior. *Journal of Consumer Protection and Food Safety, 16*(4), 305–314.

Rachwa-Rosiak, D., Nebesny, E., & Budryn, G. (2015). Chickpeas—composition, nutritional value, health benefits, application to bread and snacks: A review. *Critical Reviews in Food Science and Nutrition, 55*(8), 1137–1145.

Rahman, M. M., & Lamsal, B. P. (2021). Ultrasound-assisted extraction and modification of plant-based proteins: Impact on physicochemical, functional, and nutritional properties. *Comprehensive Reviews in Food Science and Food Safety, 20*(2), 1457–1480.

Ran, X., Lou, X., Zheng, H., Gu, Q., & Yang, H. (2022). Improving the texture and rheological qualities of a plant-based fishball analogue by using konjac glucomannan to enhance crosslinks with soy protein. *Innovative Food Science & Emerging Technologies, 75*, 102910.

Rusu, L., Grigoraş, C. G., Simion, A. I., Suceveanu, E. M., Istrate, B., & Harja, M. (2022). Biosorption Potential of Microbial and Residual Biomass of Saccharomyces pastorianus Immobilized in Calcium Alginate Matrix for Pharmaceuticals Removal from Aqueous Solutions. *Polymers, 14*(14), 2855.

Sá, M. J., & Serpa, S. (2020). COVID-19 and the promotion of digital competences in education. *Universal Journal of Educational Research, 8*(10), 4520–4528.

Sánchez-Morán, H., Ahmadi, A., Vogler, B., & Roh, K. H. (2019). Oxime cross-linked alginate hydrogels with tunable stress relaxation. *Biomacromolecules, 20*(12), 4419–4429.

Sanyal, R., Pradhan, B., Jawed, D. M., Tribhuvan, K. U., Dahuja, A., Kumar, M., . . . Bishi, S. K. (2023). Spatio-temporal expression pattern of Raffinose Synthase genes determine the levels of Raffinose Family Oligosaccharides in peanut (Arachis hypogaea L.) seed. *Scientific Reports*, *13*(1), 795.

Saremnezhad, S., Soltani, M., Faraji, A., & Hayaloglu, A. A. (2021). Chemical changes of food constituents during cold plasma processing: A review. *Food Research International*, *147*, 110552.

Sarwar, N., Ahmed, T., Rahman, N., Nayik, G. A., & Asgari, S. (2023). Industrial applications of cereals. In *Cereal Grains* (pp. 123–146). CRC Press.

Sha, L., & Xiong, Y. L. (2020). Plant protein-based alternatives of reconstructed meat: Science, technology, and challenges. *Trends in Food Science & Technology*, *102*, 51–61.

Sharafodin, H., & Soltanizadeh, N. (2022). Potential application of DBD plasma technique for modifying structural and physicochemical properties of soy protein isolate. *Food Hydrocolloids*, *122*, 107077.

Sima, J., Wang, J., Song, J., Du, X., Lou, F., Pan, Y., . . . Zhao, G. (2023). Dielectric barrier discharge plasma for the remediation of microplastic-contaminated soil from landfill. *Chemosphere*, 137815.

Šimončicová, J., Kryštofová, S., Medvecká, V., Ďurišová, K., & Kaliňáková, B. (2019). Technical applications of plasma treatments: Current state and perspectives. *Applied Microbiology and Biotechnology*, *103*, 5117–5129.

Smetana, S., Profeta, A., Voigt, R., Kircher, C., & Heinz, V. (2021). Meat substitution in burgers: Nutritional scoring, sensorial testing, and life cycle assessment. *Future Foods*, *4*, 100042.

Song, J., Jiang, B., Wu, Y., Chen, S., Li, S., Sun, H., & Li, X. (2019). Effects on surface and physicochemical properties of dielectric barrier discharge plasma-treated whey protein concentrate/wheat cross-linked starch composite film. *Journal of Food Science*, *84*(2), 268–275.

Soni, M., Maurya, A., Das, S., Prasad, J., Yadav, A., Singh, V. K., . . . Dwivedy, A. K. (2022). Nanoencapsulation strategies for improving nutritional functionality, safety and delivery of plant-based foods: Recent updates and future opportunities. *Plant Nano Biology*, *1*, 100004.

Stryczewska, H. D., & Boiko, O. (2022). Applications of plasma produced with electrical discharges in gases for agriculture and biomedicine. *Applied Sciences*, *12*(9), 4405.

Venkataratnam, H., Cahill, O., Sarangapani, C., Cullen, P. J., & Barry-Ryan, C. (2020). Impact of cold plasma processing on major peanut allergens. *Scientific Reports*, *10*(1), 1–11.

Venkateswara Rao, M., CK, S., Rawson, A., & DV, C. (2021). Modifying the plant proteins techno-functionalities by novel physical processing technologies: A review. *Critical Reviews in Food Science and Nutrition*, 1–22.

Wang, J., Li, J., Liu, W., Zeb, A., Wang, Q., Zheng, Z., . . . Liu, L. (2023). Three typical microplastics affect the germination and growth of amaranth (*Amaranthus mangostanus* L.) seedlings. *Plant Physiology and Biochemistry*, *194*, 589–599.

Wang, L., Xue, J., & Zhang, Y. (2019). Preparation and characterization of curcumin loaded caseinate/zein nanocomposite film using pH-driven method. *Industrial Crops and Products*, *130*, 71–80.

Wang, S., Yang, J., Shao, G., Liu, J., Wang, J., Yang, L., . . . Jiang, L. (2020). pH-induced conformational changes and interfacial dilatational rheology of soy protein isolated/soy hull polysaccharide complex and its effects on emulsion stabilization. *Food Hydrocolloids*, *109*, 106075.

Wang, Y., & Luo, Y. (2023). Colloidal nanoparticles prepared from zein and casein: Interactions, characterizations and emerging food applications. *Food Science and Human Wellness*, *12*(2), 337–350.

Waskow, A., Avino, F., Howling, A., & Furno, I. (2022). Entering the plasma agriculture field: An attempt to standardize protocols for plasma treatment of seeds. *Plasma Processes and Polymers*, *19*(1), 2100152.

Wen, K. W., Wang, L., Menke, J. R., & Damania, B. (2022). Cancers associated with human gammaherpesviruses. *The FEBS Journal*, *289*(24), 7631–7669.

Ye, M. J., Xu, Q. L., Tang, H. Y., Jiang, W. Y., Su, D. X., He, S., . . . Yuan, Y. (2020). Development and stability of novel selenium colloidal particles complex with peanut meal peptides. *LWT*, *126*, 109280.

Yepez, X., Illera, A. E., Baykara, H., & Keener, K. (2022). Recent advances and potential applications of atmospheric pressure cold plasma technology for sustainable food processing. *Foods*, *11*(13), 1833.

Yi, X., Chen, Y., Ding, B., Ma, K., Li, Z., & Luo, Y. (2023). High internal phase Pickering emulsions prepared by globular protein-tannic acid complexes: A hydrogen bond-based interfacial crosslinking strategy. *Journal of Molecular Liquids*, *370*, 121025.

Yu, J. J., Zhang, Y. F., Yan, J., Li, S. H., & Chen, Y. (2021). A novel glycoprotein emulsion using high-denatured peanut protein and sesbania gum via cold plasma for encapsulation of β-carotene. *Innovative Food Science & Emerging Technologies*, *74*, 102840.

Yust, M. M., Pedroche, J., Giron-Calle, J., Alaiz, M., Millán, F., & Vioque, J. (2003). Production of ace inhibitory peptides by digestion of chickpea legumin with alcalase. *Food Chemistry*, *81*(3), 363–369.

Zang, P., Gao, Y., Chen, P., Lv, C., & Zhao, G. (2022). Recent Advances in the Study of Wheat Protein and Other Food Components Affecting the Gluten Network and the Properties of Noodles. *Foods*, *11*(23), 3824.

Zhang, W., Zhao, P., Li, J., Wang, X., Hou, J., & Jiang, Z. (2022). Effects of ultrasound synergized with microwave on structure and functional properties of transglutaminase-crosslinked whey protein isolate. *Ultrasonics Sonochemistry*, *83*, 105935.

Zou, H., Luo, J., Guo, Y., Liu, Y., Wang, Y., Deng, L., & Li, P. (2022). RNA-binding protein complex LIN28/MSI2 enhances cancer stem cell-like properties by modulating Hippo-YAP1 signaling and independently of Let-7. *Oncogene*, *41*(11), 1657–1672.

10 Chemical Modification of Plant Proteins

Suhail Anees, Showkat Ahmad Ganie and Rabia Hamid

10.1 INTRODUCTION

Plant proteins have grown in popularity from last few years. This growth is being driven by consumers with their increased awareness and understanding of food ingredients because plants offer them eco-friendly and sustainable food sources. The estimated annual growth rate of plant protein sales is 4.93% from 2018 to 2023, and the sales are expected to reach 45 billion dollars by 2023, which were 35 billion dollars in 2018. Proteins are crucial biomolecules with significant nutritional and functional importance in food systems, and the growing demand for healthier and more sustainable protein choices has accelerated the expansion of the plant-based protein industry, which includes legume, plant seed, cereal, and leaf proteins. Plant protein's nutritional value gives rise to benefits including good health and well-being, weight control, a low glycemic index, improved cardiac health, core strength, and athletic performance (Miguéns, 2022). Because of various concerns about food security and its impact on the environment, there is a global trend toward shifting our focus from animal to plant proteins, which can increase food production while decreasing environmental footprint (Aiking and de Boer, 2020). Plant proteins are rapidly replacing animal proteins as the most important protein source of the future. Plant proteins pose a challenge as compared to animal proteins because they have low water solubility and are primarily storage proteins. The approach to overcoming these problems is to modify these proteins. The proteins have the unique ability to regulate nearly all biological functions, and they are used as powerful biomacromolecules in many biomaterial processes, such as potent therapeutics, cell culture, and disease detection (Lagassé et al., 2017). Proteins, which are in the form of enzymes, catalyze reactions at high rates and with high selectivity. The essential amino acids are crucial for maintaining human health, and high-quality proteins offer a sufficient supply of these easily digestible amino acids. The amount of the amino acids in a protein can indicate its quality. Proteins from grains, seeds, legumes, nuts, pulses, and vegetables have low levels of essential amino acids like lysine, cysteine, methionine, and threonine, whereas soy proteins are referred to as complete proteins because they contain not only all nine essential amino acids but also a small amount of nonessential amino acids (Jäger, 2017). Plant proteins play a variety of roles, including structural, functional, and enzymatic functions, and they aid in biosynthesis and transport. Because these proteins function as storage proteins, they aid in plant growth and nutritional requirements. To meet the high demand, we must shift our focus to plant-based proteins, which have quality and functionality higher than animal proteins. There are various methods of protein modification, such as physical, chemical, and biological, and it is critical to understand the nature of the polymer, whether it is isolated or bound, as well as the structural properties of the proteins' ingredients, so that the desired method can be used to obtain the modified target property of the protein.

Chemical modification of plant proteins used as food proteins has gained popularity due to its low efficiency and low cost. Chemically modifying proteins involves either adding a new functional group or removing constituents from the native protein structure. The chemical modification of the protein is driven by enzymes which modify its structure by adding chemical group to the specific amino acids in the molecule. It enhances the functional diversity of the proteome by the covalent addition of functional groups. The pathophysiological consequences of diseases are primarily regulated at the cellular level by switching the downstream signaling network. Chemical conjugation affects

DOI: 10.1201/9781003369790-10

the cellular processes that change the tertiary structure of proteins by cellular differentiation, protein–protein interaction, and protein degradation. Chemical modification of proteins is an effective method for modulating their macromolecular function. A variety of post-translational modifications drive the changes in the native protein, which in turn mediate protein activity. The other methods for modifying proteins, such as fluorescent or affinity tagging, allow for complicated protein determination and tracking both in vitro as well as in vivo. These protein conjugates can also be used in the treatment of a variety of diseases (Liebscher and Bordusa, 2019; de Freitas Junior and Morgado-Díaz, 2016). The chemical modification of protein is usually achieved by post-translational modifications, and this generates the protein chemodiversity which is often responsible for the great biodiversity found in nature. These modifications include, among others, acylation, methylation, phosphorylation, sulfation, farnesylation, ubiquitination, and glycosylation, and they play an important role in key biological processes such as trafficking, differentiation, migration, and signaling (Chuh and Pratt, 2015). Protein modifications such as carbonylation, citrullination, glycosylation, and S-nitrosylation at specific sites are more pathologically significant because they can lead to abnormal cellular function and disease progression (Arora and Katyal, 2019). Chemical modification of proteins is a game-changing technique in research and development. Natural deteriorations modify some of the most reactive side chains through oxidations, reductions, and nucleophilic and electrophilic substitutions. Peptide bond scissions, racemizations, and β-eliminations are deteriorations that occur when proteins react with chemicals (Reddy et al., 2020). In both chemo- and regioselectivity, reaction at an individual amino acid or site among many reactive carboxylic acids, amides, amines, alcohols, and thiols is an important challenge. The conditions required for potential transformations should be biologically relevant and should not disrupt protein structure or function and a primary driver of pharmaceutical industry advancement in response to drug delivery challenges. The properties of therapeutic development continue to expand as our knowledge of current methods grows, and one of the rapidly expanding areas of interest is chemically altered peptide and protein therapeutics (Holz, 2023).

10.2 PLANT PROTEINS

The demand for plant protein-based products has recently increased and is expected to grow significantly over the next decade. Plant proteins perform a variety of enzymatic, structural, and functional functions such as photosynthesis, biosynthesis, and transport. The development of seedlings in plants is due to these proteins, which act as storage mediums for their growth and nutrition. These proteins perform a variety of functions from composition to structural functions, via folding that ranges from compact to disordered. The essential amino acids needed for human health are present in enough amounts in the plant proteins derived from various plant sources. Plant proteins are becoming increasingly popular due to their biodegradable, diverse, and edible and nonedible properties (Kumar et al., 2022a). As reported by various studies, the benefit of plant protein as a functional food has yielded mixed results due to heterogeneity. The bioactive components found in plants are thought to be due to whole food sources rather than protein. Flavonoids and carotenoids, for example, have numerous health benefits. Plant-based proteins serve a variety of functions in nutrition and industry, including serving as functional ingredients in food formulations, emulsion stabilizers, and binding agents for water and fat (Jafari, 2020). They also have antimicrobial, anti-inflammatory, and antioxidant properties (Warnakulasuriya and Nickerson, 2018).

There are various factors which contribute to the popularity of plant proteins such as plant protein's potential health benefits in comparison to animal protein's contribution to adverse health effects, ethical issues concerning animal treatment, and the general view that protein acts as a health beneficial nutrient (Hertzler, 2020). There are two qualities in proteins which make them complete for human nutrition, and the amino acid requirement is categorized into indispensable and dispensable in order to know the essentiality of these amino acids. First, they should contain an adequate amount of indispensable amino acids and, second, they could be easily absorbed and digested by the humans. Table 10.1 shows the indispensable and dispensable amino acids.

TABLE 10.1
Amino Acids in Human Diet

Indispensable Amino Acids	Dispensable Amino Acids
Histidine	Alanine
Isoleucine	Aspartic Acid
Leucine	Asparagine
Lysine	Glutamic acid
Methionine	Serine
Phenylalanine	
Threonine	
Tryptophan	

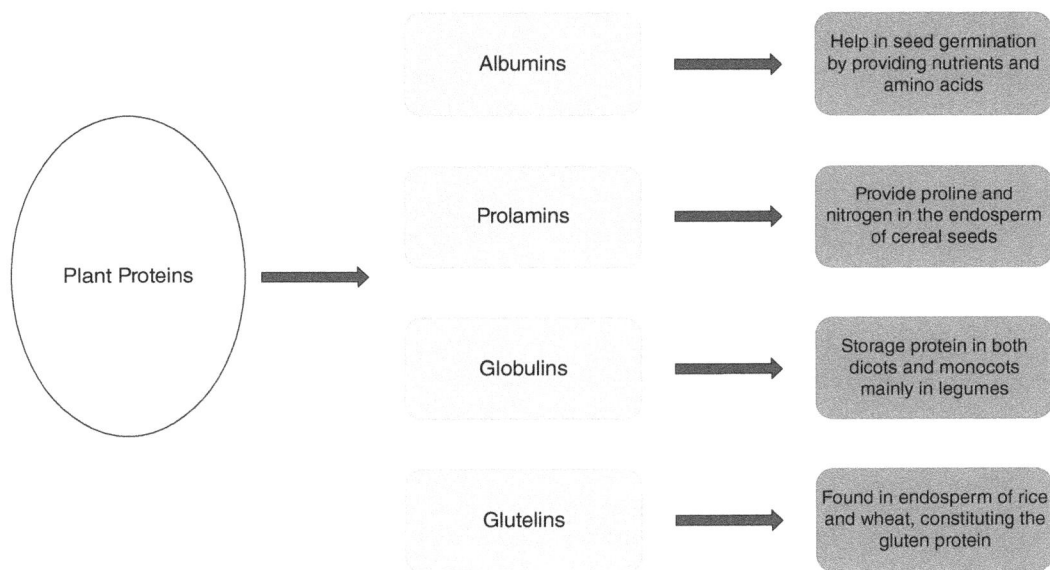

FIGURE 10.1 Types of plant proteins and their main functions.

10.3 TYPES OF PLANT PROTEINS

The plant proteins are classified into four main types such as albumins, globulins, prolamins, and glutelins based on Osborne classification (Osborne, 1924). These are generally seed storage proteins. The storage protein's function is to store amino acids required for the new seedling's growth and development. The albumins are water soluble; globulins are dilute saline solution soluble, prolamins are alcohol soluble, and glutelins are acid or alkali solution soluble. Figure 10.1 represents types of plant proteins and their main functions.

10.3.1 ALBUMINS

Albumin proteins are essential seed storage proteins that aid in seed germination by providing nutrients and amino acids from the early to late stages of development. These have a three-dimensional structure that is rich in alpha helices, which accounts for their antibacterial and antifungal activity (Souza, 2020). These have a low molecular mass and a high amino acid content, including arginine, glutamine, asparagine, and cysteine, as well as high stability due to the presence of disulfide bonds. These are mostly found in the seeds of monocotyledons and dicotyledons. Plants have several preproalbumin genes with no introns, and these preproalbumins form an endoplasmic reticulum signal domain and pro domain, which is followed by an albumin domain. When the endoplasmic reticulum signal is removed from preproalbumin, it becomes proalbumin, and when this proalbumin is proteolytically processed, it becomes albumin (Mylne et al., 2014). During biosynthesis, these proteins undergo post-translational modification, which is primarily proteolysis (Monsalve et al., 1990).

10.3.2 PROLAMINS

Prolamins are the second-most prevalent protein seed storage, and they pile up in the endosperm of cereal seeds and are important sources of essential amino acids in the diet. Prolamins proteins form the intracisternal granules which are held together by hydrophobic interactions after they are transported to ER lumen after translation (Kawakatsu and Takaiwa, 2019). The common cereals are members of the Poaceae family, also known as grasses, and their seeds are a staple food in both human and animal diets. A large portion of these seeds contain prolamins, a protein family rich in proline and amide nitrogen derived from glutamine. Due to chromosomal position and sequence divergence, gene amplification events become possible, and the loss of older gene in various subfamilies ends up creating divergence, and these modifications in protein affect their physicochemical characteristics; organelle location; and the amino acid makeup in rice, sorghum, wheat, barley, and brachypodium, representing different subfamilies of the Poaceae (Xu and Messing, 2009).

10.3.3 GLOBULINS

The globulins are seed storage proteins that are not only found primarily in legumes but are also found in higher plants such as monocots, dicots, ferns, and gymnosperms. The amino acid composition of globulins reveals that they are low in sulfur-containing amino acids, with methionine being the least abundant (Krishnan and Coe, 2001). Seven proteins were discovered in a study on Brachypodium distachyon using SDS page and mass spectroscopy, six of which were globulins. The seed proteins isolated from three hexaploid accessions were identified as Bd4, Bd14, and Bd17 globulins (Laudencia-Chingcuanco and Vensel, 2008). Globulins are classified as 11S legumin type and 7S vicilin type, which evolved from a one-domain germin predecessor or through horizontal gene transfer, and these globulins are known for their nutritional quality, resistance to microbes and oxidative stress, and involvement in sucrose binding and desiccation (Kesari, 2017).

10.3.4 GLUTELINS

Glutelin is a major storage protein found in the endosperm of various cereals, especially rice, and shares homology with leguminous 11–12S globulins. Glutelins are formed in the rough endoplasmic reticulum and then transported to ER lumen by hydrophobic interactions with pre-proglutelins, a 57 kDa protein precursor (Wakasa, 2009). Glutenin is the most commonly found glutelin in wheat, but glutelins can also be found in barley and rye (Shang et al., 2005). Many similarities exist between prolamins and glutelins, including amino acids such as proline and glutamine, as well as repetitive motifs and non-repetitive domains in the N- and C-terminals. The glutelins are made up of four protein subfamilies that are encoded by multiple genes: GluA, GluB, GluC, and GluD. When glutenins are digested by protease enzymes, they yield a good number of amino acids and peptides, and the biological processes by which they are digested are similar to those found in human digestive organs, while embryo germination occurs in rice and by microorganisms in the wine production

TABLE 10.2
Different Types of Protein Modifications

S. No	Physical Modifications	Chemical Modifications	Biological Modifications
1.	Sonication	Glycation	Enzymatic
2.	Heating	Phosphorylation	Protein fermentation
3.	High-pressure treatment	Acylation	
4.	Extrusion	Deamination	
5.	Ultrafiltration	Succinylation	
6.	Electron beam radiation	Glycosylation	
7.	Gamma radiation		
8.	UV radiation		

process (Takahashi et al., 2019). In one study, polyploid breeding enhanced the protein nutritional value and quality of rice seeds, and it was later discovered that polyploidization increased glutelin content by enhancing its biosynthesis, transport, and deposition (Gan, 2021).

10.4   MODIFICATION APPROACHES OF PROTEINS

Protein modification refers to physical, chemical, or biological changes made to functional groups or molecular structure in order to improve biological activity. It makes plant proteins more viable and multifunctional in the food system by altering their properties. Proteins are extremely complex biomolecules, and it is critical to understand the nature of these polymers, whether bound or in isolation, as well as their structural properties and mechanism, in order to achieve the modified target. Physical methods of protein modification include heating, high-pressure treatment, sonication, extrusion, and ultrafiltration, while biological methods include enzymatic and protein fermentation modification. Glycation, phosphorylation, acylation, and deamination are examples of chemical modifications that occur after translation. Chemical modification refers to reactions in which existing bonds are broken or new bonds are formed in order to change the original protein structure by utilizing the protein side chains in a chemical reaction that alters the property and function of a protein. The ultimate goal is to alter the net charge of the corresponding protein (Panyam and Kilara, 1996). In comparison to native proteins, the chemically modified protein demonstrates improved functionality. Here, we look at some of the chemical modifications of proteins, which result in high-quality products.

10.5   POST-TRANSLATIONAL MODIFICATIONS

Protein modification is a hot topic in chemical biology. The chemical change in the structure and function of proteins leads to various intra and extracellular events thereby modifying the protein. These chemical changes are known as posttranslational modifications because they take place after the protein biosynthesis step, i.e., translation (Baslé, 2010). Posttranslational modifications improve protein features, functions, and diversity while also influencing protein interactions, stability, and enzyme kinetics. These posttranslational modifications are covalent modifications which include the addition of different chemical groups to the side chain of amino acids.

10.5.1   SELECTIVE CHEMICAL MODIFICATION

Selective modification allows for the modification of specific chemical groups within a molecule while leaving others unmodified, allowing for the modification of specific properties or functionalities while preserving the overall structure and integrity of the molecule. Using a protecting group strategy, a chemical reagent is selectively reacted with a specific amino acid residue or group within a protein. Cysteine is widely used in the selective protein modification process and has long been used for such modifications due to its unique reactivity and because of its side chain thiol's high nucleophilicity and the comparable lack of free sulfhydryl groups on native proteins. Apart from nonselective cysteine modification methods, the progression of more selective protocols has managed to grow at an exponential rate. The methods for site-selective cysteine modification begin with the simple disulfide exchange reaction to form a mixture of disulfides and a similar reaction using diselenides to generate more sustainable SeS-linked conjugates; it continues with the transformation of potentially cleavable disulfide linkages to more stable thioethers via disulfurization of disulfides and concludes with the development of complementary cys-elimination methods that allow access to dehydroalanine, which is good Michael acceptor, and from this thioethers can be obtained directly upon addition of thiol nucleophiles (Chalker et al., 2009).

10.5.2 GLYCATION

Glycation, also known as glycosylation, is a food grade reaction that is used to enhance the capabilities because it does not require any exogenous chemicals and is considered safe (Doost et al., 2019). Glycation is a nonenzymatic reaction that occurs between sugars and proteins and modifies the plant protein. It is a post-translational modification that takes place between reducing sugars with a free aldehyde or ketone group, such as glucose, and the N-terminal and lysine side chains (Rabbani, 2020). Glycation has remained an important interest in both research and commercial purposes because it improves plant's crop relevance (Leonova, 2020; Chaplin et al., 2019). Cross-linking enzymes such as transglutaminase or laccase cause glycation, which occurs chemically via Maillard reactions. The best method for improving the functional properties of proteins is glycation with carbohydrates via the first step of the Maillard reaction under the influence of heat. It is the nonenzymatic reaction described by Louis-Camille Maillard between the free amino groups of a protein and the carbonyl functions of reducing carbohydrates. Because it occurs spontaneously under natural conditions, this method outperforms other chemical modification methods. Because reducing sugars contain an aldehyde or ketone group, they aid in the reaction with the protein substrate on the N-terminal and lysine side chains. The main sugars found in higher plants that aid in glycation are glucose, galactose, and fructose. When glucose reacts with the protein's N-terminal amino groups and lysine side-chain amino groups, Schiff's base is formed. Amadori rearrangement produces the fructosamines: N-1-deoxyfructosyl N-terminal amino groups and Nε-fructosyl-lysine (FL). Because of their lower pKa, N-terminal amino groups react faster with reducing sugars than lysine side chain amino groups (Rabbani and Thornalley, 2012). Glycation is a nonenzymatic process that causes numerous health problems in the form of diabetes mellitus and has been linked to certain diseases and ageing. Various studies have shown that during photosynthetic metabolism, which is essential to plants, glycating agents are produced, and the plant proteome is continuously exposed to glycating agents, resulting in the formation of advanced glycating agents (AGEs) (Soboleva, 2017). AGEs are heterogeneous compounds formed by the reaction of sugar saccharide with proteins or lipids. They can form as a result of both endogenous and exogenous mechanisms. The main driving force that catalyzes the formation of AGEs is reactive oxygen species. These can influence gene expression and cellular signaling by generating reactive oxygen species and interacting with specific receptors. When their concentration rises in the diet and circulation, they cause a variety of diseases (Chen et al., 2018). These AGEs cause oxidative stress, which activates stress-induced transcription factors, resulting in the production of inflammatory mediators such as cytokines and acute-phase proteins (Uribarri, 2007). Figure 10.2 shows the mechanism of glycation and its application in food processing.

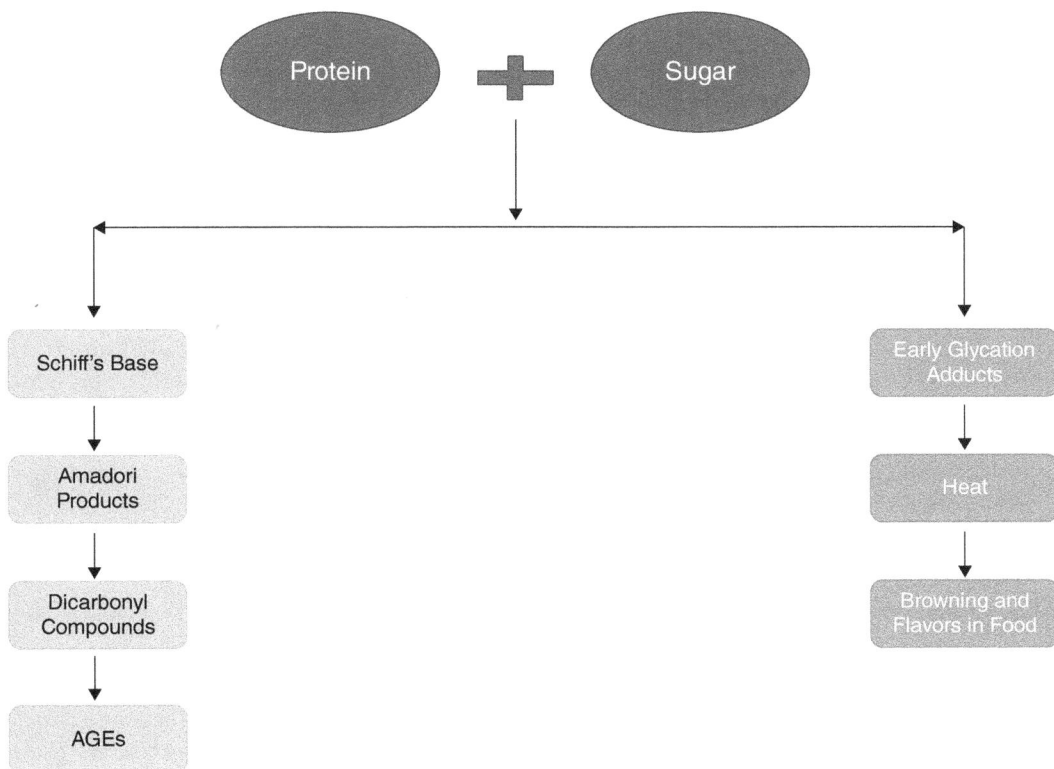

FIGURE 10.2 Mechanism of glycation and its application in food processing.

10.5.3 ACYLATION

Protein S-acylation is a reversible process that adds fatty acids to cysteine residues. This process regulates various aspects of protein function such as intracellular trafficking, protein interactions, protein stability, and protein conformation, and it is controlled by palmitoyl acyltransferases with a conserved amino acid sequence (Chen et al., 2021a). It is not well known that post-translational fatty-acid-based protein modification occurs in plants. It is also known as S-palmitoylation, which occurs when a 16-carbon palmitate is covalently attached to a cysteine residue through a thioester bond. S acylation can occur on both transmembrane and soluble proteins, and acylated soluble proteins aid in trafficking, regulation, and signaling (Blaskovic, 2013). Protein S-acylation is a significant lipid modification that is reversible in nature and catalyzed by acyl protein thioesterases, assisting in and out of microdomains found in many s-acylated proteins and their regulation (Lu and Hofmann, 2006). It is catalyzed by the DHHC domain (an enzyme that adds a palmitoyl group to the proteins), which contains Protein S-Acyl Transferases (PATs), and many of these domains are found in plant databases. Arabidopsis, for example, contains 23 DHHC PATs (AtPATs) (Hemsley et al., 2005). There are many S-acylated proteins found in plants, but one of the most important is thioesterase, which catalyzes the de S-acylation of many substrates (Zheng, 2019). The addition of an acyl moiety mediates various functions required for protein function, such as subcellular localization, protein–protein interaction, or protein–membrane interaction. These covalent modifications in proteins aid in modulating protein functions, resulting in specific tasks.

10.5.4 DEAMIDATION

Deamidation is a chemical reaction that converts a functional group, typically an amide group of an amino acid, into another functional group. This happens when a functional group from an amino acid's side chain, such as asparagine, is converted into another amino acid called aspartic acid. The protein deamidation includes the conversion of amide groups to carboxylic groups with the release of ammonia by hydrolysis. The only enzymes reported for protein deamidation are transglutaminase, protease, and peptidoglutaminase (PGase). The products produced by chemical deamidation are determined by factors such as temperature, pH, buffer composition, and the sequence/size of amino acids in the deamidated substrate. Protein solubility and emulsifying properties are enhanced by deamidation, which turns protein amide groups to carboxyl groups by imparting negative charges, lowering the protein's isoelectric point. Sometimes deamidation causes the conversion of asparagine residues to acidic residues (isoAsp), impacting protein structural stability, and some proteins are targeted for degradation by the ubiquitin–proteasome (Nowak et al., 2018). Protein deamination improves protein function, which has significant advantages over other treatments. Enzymatic protein–glutaminase deamidation can improve the chemical and physical properties of plant-based proteins, potentially allowing for the development of food derived from proteins obtained from plants which outperform the proteins that come from other sources (Liu et al., 2022).

10.5.5 PHOSPHORYLATION

Protein phosphorylation is a significant and common post translational modification of proteins mediated by kinases and phosphatases. It takes place in the cell's cytosol or nucleus, and these reactions are catalyzed by protein kinases, which are involved in cell regulation processes such as apoptosis, differentiation, and proliferation. It is a reversible mechanism in which a phosphate moiety is transferred from adenosine triphosphate (ATP) to the acceptor residue and results in the formation of adenosine diphosphate (ADP). The phosphate group is added to the polar group R of the respective amino acid and is mediated by protein kinase C. The phosphate group changes the protein's conformation from hydrophobic apolar to hydrophilic polar, and the phosphorylated protein can now interact with other proteins (Sacco, 2012). Phosphorylation controls many subcellular processes in all living cells, including eukaryotes and prokaryotes, and phosphorylation of proteins on serine, threonine, and tyrosine, which is reversible in nature (Dworkin, 2015). Sodium tripolyphosphate (STP), sodium trimetaphosphate (STMP), and POCl$_3$ are the primary phosphorylating agents in plant proteins. Nowadays, there is a lot of interest in phosphorylation of cereals, and STMP phosphorylated rice proteins outperform untreated protein (Hu et al., 2019).

The amino acids containing side chain groups such as -NH, -OH, or -SH are capable of phosphorylation as shown in Table 10.3.

10.5.6 SUCCINYLATION

Succinylation is the transfer of a succinyl group to a protein molecule at its lysine (K) residue, and it is an essential step in regulating plant growth and development (Hashiguchi and Komatsu, 2016). During rice seed germination, it leads the proteomic pool changes, impacting metabolic mechanisms such as carbon fixation, glycolysis, gluconeogenesis, and the citric acid cycle. Succinylation aids in the regulation of various biological processes as well as the adaptation of plants to various environmental stresses (Zhen, 2016). Succinylation is a very important protein regulatory mechanism that is important for cell energy metabolism. Furthermore, research on the mitochondrial specificity of desuccinylase (SIRT5) suggests that succinylation may be associated in several diseases where mitochondrial dysfunction is frequent, including acute lung injury and age-related diseases (Galam, 2015). Because various plant-based proteins are lost as waste during food processing, succinylation can aid in their proper utilization. For example, mung-bean protein, a by-product of

TABLE 10.3
Commonly Phosphorylated Amino Acids

S. No	Protein	Side Chain Group
1.	Tryptophan	-NH
2.	Threonine	-OH
3.	Cysteine	-SH
4.	Asparagine	-NH
5.	Serine	-OH
6.	Tyrosine	-OH
7.	Aspartic Acid	-OH
8	Glutamic Acid	-OH
9.	Histidine	-NH
10.	Lysine	-NH
11.	Arginine	-NH
12.	Glutamine	-NH

mung-bean vermicelli production, has a high protein content but is no longer utilized due to poor functional properties and a bitter flavor. This can be reduced by succinylation, which contributes negatively charged succinyl groups which makes it functional (Charoensuk et al., 2018). When compared to other posttranslational modifications, succinylation switches a protein's positive charge to negative charge and increases mass. Numerous protein succinylation sites can be found in various tissues and species. Nonenzymatic succinylation is mediated by succinyl coA in the cell, and succinyl CoA, a TCA cycle metabolite, aids in the supply of succinyl groups. Succinyl coA is a short reactive chain CoA thioester that keeps the mitochondrial matrix at steady-state concentrations. When albumin protein is combined with succinyl coA, the succinylation increases, while the pH of the mitochondria aids in nonenzymatic regulation (Wagner and Payne, 2013). Nonenzymatic succinylation of BSA and ovalbumin by succinyl coA confirms that succinylation is dependent on intracellular succinyl coA. The enzyme succinyl–CoA synthetase generates succinyl coA from the TCA cycle, lipids, and amino acids (Burch et al., 2018). Succinylation is important in cancer, but its role in other diseases is unknown. Hypersuccinylation (i.e., SIRT5 knockout) causes hypertrophic cardiomyopathy, as evidenced by an increased heart weight to body weight ratio (Sadhukhan et al., 2016). SIRT5 belongs to a protein family that functions as nicotinamide adenine dinucleotide-dependent lysine deacetylases and desuccinylases. The SIRT5 protein has a desuccinylase function, and its in vitro desuccinylase activity exceeds its deacetylase activity. SIRT5 was knocked out of mice to confirm its desuccinylation activity, and the results indicated that the SIRT5 knockout increased succinylation on CPS1 (carbamoyl phosphate synthase), an enzyme in the urea cycle. This research contributes to our understanding of SIRT5 as a lysine desuccinylase that catalyzes the expulsion of a succinyl group while using NAD+ as a cofactor (Du et al., 2011).

10.5.7 GLYCOSYLATION

Protein glycosylation is a common posttranslational modification of secretory and membrane proteins found in all aspects of life. Glycosylation is the organized bonding of carbohydrates to proteins or lipids. This process necessitates the use of enzymes such as glycosyltransferases, which are found in the smooth endoplasmic reticulum and the Golgi apparatus. The protein moiety is covalently attached to various types of glycans such as carbohydrates, saccharides, or sugars. A conjugate is formed by the reaction of a carbohydrate (or 'glycan,' i.e., a glycosyl donor and a glycosyl acceptor.

Glycation differs from glycosylation in that glycosylation is an enzyme-catalyzed reaction, whereas glycation is a nonenzymatic reaction. Glycosylation occurs on more than half of all proteins and several lipids in biological systems. N- and O-linked glycosylation are the two types of glycosylation that occur in eukaryotic as well as Gram-negative and Gram-positive bacteria. In bacteria, O-glycosylation is more common than N-glycosylation. The initial sugar residue of N-glycans in eukaryotic cells is normally β-GlcNAc, whereas the first residue of O-glycans can be β-GlcNAc, α-GalNAc, α-Man, or other monosaccharide units. N-glycosylation is the attachment of a nitrogen atom N4 to asparagine residues, which occurs on secreted or membrane-bound proteins found in eukaryotes but not in most bacteria. N-glycosylation occurs in two stages: First, lipid-linked oligosaccharide is assembled, and then the oligosaccharide is transferred to asparagine residues of polypeptide chains. Attachment of glycans to asparagine is an important modification in eukaryotes and a highly conserved process. As a result, the N-glycosidic linkage between the asparagine side-chain amide and the oligosaccharide is formed (Breitling and Aebi, 2013). N-glycans also have an impact on cellular proteins function and activity, and their terminal residues are important in the quality control of protein folding (Skropeta, 2009). O-glycosylation is the modification which covalently attaches oligosaccharides to secreted proteins in certain threonine or serine residues, which occurs in the Golgi compartment. Protein localization and trafficking depend largely on O-linked glycans thereby contain N-acetylgalactosamine (GalNAc) as their core. O-linked glycans contain glycoproteins that are O-glycosylated, such as mucins (MUCs), immunoglobulins, and caseins. O-linked glycans are produced in mucins or found in milk secretions and are released and consumed by gut microbiota. O-glycans obtained from milk glycopeptides, such as those found in GMP, have the potential to be used in foods that promote a healthy gut microbiota and act as emerging prebiotics (González-Morelo et al., 2020). These O-glycans can act as signal molecules, assisting in cell secretion and providing better resistance to proteolysis (Boutrou, 2008). Glycosylation is significant in improving dietary protein functioning because carbohydrate inclusion in proteins makes the products more hydrophilic, and glycosylated casein improved its solubility (Ma et al., 2020). Figure 10.3 represents overall insight of glycosylation.

10.6 APPLICATION OF CHEMICAL MODIFICATIONS IN PRODUCING FOOD

The task after achieving highly efficient plant proteins is to modify these components into nutritious and delicious foods. Plants aid in the reduction of various lipids, the maintenance of good health, and the fulfilment of individual's protein requirements. Plant sources such as legumes, cereals, seeds, and dry fruits have all been studied for their potential use as protein supplements (Sá et al., 2020b). Natural biopolymers such as proteins which are biodegradable and environmentally friendly can be used as food coating and packaging. The most commonly used proteins in the production of edible films that are also water soluble are globular proteins. The main protein ingredients used to make plant-based meat are wheat, pea, and soy, but consumers have yet to accept legume, soy, and pea proteins due to their bitter taste (Sha and Xiong, 2020). Covalent and hydrogen bonding aid in complex formation, and these globular proteins can be denatured by physical modifications such as heat, allowing them to be used in film formation.

Glycation is the most desired chemical modification of plant protein for food applications because it produces no side chemicals and is thought to be beneficial for the modification of proteins derived from plants with regard to the preferences of consumers and for production and marketing. Glycation is an acceptable food reaction that is commonly used to improve protein functions since it is safe and does not usually involve the addition of external substances. The structural properties of various plant proteins are altered by glycation, which increase features and functions, bioactivity, solubility, and emulsification in oat and pea protein isolates (Zhong, 2019). Succinylation is an important method for obtaining modified protein products, when the male date palm pollen protein is succinylated, it improves its thermal stability and solubility. Non-diary analogs were discovered through succinylation, and these analogs, along with succinylated proteins, were combined

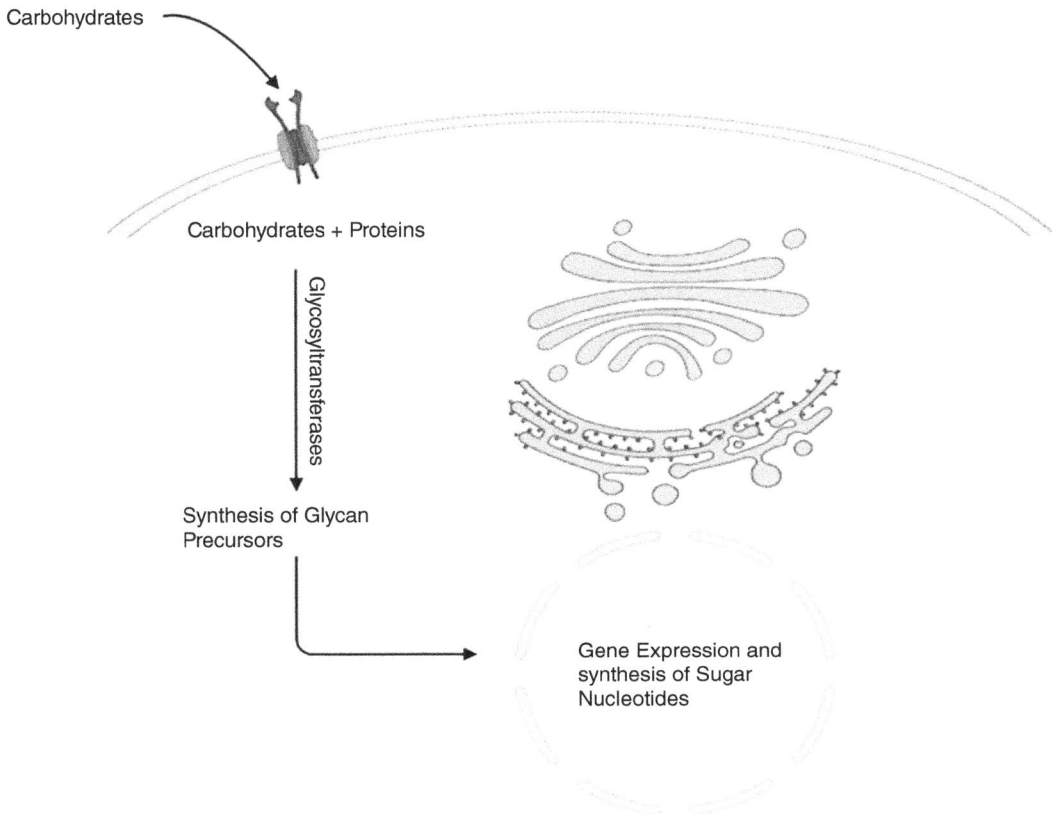

FIGURE 10.3 Overall insight of glycosylation.

with other ingredients to create a diary-free product that could be used in beverages (Stiles, 2021). Deamidation affects protein functionality and improves biological functions such as gluten solubilization, emulsification property, and foaming property, and it also increases gluten hydrophobicity, which is most likely caused by a conformational change, thereby increasing surface activity (Matsudomi, 1982). Deamidation is regarded as a safe method of protein modification in the food system because cereals and legumes contain a high concentration of glutamine and asparagine, making this a good method for their food system application. The effect of acetylation of acetic anhydride on the chemical composition and functional properties of pumpkin protein concentrate was studied in a study, and it was discovered that acetylation generally improves the functional characteristics of pumpkin protein, primarily protein solubility, water-binding ability, and oil-absorption capabilities. This acetylation of pumpkin protein concentrate resulted in a significant improvement in water-binding and oil-absorption capacity, allowing it to be used in the food industry (Miedzianka et al., 2021). Phosphorylation was used to improve the emulsifying activity of peanut and soybean protein isolates. Soybean and peanut protein isolates were phosphorylated with sodium trimetaphosphate (STMP), which influenced and caused changes in functional properties. Phosphorylation of peanut and soybean proteins with sodium trimetaphosphate resulted in isolates with broad functional applications in the food system. Plant-based proteins have been shown to improve phosphorylation by increasing electronegativity and electrostatic repulsion among various protein moieties (Sánchez-Reséndiz, 2018). In order to understand the relationship between N-glycosylation and fruit ripening, a study was conducted in which the silencing of RNA interference in α-mannosidase and β-d-N-acetylhexosaminidase, two N-glycoprotein-modifying enzymes

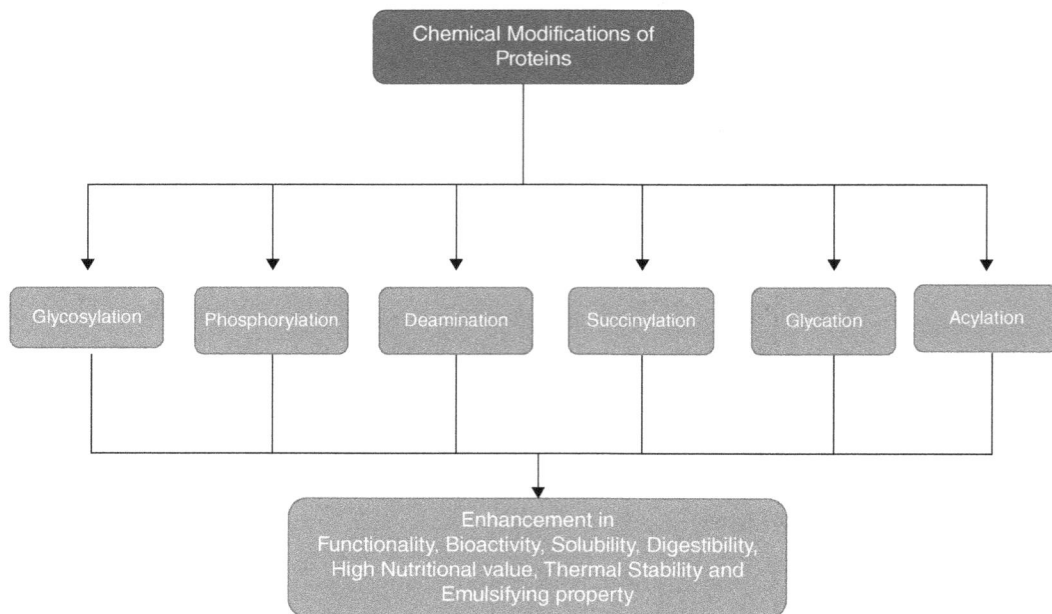

FIGURE 10.4 Effect of protein modifications on the functional properties of proteins, resulting in a highly efficient and improved functionality.

found in capsicum annuum fruits, delayed fruit deterioration and aided in fruit ripening (Ghosh, 2011). Plants can perform human-like complex glycosylation, and manipulating this pathway allows the production of proteins with human-type oligosaccharides (Bosch, 2013). Figure 10.4 shows effect of protein modifications on the functional properties of proteins, resulting in highly efficient and improved functionality.

10.7   EFFECT OF CHEMICAL PROCESSING ON PLANT PROTEINS

Protein chemical processing is used in a variety of applications, including drug development and food processing. Protein processing includes the addition or removal of specific components, such as amino acids, to improve nutritional value or extend food shelf life. Protein processing aims to make protein-rich foods safe, palatable, and easy to consume. The quality of plant protein refers to the amino acid composition, digestibility, and bioavailability; digestibility is determined by the structure and processing of the plant protein. Organic proteins must have the solubility in order to function as essential components in food, and it is a key factor in their emulsifying and foaming properties, which qualify them for use in the food industry (Kumar et al., 2022b). Different strategies are required because native and nonfunctional proteins have various posttranslational states. In order to improve the foaming and emulsifying characteristics of protein through denaturation, we must separate nonfunctional proteins from native proteins, as this will not affect the proteins which are already denaturated and have lost the function. As a result, different methods are required to transform the protein in order to improve its properties (Dhalleine et al., 2016). In food processing industry, the commonly modification techniques used for the protein modification include chemical hydrolysis with acid or base and enzymatic proteolysis. Antinutritional factors (ANFs) such as phytates, tannins, lectins, and inhibitors of trypsin make plant proteins less digestible, and their inactivation by food processing improves plant protein quality. Thus, plant protein processing facilitates the use of plant proteins as an alternative to animal proteins (Sá et al., 2020a). In order to improve

TABLE 10.4

Protein Modification and Their Applications in Food

S. No	Chemical Modification	Application	Reference
1.	Glycosylation	Improves solubility, rheological, emulsifying, foaming, film-forming properties, thermal stability, antioxidant activity, and antibacterial properties	Zhang et al., 2019
2.	Phosphorylation	Gelling and foaming agents in functional food models to improve the protein function and processing compatibilities	Hu et al., 2022
3.	Glycation	Glycation can improve the techno-functional characteristics of proteins, and glycated products show superior performance as emulsifiers, foam stabilizers, film-forming biopolymers, and encapsulation agents compared to their non-glycated forms.	Kutzli et al., 2021
4.	Deamination	Deamidation increases the characteristics of food proteins originating from cereals, legumes, milk, and other sources, aiding in oral perception; improving gastrointestinal digestion and bioavailability; and protecting and/or delivering bioactive nutrients.	Chen et al., 2021b
5.	Succinylation	Proteins that are succinylated include legume, seed, cereal, and leaf proteins, and this improves their encapsulation, oral administration, film-forming, thermal stability, solubility, emulsification, and foaming capabilities.	Basak and Singhal, 2022
6.	Acylation	Glycation increases the nutritional value of a protein while also improving its thermal stability, solubility, water-binding and oil-absorption capacity, and foaming capabilities.	Miedzianka et al., 2021

internal disadvantages, native proteins can be glycosylated with polysaccharides. Solubility, emulsifying properties, thermal stability, and antioxidant properties of native proteins can be managed to improve, allowing these proteins to be used more effectively in the food system (Zhang et al., 2019). Plants are easily accessible for glyco-engineering approaches and can thus be used to produce recombinant proteins.

10.8 FUTURE PERSPECTIVES

Plant-based protein consumption has skyrocketed in the last decade. Apart from health benefits, this trend will see a significant increase in the future due to strong factors of sustainability and less environmental consequences of plant protein production than animal protein (Akharume et al., 2021). The use of genetics to enhance plant protein quantity while also making it commercially available is a promising future direction. The development of plants with high levels of essential amino acids is dependent on public acceptance of genetically modified organisms. The United States Department of Agriculture (USDA) has acknowledged LY038 maize, which has an almost double lysine content without changing the grain protein content, as safe and accepted for commercial use in many countries (Tien Lea, 2016). China and the United States are in the process of patenting novel methods and products related to modified plant-based proteins. Amino acids that humans cannot synthesize are obtained by modifying soy protein. Industries and businesses are racing to modify proteins that contain an adequate number of amino acids so that they can be incorporated into food systems (Heredia-Leza et al., 2022). The genome sequencing of complete plant genomes, as well as the accessibility of many other genome-editing technologies, will aid in the effective modification of N-glycan-processing reactions in several plant species, allowing plant proteins to be used in the food system (Quétier, 2016). Glyco-engineering of biopharmaceutical expression systems is a new and emerging field that can be utilized to achieve particularly active and potential products, and various systems are now being explored to meet the high expectations of commercial enterprises (Jacobs and Callewaert, 2009). In the future, plant-based proteins should be considered not only in terms of chemical analysis, but also in terms of product stability, consumer acceptance, and physiological data for commercial and industrial applications.

10.9 CONCLUSION

Plant-based proteins have become a main product to be used in mainstream applications such as commercial and industrial purposes in recent years, and enhancing the taste, texture, and proper processing of proteins derived from plants can increase their functional properties and consumer acceptance, making them a good alternative to animal-based proteins. Chemical modification of plants can be accomplished by modifying some components of protein structure using chemical methods. The main goal of plant protein modification is to change the chemical properties of plant-based proteins in order to improve their functional properties such as solubility, stability, and digestibility. This modification makes plant proteins more appealing to consumers than animal proteins.

The food processing industries are keenly interested in the physicochemical, structural, biochemical, and functional properties of plant proteins, which can be modified via enzymatic and chemical means. It is believed that modified plant-based proteins will open new doors and replace animal counterparts, revolutionizing the food industry.

REFERENCES

Aiking, H. and de Boer, J., 2020. The next protein transition. *Trends in Food Science & Technology*, 105, pp. 515–522.

Akharume, F.U., Aluko, R.E. and Adedeji, A.A., 2021. Modification of plant proteins for improved functionality: A review. *Comprehensive Reviews in Food Science and Food Safety*, 20(1), pp. 198–224.

Arora, S. and Katyal, A., 2019. Protein modifications and lifestyle disorders. In *Protein Modificomics* (pp. 87–108). Academic Press.

Basak, S. and Singhal, R.S., 2022. Succinylation of food proteins-a concise review. *LWT*, 154, p. 112866.

Baslé, E., Joubert, N. and Pucheault, M., 2010. Protein chemical modification on endogenous amino acids. *Chemistry & Biology*, 17(3), pp. 213–227.

Blaskovic, S., Blanc, M. and van der Goot, F.G., 2013. What does S-palmitoylation do to membrane proteins? *The FEBS Journal*, 280(12), pp. 2766–2774.

Bosch, D., Castilho, A., Loos, A., Schots, A. and Steinkellner, H., 2013. N-glycosylation of plant-produced recombinant proteins. *Current Pharmaceutical Design*, 19(31), pp. 5503–5512.

Boutrou, R., Jardin, J., Blais, A., Tomé, D. and Léonil, J., 2008. Glycosylations of κ-casein-derived caseinomacropeptide reduce its accessibility to endo-but not exointestinal brush border membrane peptidases. *Journal of Agricultural and Food Chemistry*, 56(17), pp. 8166–8173.

Breitling, J. and Aebi, M., 2013. N-linked protein glycosylation in the endoplasmic reticulum. *Cold Spring Harbor Perspectives in Biology*, 5(8), p. a013359.

Burch, J.S., Marcero, J.R., Maschek, J.A., Cox, J.E., Jackson, L.K., Medlock, A.E., Phillips, J.D. and Dailey Jr, H.A., 2018. Glutamine via α-ketoglutarate dehydrogenase provides succinyl-CoA for heme synthesis during erythropoiesis. *Blood, the Journal of the American Society of Hematology*, 132(10), pp. 987–998.

Chalker, J. M., Bernardes, G. J., Ya, L. and Davis, B. G., 2009. Chemical modification of proteins at cysteine: Opportunities in chemistry and biology. *Chemistry: An Asian Journal*, 4, pp. 630–640.

Chaplin, A.K., Chernukhin, I. and Bechtold, U., 2019. Profiling of advanced glycation end products uncovers abiotic stress-specific target proteins in Arabidopsis. *Journal of Experimental Botany*, 70(2), pp. 653–670.

Charoensuk, D., Brannan, R.G., Chanasattru, W. and Chaiyasit, W., 2018. Physicochemical and emulsifying properties of mung bean protein isolate as influenced by succinylation. *International Journal of Food Properties*, 21(1), pp. 1633–1645.

Chen, J.H., Lin, X., Bu, C. and Zhang, X., 2018. Role of advanced glycation end products in mobility and considerations in possible dietary and nutritional intervention strategies. *Nutrition & Metabolism*, 15(1), pp. 1–18.

Chen, J.J., Fan, Y. and Boehning, D., 2021a. Regulation of dynamic protein S-acylation. *Frontiers in Molecular Biosciences*, 8, p. 656440.

Chen, X., Fu, W., Luo, Y., Cui, C., Suppavorasatit, I. and Liang, L., 2021b. Protein deamidation to produce processable ingredients and engineered colloids for emerging food applications. *Comprehensive Reviews in Food Science and Food Safety*, 20(4), pp. 3788–3817.

Chuh, K.N. and Pratt, M.R., 2015. Chemical methods for the proteome-wide identification of posttranslationally modified proteins. *Current Opinion in Chemical Biology*, 24, pp. 27–37.

de Freitas Junior, J.C.M. and Morgado-Díaz, J.A., 2016. The role of N-glycans in colorectal cancer progression: Potential biomarkers and therapeutic applications. *Oncotarget*, 7(15), p. 19395.

Dhalleine, C., Passe, D. and Roquette Freres, S.A., 2016. *Process for manufacturing soluble and functional plant proteins, products obtained and uses.* U.S. Patent 9, pp. 017, 259.

Doost, A.S., Nasrabadi, M.N., Kassozi, V., Dewettinck, K., Stevens, C.V. and Van der Meeren, P., 2019. Pickering stabilization of thymol through green emulsification using soluble fraction of almond gum–Whey protein isolate nano-complexes. *Food Hydrocolloids*, 88, pp. 218–227.

Du, J., Zhou, Y., Su, X., Yu, J.J., Khan, S., Jiang, H., Kim, J., Woo, J., Kim, J.H., Choi, B.H. and He, B., 2011. Sirt5 is a NAD-dependent protein lysine demalonylase and desuccinylase. *Science*, 334(6057), pp. 806–809.

Dworkin, J., 2015. Ser/Thr phosphorylation as a regulatory mechanism in bacteria. *Current Opinion in Microbiology*, 24, pp. 47–52.

Galam, L., Failla, A., Soundararajan, R., Lockey, R.F. and Kolliputi, N., 2015. 4-Hydroxynonenal regulates mitochondrial function in human small airway epithelial cells. *Oncotarget*, 6(39), p. 41508.

Gan, L., Huang, B., Song, Z., Zhang, Y., Zhang, Y., Chen, S., Tong, L., Wei, Z., Yu, L., Luo, X. and Zhang, X., 2021. Unique glutelin expression patterns and seed endosperm structure facilitate glutelin accumulation in polyploid rice seed. *Rice*, 14(1), pp. 1–19.

Ghosh, S., Meli, V.S., Kumar, A., Thakur, A., Chakraborty, N., Chakraborty, S. and Datta, A., 2011. The N-glycan processing enzymes α-mannosidase and β-D-N-acetylhexosaminidase are involved in ripening-associated softening in the non-climacteric fruits of capsicum. *Journal of Experimental Botany*, 62(2), pp. 571–582.

González-Morelo, K.J., Vega-Sagardía, M. and Garrido, D., 2020. Molecular insights into O-linked glycan utilization by gut microbes. *Frontiers in Microbiology*, 11, p. 591568.

Hashiguchi, A. and Komatsu, S., 2016. Impact of post-translational modifications of crop proteins under abiotic stress. *Proteomes*, 4(4), p. 42.

Hemsley, P.A., Kemp, A.C. and Grierson, C.S., 2005. The tip growth defective1 S-acyl transferase regulates plant cell growth in Arabidopsis. *The Plant Cell*, 17(9), pp. 2554–2563.

Heredia-Leza, G.L., Martínez, L.M. and Chuck-Hernandez, C., 2022. Impact of hydrolysis, acetylation or succinylation on functional properties of plant-based proteins: Patents, regulations, and future trends. *Processes*, 10(2), p. 283.

Hertzler, S.R., Lieblein-Boff, J.C., Weiler, M. and Allgeier, C., 2020. Plant proteins: Assessing their nutritional quality and effects on health and physical function. *Nutrients*, 12(12), p. 3704.

Holz, E., Darwish, M., Tesar, D.B. and Shatz-Binder, W., 2023. A review of protein-and peptide-based chemical conjugates: Past, present, and future. *Pharmaceutics*, 15(2), p. 600.

Hu, Y., Du, L., Sun, Y., Zhou, C. and Pan, D., 2022. Recent developments in phosphorylation modification on food proteins: Structure characterization, site identification and function. *Food Hydrocolloids*, p. 108390.

Hu, Z., Qiu, L., Sun, Y., Xiong, H. and Ogra, Y., 2019. Improvement of the solubility and emulsifying properties of rice bran protein by phosphorylation with sodium trimetaphosphate. *Food Hydrocolloids*, 96, pp. 288–299.

Jacobs, P.P. and Callewaert, N., 2009. N-glycosylation engineering of biopharmaceutical expression systems. *Current Molecular Medicine*, 9(7), pp. 774–800.

Jafari, S.M., Doost, A.S., Nasrabadi, M.N., Boostani, S. and Van der Meeren, P., 2020. Phytoparticles for the stabilization of Pickering emulsions in the formulation of novel food colloidal dispersions. *Trends in Food Science & Technology*, 98, pp. 117–128.

Jäger, R., Kerksick, C.M., Campbell, B.I., Cribb, P.J., Wells, S.D., Skwiat, T.M., Purpura, M., Ziegenfuss, T.N., Ferrando, A.A., Arent, S.M. and Smith-Ryan, A.E., 2017. International society of sports nutrition position stand: Protein and exercise. *Journal of the International Society of Sports Nutrition*, 14(1), p. 20.

Kawakatsu, T. and Takaiwa, F., 2019. Rice proteins and essential amino acids. In *Rice* (pp. 109–130). AACC International Press.

Kesari, P., Sharma, A., Katiki, M., Kumar, P., Gurjar, B.R., Tomar, S., Sharma, A.K. and Kumar, P., 2017. Structural, functional and evolutionary aspects of seed globulins. *Protein and Peptide Letters*, 24(3), pp. 267–277.

Krishnan, H.B. and Coe, E.H., 2001. Seed storage proteins. *Encyclopedia of Genetics*, pp. 1782–1787.

Kumar, M., Tomar, M., Potkule, J., Punia, S., Dhakane-Lad, J., Singh, S., Dhumal, S., Pradhan, P.C., Bhushan, B., Anitha, T. and Alajil, O., 2022a. Functional characterization of plant-based protein to determine its quality for food applications. *Food Hydrocolloids*, 123, p. 106986.

Kumar, M., Tomar, M., Punia, S., Dhakane-Lad, J., Dhumal, S., Changan, S., Senapathy, M., Berwal, M.K., Sampathrajan, V., Sayed, A.A. and Chandran, D., 2022b. Plant-based proteins and their multifaceted industrial applications. *LWT*, 154, p. 112620.

Kutzli, I., Weiss, J. and Gibis, M., 2021. Glycation of plant proteins via Maillard reaction: Reaction chemistry, technofunctional properties, and potential food application. *Foods*, 10(2), p. 376.

Lagassé, H.D., Alexaki, A., Simhadri, V.L., Katagiri, N.H., Jankowski, W., Sauna, Z.E. and Kimchi-Sarfaty, C., 2017. Recent advances in (therapeutic protein) drug development. *F1000Research*, 6.

Laudencia-Chingcuanco, D.L. and Vensel, W.H., 2008. Globulins are the main seed storage proteins in Brachypodium distachyon. *Theoretical and Applied Genetics*, 117(4), pp. 555–563.

Leonova, T., Popova, V., Tsarev, A., Henning, C., Antonova, K., Rogovskaya, N., Vikhnina, M., Baldensperger, T., Soboleva, A., Dinastia, E. and Dorn, M., 2020. Does protein glycation impact on the drought-related changes in metabolism and nutritional properties of mature pea (Pisum sativum L.) seeds? *International Journal of Molecular Sciences*, 21(2), p. 567.

Liebscher, S. and Bordusa, F., 2019. Site-specific modification of proteins via trypsiligase. *Bioconjugation: Methods and Protocols*, pp. 95–115.

Liu, X., Wang, C., Zhang, X., Zhang, G., Zhou, J. and Chen, J., 2022. Application prospect of protein-glutaminase in the development of plant-based protein foods. *Foods*, 11(3), p. 440.

Lu, J.Y. and Hofmann, S.L., 2006. Inefficient cleavage of palmitoyl-protein thioesterase (PPT) substrates by aminothiols: Implications for treatment of infantile neuronal ceroid lipofuscinosis. *Journal of Inherited Metabolic Disease: Official Journal of the Society for the Study of Inborn Errors of Metabolism*, 29(1), pp. 119–126.

Ma, B., Guan, X., Li, Y., Shang, S., Li, J. and Tan, Z., 2020. Protein glycoengineering: An approach for improving protein properties. *Frontiers in Chemistry*, 8, p. 622.

Matsudomi, N., Kato, A. and Kobayashi, K., 1982. Conformation and surface properties of deamidated gluten. *Agricultural and Biological Chemistry*, 46(6), pp. 1583–1586.

Miedzianka, J., Zambrowicz, A., Zielińska-Dawidziak, M., Drożdż, W. and Nemś, A., 2021. Effect of acetylation on physicochemical and functional properties of commercial pumpkin protein concentrate. *Molecules*, 26(6), p. 1575.

Miguéns-Gómez, A., Sierra-Cruz, M., Rodríguez-Gallego, E., Beltrán-Debón, R., Blay, M.T., Terra, X., Pinent, M. and Ardévol, A., 2022. Effect of an acute insect preload vs. an almond preload on energy intake, subjective food consumption and intestinal health in healthy young adults. *Nutrients*, 14(7), p. 1463.

Monsalve, R.I., Menéndez-Arias, L., López-Otin, C. and Rodríguez, R., 1990. β-Turns as structural motifs for the proteolytic processing of seed proteins. *FEBS Letters*, 263(2), pp. 209–212.

Mylne, J.S., Hara-Nishimura, I. and Rosengren, K.J., 2014. Seed storage albumins: Biosynthesis, trafficking and structures. *Functional Plant Biology*, 41(7), pp. 671–677.

Nowak, C., Tiwari, A. and Liu, H., 2018. Asparagine deamidation in a complementarity determining region of a recombinant monoclonal antibody in complex with antigen. *Analytical Chemistry*, 90(11), pp. 6998–7003.

Osborne, T.B., 1924. *The Vegetable Proteins*. Longmans, Green and Company.

Panyam, D. and Kilara, A., 1996. Enhancing the functionality of food proteins by enzymatic modification. *Trends in Food Science & Technology*, 7(4), pp. 120–125.

Quétier, F., 2016. The CRISPR-Cas9 technology: Closer to the ultimate toolkit for targeted genome editing. *Plant Science*, 242, pp. 65–76.

Rabbani, N., Al-Motawa, M. and Thornalley, P.J., 2020. Protein glycation in plants—An under-researched field with much still to discover. *International Journal of Molecular Sciences*, 21(11), p. 3942.

Rabbani, N. and Thornalley, P.J., 2012. Glycation research in amino acids: A place to call home. *Amino Acids*, 42(4), pp. 1087–1096.

Reddy, N.C., Kumar, M., Molla, R. and Rai, V., 2020. Chemical methods for modification of proteins. *Organic & Biomolecular Chemistry*, 18(25), pp. 4669–4691.

Sá, A.G.A., Moreno, Y.M.F. and Carciofi, B.A.M., 2020a. Food processing for the improvement of plant proteins digestibility. *Critical Reviews in Food Science and Nutrition*, 60(20), pp. 3367–3386.

Sá, A.G.A., Moreno, Y.M.F. and Carciofi, B.A.M., 2020b. Plant proteins as high-quality nutritional source for human diet. *Trends in Food Science & Technology*, 97, pp. 170–184.

Sacco, F., Perfetto, L., Castagnoli, L. and Cesareni, G., 2012. The human phosphatase interactome: An intricate family portrait. *FEBS Letters*, 586(17), pp. 2732–2739.

Sadhukhan, S., Liu, X., Ryu, D., Nelson, O.D., Stupinski, J.A., Li, Z., Chen, W., Zhang, S., Weiss, R.S., Locasale, J.W., Auwerx, J. and Lin, H., 2016. Metabolomics-assisted proteomics identifies succinylation and SIRT5 as important regulators of cardiac function. *Proceedings of the National Academy of Sciences*, 113, pp. 4320–4325.

Sánchez-Reséndiz, A., Rodríguez-Barrientos, S., Rodríguez-Rodríguez, J., Barba-Dávila, B., Serna-Saldívar, S.O. and Chuck-Hernández, C., 2018. Phosphoesterification of soybean and peanut proteins with sodium trimetaphosphate (STMP): Changes in structure to improve functionality for food applications. *Food Chemistry*, 260, pp. 299–305.

Sha, L. and Xiong, Y.L., 2020. Plant protein-based alternatives of reconstructed meat: Science, technology, and challenges. *Trends in Food Science & Technology*, 102, pp. 51–61.

Shang, H.Y., Wei, Y.M., Long, H., Yan, Z.H. and Zheng, Y.L., 2005. Identification of LMW glutenin-like genes from *Secale sylvestre* host. *Russian Journal of Genetics*, 41(12), pp. 1372–1380.

Skropeta, D., 2009. The effect of individual N-glycans on enzyme activity. *Bioorganic & Medicinal Chemistry*, 17(7), pp. 2645–2653.

Soboleva, A., Vikhnina, M., Grishina, T. and Frolov, A., 2017. Probing protein glycation by chromatography and mass spectrometry: Analysis of glycation adducts. *International Journal of Molecular Sciences*, 18(12), p. 2557.

Souza, P.F., 2020. The forgotten 2S albumin proteins: Importance, structure, and biotechnological application in agriculture and human health. *International Journal of Biological Macromolecules*, 164, pp. 4638–4649.

Stiles, A., Homyak, C., Astor, S., Smith, B. and Ripple Foods Pbc, 2021. *Non-dairy Analogs with Succinylated Plant Proteins and Methods Using Such Products*. U.S. Patent Application 17/339, 180.

Takahashi, K., Kohno, H., Kanabayashi, T. and Okuda, M., 2019. Glutelin subtype-dependent protein localization in rice grain evidenced by immunodetection analyses. *Plant Molecular Biology*, 100(3), pp. 231–246.

Tien Lea, D., Duc Chua, H. and Quynh Lea, N., 2016. Improving nutritional quality of plant proteins through genetic engineering. *Current Genomics*, 17(3), pp. 220–229.

Uribarri, J., Cai, W., Peppa, M., Goodman, S., Ferrucci, L., Striker, G. and Vlassara, H., 2007. Circulating glycotoxins and dietary advanced glycation end products: Two links to inflammatory response, oxidative stress, and aging. *The Journals of Gerontology Series A: Biological Sciences and Medical Sciences*, 62(4), pp. 427–433.

Wagner, G.R. and Payne, R.M., 2013. Widespread and Enzyme-independent Nε-Acetylation and Nε-Succinylation of proteins in the chemical conditions of the mitochondrial matrix. *Journal of Biological Chemistry*, 288(40), pp. 29036–29045.

Wakasa, Y., Yang, L., Hirose, S. and Takaiwa, F., 2009. Expression of unprocessed glutelin precursor alters polymerization without affecting trafficking and accumulation. *Journal of Experimental Botany*, 60(12), pp. 3503–3511.

Warnakulasuriya, S.N. and Nickerson, M.T., 2018. Review on plant protein–polysaccharide complex coacervation, and the functionality and applicability of formed complexes. *Journal of the Science of Food and Agriculture*, 98(15), pp. 5559–5571.

Xu, J.H. and Messing, J., 2009. Amplification of prolamin storage protein genes in different subfamilies of the Poaceae. *Theoretical and Applied Genetics*, 119(8), pp. 1397–1412.

Zhang, Q., Li, L., Lan, Q., Li, M., Wu, D., Chen, H., Liu, Y., Lin, D., Qin, W., Zhang, Z. and Liu, J., 2019. Protein glycosylation: A promising way to modify the functional properties and extend the application in food system. *Critical Reviews in Food Science and Nutrition*, 59(15), pp. 2506–2533.

Zhen, S., Deng, X., Wang, J., Zhu, G., Cao, H., Yuan, L. and Yan, Y., 2016. First comprehensive proteome analyses of lysine acetylation and succinylation in seedling leaves of Brachypodium distachyon L. *Scientific Reports*, 6(1), pp. 1–15.

Zheng, L., Liu, P., Liu, Q., Wang, T. and Dong, J., 2019. Dynamic protein S-acylation in plants. *International Journal of Molecular Sciences*, 20(3), p. 560.

Zhong, L., Ma, N., Wu, Y., Zhao, L., Ma, G., Pei, F. and Hu, Q., 2019. Characterization and functional evaluation of oat protein isolate-*Pleurotus ostreatus* β-glucan conjugates formed via Maillard reaction. *Food Hydrocolloids*, 87, pp. 459–469.

11 Processing Techniques for Scaling Up of Modified Plant Proteins

Atefeh Karimidastjerd, Gizem Sevval Tomar,
Rukiye Gundogan and Asli Can Karaca

11.1 INTRODUCTION

Globally, the population has been increasing regardless of any natural disasters, wars, pandemics, and other threats. On the other hand, water, soil, and food resources have been decreasing each year. According to some researchers, the world's population will reach 10 billion by 2050, which suggest that just like other food sources, the food protein supply may be scarce. Therefore, it is obvious that there must be a focus on the consumption of proteins from new sources as alternatives to animal-based proteins. Proteins from plant-based sources, such as vegetables, seeds, grains, leaves, legumes, cereals, and other sources including algae and insects, are currently being evaluated (Kamani et al., 2022; Sá et al., 2020). Among these protein-source alternatives, plant-based proteins are indicated to have several advantages. Ethnical, consumer acceptance behaviour, religious concerns, availability, biodiversity, and health aspects are some of the advantages which can be maintained (Onwezen et al., 2021). Proteins from plant-based sources can be categorized into eight different main groups as given here:

1. Cereals
2. Legumes
3. Editable seeds
4. Nuts
5. Oilseeds
6. Tubers
7. Macroalgae
8. Microalgae (Day et al., 2022; Nasrabadi et al., 2021)

Among these various plant-based protein sources, cereals and legumes have been consumed and accepted by large segments of consumers around world. They are high in amount of protein (18–32%) and bioactive peptides; show techno-functional properties such as water holding, fat binding, foaming, emulsifying, and gelation; and have potential health benefits (Singhal et al., 2016). Owing to high availability and low costs of native plant proteins, their application in food industry is extending. Such plant-based proteins are potentially useful as alternative materials for packaging or agricultural purposes (Bräuer et al., 2007). Moreover, due to their functional properties, it is desirable to make them accessible for various processes and application in foods (Onwezen et al., 2021; Sá et al., 2020; Akharume et al., 2021). Beside these advantages and positive aspects, some disadvantages for plant-based proteins have also been reported. In comparison to animal-based proteins, lower nutritive value, protein digestibility, and functionality are reported for plant-based proteins (Figure 11.1) (Munialo et al., 2022).

The knowledge of the physio-chemical properties of plant-based protein products may aid in predicting how their functionality supports product formulation and quality (Akharume et al., 2021; Paulson et al., 1984). Based on increased incorporation of plant protein ingredients such as flour, isolates, and concentrates in traditional and novel foods, more studies are required to look at the

DOI: 10.1201/9781003369790-11

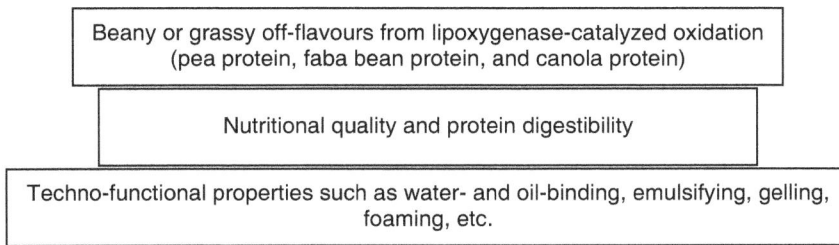

FIGURE 11.1 Some challenges linked to the application of plant-based proteins in food industry. (Munialo et al., 2022; Sá et al., 2020.)

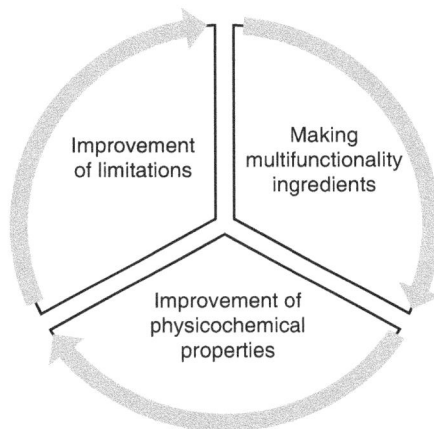

FIGURE 11.2 The major modification outcomes of plant-based proteins. (Nasrabadi et al., 2021.)

way these proteins' structure changes and interacts with other components in the food matrix. For example, flavours and aromas can be bound to plant-based proteins impacting their overall perception in foods. This binding can occur through either covalent or non-covalent bonds (more lysine, arginine, and cysteine amino acid content could potentially result in higher flavour-binding ability). In reduced-fat food, this phenomenon is considered to be undesirable (Wang and Arntfield, 2017). Another example in juices and beer is the reaction between polyphenols and proteins, which results in haze formation. During food processing, plant-based proteins react via covalent bonds with plant metabolites (polyphenols as well as sulphur compounds) (Keppler et al., 2020). Parameters which cause these changes can be listed as, first, environment factors such as pH, temperature, and ionic strength of solution and, second, food-processing techniques including cooking, dehulling, soaking, germination, microwave, irradiation, fermentation, and extrusion (Wang and Arntfield, 2017; Akharume et al., 2021; Sá et al., 2020). Different protein extraction steps are designed to be beneficial in terms of protein quality by inactivating the compounds that lower the protein digestibility of plant proteins and recover more purified proteins. Alongside isolation steps, pre-treatment steps such as soaking, germination, and fermentation aim to improve techno-functionality of plant-based proteins. For example the treatment of soy protein was reported to increase emulsion viscosity and stability as well as foam volume and stability (Ismail et al., 2020).

To improve the nutritional, techno-functional, and organoleptic properties of plant-based proteins, protein modification via physical, chemical, biological and other methods has been developed (Figure 11.2) (Munialo et al., 2022). The term "protein modification" means the process of modifying and changing the molecular structure or a few chemical groups of a protein by specific methods for the targeted purpose (Nasrabadi et al., 2021; Akharume et al., 2021).

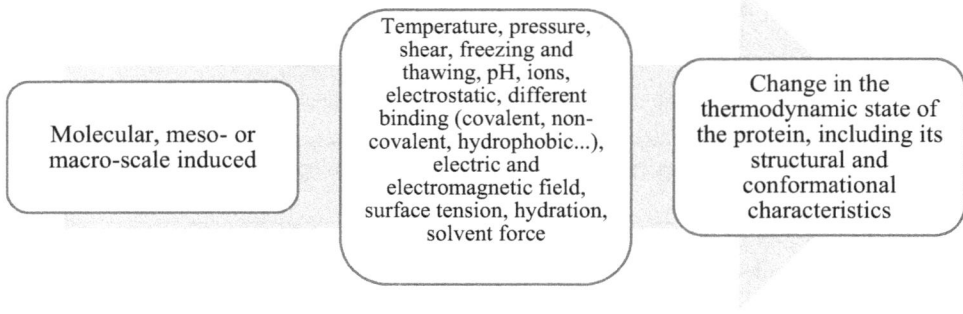

FIGURE 11.3 The modification processes' pathway to develop a modified plant-based protein ingredient. (Munialo et al., 2022.)

The protein modification methods can be classified into physical, chemical, biological and reaction with substances, amyloid fibrillization, and dual modification (Nasrabadi et al., 2021; Akharume et al., 2021). To select the most suitable modification method, it is important to have information on: (i) the nature of native plant-based protein (whether in isolation or bound with other polymers) and (ii) the accepted structural properties of the protein ingredient (Akharume et al., 2021). Several mechanisms that cause improvements in the techno-functional properties of plant-based protein ingredients are summarized next (Akharume et al., 2021; Bräuer et al., 2007).

- An improvement in the solubility requires
  - Size reduction
  - Electrostatic repulsion
  - An increase in hydrophilicity by conjugation with a more hydrophilic polymer
  - Manipulating the protein isoelectric point and net charge
- An improvement in the emulsifying and foaming activities promotes
  - A balance of its hydrophobic–hydrophilic property
  - Improvement of solubility
  - Conjugation with other polymers' disruption of agglomeration using ultrasonication and modifications that reveal the hydrophobic core of the proteins
- An improvement of mechanical integrity requires
  - The acylated protein mater (using palmitic acid chloride and alkenyl-substituted succinic anhydrides)
  - Long hydrophobic alkyl chains

Overall, selecting one general modification method for all types of plant-based proteins couldn't be logical. Proteins are complex macromolecules, and their modification approach should be unique, depending on the application and functionality of interest. A modification method by means of different parameters (Figure 11.3) could change the size, surface charge, hydrophobic/hydrophilic ratio, and molecular flexibility of plant-based proteins (Akharume et al., 2021).

11.2 EXTRACTION METHOD AND VARIOUS MODIFICATION TECHNIQUES FOR PLANT-BASED PROTEINS

11.2.1 Extraction Method

The processes that are mainly used for production of plant-based protein concentrates (50–80% protein) and isolates (>80% protein) can be classified into aqueous and dry fractionation. Plants with high oil content, such as pulses and oilseeds, first undergo oil extraction. A general five-step food protein processing procedure includes: (1) pre-treatment (soaking and/or milling), (2) defatting by organic solvent, (3) protein solubilization and recovery (NaOH or water solvent and centrifugation), (4) protein purification (membrane filtration), and (5) drying. Ultrafiltration in the membrane filtration step is applied to the protein to from either a concentrate or an isolate depending on protein content. An additional step of acid precipitation of proteins at their isoelectric point to obtain proteins with higher yield and facilitate the concertation process may also be used (Munialo et al., 2022; Ismail et al., 2020). Isolation method and protein original backbone are two factors that determine techno-functional properties of plant-based proteins. To apply isolated plant-based proteins as functional ingredients, these factors should be taken into consideration.

11.2.2 Various Processing Techniques for Modification of Plant Proteins

After isolation, plant-based protein modification is required for desired techno-functional properties by using various processes in the molecular structure or some chemical groups of the protein (Nasrabadi et al., 2021). In addition to imparting techno-functional properties to proteins, a reduction of protein allergenicity can also be achieved by modification. These approaches may improve the utilization of plant-based proteins (Fadimu et al., 2023). Table 11.1 lists various plant-based protein modification methods and their processing parameters. In recent years, studies have been carried out on nonthermal innovative and green technologies. These methods are preferred since they cause minimal damage to the nutritional and sensory properties of the protein (Pan et al., 2022). In this section, several novel protein modification techniques including cold atmospheric plasma, sonication, pulsed ultrasound, high pressure, and conventional heat treatment application will be discussed.

11.2.2.1 Cold Atmospheric Plasma

Cold atmospheric plasma treatment is the application of plasma, the fourth state of matter, over a wide temperature and pressure range. This high energy applied causes the breaking of covalent bonds in the protein, causing various chemical reactions to change the characteristics of proteins (Nasrabadi et al., 2021). Plasma production is usually achieved with the help of an electric discharge. This process involves the concentration of the kinetic energy of the electrons as a result of increased collision of ions, radicals, and electrons in the gas (Sonawane and Patil, 2020). It can be produced by different systems such as cold plasma, corona discharges, microwave plasma, radio frequency plasma, inductively coupled plasma, capacitance-coupled plasma, electron cyclotron resonance plasma, and dielectric barrier discharge plasma. Among these, the most widely used systems in the food industry are dielectric barrier discharge and jet plasma due to their simple and versatile use (Mollakhalili-Meybodi et al., 2021). The important parameters in the cold plasma process are the type of process gas, the flow rate of the plasma, the design of the plasma device, and processing time (Saremnezhad et al., 2021). The cold plasma treatment causes the oxidation of the backbone of the protein and amino acids in the presence of reactive species. This can cause protein–protein cross-links or protein degradation (Fadimu et al., 2023). With the oxidation process, properties of the protein such as solubility, foaming, emulsifying ability, and allergenicity are changed in a positive way (Basak and Annapure, 2022; Saremnezhad et al., 2021). The main advantages the cold atmospheric plasma treatment can be listed as the features such as no need for the use of chemicals, low operating temperature, low cost, and little damage to nutritional properties (Pan et al., 2022).

11.2.2.1.1 Disadvantages

The main disadvantage of cold plasma treatment is that it causes a decrease in the thermal stability of proteins. At the same time, the scarcity of information on its industrial use is one of the most important reasons limiting the widespread use of the method (Basak and Annapure, 2022).

TABLE 11.1

Various Plant-based Protein Modification Methods and Their Processing Limitations and Parameters

	Physical	Chemical	Biological	Reactions with Organic Substance	Other Methods
Different Methods	• Thermal treatment • Irradiation • Pulsed-electric field • High-pressure treatment • Extrusion • Sonication • Cold atmospheric plasma process • Ultrafiltration	• Glycation • Phosphorylation • Acetylation • Deamidation • Cationization • pH shifting	• Fermentation • Enzymatic treatment	• Polysaccharide • Protein • Phenolic • Surfactant	• Amyloid fibrillization • Dual modification
Processing Limitations	• Adjusting method to achieve desired properties in product	• Producing toxic by-product • Safety • Cost of reactors/catalyst	• High cost of enzymes/cultures • Complex factors to control	• Cost	• Novelty and newly adoption
Processing Parameters	• Time • Temperature (efficiency of heating method) • Pressure • Radiation dose • Membrane cut-off	• Time • Temperature • Reaction speed • Chemical substance efficiency • Sustainability	• Time • Temperature • Cost • pH	• Time • Temperature • Sustainability • Cost	• Time • Temperature • Fluorescence • pH • Solvent concentration
Industrial -Level Potential	• High	• Low	• Moderate	• High	• Moderate

Source: Akharume et al., 2021; Nasrabadi et al., 2021; Kamani et al., 2022

Ji et al. (2018) investigated the effects of cold plasma treatment and dielectric barrier discharge on the physicochemical and functional properties of peanut proteins. The solubility, emulsion stability, and water-holding capacity of the modified peanut protein isolate were reported to improve after cold plasma treatment. The distance between the electrodes used was 8 mm. The applied plasma treatment was applied for 1, 2, 3, and 4 min at 35 V and 2 ± 0.2 A power. It was observed that the primary structure of the protein did not change after the procedure, whereas the secondary structure was altered. Properties such as solubility and emulsion stability were improved as a result of the protein structure being opened and the binding of water micelles to protein molecules with the applied cold plasma treatment. Cold plasma treatment was suggested as an effective method for modifying peanut protein (Ji et al., 2018).

In another study, the combination of cold plasma treatment and high-power sonication was evaluated in the modification of mung bean protein. High power sonication at 250 J/mL for 2 min and atmospheric cold plasma treatment at 80 kV for 5 min were applied to different concentrations of mung bean protein isolate. It has been reported that 16% protein dispersions treated with cold plasma show six times more storage stability compared to the control sample. The combination of the two methods was indicated to improve gel hardness. In addition, cold-plasma-treated samples showed greater hydrophobic bonding capacity than high-power-sonication-treated samples (Rahman and Lamsal, 2023).

Tahsiri et al. (2023) conducted a study on the improvement of the functional properties of films produced with wild almond protein isolate by using Iranian gum and cold plasma treatment. The films were subjected to cold plasma treatment for 5, 10, and 15 min. The cold plasma treatment applied for 10 min was indicated to improve the mechanical properties of the films. The process applied for 15 min was observed to have adverse effects on the film properties. The microstructures of the surface morphology in the processed films increased the surface roughness of longer processing times. At the same time, the cold plasma treatment applied for 10 min was reported to increase the thickness of the films (Tahsiri et al., 2023).

Zhang et al. (2021) evaluated the atmospheric cold plasma treatment in order to improve the gelling properties of pea protein. As a result of the atmospheric cold plasma treatment applied to the pea protein, the structure of the protein was reported to change and a network structure was formed. In addition, gel formation was performed below the protein denaturation temperature (70–90°C). Gels formed at 80–90°C were indicated to show improved mechanical properties. Atmospheric cold plasma treatment has been reported to be a promising approach for the modification of plant-based proteins (Zhang et al., 2021).

In another study conducted in 2023, synergistic tartaric acid was used in combination with cold plasma for the preparation of fibrotic pea protein isolate. The development of foam stabilizers based on hypoallergenic pea protein isolate with high foam stability is of great importance in terms of use in products such as ice cream, beer, and cake in the food industry. In this application, tartaric acid was used to carry out acid deamidation. Deamidation is a modification method used to change protein or peptide molecules. By using these two methods together, it is predicted that protein hydrolysis can be regulated, and the deamidation rate will be increased. The combined applied process was reported to improve protein deamidation efficiency, solubility, foamability, and stability. The application of cold plasma treatment was reported to change the composition and primary structure of the protein by destroying the subunits and disulphide bonds of the proteins (Qu et al., 2023).

11.2.2.2 Extrusion

Extrusion is a technique in which a material or food is passed through a die or hole of the desired cross-section with high pressure, temperature, and shear force (Prabha et al., 2021). This process causes unfolding, denaturation, or realignment of protein molecules, improving their functional properties. In particular, antinutritional substances are destroyed by this process, which contributes to the improvement of the digestibility of the protein (Nasrabadi et al., 2021). This method is particularly used for the production of meat analogues from plant-based proteins. The high pressure, temperature, and shear force used during the process ensure the formation of a meat-like structure in the plant protein. High-humidity (50–80%) extrusion is used to produce plant proteins with meat-like properties, while low-moisture (<30%) extrusion is used to obtain harder and less-soluble proteins (Akharume et al., 2021). Apart from moisture content, some other components that affect extrusion are shear force, mechanical energy, barrel temperature, screw speed, and feed rate (Figure 11.4). These parameters can affect the quality of the extruded plant protein (J. Zhang et al., 2018). In the extruder where the extrusion process is carried out, the motor consists of one or two screws driven by the motor, the transition zone, and the sleeve with the cooling mold (Cornet et al., 2022).

The working principle in extrusion involves feeding the raw materials into the extruder sleeve and then transporting the food through the extruder. Down the barrel, smaller flights restrict volume and increase resistance to food movement. As it travels through the barrel, the screw kneads the material into a semi-solid, plasticized mass. If the food is heated above 100°C, the process is

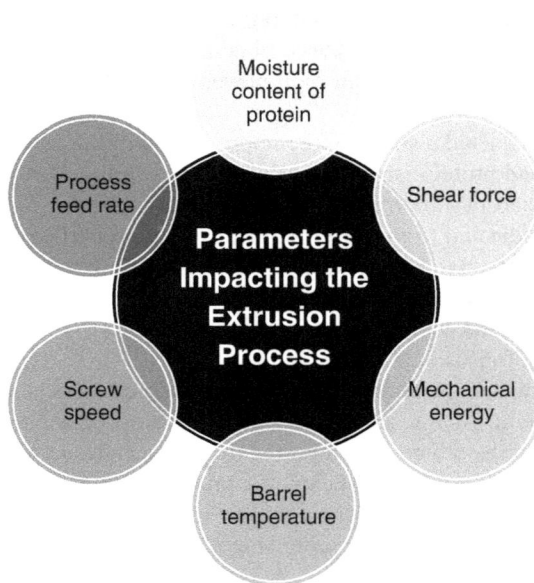

FIGURE 11.4 Parameters impacting the extrusion process.

called extrusion cooking. Here, frictional heat and any additional heating process used cause the temperature to rise rapidly. The food then moves into the barrel section, where the pressure and shear are further increased due to the smaller flight. Finally, as the food exits the die under pressure, it is forced through one or more of the limited openings (moulds) at the discharge end of the barrel. Finally, the food becomes desirable and cools rapidly (Choton et al., 2020).

Since the extrusion process does not change the main chemical bonds, it changes the interactions responsible for the initial conformation of the protein. With the effect of high temperature applied during extrusion, hydrogen bonds are broken, resulting in unfolding of the protein structure (Nasrabadi et al., 2021). In addition, the combination of the applied temperature and the mechanical shift of the screw speed results in the reduction of heat-sensitive antinutritional compounds such as phytic acid, trypsin inhibitors, and tannins. This also improves the digestibility of proteins (Fernando, 2022). During the extrusion process, proteins undergo four major conformational changes which are molecular chain opening, association, aggregation, and crosslinking by oxidation (Figure 11.5). No significant changes in the protein structure are indicated to occur in the feeding zone of the extruder where the extrusion process takes place. In the mixing zone, the molecular chain of the protein moving along the flow direction expands, and the hydrophobic amino acids in the molecular chain are exposed. If the melting region has low density, protein–protein and protein–water interaction may increase and protein aggregation can be observed. If the melting region is high density, it causes the protein to break down and the particle size to decrease. In the cooling zone, protein molecules are rearranged and crosslinked, resulting in the formation of a fibrous structure (Zhang et al., 2022).

11.2.2.2.1 Disadvantages

Despite the properties that the extrusion process brings to proteins, it also has several disadvantages. Lysine, the essential and limiting amino acid, can degrade during the extrusion process (Gulati et al., 2020). Furthermore, complex physical, chemical, and biological reactions can occur during extrusion, and these reactions are difficult to control. In addition to these, undesirable colour changes may occur in the final product due to the high temperatures used. In order to achieve

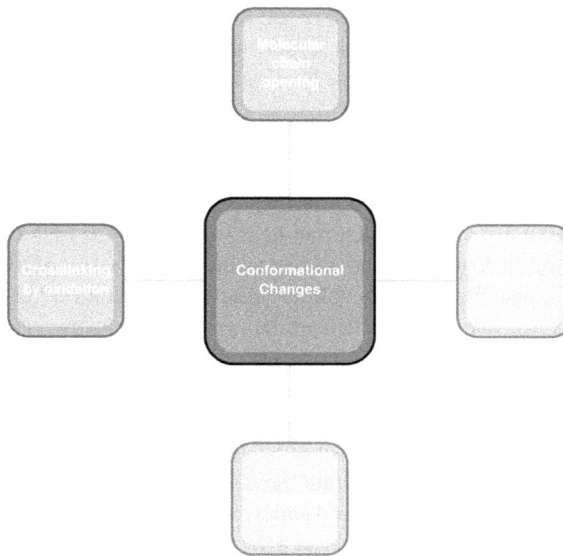

FIGURE 11.5 Four basic conformational changes that occur in the structure of the protein in the extrusion process.

quality optimization, it is essential to control the conformational changes of the main components by dividing the extrusion process into several parts (J. Zhang et al., 2018).

Gao et al. (2022) evaluated the extrusion process in order to improve the functional properties of rice proteins. Rice protein was extruded at different screw speeds (100–250 rpm), extrusion temperatures (90–150°C), and moisture contents (25–40%). The treated rice proteins were compared with the untreated control sample. The water-holding capacity of rice protein was observed to show the highest results at 200 rpm, 130°C, and 25% moisture content. On the other hand, emulsion stability was observed to show the highest results at 200 rpm, 130°C, and 35% moisture content. The microstructure was reported to change drastically after the extrusion process, resulting in the production of tighter structured proteins (Gao et al., 2022).

In another study examining the modification of rice protein by the extrusion process, the production of meat analogues from rice protein mixed with soy protein in various proportions by low-moisture extrusion process was evaluated. Rice protein and soy protein were mixed in different ratios of 25:75, 50:50, 75:25, and 100:0 (w/w). In addition, corn starch and wheat gluten were added to the formulations at 29% and 13% (w/w) by total weight of all components, respectively. The feed rate was set at 100 g/min, the screw speed was 350 rpm, and the die pressure was 100 bar in the extrusion process. The water flow rate of the water pump was 1.5 mL/min. The temperature profile of the extruder drum is 20°C in the feed zone and 195°C in the die. With the addition of rice protein isolate, there was a decrease in the water absorption capacity of total soluble substances due to hydrophobicity, formation of agglomerated bodies, and disulphide crosslinking of glutelin. The combined production of soya bean and rice protein resulted in a better structure in the analogue (Lee et al., 2022).

A study conducted in 2019 evaluated the effect of high-moisture extrusion of soy protein concentrate. The formation on anisotropic structures was investigated in terms of extruder temperature (100–143°C), extruder pressure (1.7–2.7 MPa), and specific mechanical energy input (85–350 kJ/kg). The increase in temperature was observed to be the most effective parameter in the formation of the anisotropic structure. In addition, the increase in temperature was indicated to cause a decrease in viscosity by changing the protein flow behaviour (Pietsch et al., 2019). Ferawati et al. (2021)

evaluated the potential of local yellow pea and broad bean protein isolates/concentrate as meat analogue products. It was concluded that the most important process parameters were moisture content,
extrusion temperature, and screw speed. These local protein sources were reported to be suitable for
producing meat analogues (Ferawati et al., 2021).

11.2.2.3 Sonication

Ultrasonication or ultrasound treatment is a non-thermal protein modification method using high-
frequency sound waves that are above the human hearing limit of 20 kHz (Barbhuiya et al., 2021).
The expansion and compression of the sound waves produced as a result of ultrasound waves change
the ambient pressure and cause the formation of bubbles called cavitation (Sá et al., 2022). Cavitation
generally includes three basic aspects. The first of these is that the formed cavitation bubbles shrink
until they reach a critical size and then collapse violently. As a result of the collapse, high temperature, pressure, strong hydrodynamic shear forces, and turbulence occur. The mass transfer of
ultrasound process increases as a result of temperature, pressure, and turbulence formation. Second,
physical effects such as cavitation bubble formation, microturbulence, and high shear stress occur
with explosion. These physical effects accelerate the chemical reactions and increase the yield. The
third includes the formation of free radicals and superoxide. The radicals formed react with proteins
to form crosslinks (Wen et al., 2019). With these effects, which occur as a result of cavitation, the
structure of proteins can be modified resulting in changes in protein functionality. In the application of the sonication process, frequency, power density, application time, processing temperature,
and the structure of the protein to which the process will be applied are important parameters for
the effectiveness of the process. Protein modification by sonication is mainly performed to improve
foaming and emulsification properties. With the ultrasound process, a better dispersion of the protein particles is achieved, which contributes to the improvement of interfacial properties. Reducing
the droplet size provides better adsorbing of the protein to the oil–water interface, improving the
emulsion properties. In addition, functional properties such as solubility and gelation and digestibility of proteins can be improved by sonication (Pan et al., 2022).

Ultrasound is an innovative and environmentally friendly technique that uses non-toxic, green
solvents (Rahman and Lamsal, 2021). Moreover, this technique has simple, fast, low cost, energy-
saving, and aseptic working conditions (Gharibzahedi and Smith, 2020).

11.2.2.3.1 Disadvantages

Despite the advantages of using the sonication technique, there are also several disadvantages.
These are various disadvantages such as the mode of action of the process, the dependence of
penetration on water and air content, the potential negative effects of free radicals formed, and the
possibility of textural changes (Yousefi and Abbasi, 2022).

Li et al. (2020) investigated the changes in protein solubility and structural properties using
ultrasound-assisted pH-shifting technique in rapeseed protein isolates. Fixed and sweeping modes,
which are two different operating modes of ultrasound, were used. Compared with the control
sample, the amount of free sulfhydryl increased, and the amount of α-helix decreased in the treated
samples. The results show that ultrasonic treatment and operating modes greatly change the secondary and tertiary structures of the protein structure, improving its solubility properties. Ultra + pH
mode at a fixed frequency of 20 kHz was the parameter that had the most effect on the solubility of
the protein. At the same time, the surface hydrophobicity and zeta potential were increased compared to the unmodified protein (Li et al., 2020).

Zhu et al. (2018) conducted a study examining the effect of sonication on walnut protein isolates.
For this purpose, aqueous walnut protein suspensions were sonicated at varying power levels (200,
400, or 600 W) and durations (15 or 30 min). The authors observed that the sonication process
did not break the covalent bonds in the protein's structure but caused changes in the secondary
structure. The applied sonication process was effective on protein structure, aggregation status,
emulsifying properties, and solubility. With this process, the solubility of the protein increased,

and the number of large aggregates decreased, resulting in improved emulsifying properties. It has been reported that the cause of these effects is the effect of ultrasonic waves on the physical bonds between globular protein molecules (Zhu et al., 2018).

Mir et al. (2019) investigated the physicochemical, molecular, and thermal properties of protein isolates treated with high-intensity ultrasound from albumin (*Chenopodium album*) seed. Ultrasound was applied to the proteins at 20 kHz for 5, 15, 25, and 35 min. With the applied ultrasound process, the color properties, whiteness index, solubility, foaming capacity, and stability of the protein were significantly increased. It was thought that the reason for the improvement in color of the album protein isolates after the treatment might be due to the cavitation changing the pigments in the foodstuffs. The most effective reduction in protein particle size was observed in samples treated for 25 min (Mir et al., 2019).

In a study conducted in 2021, the effect of ultrasound treatment on the physicochemical properties of water-soluble protein obtained from Moringa oleifera seed was evaluated. It was stated that solubility, foaming properties, and emulsifying properties of the modified protein first increased and then decreased with the increase of ultrasonic power. Ultrasonic processing was reported to change the secondary and tertiary structures of the protein (Tang et al., 2021).

Khan et al. (2021) investigated the effect of ultrasonication on the structural and functional properties of Sea buckthorn seed protein concentrate in their study. Ultrasonication was observed to carry out with a fixed power supply of 500 W/cm^2 for varying times (0, 5, 15, 25, and 35 min). The solubility and emulsifying properties of all processed protein concentrates were indicated to be improved. With the process used, the decrease in the percentage of α-helix and the increase in the structure of the β-sheet were confirmed by FTIR analysis. Increasing the ultrasonication duration was observed to result in a decrease in protein particle size (Khan et al., 2021).

In another study, the effect of high-intensity ultrasound treatment on the physicochemical, interfacial, and gel properties of chickpea protein was investigated. High-intensity ultrasound treatment was observed to reduce the particle size of chickpea protein, improving its emulsifying and gel properties. Protein solubility was reported to increase by 26.7% by decreasing the particle size. The foaming capacity was indicated to increase from 62.1% to 136.7%, and the final treatment was indicated to increase the surface free sulfhydryl content and surface hydrophobicity (Setyaningsih et al., 2016).

Cui et al. (2020) conducted a study in which the effect of ultrasonic treatment on soy protein isolate–glucose conjugates were examined. The degree of Maillard reaction was observe to increase with the increase in the intensity of the applied ultrasound treatment. Increasing the Maillard reaction, on the other hand, was observed to have an effect on protein flexibility. After the process, the particle structure of the protein was reported to become smaller, and a more flexible structure was obtained. A significant correlation was found between molecular flexibility and emulsion properties. This finding emphasizes the role of protein flexibility in interfacial properties (Cui et al., 2020).

Pulsed ultrasonic treatment involves applying the ultrasound process as intermittent pulses which contribute to making the process more efficient and saving energy. Jin et al. (2021) evaluated the effects of pulsed ultrasonic treatment on buckwheat protein isolates. The treatment was indicated to improve the digestibility of protein isolates at 20 kHz power, 10 s on time, 5 s pulsed, 60% amplitude, and 10 min processing time. According to the analyses evaluating the tertiary structure, the sonication process was reported to expose the hydrophobic core and was reported to cause the intramolecular crosslinks to break. The applied process was indicated to cause the macromolecules to break down into smaller pieces, thus improving the surface activity properties of proteins (Jin et al., 2021).

Fu et al. (2022) investigated the effect of pulsed ultrasonic treatment on the structural and functional properties of cottonseed protein isolate. Pulsed ultrasound treatment was applied at different power values (0, 200, 400, 600, and 800 W) and durations (15 and 30 min). As a result of the applied ultrasound process, the particle size of the protein was observed to decrease. As a result of the process, changes occurred in the tertiary structure of the protein. Also, α-helix, β-sheet, and random coil contents decreased while β-turn contents increased. In addition, free sulfhydryl groups increased and fluorescence intensity decreased. In addition, the functional properties of the protein such as solubility, foaming ability, and emulsifying ability were reported to have improved (Fu et al., 2022).

In a study conducted in 2020, the structural changes and digestibility of the pulsed ultrasound treatment on almond milk protein were tested. After the pulsed ultrasound treatment was applied for 4 min, minor losses in the α-helix structure were observed. It was reported that the conformational changes observed in almond protein were very similar to the secondary structure changes in walnut and black bean proteins. It was observed that protein digestibility increased with increasing processing time (Vanga et al., 2020). In another study, quinoa proteins were modified by applying pulsed ultrasound. Ultrasound treatment was applied to the samples with on–off pulses of 10 s/10 s, 5 s/1 s, and 1 s/5 s at 5, 10, 20, and 30 min. After the applied process, the solubility of the proteins was indicated to increase by 48%. At the same time, the secondary and tertiary structures of quinoa proteins were reported to change, and these changes were reported to cause an increase in solubility and particle size (Vera et al., 2019).

Mozafarpour et al. (2022) carried out the modification of pea protein isolate by pulsed ultrasound process. Pulsed ultrasound was performed in 5, 10, and 20 min with 2 s on and 2 s off. When the applied process was performed at lower protein concentrations, stronger gels were indicated to form. In addition, the emulsification properties were reported to improve with the applied process (Mozafarpour et al., 2022).

11.2.2.4   High-Pressure Processing

High-pressure application is a non-thermal process in which proteins are modified by exposing food materials to high pressure (100–1,000 MPa) (Barbhuiya et al., 2021). The high-pressure process is based on Le Chatelier's principle, which states that any phenomenon that causes a decrease in volume increases with pressure (Marciniak et al., 2018). Process flow of high-pressure processing is indicated in Figure 11.6. This process causes the protein to physically firm, creating conformational changes in the protein structure. This contributes to the improvement of various properties of protein such as water-holding capacity, digestibility, foaming, emulsion and gelling (Barbhuiya et al., 2021). The effectiveness of high-pressure treatment is highly dependent on the applied pressure, the processing time, the solution conditions, and the protein properties (Pan et al., 2022). Covalent bonds in the structure of proteins are not affected by high pressure, especially at low temperatures. Therefore, the primary structure of the proteins is not destroyed in this process. However, electrostatic interactions occurring in the process cause changes in the secondary, tertiary, and quaternary structures of the protein. Changes in processes applied below 300 MPa can be reversible; however, the effects at higher pressures are irreversible (Fadimu et al., 2022).

11.2.2.4.1   Disadvantages

High-pressure application has advantages such as not causing environmental pollution and providing less nutrient and taste loss. However, this process has limitations such as the need for expensive equipment, high energy consumption, and inadequacy in large-scale production (Fadimu et al., 2023). In addition, although protein degradation is usually desired, applying severe pressure can lead to protein–protein and peptide–peptide aggregation (Marciniak et al., 2018).

Chao et al. (2018) modified pea proteins by applying high-pressure treatment and investigated the changes in physicochemical and functional properties. The applied process was carried out at pressures of 200, 400, and 600 MPa. After the applied process, high-molecular-weight protein aggregates were observed to form. After the treatment, the emulsification and foaming capacities of the proteins were observed to increase. Pressure treatment applied at 600 MPa was indicated to cause the highest level of denaturation in the samples (Chao et al., 2018).

In a study conducted in 2018, kidney beans were modified by applying a high-pressure treatment at 200, 400, and 600 MPa for 15 min. With the pressure treatment applied at 600 MPa, the water-holding capacity and foaming and emulsification properties were reported to be greatly improved.

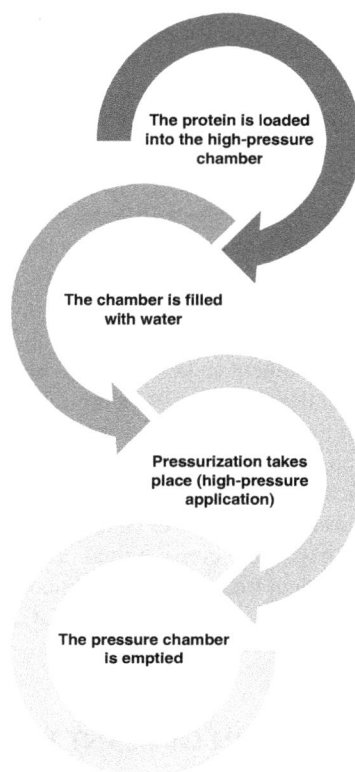

FIGURE 11.6 High-pressure process flow.

Pressure treatments below 600 MPa was observe to have little effect on protein structure. High-pressure treatment was reported to be an effective process for the improvement of plant-based proteins (Ahmed et al., 2018).

Yang et al. (2018) studied the effect of high-pressure treatment on broad bean proteins. Insoluble pod protein aggregates were reported to be separated with high-pressure treatment. In addition, hydrophobic interactions were reported to occur between protein molecules, resulting in changes in the tertiary and quaternary structures of the protein. As a result of these changes, protein solubility and foaming capacity were reported to have increased (Yang et al., 2018).

11.2.2.5 Conventional Heat Treatment

Conventional heat treatment is among the oldest and most common methods applied to change the structural and functional properties of plant-based proteins. Heat treatment applied at lower temperatures contributes to gaining the desired functional properties by unfolding the protein structure. However, the heat treatment applied at high temperatures causes irreversible changes in the protein structure and results in protein denaturation. This causes the nutritional and sensory properties of the protein to be negatively affected. During the heat treatment of a protein solution, the chains, internal sulfhydryl groups, and hydrophobic side chains embedded in the core of the protein are exposed more as a result of the polypeptide opening. These structural changes contribute to the improvement of the functional properties of plant-derived proteins. In addition to the use of heat treatment alone, its use in combination with other modification methods is widely preferred (Nasrabadi et al., 2021).

11.2.2.5.1 Disadvantages

Heat treatment, which is one of the best-known and easy-to-apply techniques for protein modification, has disadvantages such as long processing time, high energy and water consumption, and nutritional and sensory losses. In addition, uncontrolled Maillard reaction may result in the formation of toxic compounds (Sá et al., 2022).

Jiang et al. (2016) examined the effect of heat treatment on legume proteins. The most important reason limiting the use of broad bean protein in food products is the unpleasant bean taste. This problem can be eliminated by heat treatment. With the applied heat treatment, enzymes that are harmful for flavour are neutralized, and more pleasant, sweet products are obtained. The best results were obtained by applying the heat treatment at 170°C for 1.5 min. Excessive application of heat treatment was indicated to cause a decrease in protein solubility (Jiang et al., 2016).

Sarkar et al. (2016) applied heat treatment to tomato seed protein isolate and reported that heat treatment above 80°C caused protein denaturation (Sarkar et al., 2016). In another study conducted in 2018, the effects of conventional and microwave treatments applied to soy milk were evaluated. Soy milk is highly preferred because it is an excellent source of plant-based protein and low in calories. However, it contains protease inhibitors that reduce its nutritional value and digestibility. Microwave processing conditions were applied at 45 GHz and at temperatures of 70°C, 85°C, and 100°C for 2, 5, and 8 min, respectively, while conventional heat treatment was applied at the same temperatures for 10, 20, and 30 min. Heat treatment in combination with microwave treatment was observed to increase digestibility by 7% whereas traditional heat treatment was reported to improve digestibility by 11% compared to control (Vagadia et al., 2018).

11.3 DESIGN AND CONSTRUCTION OF PLANT PROTEIN-BASED MANUFACTURING FACILITY

Scale-up of well-established extraction and modification methods to industrial scales is the final challenge for the production of plant-based proteins ingredients. Therefore, the specific process parameters should be kept constant. Time, cost, and ease to apply in small- and large-scale ease the scale-up procedure. Scale-up of plant-based proteins' production is relatively satisfying since novel and green technologies have been promoted. The main point in scale-up is to guarantee sustainability to ensure optimized variable parameters.

There are newly established companies which produce plant-based foods and proteins. These companies are challenged to provide consumers high-quality plant proteins in large scale. To handle this, first laboratory and then pilot scale of production are recommended to be settled. Successful laboratory process will develop into "as-is" in a manufacturing environment. A pilot plant simulates the conditions of a full-sized manufacturing facility, enabling companies to examine effecting factors and how to control process. In pilot simulation step, it is possible to try out different steps, techniques, and equipment. While a pilot plant may initially upscale production at a university, or in a non-food-grade plant, and then move to a food-grade pilot plant, there are two main advantages to work with a food-grade pilot plant from the start: (i) It can reduce the number of transfers and (ii) the final product will be produced on the basis of regulations for safe food product or ingredient from the start. For most companies, extrusion is central to the production of plant-based innovative foods and ingredients. The advantage of extrusion is that during this process, all the ingredients can be mixed and cooked under pressure simultaneously (Moses, 2023). Mixing ingredients under pressure and cooking them helps to reach desirable texture and appearance by denaturation, cross-linking and hydration of ingredients. Moreover, microbial contamination and anti-nutritional components like phytate, lectins and etc... can be significantly reduced (Moses, 2023).

Using computer modelling programs can help quickly and cost-effectively simulate different processing conditions. Some of process parameters can come from laboratory experiences, while others are collected from the pilot runs. Modelling can also be used for optimizing each step of producing process. For developing a plant-based ingredient for use in food, only testing it with consumers is possible. So, final product must come from a food grade and safe facility. This means sensory analysis panels or clinical trials with volunteers should also be considered (Kamani et al., 2022).

11.4 DOWNSTREAM PROCESSING OF PLANT PROTEIN-BASED MANUFACTURING FACILITY

Transgenic plants are produced by DNA modification by adding one or more genes to them. These are obtained by the gene gun method or *Agrobacterium tumefaciens*-based transformation method. The gene gun method is applied by shooting the plant tissue with a high-pressure gun of DNA fixed on gold or tungsten. After DNA is separated from coated metals, it connects with the plant genome. On the other side, Agrobacterium tumefaciens infects plant cells with its DNA (Rani and Usha, 2013). The tendency to obtain recombinant proteins for industrial and medical use is increasing. Transgenic plants have been found to be safe and inexpensive in producing recombinant proteins (Conley et al., 2011). The use of plants in protein expression systems has many advantages such as low cost, low probability of contamination with human pathogens, and easy scalability (Menary et al., 2020; Shanmugaraj et al., 2021).

Antimicrobial peptides can be produced by biotechnological techniques. Due to their high resistance to fungal and microbial pathogens, they increase crop yields (Montesinos, 2007). Onions (*Allium cepa* L.) are susceptible to fungal infections such as *Aspergillus niger*. The thionine protein has antimicrobial properties against a variety of pathogens, including *A. niger*. The thionine gene is loaded onto the chitosan nanoparticle as a carrier so that the thionine gene is transferred to the onion. After thionine protein expression, *A. niger* spore germination was inhibited by 52%. This revealed that the antimicrobial peptide form of the thionine protein showed antifungal activity (Tawfik et al., 2022). In another study, a potato antimicrobial peptide called Snaking-1 was used for rice prevention against sheath blight disease caused by the fungus *Rhizoctonia solani* by Das et al. (2021). Transgenic rice capable of producing the Snaking-1 peptide was produced, and this showed increase in protection against sheath blight disease.

Plant molecular farming (PMF) is a cost-effective and scalable way to produce recombinant proteins, enzymes, and secondary metabolites in plants for pharmaceutical and industrial platforms. Recombinant proteins are commonly produced proteins for increasing the yield of native material in heterogeneous systems. Tobacco, potato, tomato, alfalfa, safflower, carrot, lettuce, strawberry, duckweed, maize, wheat, and rice are plants used in recombinant protein production (Sing et al., 2021; Coates et al., 2022). Seeds are bioreactors to produce a high-protein and low-moisture recombinant proteins. The recombinant proteins, antibodies, antigens, and others can accumulate in seeds and remain stable for a long time. The relatively homogeneous seeds are materials for further processing and purification (Lau and Sun, 2009). Robić et al. (2010) studied the effect of pH and ionic strength on the production of recombinant β-glucuronidase (rGUS) in soya bean seeds as a bioreactor. Recombinant protein concentrations of soya bean and maize as bioreactors were compared. The most efficient rGUS production was observed at pH 5.5 without salt. Recombinant protein from soya bean seeds had lower protein co-extracted compounds compared to maize seeds (Robić et al., 2010). The use of tobacco in obtaining recombinant protein is extensively shown in recent researches (Sing et al., 2021; Coates et al., 2022). However, recombinant proteins may be found to have insufficient yield and low activity. Various researches are underway, including targeted genome editing, deconstructed vectors, virus-like particles, and humanized glycosylation, to meet industrial production (Liu and Timko, 2022). PMF is based on many protein expression methods including the nuclear transformation of crops growing in a field, plastid transformation of crops, the transient transformation of crops, and the transformation of a hydroponically grown plant species such that protein is secreted into the medium and recovered (Menary et al., 2020). The soil-borne pathogen Agrobacterium tumefaciens is used as a vector for the nuclear transformation of various plant species, including mono- and dicotyledons. It has a Ti-plasmid that integrates T-DNA into the host cell, which leads to transformation. Although this method is widely used, it has a disadvantage as it has a high production time compared to other plant-based expression techniques (Desai et al., 2010). Vacuum infiltration of leaves of plants in the presence of A. tumefaciens causes transient transformation of cells, and a huge number of proteins are expressed with this method (Desai et al., 2010). Joh et al. (2005) studied transient expression followed by agroinfiltration of plants in recombinant protein production. Agrobacterium tumefaciens involved in the beta-glucuronidase (rGUS) gene was infiltrated into lettuce leaf by vacuum. GUS production in lettuce leaves was the highest in darkness after 72 h at 22°C. Agro-filtration was found as a potential scale-up to obtain recombinant protein. Recombinant proteins having high activity and affinity can be produced in rice endosperm. Rice endosperm is a good source for the storage of recombinant proteins due to its stability in structure and composition. Because the cost of purification is high, various buffers can be used to separate target molecules from rice endosperm (Zhu et al., 2022).

The downstream process (DSP) is known as a method that enables the recovery of the pure component from the biological matrix. In recombinant protein extraction, DSP accounts for the bulk of the total cost. The reduction steps of DSP and the use of cheap materials can result in lower costs. Protein extraction and purification are the main steps of DSP (Lojewska et al., 2016). The recovery of products from leafy and seed-based systems for downstream processing starts with homogenization or aqueous extraction of raw material. This is followed by solid–liquid separation and conditioning. Before the homogenization step, fractionation is required to lower solid content and increase the level of recombinant protein in seed-based systems (Wilken and Nikolov, 2012). Zhang et al. (2009) studied the enrichment of recombinant collagen-related protein using wet and dry milling. Thus, impurities from corn could be minimized at the extraction stage. In bioreactor-based systems, separation of biomass for harvesting occurs by using a centrifuge or membrane filtration. Other steps are the same for leafy and seed-based systems. Adsorption chromatography is the main purification method of recombinant proteins. Thanks to the removal of impurities, protein yield and quality are increased (Wilken and Nikolov, 2012).

Downstream processing is complicated by the sensitivity of samples to organic solvents, pH, or high temperatures. It also contains impurities and relatively low-product concentrations. Therefore, aqueous two-phase systems (ATPSs) or aqueous two-phase extraction (ATPE) are being explored for use in the selective, convenient, and stable DSP processes for the separation and purification of sensitive molecules (Glyk et al., 2015). ATPS is a recovery method that will be used for the extraction and purification of biological materials such as protein, enzymes, and genetic material. The design of protein recovery using ATPS consists of several steps, including physicochemical characterization of the raw material, the determination of the ATPS type and parameters, and the determination of the effect of the parameters on the purity of the sample (Benavides and Rito-Palomares, 2008). Vazquez-Villegas et al. (2015) studied the extraction of chlorophyl-free leaf protein concentrates using ATPS. A two-level factorial design with five different factors was made, and the recovery of alfalfa (Medicago sativa) protein in ATPS was examined. They have found that freeze-dried soluble protein consisted of 51% of protein. They indicated that the integration of ATPS with proteomic tools such as electrophoresis and analytical techniques can enable the protein to be separated from the crude extract according to its hydrophobicity or molecular size (Benavides and Rito-Palomares, 2008). In order to find the protein profile of corn germ, ATPSs containing polyethylene glycol (PEG) 1890 and salt were preferred for protein separation by Gu and Glatz (2007). Thus, the hydrophobic resolution of mixtures was obtained. When ATPS is accompanied by two-dimensional gel electrophoresis, the protein's molecular weight, isoelectric point, and expression levels can be obtained by mapping the protein. As a result of this research, 3D mapping and ATPS can be valuable methods for determining the optimal purification of protein from a host and selecting the best host and extraction condition while obtaining a particular recombinant protein to ease the downstream process. The deficiency in purification and low yield of plant-based recombinant proteins can be solved with some fusion protein techniques. Therefore, the expression of transgenic plants is limiting in industrial use. For instance, the recovery of fused proteins with density-based separation is provided by zera, which is the domain of the maize seed storage protein c-zein, leading to the formation of storage protein structures. This new technique results in the collection of various recombinant proteins, which in addition to leads to the formation of protein bodies. Therefore, this method provides an easy and cheap means of protein purification without using chromatographic methods (Conley et al., 2011).

11.5   CONCLUSION AND FUTURE PERSPECTIVES

Beside the increasing consumption of plant-based diets and foods all around the world, the application of isolated plant-based proteins in different food products and formulations has been continuously increasing. Environmental issues such as carbon emission, animal welfare organizations' awareness activities, and increasing health problems related to the consumption of animal-based foods are known as the main reasons of rising demand for plant-based foods and protein ingredients. Compared to animal-based proteins, plant-based ones have some challenging aspects according to their low techno-functional properties, lack of essential amino acids (nutritional concerns) and a good number of organoleptic properties. Due to these aspects, plant-based modification methods have been considered as an effective strategy to enhance and improve their techno-functional, nutritional, and sensory values. Similar to significant role of different isolation methods on the structure of plant-based proteins, various modification methods are also crucial. Modification methods applied to plant proteins result in the structural (e.g., sulfhydryl and disulphide bonds, secondary structures, molecular weight distribution) and morphological changes. To select the best method, advantages and disadvantages of all various methods should be discussed widely. Physical modification methods including heat treatments and high pressure have been already used in food processing for a long time. The combination of these two methods with ultrasonication and micro fluidization is being widely used in food processing to stabilize plant-based protein emulsion products and beverages. Most chemical modifications create toxic chemical by-products which give rise to certain safety and consumers' regulatory concerns. Glycation is the chemical modification method which does not need external chemicals and was known as the safest method among other chemical modification methods. Two main biological modifications, enzymes' treatment and fermentation, are known to be environmentally friendly as no toxic by-products are produced during the process. However, the cost of enzymes and cultures and as well the process parameters' control to produce a desirable modified plant-based protein must be considered. During biological modification as an action of enzymes, protein hydrolysates are changed into peptides which may lead to strong bitter and intense after-taste. Therefore, limited hydrolysis by narrow enzyme hydrolyzation can be applied to develop functional plant-based protein hydrolysate ingredients. Amyloid fibrillization is launched as a novel modification method, and research about its biosafety is still ongoing. Dual modification is reported as a suitable method to enhance functional plant-based protein properties such as emulsifying, water-holding, and protein solubility, but yet significant limitations for its facility to become an industrial technique have been reported. Among dual modification methods, enzyme cross-linking combinations with ultrasound and glycosylation were commonly used techniques. Industrial modification method must be capable of sustaining development goals so the energy and cost efficiency of these techniques need to be evaluated. These factors have certain roles in the feasibility and implementation evaluation at a large scale. To optimize and understand the process parameters' response, surface methodology and simulation tools as *in silico* (dynamic molecular modelling) and also computational fluid dynamics models can be utilized.

In conclusion, various modification methods can be applied on the basis of expected molecular perspective and desirable functionality of plant-based protein ingredients from an industrial viewpoint. In addition, factors for scaling up modification methods such as cost and energy efficiency, food safety regulations, and clean-labelling need to be evaluated. It seems among different modification methods, physical modifications such as ultrasonication, micro fluidization, high pressure, conventional heating, and extrusion and also among biological methods, fermentation and enzymatic reactions reveal the most environmentally friendly industrial level of modification methods for scaling up the production of modified plant proteins. The only concerns regarding biological modification methods in large-scale processes are the cost of enzymes and cultures. While dual modification methods are reported to improve the techno-functional properties of plant-based proteins, more research is required in industrial processing scale.

REFERENCES

Ahmed, J., Al-Ruwaih, N., Mulla, M., & Rahman, M. H. (2018). Effect of high pressure treatment on functional, rheological and structural properties of kidney bean protein isolate. *LWT*, *91*, 191–197. https://doi.org/10.1016/j.lwt.2018.01.054.

Akharume, F. U., Aluko, R. E., & Adedeji, A. A. (2021). Modification of plant proteins for improved functionality: A review. *Comprehensive Reviews in Food Science and Food Safety*, *20*(1), 198–224. https://onlinelibrary.wiley.com/doi/full/10.1111/1541-4337.12688.

Barbhuiya, R. I., Singha, P., & Singh, S. K. (2021). A comprehensive review on impact of non-thermal processing on the structural changes of food components. *Food Research International*, *149*, 110647. https://doi.org/10.1016/j.foodres.2021.110647.

Basak, S., & Annapure, U. S. (2022). Recent trends in the application of cold plasma for the modification of plant proteins-a review. *Future Foods*, 100119. https://doi.org/10.1016/j.fufo.2022.100119.

Benavides, J., & Rito-Palomares, M. (2008). Practical experiences from the development of aqueous two-phase processes for the recovery of high value biological products. *Journal of Chemical Technology and Biotechnology*, *83*(2), 133–142. https://doi.org/10.1002/jctb.1844.

Bräuer, S., Meister, F., Gottlöber, R. P., & Nechwatal, A. (2007). Preparation and thermoplastic processing of modified plant proteins. *Macromolecular Materials and Engineering*, *292*(2), 176–183. https://doi.org/10.1002/mame.200600364.

Chao, D., Jung, S., & Aluko, R. E. (2018). Physicochemical and functional properties of high pressure-treated isolated pea protein. *Innovative Food Science & Emerging Technologies*, *45*, 179–185. https://doi.org/10.1016/j.ifset.2017.10.014.

Choton, S., Gupta, N., Bandral, J. D., Anjum, N., & Choudary, A. (2020). Extrusion technology and its application in food processing: A review. *The Pharma Innovation Journal*, *9*(2), 162–168. https://doi.org/10.22271/tpi.2020.v9.i2d.4367.

Coates, R. J., Young, M. T., & Scofield, S. (2022). Optimising expression and extraction of recombinant proteins in plants. *Frontiers in Plant Science*, *13*. https://doi.org/10.3389/fpls.2022.1074531.

Conley, A. J., Joensuu, J. J., Richman, A., & Menassa, R. (2011). Protein body-inducing fusions for high-level production and purification of recombinant proteins in plants. *Plant Biotechnology Journal*, *9*(4), 419–433.

Cornet, S. H., Snel, S. J., Schreuders, F. K., van der Sman, R. G., Beyrer, M., & van der Goot, A. J. (2022). Thermo-mechanical processing of plant proteins using shear cell and high-moisture extrusion cooking. *Critical Reviews in Food Science and Nutrition*, *62*(12), 3264–3280. https://doi.org/10.1080/10408398.2020.1864618.

Cui, Q., Zhang, A., Li, R., Wang, X., Sun, L., & Jiang, L. (2020). Ultrasonic treatment affects emulsifying properties and molecular flexibility of soybean protein isolate-glucose conjugates. *Food Bioscience*, *38*, 100747. https://doi.org/10.1016/j.fbio.2020.100747.

Das, K., Datta, K., Sarkar, S. N., & Datta, S. K. (2021). Expression of antimicrobial peptide snakin-1 confers effective protection in rice against sheath blight pathogen, Rhizoctonia solani. *Plant Biotechnology Reports*, *15*(1), 39–54. https://doi.org/10.1007/s11816-020-00652-3.

Day, L., Cakebread, J. A., & Loveday, S. M. (2022). Food proteins from animals and plants: Differences in the nutritional and functional properties. *Trends in Food Science & Technology*, *119*, 428–442. https://doi.org/10.1016/j.tifs.2021.12.020.

Desai, P. N., Shrivastava, N., & Padh, H. (2010). Production of heterologous proteins in plants: Strategies for optimal expression. *Biotechnology Advances*, *28*(4), 427–435. https://doi.org/10.1016/j.biotechadv.2010.01.005.

Fadimu, G. J., Le, T. T., Gill, H., Farahnaky, A., Olatunde, O. O., & Truong, T. (2022). Enhancing the biological activities of food protein-derived peptides using non-thermal technologies: A review. *Foods*, *11*(13), 1823. https://doi.org/10.3390/foods11131823.

Fadimu, G. J., Olatunde, O. O., Bandara, N., & Truong, T. (2023). Reducing allergenicity in plant-based proteins. In *Engineering plant-based food systems* (pp. 61–77). Academic Press. https://doi.org/10.1016/B978-0-323-89842-3.00012-9.

Ferawati, F., Zahari, I., Barman, M., Hefni, M., Ahlström, C., Witthöft, C., & Östbring, K. (2021). High-moisture meat analogues produced from yellow pea and Faba bean protein isolates/concentrate: Effect of raw material composition and extrusion parameters on texture properties. *Foods*, *10*(4), 843. https://doi.org/10.3390/foods10040843.

Fernando, S. (2022). Pulse protein ingredient modification. *Journal of the Science of Food and Agriculture*, *102*(3), 892–897. https://doi.org/10.1002/jsfa.11548.

Fu, J., Ren, Y., Jiang, F., Wang, L., Yu, X., & Du, S. K. (2022). Effects of pulsed ultrasonic treatment on the structural and functional properties of cottonseed protein isolate. *LWT*, *172*, 114143. https://doi.org/10.1016/j.lwt.2022.114143.

Gao, Y., Sun, Y., Zhang, Y., Sun, Y., & Jin, T. (2022). Extrusion modification: Effect of extrusion on the functional properties and structure of rice protein. *Processes*, *10*(9), 1871. https://doi.org/10.3390/pr10091871.

Gharibzahedi, S. M. T., & Smith, B. (2020). The functional modification of legume proteins by ultrasonication: A review. *Trends in Food Science & Technology*, *98*, 107–116. https://doi.org/10.1016/j.tifs.2020.02.002.

Glyk, A., Scheper, T., & Beutel, S. (2015). PEG–salt aqueous two-phase systems: An attractive and versatile liquid–liquid extraction technology for the downstream processing of proteins and enzymes. *Applied Microbiology and Biotechnology*, *99*(16), 6599–6616. https://doi.org/10.1007/s00253-015-6779-7.

Gu, Z., & Glatz, C. E. (2007). A method for three-dimensional protein characterization and its application to a complex plant (corn) extract. *Biotechnology and Bioengineering*, *97*(5), 1158–1169. https://doi.org/10.1002/bit.21310.

Gulati, P., Brahma, S., & Rose, D. J. (2020). Impacts of extrusion processing on nutritional components in cereals and legumes: Carbohydrates, proteins, lipids, vitamins, and minerals. In *Extrusion cooking* (pp. 415–443). Woodhead Publishing. https://doi.org/10.1016/B978-0-12-815360-4.00013-4.

Ismail, B. P., Senaratne-Lenagala, L., Stube, A., & Brackenridge, A. (2020). Protein demand: Review of plant and animal proteins used in alternative protein product development and production. *Animal Frontiers*, *10*(4), 53–63. https://doi.org/10.1093/af/vfaa040.

Jhansi Rani, S., & Usha, R. (2013). Transgenic plants: Types, benefits, public concerns and future. *Journal of Pharmacy Research*, *6*(8), 879–883. https://doi.org/10.1016/j.jopr.2013.08.008.

Ji, H., Dong, S., Han, F., Li, Y., Chen, G., Li, L., & Chen, Y. (2018). Effects of dielectric barrier discharge (DBD) cold plasma treatment on physicochemical and functional properties of peanut protein. *Food and Bioprocess Technology*, *11*, 344–354. https://doi.org/10.1007/s11947-017-2015-z.

Jiang, Z. Q., Pulkkinen, M., Wang, Y. J., Lampi, A. M., Stoddard, F. L., Salovaara, H., Piironen, V., & Sontag-Strohm, T. (2016). Faba bean flavour and technological property improvement by thermal pre-treatments. *LWT-Food Science and Technology*, *68*, 295–305. https://doi.org/10.1016/j.lwt.2015.12.015.

Jin, J., Okagu, O. D., Yagoub, A. E. A., & Udenigwe, C. C. (2021). Effects of sonication on the in vitro digestibility and structural properties of buckwheat protein isolates. *Ultrasonics Sonochemistry*, *70*, 105348. https://doi.org/10.1016/j.ultsonch.2020.105348.

Joh, L. D., Wroblewski, T., Ewing, N. N., & VanderGheynst, J. S. (2005). High-level transient expression of recombinant protein in lettuce. *Biotechnology and Bioengineering*, *91*(7), 861–871. https://doi.org/10.1002/bit.20557.

Kamani, M. H., Semwal, J., & Khaneghah, A. M. (2022). Functional modification of grain proteins by dual approaches: Current progress, challenges, and future perspectives. *Colloids and Surfaces B: Biointerfaces*, *211*, 112306. https://doi.org/10.1016/j.colsurfb.2021.112306.

Keppler, J. K., Schwarz, K., & van der Goot, A. J. (2020). Covalent modification of food proteins by plant-based ingredients (polyphenols and organosulphur compounds): A commonplace reaction with novel utilization potential. *Trends in Food Science & Technology*, *101*, 38–49. https://doi.org/10.1016/j.tifs.2020.04.023.

Khan, Z. S., Sodhi, N. S., Dhillon, B., Dar, B., Bakshi, R. A., & Shah, S. F. (2021). Seabuckthorn (Hippophae Rhamnoides L.), a novel seed protein concentrate: Isolation and modification by high power ultrasound and characterization for its functional and structural properties. *Journal of Food Measurement and Characterization*, *15*(5), 4371–4379. Retrieved March 23, 2023, from https://link.springer.com/article/10.1007/s11694-021-01020-7.

Lau, O. S., & Sun, S. S. M. (2009). Plant seeds as bioreactors for recombinant protein production. *Biotechnology Advances*, *27*(6), 1015–1022. https://doi.org/10.1016/j.biotechadv.2009.05.005.

Lee, J. S., Oh, H., Choi, I., Yoon, C. S., & Han, J. (2022). Physico-chemical characteristics of rice protein-based novel textured vegetable proteins as meat analogues produced by low-moisture extrusion cooking technology. *LWT*, *157*, 113056. https://doi.org/10.1016/j.lwt.2021.113056.

Li, Y., Cheng, Y., Zhang, Z., Wang, Y., Mintah, B. K., Dabbour, M., Jiang, H., He, R., & Ma, H. (2020). Modification of rapeseed protein by ultrasound-assisted pH shift treatment: Ultrasonic mode and frequency screening, changes in protein solubility and structural characteristics. *Ultrasonics Sonochemistry*, *69*, 105240. https://doi.org/10.1016/j.ultsonch.2020.105240.

Liu, H., & Timko, M. P. (2022). Improving protein quantity and quality-the next level of plant molecular farming. *International Journal of Molecular Sciences, 23*(3). https://doi.org/10.3390/ijms23031326.

Łojewska, E., Kowalczyk, T., Olejniczak, S., & Sakowicz, T. (2016). Extraction and purification methods in downstream processing of plant-based recombinant proteins. *Protein Expression and Purification, 120*, 110–117. https://doi.org/10.1016/j.pep.2015.12.018.

Marciniak, A., Suwal, S., Naderi, N., Pouliot, Y., & Doyen, A. (2018). Enhancing enzymatic hydrolysis of food proteins and production of bioactive peptides using high hydrostatic pressure technology. *Trends in Food Science & Technology, 80*, 187–198. https://doi.org/10.1016/j.tifs.2018.08.013.

Menary, J., Hobbs, M., de Albuquerque, S. M., Pacho, A., Drake, P. M. W., Prendiville, A., Ma, J. K. C., & Fuller, S. S. (2020). Shotguns vs lasers: Identifying barriers and facilitators to scaling-up plant molecular farming for high-value health products. *PLoS ONE, 15*(3). https://doi.org/10.1371/journal.pone.0229952

Mir, N. A., Riar, C. S., & Singh, S. (2019). Physicochemical, molecular and thermal properties of high-intensity ultrasound (HIUS) treated protein isolates from album (Chenopodium album) seed. *Food Hydrocolloids, 96*, 433–441. https://doi.org/10.1016/j.foodhyd.2019.05.052.

Mollakhalili-Meybodi, N., Yousefi, M., Nematollahi, A., & Khorshidian, N. (2021). Effect of atmospheric cold plasma treatment on technological and nutrition functionality of protein in foods. *European Food Research and Technology, 247*, 1579–1594. https://doi.org/10.1007/s00217-021-03750-w.

Montesinos, E. (2007). Antimicrobial peptides and plant disease control. *FEMS Microbiology Letters, 270*(1), 1–11. https://doi.org/10.1111/j.1574-6968.2007.00683.x

Moses, T. (2023). "Plant-based protein manufacturing: Scaling-up extrusion." [accessed March 05, 2023]. www.crbgroup.com/insights/food-beverage/plant-based-protein-manufacturing.

Mozafarpour, R., Koocheki, A., & Nicolai, T. (2022). Modification of grass pea protein isolate (Lathyrus sativus L.) using high intensity ultrasound treatment: Structure and functional properties. *Food Research International, 158*, 111520. https://doi.org/10.1016/j.foodres.2022.111520.

Munialo, C. D., Stewart, D., Campbell, L., & Euston, S. R. (2022). Extraction, characterisation and functional applications of sustainable alternative protein sources for future foods: A review. *Future Foods*, 100152. https://doi.org/10.1016/j.fufo.2022.100152.

Nasrabadi, M. N., Doost, A. S., & Mezzenga, R. (2021). Modification approaches of plant-based proteins to improve their techno-functionality and use in food products. *Food Hydrocolloids, 118*, 106789. https://doi.org/10.1016/j.foodhyd.2021.106789.

Onwezen, M. C., Bouwman, E. P., Reinders, M. J., & Dagevos, H. (2021). A systematic review on consumer acceptance of alternative proteins: Pulses, algae, insects, plant-based meat alternatives, and cultured meat. *Appetite, 159*, 105058. https://doi.org/10.1016/j.appet.2020.105058.

Pan, J., Zhang, Z., Mintah, B. K., Xu, H., Dabbour, M., Cheng, Y., Dai, C., He, R., & Ma, H. (2022). Effects of nonthermal physical processing technologies on functional, structural properties and digestibility of food protein: A review. *Journal of Food Process Engineering, 45*(4), e14010. https://doi/full/10.1111/jfpe.14010.

Paulson, A. T., Tung, M. A., Garland, M. R., & Nakai, S. (1984). Functionality of modified plant proteins in model food systems. *Canadian Institute of Food Science and Technology Journal, 17*(4), 202–208. https://doi.org/10.1016/S0315-5463(84)72558-2.

Pietsch, V. L., Bühler, J. M., Karbstein, H. P., & Emin, M. A. (2019). High moisture extrusion of soy protein concentrate: Influence of thermomechanical treatment on protein-protein interactions and rheological properties. *Journal of Food Engineering, 251*, 11–18. https://doi.org/10.1016/j.jfoodeng.2019.01.001.

Prabha, K., Ghosh, P., Abdullah, S., Joseph, R. M., Krishnan, R., Rana, S. S., & Pradhan, R. C. (2021). Recent development, challenges, and prospects of extrusion technology. *Future Foods, 3*, 100019. https://doi.org/10.1016/j.fufo.2021.100019.

Qu, Z., Chen, G., Wang, J., Xie, X., & Chen, Y. (2023). Preparation, structure evaluation, and improvement in foaming characteristics of fibrotic pea protein isolate by cold plasma synergistic organic acid treatment. *Food Hydrocolloids, 134*, 108057. https://doi.org/10.1016/j.foodhyd.2022.108057.

Rahman, M. M., & Lamsal, B. P. (2021). Ultrasound-assisted extraction and modification of plant-based proteins: Impact on physicochemical, functional, and nutritional properties. *Comprehensive Reviews in Food Science and Food Safety, 20*(2), 1457–1480. https://doi.org/10.1111/1541-4337.12709.

Rahman, M. M., & Lamsal, B. P. (2023). Effects of atmospheric cold plasma and high-power sonication on rheological and gelling properties of mung bean protein dispersions. *Food Research International, 163*, 112265. https://doi.org/10.1016/j.foodres.2022.112265.

Robić, G., Farinas, C. S., Rech, E. L., & Miranda, E. A. (2010). Transgenic soybean seed as protein expression system: Aqueous extraction of recombinant β-glucuronidase. *Applied Biochemistry and Biotechnology, 160*(4), 1157–1167. https://doi.org/10.1007/s12010-009-8637-5.

Sá, A. G. A., Laurindo, J. B., Moreno, Y. M. F., & Carciofi, B. A. M. (2022). Influence of emerging technologies on the utilization of plant proteins. *Frontiers in Nutrition*, 9, 120. https://doi.org/10.3389/fnut.2022.809058.

Sá, A. G. A., Moreno, Y. M. F., & Carciofi, B. A. M. (2020). Food processing for the improvement of plant proteins digestibility. *Critical Reviews in Food Science and Nutrition*, 60(20), 3367–3386. https://doi.org/10.1080/10408398.2019.1688249.

Saremnezhad, S., Soltani, M., Faraji, A., & Hayaloglu, A. A. (2021). Chemical changes of food constituents during cold plasma processing: A review. *Food Research International*, 147, 110552. https://doi.org/10.1016/j.foodres.2021.110552.

Sarkar, A., Kamaruddin, H., Bentley, A., & Wang, S. (2016). Emulsion stabilization by tomato seed protein isolate: Influence of pH, ionic strength and thermal treatment. *Food Hydrocolloids*, 57, 160–168. https://doi.org/10.1016/j.foodhyd.2016.01.014.

Setyaningsih, W., Saputro, I. E., Palma, M., & Barroso, C. G. (2016). Pressurized liquid extraction of phenolic compounds from rice (Oryza sativa) grains. *Food Chemistry*, 192, 452–459. https://doi.org/10.1016/j.foodchem.2015.06.102.

Shanmugaraj, B., Bulaon, C. J. I., Malla, A., & Phoolcharoen, W. (2021). Biotechnological insights on the expression and production of antimicrobial peptides in plants. *Molecules*, 26(13). https://doi.org/10.3390/molecules26134032

Singh, A. A., Pillay, P., & Tsekoa, T. L. (2021). Engineering approaches in plant molecular farming for global health. *Vaccines*, 9(11). https://doi.org/10.3390/vaccines9111270.

Singhal, A., Karaca, A. C., Tyler, R., & Nickerson, M. (2016). Pulse proteins: From processing to structure-function relationships. *Grain Legumes*, 55. https://doi.org/10.5772/64020.

Sonawane, S. K., Marar, T., & Patil, S. (2020). Non-thermal plasma: An advanced technology for food industry. *Food Science and Technology International*, 26(8), 727–740. https://doi.org/10.1177/1082013220929474

Tahsiri, Z., Hedayati, S., & Niakousari, M. (2023). Improving the functional properties of wild almond protein isolate films by Persian gum and cold plasma treatment. *International Journal of Biological Macromolecules*, 229, 746–751. https://doi.org/10.1016/j.ijbiomac.2022.12.321.

Tang, S. Q., Du, Q. H., & Fu, Z. (2021). Ultrasonic treatment on physicochemical properties of water-soluble protein from Moringa oleifera seed. *Ultrasonics Sonochemistry*, 71, 105357. https://doi.org/10.1016/j.ultsonch.2020.105357.

Tawfik, E., Hammad, I., & Bakry, A. (2022). Production of transgenic Allium cepa by nanoparticles to resist Aspergillus niger infection. *Molecular Biology Reports*, 49(3), 1783–1790. https://doi.org/10.1007/s11033-021-06988-5

Vagadia, B. H., Vanga, S. K., Singh, A., Gariepy, Y., & Raghavan, V. (2018). Comparison of conventional and microwave treatment on soymilk for inactivation of trypsin inhibitors and in vitro protein digestibility. *Foods*, 7(1), 6. https://doi.org/10.3390/foods7010006.

Vanga, S. K., Wang, J., Orsat, V., & Raghavan, V. (2020). Effect of pulsed ultrasound, a green food processing technique, on the secondary structure and in-vitro digestibility of almond milk protein. *Food Research International*, 137, 109523. https://doi.org/10.1016/j.foodres.2020.109523.

Vazquez-Villegas, P., Acuna-González, E., Mejía-Manzano, L. A., Rito-Palomares, M., & Aguilar, O. (2015). Production and optimization of a chlorophyl-free leaf protein concentrate from alfalfa (medicago sativa) through aqueous two-phase system. *Revista Mexicana de Ingeniería Química*, 14(2), 383–392.

Vera, A., Valenzuela, M. A., Yazdani-Pedram, M., Tapia, C., & Abugoch, L. (2019). Conformational and physicochemical properties of quinoa proteins affected by different conditions of high-intensity ultrasound treatments. *Ultrasonics Sonochemistry*, 51, 186–196.

Wang, K., & Arntfield, S. D. (2017). Effect of protein-flavour binding on flavour delivery and protein functional properties: A special emphasis on plant-based proteins. *Flavour and Fragrance Journal*, 32(2), 92–101. https://doi.org/10.1002/ffj.3365

Wen, C., Zhang, J., Yao, H., Zhou, J., Duan, Y., Zhang, H., & Ma, H. (2019). Advances in renewable plant-derived protein source: The structure, physicochemical properties affected by ultrasonication. *Ultrasonics Sonochemistry*, 53, 83–98. https://doi.org/10.1016/j.ultsonch.2018.12.036.

Wilken, L. R., & Nikolov, Z. L. (2012). Recovery and purification of plant-made recombinant proteins. *Biotechnology Advances*, 30(2), 419–433. https://doi.org/10.1016/j.biotechadv.2011.07.020.

Yang, J., Liu, G., Zeng, H., & Chen, L. (2018). Effects of high-pressure homogenization on Faba bean protein aggregation in relation to solubility and interfacial properties. *Food Hydrocolloids*, 83, 275–286. https://doi.org/10.1016/j.foodhyd.2018.05.020.

Yousefi, N., & Abbasi, S. (2022). Food proteins: Solubility & thermal stability improvement techniques. *Food Chemistry Advances*, 100090. https://doi.org/10.1016/j.focha.2022.100090.

Zhang, C., Glatz, C. E., Fox, S. R., & Johnson, L. A. (2009). Fractionation of transgenic corn seed by dry and wet milling to recover recombinant collagen-related proteins. *Biotechnology Progress*, *25*, 1396–1401.

Zhang, J., Liu, L., Liu, H., Yoon, A., Rizvi, S. S., & Wang, Q. (2019). Changes in conformation and quality of vegetable protein during texturization process by extrusion. *Critical Reviews in Food Science and Nutrition*, *59*(20), 3267–3280. https://doi.org/10.1080/10408398.2018.1487383.

Zhang, S., Huang, W., Feizollahi, E., Roopesh, M. S., & Chen, L. (2021). Improvement of pea protein gelation at reduced temperature by atmospheric cold plasma and the gelling mechanism study. *Innovative Food Science & Emerging Technologies*, *67*, 102567. https://doi.org/10.1016/j.ifset.2020.102567.

Zhang, Z., Zhang, L., He, S., Li, X., Jin, R., Liu, Q., Chen, S., & Sun, H. (2022). High-moisture extrusion technology application in the processing of textured plant protein meat analogues: A review. *Food Reviews International*, 1–36. https://doi.org/10.1080/87559129.2021.2024223

Zhu, Q., Tan, J., & Liu, Y. G. (2022). Molecular farming using transgenic rice endosperm. *Trends in Biotechnology*, *40*(10), 1248–1260. https://doi.org/10.1016/j.tibtech.2022.04.002.

Zhu, Z., Zhu, W., Yi, J., Liu, N., Cao, Y., Lu, J., Decker, E.A. & McClements, D. J. (2018). Effects of sonication on the physicochemical and functional properties of walnut protein isolate. *Food Research International*, *106*, 853–861. https://doi.org/10.1016/j.foodres.2018.01.060.

12 Applications of Modified Plant Protein-Based Future Foods

Koyel Kar, Sailee Chowdhury, Priyanka Chakraborty and Kamalika Mazumder

12.1 INTRODUCTION

The next very important macronutrient used for individual existence, which remains after calories, is protein. Since conventional animal protein supplies involve a lot of ground supplies, protein production is a major concern (Calicioglu et al. 2019). Protein malnutrition, which is characterized by dietary protein deficiencies or imbalances, has a significant impact on a person's digestion, alignment, and roles, besides scientific consequences. Despite being scarce in the sphere of technologically advanced countries, protein malnourishment continues to be the leading cause of childhood impermanence and sickness worldwide (Torres-León et al. 2018).

To sustainably feed humans and provide them with the nutrients and energy they need, land herbs have prolonged been a significant component of their intake. Plant-based proteins are still not widely used directly despite being more plentiful and more affordable than animal proteins in comparison. Since animal protein is produced, there is an excessive amount of food distributed because of environmental pressure. Regarding ground usage, if the identical quantity of herb-based proteins were used for individual utilization exclusively, fewer than 10% of the soil space would remain necessary towards growing staple harvests in place of fodder harvests to produce the equivalent quantity of animal proteins. In addition, the production of animal proteins needs roughly 100 times additional water than the manufacture of herb-based proteins of the same amount (Jiménez-Munoz et al. 2021).

Given the world's speedily expanding populace, ensuring nutrition as well as diet safety continue to be a massive task intended for cooking production. Due to populace growth and other factors, such as changing socio-demographics, the demand for nutrient-rich food products will eventually put a significant strain on the world's resources. The increased demand for protein that results from this population growth is also necessarily propelled by socioeconomic fluctuations akin to enhanced growth, and an increasing wage, combined with elderly populaces. It remains critical to comprehend how crucial protein is to a nutritious intake and healthful aging. Monetary expansion along with growing development, especially in middle- and low-income nations, has significantly altered the dietary habits of the populace. Even the World Health Organization (WHO), along with the World Cancer Research Fund (WCRF), endorses herbal-based diets (Lehikoinen & Salonen 2019). Factory-based proteins make up a more promising explanation owing to their extended account of harvest usage as well as farming, decreasing the expense of making, along with an easygoing approach around several sections of the people. Herb proteins remain as well beyond being biologically viable (Willett et al. 2019). On the other hand, plant proteins have lower protein quality and relatively poor functionality, which is defined as poor solubility, frothing, emulsification, and gelation characteristics. This limits their use in food products. Plant protein science remains still within the problem of infancy as compared to animal proteins, including dairy. It is urgently necessary to make advancements in the creation of plant-based food components and plant protein ingredient development to close this gap. A roadmap for advancing plant protein science and technology is presented in this chapter, which aims to pique readers' interest while concentrating on the advancement of plant protein components combined with the creation of new foods. The condition

DOI: 10.1201/9781003369790-12

of the art in each field is briefly discussed, and recent trends in research are highlighted. The goal of this analysis is to introduce new concepts and promote innovative thinking, not to duplicate what has already been done. Even though the emphasis of this assessment of plant proteins remains in the perspective of developed countries, it remains crucial to acknowledge that around 1 billion individuals might endure protein defects (McClements & Grossmann 2021).

12.2 ARRANGEMENT AND PROPERTIES OF PROTEINS FROM PLANT SOURCES

The quality of the protein remains just as important for various aspects of health as the quantity of protein consumed (Millward et al. 2008). The essential amino acids needed for protein synthesis are present in sufficient amounts in high-quality proteins that are simple to digest. The most crucial element influencing protein features is its amino acid structure. The amino acid composition of proteins derived from various plant supplies varies, displaying a range of attributes and well-being benefits. Several essential amino acids, such as lysine, sulfur-containing amino acids (cysteine and methionine), and threonine, are typically lacking or low in plant proteins derived from grains, seeds, legumes, nuts, pulses, and vegetables. Even though soy proteins are sometimes referred to as "complete" proteins, they only contain 85% of milk's essential amino acids (Mattila et al. 2018). Cutting-edge, the main distinction among the essential amino acid substance of plant and animal proteins is in the content of lysine, except soybeans, which are low in sulfur amino acids, and maize, which is low in tryptophan. Thus, to meet the body's need for amino acids, a number of nutritionists have suggested combining grains with legumes or else soy, which don't contain sulfur-containing amino acids. In contrast to kidney beans and lentils, the protein fractions from wheat gluten, peanuts, wheat flour/bread, and soy protein isolates had a higher rate of digestion (94–99%) (Gilbert et al. 2011). Table 12.1 lists the amino acid composition of various plant proteins.

12.3 HERBAL PROTEIN EXTRACTION

Protein fractionation, as well as isolation, has traditionally been accomplished through different methods (Hinderink et al. 2021).

12.3.1 CHEMICAL EXTRACTION TECHNIQUES

12.3.1.1 Aqueous Two-Phase System (ATPS)

Owing to the properties of the aqueous two-phase system (ATPS), including the hydrophobicity of the phase method, the electrical potential among phases, and molecular size, combined with the bio-affinity of the protein, it remains currently used for effective protein extraction (Kumar et al. 2021).

TABLE 12.1

Amino Acid Composition of Some Plant Proteins (Devi et al. 2015)

Plant/Corp	Soy	Wheat	Rice	Potato	Lupin	Cotton Seed	Lentil	Peanut	Kidney Bean	Corn
Threonine	2.3	1.8	2.3	4.1	1.6	2.9–3.1	3.0	–	3.17–3.77	1.8
Methionine	0.3	0.7	0.3	1.3	0.2	1.3	0.8	0.89–1.13	0.72–1.62	1.1
Histidine	1.5	1.4	1.6	1.4	1.2	2.6–2.8	2.5	2.10–2.33	2.61–2.94	1.1
Lysine	3.4	1.1	4.7	4.8	2.1	4.2–4.6	7.3	3.01–3.72	4.91–6.48	1
Valine	2.2	2.3	2.7	3.7	1.4	4.2–4.6	4.5	3.18–4.02	4.58–5.38	2.1
Aspartic acid	–	–	–	–	–	7.4–8.5	13.7	8.37–11.73	9.02–10.86	–

12.3.1.2   Subcritical Water Extraction (SWE)

Using hot water between the temperature where it normally boils (100°C) and the temperature where it becomes critical (374°C), applying elevated pressure to keep the aquatic in a liquid state at those temperatures, is the main basis (Varadavenkatesan et al. 2021).

12.3.2   Enzyme-Aided Extraction

The method is established on the basis of the action of enzymes that break down the major cell-wall constituents (Kumar et al. 2021).

12.3.2.1   Microwave-Aided Extraction (MAE)

It employs electromagnetic signals through a frequency range of 300 MHz to 300 GHz (Pojić et al. 2018).

12.3.2.2   Ultrasound-Aided Extraction (UAE)

The nutrition pattern, extraction solvent, contact period and temperature, ultrasound frequency, energy, and profusion, as well as the sort of equipment being utilized, are just a few of the variables that can affect UAE's performance (Rahman & Lamsal 2020).

12.3.2.3   High Hydrostatic Pressure-Assisted Extraction

The intent of an isostatic pressure between 100 and 1000 MPa, which is instantly and consistently spread across a liquid, typically water, is the foundation of high-level hydrostatic pressure (HHP) expertise (Shouqin et al. 2005).

12.4   APPROACHES FOR MODIFICATION OF PLANT PROTEIN

The practice of changing a protein's molecular structure or a few chemical groups using specific techniques to increase its bioactivity and technological functionality is referred to as "protein modification."

12.4.1   Physical Modification

Simple procedures that don't rely on chemicals or enzymes can increase the functioning of proteins through physical means without using chemicals thus diminishing adverse effects. In the *heating method*, the protein is encouraged to unfold under mild heat conditions, which results in a transitional molten globule condition (Yang et al. 2018). *Sonication* uses high sound waves (>16 kHz) that are inaudible to the human ear for the breaking of non-covalent bonds (Yang et al. 2018). *Extrusion* involves high pressure of 1.5–30.0 MPa, heat, and mechanical shear (Matsuyama et al. 2021). *Ball mill treatment* has been used to make plant-derived proteins' particle sizes smaller, ensuring greater solubility (Liu et al. 2022). *Ultrafiltration* is a nonthermal technique based on applying pressure or an electric field to a membrane to force a fluid through its pores (Aryee et al. 2018). Modification is also executed using *high pressure* varying from 200 to 800 MPa for 5 min (Gharibzahedi & Smith 2021). The *ultraviolet radiation method* is executed by passing the liquid through UV tubes (200–280 nm) with the aid of laminar or turbulent flows (Kumar et al. 2020). Figure 12.1 represents a protein structure that is treated and untreated by pressure (Sim et al. 2021).

12.4.2   Chemical Modification

Food proteins are reformed by chemical means because of their competence, economics, and ease of comprehension. The chemical reformation method is executed by either adding a moiety or group abolition from protein. This method is always accompanied by suitable and clean labels following

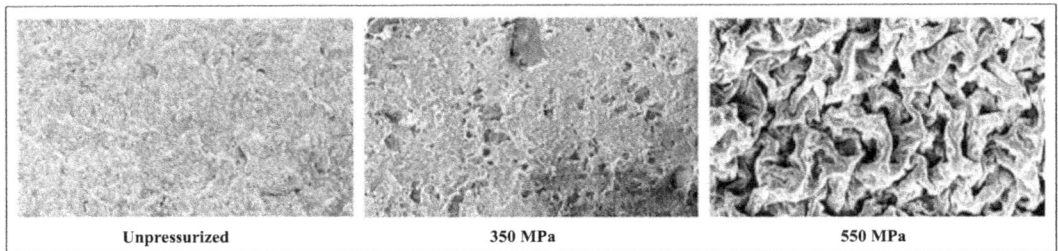

FIGURE 12.1 Representation of a protein structure that is treated and untreated by pressure. (Sim et al. 2021.)

regulations as chemicals are used in this method and by-products containing chemicals are fabricated. (Zhang et al. 2020). Various types of chemical reformation techniques have been discussed. *Glycation* is a type of chemical reaction that is used commonly to improvise the different protein characteristics. It is executed by Maillard chemical reaction or with the help of transglutaminase (Wen et al. 2020). In the *phosphorylation method*, a phosphate group is introduced into the primary structure of the protein and enhances the functional ability of the protein (Zhang et al. 2019). *Shifting of pH* also helps in protein structure modification. Any alterations in the pH of the liquid matrix may lead to structural and functional changes in the protein structure (Muneer et al. 2019). *Acylation* also helps in executing protein modification. It is a type of nucleophilic reaction that is based on acetylation and succinylation, where acylating agent acetic or succinic anhydride is used respectively (Nikbakht et al. 2021). The *deamidation method* is also used in which the amide group from glutamine and asparagine residues of a protein are transformed into carboxyl groups (Hadidi et al. 2021).

12.4.3 BIOLOGICAL MODIFICATION

Both the enzymatic and fermentation processes are considered under biological modification. The *enzymatic method* is a nontoxic method that is carried out either by enzymatic hydrolysis or enzymatic cross-linking method. In the case of *enzymatic hydrolysis*, peptide bonds are enzymatically broken down (Nikbakht et al. 2021). In the *enzymatic cross-linking method*, trans-glutaminase and laccase catalyze acyl transfer reactions occur between the γ-carboxamide group of protein-bound glutamine and lysine leading to the enzymatic formation of covalent bonds (García Arteaga et al. 2020). *Fermentation* is a traditional cost-effective biological method. Different cultures are used as starters, such as lactic acid bacteria, Bacillus strains, yeast, and mold. They have been used for the fermentation proteins. They help in the degradation of allergens and antinutritional ingredients (Xing et al. 2020). Figure 12.2 represents different approaches for protein modification.

12.5 APPLICATION OF MODIFIED PLANT PROTEIN

12.5.1 PROTEIN–POLYSACCHARIDE COMPLEX

It is the primary challenge to convert plant proteins into delicious and also nutritious food. For this purpose, protein–polysaccharide interaction and modification of flavor and taste are required. Polysaccharides are sugar polymers belonging to the family of starch, pectins, alginates, cellulose, agars, carrageenans, and gums (Stephen & Phillips 2016). Polysaccharides are considered as the major building blocks in food. They help in the formation of proper structure and stabilize it through their emulsifying, thickening, and gelling properties. When they are used in combination with proteins, due to the biopolymer interactions functionality can be improved. For plant-based milk, ice cream, and pudding, protein–polysaccharide interactions can be designed. Literature

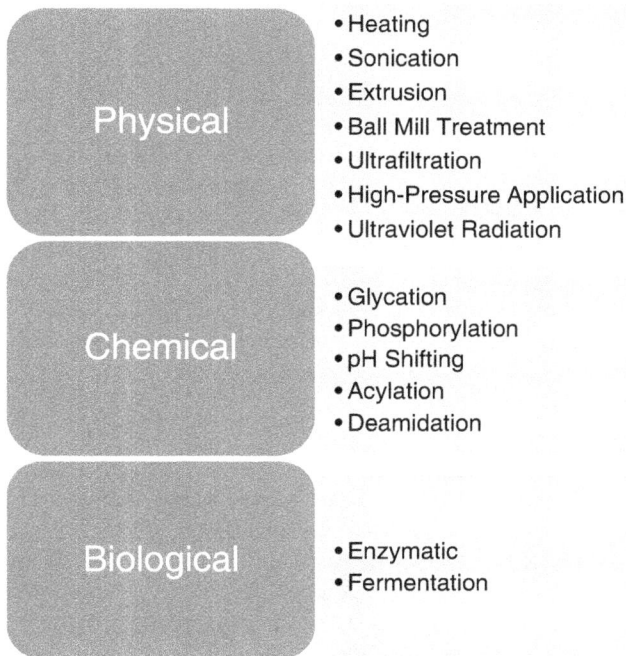

FIGURE 12.2 Different approaches for protein modification.

showed that in the last five years, many plant proteins with polysaccharides were found. When polysaccharides are added to proteins, the solubility of the complex is increased. It is due to the change of surface charge during complexation (Yildiz et al. 2018). The change in isoelectric point is used for preparing high-protein beverages. Commercially available plant protein when isolated gets denatured. More research is required to properly understand how to overcome these. Some factors are also to be considered like pH, concentrations, the ratio of biopolymer, and the strength of ions in the system.

12.5.2 Formation of Fibrous Structure of Protein

During cooking, proteins thermally unfold and cross-link and form continuous gel structures (Sha & Xiong 2020). Hydrogen bonding is increased during cooling. So, it makes the gel firmer and more elastic. Adhesiveness, visco-elasticity, and juiciness are increased due to this property. In the industry, plant proteins from products like soy, pea, and wheat are widely used due to their availability, cost-effectiveness, and processing functionality. Globular plant proteins cannot form fibrous textures like meat. Globules change into fibers by extrusion and fiber spinning, and the products are produced with a meat-like surface. Mung bean and chickpea isolates have good gelling properties. They can form porous networks which are also heterogeneous. Mung bean flour is extruded to make meat-like products. Gluten present in wheat has film-forming characteristics which may produce meat-like fibers (Kyriakopoulou et al. 2021). Wheat gluten, when combined with legume proteins, produces meat-like chewiness. Animal and plant proteins have physicochemical differences. To produce the complex structure of meat fibers, potential methods of the thermo-mechanical process should be developed. Recent studies showed that Maillard-reacted beef bone hydrolysate combined with soy protein and wheat gluten increased the amount of protein of the meat-like food components and the sensory profiles of the meat-like foods (Chiang et al. 2021).

12.5.3 Formation of Gel-Like Structure

Different methods can be applied to form a gel. The different methods are heating gelation, cold gelation, change of pH, the addition of salts, cross-linking with enzymes, and pressure-induced gelation. Various types of gels like hydrogels, oleo gels (oil gels), aerogels, and emulsion gels are formed. Tofu and bean curds are made from peas and various legumes. Another achievement is to develop plant-based yogurt and cheese analogs. Zein is used to increase stretchability in plant-based cheese. Plant-based yogurt can be produced by the traditional fermentation of plant-based milk (Margolis et al. 2019). To increase plant-based animal substitutes, new products should be modified. There are several challenges to designing future plant-based foods. Imitating the texture of animal-based fats is very difficult. Solid fats containing saturated fats increase the risk of cardiovascular disease and other health complications. Oleogels reduce postprandial insulinemia and lipidemia. Oleogels modify plant-based lipids to have the same texture as animal fat.

12.5.4 Plant Protein Flavors

Flavor increases the eating quality. The development of meat-like flavor from nonmeat sources such as plant proteins can be produced by conducting the Maillard reaction. In this process, free amino compounds and reducing sugars are reacted together to produce melanoidin (Van Ba et al. 2012). The plant proteins generate amino acids and peptides via enzymatic hydrolysis and produce meaty flavors. Oxygen, nitrogen, and sulfur-containing heterocyclic compounds play a significant role in the development of flavor in thermally processed foods. Cysteine and reducing sugars react with each other, and it is the main pathway to forming meat-like flavors for most food products. The Strecker degradation reaction forms meat-like flavored compounds. There are several plant proteins which generate meat-like flavors. There are many literature studies regarding pea protein, flaxseed protein, quinoa protein, and soybean protein which generate meat-like flavors. Further studies have to be done to avoid compounds containing sulfur. Only using the peptides and free amino acids from the plant protein hydrolysate's reaction can be done with the reducing sugars.

12.5.5 Industrial Use

Plant-based proteins are used as food substitutes, components of consumable coatings, surface active agents in foods, hydrolysates, and hydrogels for different pharmaceutical and nutraceutical uses. Protein hydrolysates also help in bearing the functional compounds.

12.5.5.1 As Food Substitutes

The human diet consists of food substitutes obtained from plant protein. These food substitutes provide extra food value and thus escalate the food value of the diet. Plant-based proteins have many roles in human health. They include the maintenance of the health of bone, escalating the muscles, the management of weight, and extra nutritional value. Protein is a biologically active compound and helps in maintaining the immune system of the body. It also helps in maintaining cardiovascular health by diminishing the level of cholesterol. Proteins are obtained from different plants like legumes, cereals, pseudocereals, seeds, and dry fruits (Almeida et al. 2020). Figure 12.3 represents plant protein as a food substitute (Satankar et al. 2021).

Lupin is a type of seed devoid of any starch. It consists of 32.2% of protein, and the oil content is around 5.95% which is very low. Currently, it is preferably used as a supplement because of its simple chemical content and ease of availability worldwide. Soya protein is a good substitute for food for those who are allergic to lactose protein. Soy flour can be used as a good source of proteins as the content of fat and sugar is low, but it has a high content of essential amino acids. It has been seen that the level of cholesterol has diminished after consuming proteins obtained from soybeans. Cereals and pulses also contain an applicable level of essential amino acids. Bakery products are

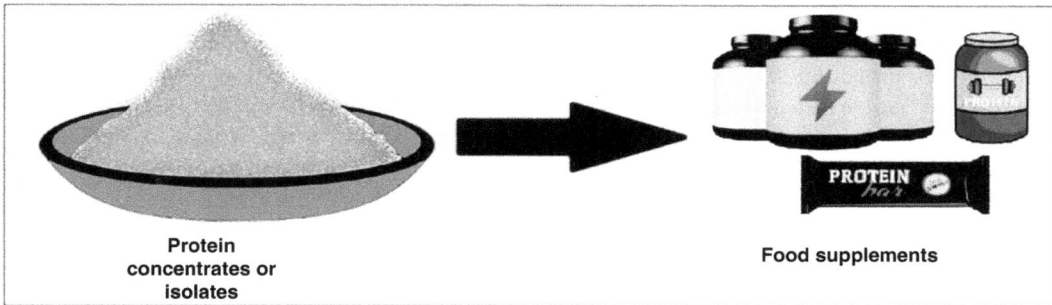

FIGURE 12.3 Plant protein as a food substitute. (Satankar et al. 2021.)

prepared from wheat flour, but when faba bean flour was added, the content of essential amino acids was seen to be escalated. These studies prove that plant proteins increase food value in terms of nutrition (Satankar et al. 2021). Cotton seed consists of protein which has a low gossypol percentage and is hence suitable as a food substitute. Research has been executed on spirulina because it contains a high percentage of carotenoids and amino acids (Grahl et al. 2020).

12.5.5.2 Plant-Derived Proteins as Edible Coating or Films

Due to their advantages over synthetic films, particularly their use as edible packaging materials, protein-based edible films have gained attention in recent years. Additionally, food products that are currently not individually packaged due to practical considerations, such as beans, nuts, and cashew nuts, can be individually wrapped using protein-based edible films. Furthermore, the interfaces between various layers of components in diverse foods can be treated with edible films made of protein. In foods like pizzas, pies, and candies, they can be adjusted to stop the migration of solute. Additionally, edible films made of protein can serve as transporters for antibacterial and antioxidant substances. Thin, covering layers consisting of edible substances that are used to package food are known as edible coatings and films. These films prevent moisture entrapment, oxygen, and solutes while maintaining the flavor and texture of the food product (Hassan et al. 2018). Over time, people began creating various coatings and films to shield food from spoiling. For instance, fruits were wax-coated in the 12th century to reduce the process of water loss, and in the 15th century, a scientist from Japan by the name of Yuba produced the first edible film made from soymilk.

Gelatin and oil were first used to cover meals in the 19th century. To stop oxidation and rancidity, foods like hazelnuts and almonds were used along with sucrose coating at this time (Satankar et al. 2021). Due to its increased concentration of nonpolar amino acids and hydrophobic properties, corn zein is ideally suited for use in the creation of edible films. Zein has more disulfide and hydrogen bonds than other proteins, which gives it superior film-forming and moisture-blocking abilities (Grahl et al. 2020). The hydrophobic globular protein wheat gluten, which is elastic and cohesive, also has great film-forming qualities (Hassan et al. 2018). Wet processing has created edible packaging from ingredients including whey protein isolate and yellow pea protein. FTIR analysis is used to identify the hydrophobic and hydrophilic surfaces of the produced film. It was discovered that yellow pea protein concentrate-based protein films have very few hydrophilic surfaces.

In a different report, mung bean protein was combined with powdered pomegranate peel to create edible films that were high in antioxidants. The films had stronger antioxidant and antibacterial activity together with higher total phenolic content (Moghadam et al. 2020). Polysaccharides are an excellent choice for edible packaging material when it comes to the packaging of edible food. Different polysaccharides like starch and carrageenan were utilized to manufacture edible films and coatings since they are ingredients from natural sources with minimal toxicity and permeability for oxygen and carbon dioxide selectively This reduced the need for conventional plastic packaging.

These qualities of the edible polysaccharide coatings and films can extend the shelf life of fruits (Mohamed et al. 2020). Protein-based edible films may be produced using one of two techniques: Deposition or formation of the surface film. Several integral factors can affect the development of protein-based edible films, like raw material types, pH, Polymer characteristics, concentration, relative humidity, and the addition of film additives. Figure 12.4 represents as a role as an edible coater (Mohamed et al. 2020).

12.5.5.3 Role as Bioactive Peptides

Herbal resources are still expanding, making them more sustainable than ecologically sound resources for the production of bioactive peptides or protein hydrolysates. The multifaceted combination of peptides, oligopeptides besides free amino acids found following the vast or else partial hydrolysis of protein peptide bonds is present in labeled protein hydrolysates. Protein hydrolysates include bioactive peptides (BAPs), which are modest protein sections getting 2–20 amino acid deposits as well as molecular quantities to a lesser extent than 6,000 Da (Bhandari et al. 2020). Numerous procedures have been used to create BAPs, such as (i) hydrolysis out of proteolytic microbes, (ii) hydrolysis beyond digestive enzymes, (iii) hydrolysis across acids or else alkalis, as well as (iv) hydrolysis beyond proteolytic enzymes of herb or microscopic origin. In addition to supplementing as a protein supply, these BAPs are associated with antitumor interest, hypoglycaemic impacts, cholesterol-reducing capability, antithrombotic impacts, antihypertensive consequence, disinfectant properties along with antioxidant interests. These BAPs are released into the blood after proteolytic hydrolysis and stay inactive inside the parent protein. Following assimilation, the BAPs immediately move into the blood, and later, after being captivated and then assimilated through the intestine, confirm their biological influence at the particular location as well as their in-vivo bioavailability capability (Adams et al. 2020). BAPs need the capability to control the renin-angiotensin structure (RAS) and kinin-nitric oxide structure to regulate blood stress in the body. Antihypertensive BAPs have been essentially separated from helianthus, garlic, then barley through using γ-zein, α-lactalbumin along with β-lactoglobulin enzymes. Angiotensin-I converting enzyme (ACE) remains the key element of the RAS, which catalyzes the transformation of the peptide hormone angiotensin-I into effective angiotensin-II. Angiotensin-II then improves blood stress with vasoconstriction. Furthermore, bradykinin created by the kinin-nitric oxide method remains a vasodilator (reduces blood tension). Antihypertensive BAPs regulate blood pressure by reducing ACE. BAPs with C terminal prolines are hypothesized to circuitously supplement nitric oxide synthase by cell signaling or else to exert their NO (nitric oxide)-motivating effect through increased bradykinin bioavailability, whereas proline cutting-edge convinced BAPs to consume a significant ligand-binding component.

Disinfectant BAPs protect animals from a variety of microorganisms, including bacteria, fungi, and protozoans, by differentiating the bacterium without altering the existing tissue. Antimicrobic BAPs comprise a greater amount of amphipathic along with cationic α-helical constructions as well as a lesser number of hydrophobic α-helices. These cationic α-helices relate with the anionic

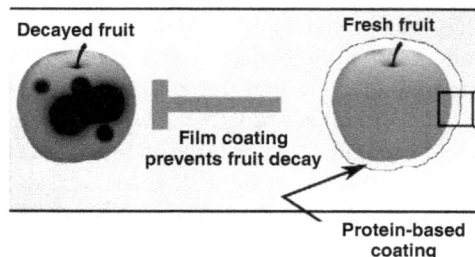

FIGURE 12.4 Role plant-derived protein as an edible coating agent. (Mohamed et al. 2020.)

phospholipids of the microbial membrane by static interfaces. This interaction involves the formation of transmembrane pores, which leads to the emigration of vital intracellular components like ions and biomolecules, impacting an ionic disparity as well as cell lysis. As the membrane of mammalian cells typically consists of phospholipids like sphingomyelin in addition to phosphatidylcholine, these BAPs are specific for the bacterial membrane since they are plentiful in acidic phospholipids (anionic). Furthermore, BAP can build a membrane on the microbial cell and incorporate hooked structures into its form, which is crucial for tissue rupturing and cell death (Adams et al. 2020). Numerous BAPs attach with minerals such as PO_4^{3-} plus Ca^2 on gastric pH + on the way to create a complicated structure that improves inorganic bioavailability and controls metal homeostasis along with advances in the purification of heavy metals. BAPs remain as well common to avoid descent coagulation within the body over their antithrombotic properties. Abnormal fibrinolysis and sensitive platelets combined with elevated concentrations of fibrinogen raise the possibility of coagulation. BAPs raise fibrinolysis and then lessen platelet accumulation (Mohamed et al. 2020). Several antioxidant BAPs get productively separated from flaxseed, soybean, cottonseed, and many other herb-centered resources. These BAPs need the property of sharing within or producing single-electron transmit responses. They comprise amino acids that are able to catalyze antioxidant actions along with transmission electrons at a biological pH (Muhialdin et al. 2020). Histidine, methionine, proline, cysteine, and various aromatic amino acids can chelate scavenged OH radicals or pro-oxidant metal ions, contributing to the antioxidant activity of BAP. Protein separations of pungent beans (Parkia speciosa) have been exposed in the process of fermentation along with simmering, as well as the BAPs produced remained assessed in support of their antiseptic as well as antioxidant actions. Fermentation drastically enhanced the antiseptic as well as antioxidant actions owing to the existence of small-molecular-weight BAPs (Muhialdin et al. 2020).

Various findings need to be reflected by the antitumor plus antineoplastic properties of lunatic, a BAP separated from cereal granules and then soybean. These BAPs are extremely useful compared to virus-related transforming genes along with chemically stimulated malignancies through reducing histone acetylation out of reticence of the enzyme histone acetyltransferase, thus suppressing melanoma cell development. Lunasin in transgenic soybean herbs demonstrated solid antioxidant, anti-incendiary, as well as antineoplastic actions as well as silenced the presence of pro-inflammatory cytokines such as interleukin-6 then -1. Likewise, BAPs inaccessible after the protein hydrolysates of amaranth seeds remained exposed to display robust antineoplastic actions through encouraging membrane breakdown, DNA disintegration, caspase-3 initiation, and phosphatidylserine translocation in breast cancer cells, subsequently in programmed cell death then antimigratory belongings. In a related study, BAPs made from wheat germ using the enzymes alcalase and pepsin had anti-cancer, ACE-inhibitory, and antioxidant activities (Moghadam et al. 2020).

It has been revealed that BAPs separated from cumin, buckwheat, and soybean have hypolipidemic as well as hypocholesterolemia properties. Low-density lipoprotein (LDL) is decreased by the subunit of the soybean 7S globulin protein by boosting the reaction of LDL receptors in refined hepatocytes (Moghadam et al. 2020). BAPs CSP1, CSP2, as well as CSP3 separated from cumin plant seeds inhibited the development of cholesterol micelles and reduced lipase movement by intermingling along with bile acids, leading to a hypocholesterolemia impact (Hassan et al. 2018). Lupin- then soy-derived proteins have been developed to decrease blood cholesterol by preventing the action of HMG CoA reductase, a rate-restricting enzyme for cholesterol biogenesis (Sim et al. 2021). Still, although the international marketplace for bioactive peptides remains gradually increasing owing to user recognition of a direct association between efficient diets and physical condition, systematic verification is necessary for the effectiveness along with protection of BAPs.

Hydrolysis affects the development of lesser protein sections along with that of fascinating physicochemical as well as useful assets. Such fundamental alterations involve a decrease in molecular mass and growth in the field of hydrophilicity owing to the growth in polar groups ($-NH_4^+$, $-CO_2^-$), as well as variation within their molecular association. Vegetable resources such as renewable as well as ecologically viable resources remain developing on the way to generate protein hydrolysates before bioactive peptides. BAPs control several metabolic activities in the interior of the body. They can be treated like nutraceuticals, medications, or else practical diet components intended for increasing individual well-being as well as avoiding ailments.

12.5.5.4 Role as Emulsifiers

An emulsion comprises two or more immiscible liquids in which one liquid diffuses into the other having a very small droplet size. There are two classes of emulsions—oil-in-water and water-in-oil. Examples of oil-in-water emulsions are mayonnaise, salad dressings, and milk, and those of water-in-oil emulsions are butter and margarine. They are not stable thermodynamically, and the phases can get separated because of high surface energy (Jiang et al. 2020). Emulsions that are food-based can be equilibrated by proteins by diminishing the surface tension between the phases. These proteins also have a food value and hence can be used to prepare equilibrated food emulsions, especially Pickering-emulsions (Jiang et al. 2020). They may be stable because of the greater adsorption of protein particles. This makes them very suitable for the encapsulation and protection of bioactive molecules and enzymes. Plant proteins can be used for the equilibration of food emulsion. They help in the formulation of margarine, discourage oxidation of lipid, and also enable the encapsulation of biologically active compounds. The solubility of cereals is very low, and so they cannot be effectively used as a food emulsifier. So, researchers have tried to modify these cereal proteins to make them suitable as an emulsifier. Nanoparticles of soya protein were contrived by the crosslinking method using glutaraldehyde to make them suitable as food emulsifiers. Lupin seed protein was able to stabilize Pickering emulsions. It had good stability and was capable enough to maintain the surface tension between the two phases (Liu et al. 2021). Peas, potatoes, and rice protein are also capable to act as a good stabilizers. Emulsions contrived from soya, whey, and gum arabic ensure stable and fabricated traits. Pea protein emulsions were also checked for their oxidating and surface tension stability. Figure 12.5 shows role of an emulsifier.

Research was executed on camellia and almond to evaluate the oxidating solidity of emulsions of walnut. These emulsions exhibited firmness and low oxidation of lipids. Proteins from grape seeds and peas are used to prepare stable oil-in-water emulsions and their stability was checked (Liu et al. 2021). They readily form a stable complex because of the formation of hydrogen bonds.

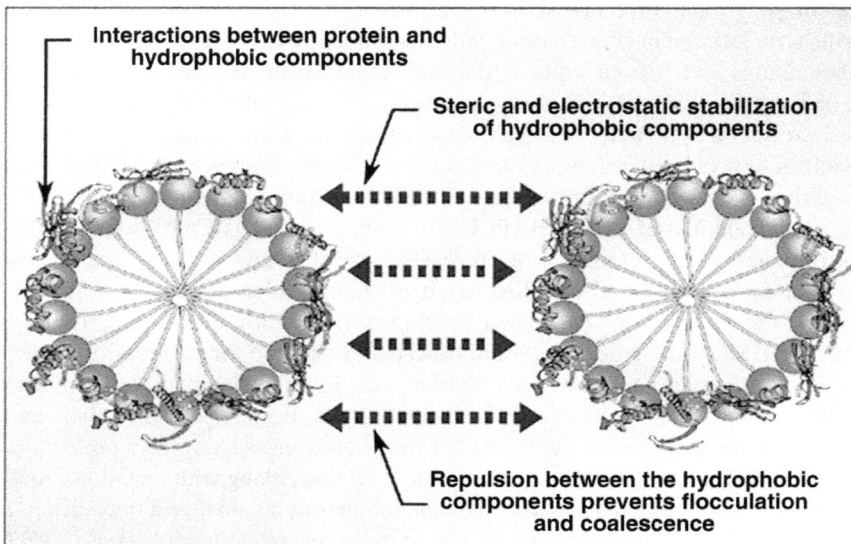

FIGURE 12.5 Role of protein as an emulsifier. (Burger & Zhang 2019.)

12.5.5.5 Role as Hydrogels

Hydrogels possess a three-dimensional structure. They are hydrophilic in nature. They chemically have crosslinked polymeric mesh works. Hydrogels can absorb water or biological fluids 1,000 times their dry weight (Katyal et al. 2020). Hydrogels, which are protein- and peptide-based, can produce responses to some external signals. The response depends on light, temperature, pH, electric field, and the presence of other molecules. These components can cause an alteration of structure chemically. These are also related to the controlled release of encapsulated substances within hydrogels. Hydrogels can be used in food additives, cell encapsulation, regenerative medicines, tissue engineering, biosensors, regulation of biological adhesions, barrier compounds, cellular immobilization, some diagnostic purposes, biomedical implants, pharmaceuticals, and so on. For the manufacturing of hydrogels, proteins are successfully used over carbohydrates. For the development of cold- and heat-set hydrogels, proteins that are globular in nature can be used as food. Hydrogels are with advanced functional properties and amphiphilic nature, and their nutritive values are very high. Soybean and whey protein are widely used to synthesize mixed hydrogels (Katyal et al. 2020). Corn fiber gum and soybean protein isolate can form a double-network hydrogel which is required to deliver thermolabile bioactive compounds like riboflavin. Pure soybean protein isolates are also used to produce hydrogels. But this type of hydrogel poorly transports riboflavin due to its high swelling ratio, and it is also degraded by pepsin. But it is potentially used in the food industry as an oral delivery medium. Thermolabile bioactive compounds are suitable for that purpose (Yan et al. 2020). Hydrogels having superior mechanical properties can be synthesized by the chemical cross-linking method. It is superior to the hydrogels formed by physical crosslinking. Chemically cross-linked hydrogels are immunogenic, and the materials used for cross-linking can produce toxicity (Yan et al. 2020). Crosslinking is the main parameter to determine water insolubility, physical integrity, and mechanical strength of the hydrogels. Hydrogels maintain the equilibrium state through their mechanical strength, diffusion, and internal transport. The study emphasizes establishing protein-based and peptide-based hydrogels due to their biodegradable and biocompatible nature. More research can be done to remove heavy metal toxicity and use it in controlled drug delivery systems by removing. Figure 12.6 represents the application of plant protein for industrial purpose.

12.5.6 As Adhesives

Scientists are looking for nonfood applications because the protein contains nonprotein contaminants and antinutritional elements. Researchers have shown how to employ bovine plasma hydrolysates to create films using proteins from soybean and sunflower. In recent years, the wood, plywood, and particle board industries have used cottonseed protein, or water-washed cottonseed meal, as an environmentally acceptable wood glue (Taniya et al. 2020). In addition, proteins possess specific

Food Substitute	Edible coating or films	As emulsifier	As hydrogels	As bioactive peptides

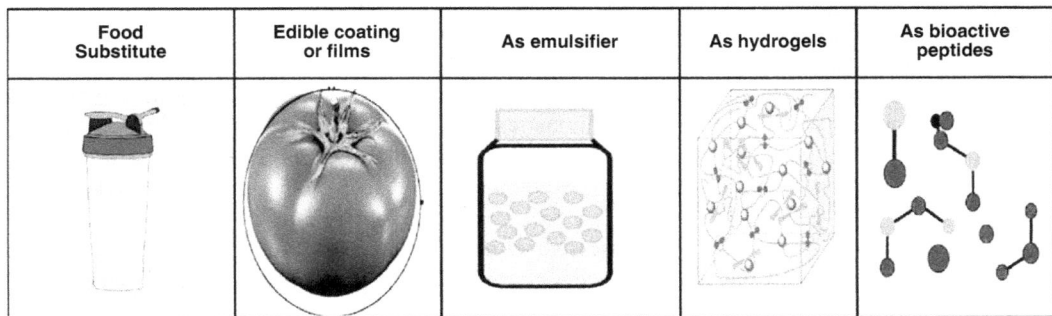

FIGURE 12.6 The application of plant protein for industrial purpose.

features that make them acceptable for use as an alternative to harmful chemical-based adhesives. Protein molecules are denatured, which causes them to unfold, exposing more hydrophobic groups to ward off water infiltration, enhancing the effectiveness of protein-based adhesives. Protein molecules' active groups make it easier for them to react with cross-linking agents to create cross-linking structures, which enhance their characteristics and effectiveness as adhesives.

Also, the unfolding of the protein structure increases the exposure of hydrophilic groups, which helps the reaction between the protein and the cross-linking agent produce a dense cross-linking structure that quickens the sticky characteristics (Tomar et al. 2021). Currently employed in the creation of wood adhesives, formaldehyde-based resins are cancer-causing and made from petroleum. There has been a trend in recent years toward the creation of superior, environmentally friendly adhesives, particularly those derived from renewable resources. Protein-based wood adhesives are extremely sustainable despite having subpar adhesive qualities. Several modifiers, including sodium dodecyl sulfonate, urea, guanidine hydrochloride, carboxylic acids, lignin, and phosphoric acid, have been used to enhance the quality and performance of these materials (Mir et al. 2020).

The benefit/cost ratio of wood in the plywood business is improved by enhancements in the quality and effectiveness of protein for adhesive production. Following the extraction of oil from their seeds, the principal by-product, press cake, contains a significant amount of camelina protein. The use of polymeric amine epichlorohydrin, a cationic polyelectrolyte, enhances the adhesive capabilities of these proteins. Small proteins are produced as a result of the treatment; however, their solubility is restricted due to their enhanced hydrophobicity.

In testing on two layers of cherry wood, the modified camelina protein outperformed the unmodified camelina protein by factors of 2 and 6, respectively, for both dry and wet adhesion bond strength. Another study compared the qualities of cottonseed protein with those of protein derived from soybeans when cottonseed protein was employed as a wood glue. Cottonseed proteins were discovered to have greater shear strength and hot-water resistance than soy proteins (Nasrabadi et al. 2019).

12.5.7 IN THE DAIRY INDUSTRY

12.5.7.1 Double-Protein Yogurt

After rapid homogenization, sterilization, and fermentation, raw cow (goat) milk or milk powder is transformed into yogurt, a quick acid-producing product. Plant protein serves as the primary nutritional foundation for double-protein yogurt. It has a distinctive flavor and a high nutritional value thanks to probiotic fermentation, which helps to promote nutrition and human health. Streptococcus thermophilus and Lactobacillus bulgaricus are currently the most commonly used starters in dairy production firms for making yogurt. Yogurt starters made from plant proteins have been the subject of further investigation in recent years. One or more functional strain combinations may be used in the fermentation of acceptable strains of plant-based yogurt (Mir et al. 2020). For the fermentation of milk and soymilk, Streptococcus thermophilus, Lactobacillus bulgaricus, and Lactobacillus acidophilus were combined. It was discovered that the acidity of the fermented milk was higher than that of the fermented soymilk. Research demonstrated that the ideal starter for soy protein yogurt differed slightly from the starter for regular milk yogurt (Muhialdin et al. 2020).

12.5.7.2 Double-Protein Beverages

People are becoming more and more obsessed with health as science and technology advance. The evolution of beverages has reached a new phase, moving from the initial scale expansion to quality improvement. As a result, the market share of carbonated beverages has continued to fall, while the consumption of natural and healthy beverages such as tea beverages, juices made from fruits and vegetables, and beverages containing plant proteins has increased. Now, plant-based cereal and nut beverages are the newcomers to the dairy sector, but creating new dairy products from cereals and nuts has certain technical obstacles. Certain grains and nuts are higher in carbohydrates and fiber than natural milk. Since milk beverages have weak suspension stability, stratified precipitation and particle suspension are common occurrences. The flavors of beverage products are also delicate and bitter. Researchers have recently carried out related research to address these issues, including the addition of thickeners and stabilizers to stabilize product quality, the addition of flavorings to enhance flavor, the enriching and strengthening of nutrients to enhance nutritional value, etc. Yang et al. investigated oat milk beverages' emulsion stabilizer composition (Yang et al. 2018)

12.5.7.3 Double-Protein Cheese

Goats and cow milk are both used to make cheese. When the proper quantities of starter and rennet are added, the protein can coagulate, release some of the whey, and eventually ferment and mature over time. Cheese absorbs and is used more readily when it is in the fermentation stage because proteins and lipids are enzymatically broken down into little molecules that are easily absorbed in the human digestive system. As a result, it is known as milk gold in the sector. In recent years, mixed cheese—a type of cheese manufactured by substituting plant protein for some animal milk's protein—has entered the market. Cheese's nutritional value can be increased while production costs are decreased by using plant protein to partially replace animal protein. The experiment of partially substituting animal protein with plant protein has emerged as a new research and development trend in the cheese industry (Yildiz et al. 2017) The most-researched double-protein cheese of them all is blended soybean cheese.

12.5.7.4 Double-Protein Milk Replacer for Calves

At 10 days after birth, the calves should be switched from receiving ordinary milk to a milk replacer (also known as artificial milk) to hasten weaning. Currently, the global dairy farming industry uses milk replacers frequently as a technical tool to develop and implement the early weaning technology of sucking calves. Dairy by-products such as skim milk, whey protein concentrate, and dry whey make up the majority of the raw materials used to make milk substitutes (Zhang et al. 2019). Low-cost, high-quality plant protein has emerged as the primary research area for the development of milk replacement protein sources as a result of increased research and technological advancements in milk replacement processing. High-quality plant protein and a large amount of milk protein both affect calf diarrhea prevention and reduction, as well as calf daily weight gain. Saving on feeding expenses has produced positive economic results (Zhang et al. 2020). Soybean, wheat, rice, and other plant proteins are the most often utilized plant proteins in calf milk substitutes.

12.5.8 REFINERY OF SEED STORAGE PROTEINS

There is an increasing demand to employ abundantly plentiful and reasonably cheap seed storage proteins as a nutritious component in food and feed. Getting functional proteins following the refinery processes is this protein source's main challenge. Because of the challenges in resolubilizing the protein and the unpredictable caking of the powder during storage and transit, functional loss is most common when the protein is utilized as a powdered ingredient (Stephen & Phillips 2016).

12.5.9 NANOPARTICLES

Enzymatic insertion of lipidic groups such as tiny fatty acids or PEGylation has been used to encapsulate bioactive compounds, usually, as microemulsions, to control the characteristics of protein-based nano-particles. These products are regarded as food-grade or have undergone testing and usage that have received medical–ethical approval. There are no signs that these uses have entered the market for food and/or feed products (Van Ba et al. 2012).

TABLE 12.2
Summary of the Application of Modified Proteins

1. Protein–polysaccharide complex
2. Formation of the fibrous structure of the protein
3. Formation of gel-like structure
4. Plant protein flavors
5. Industrial use
    a. As food substitutes
    b. Plant-derived proteins as edible coating or fibers
    c. Role as bioactive peptides
    d. Role as emulsifiers
    e. Role as hydrogels
6. As adhesives
7. In the dairy industry
    a. Double-protein yogurt
    b. Double-protein beverages
    c. Double-protein cheese
    d. Double-protein milk replacer for calves
8. Refinery of seed storage proteins
9. Nanoparticles
10. Potential in the area of food product sustainability

12.5.10 POTENTIALS IN THE AREA OF FOOD PRODUCT SUSTAINABILITY

The demand for protein as a food and nutritional component is expanding quickly in response to the growing global population and standard of living. A more sustainable economy has also increased pressure on innovations in the production, refinement, and use of proteins from sources other than those found in the existing economy. This is due to the recognition of proteins as building blocks in nonfood applications. The creation of a vision for food quality, and in particular the importance of nutritional effect, must be viewed from the perspective of the fundamental standards established by participating industries as well as by consumer desires (Wen et al. 2020). An illustration of this would be the "Pyramid of Food Innovation."

12.6 CONCLUSION

This review chapter has focused on plant protein-based technology, the development of plant protein ingredients, and future foods based on plant protein. Foods must be nutritious. The areas for further improvement include the extraction of plant protein, fractionation, and its modification. Plant proteins are becoming fast-growing in the food industry. It is advantageous for its nutritious value to fulfill the demand of the increasing population, the demand for a variety of foods, and demand for hybrid and healthier products. Plant proteins are inferior to animal-based proteins. They are affected by temperature, pH, and ionic strength. They have antinutritional compounds. So, a huge modification is required to successfully process plant protein-based foods. The advantages, as well as the challenges, were discussed in this chapter.

At the industrial level, physical and biological methods can be used for the modification of plant-based proteins for providing increased benefits with affordable costs. There is a requirement to understand the plant protein–polysaccharide interactions and plant protein-generated flavors. More research is required to develop different structuring techniques and improve nutritional value. It is also required to assess the biosafety of plant protein-based foods. This point should be emphasized that we eat the whole food and not a particular component from the food, so safety study for other ingredients is also required.

REFERENCES

Adams, C., Sawh, F., Green-Johnson, J. M., Taggart, H. J., & Strap, J. L. 2020. "Characterization of casein-derived peptide bioactivity: Differential effects on angiotensin-converting enzyme inhibition and cytokine and nitric oxide production." *Journal of Dairy Science* 103 (7): 5805–5815. https://doi.org/10.3168/jds.2019-17976.

Almeida Sá, A. G., Franco Moreno, Y. M., & Mattar Carciofi, B. A. 2020. "Plant proteins as a high-quality nutritional source for the human diet." *Trends in Food Science & Technology* 97: 170–184.

Aryee, A. N. A., Agyei, D., &Udenigwe, C. C. 2018. "The impact of processing on the chemistry and functionality of food proteins." In R. Y. Yada (Ed.), *Proteins in food processing* (2nd ed.). Woodhead Publishing.

Bhandari, D., Rafiq, S., & Gat, Y. 2020. "A review on bioactive peptides: Physiological functions, bioavailability and safety." *International Journal of Peptide Research and Therapeutics* 26: 139–150. https://doi.org/10.1007/s10989-019-09823-5.

Burger, T. G., & Zhang, Y. 2019. "Recent progress in the utilization of pea protein as an emulsifier for food applications." *Trends in Food Science & Technology* 86: 25–33. https://doi.org/10.1016/j.tifs.2019.02.007.

Calicioglu, O., Flammini, A., Bracco, S., Bellù, L., & Sims, R. 2019. "The future challenges of food and agriculture: An integrated analysis of trends and solutions." *Sustainability* 13 (11): 8012–8025. https://doi.org/10.3390/su11010222.

Chiang, J. H., Tay, W., Ong, D. S. M., Liebl, D., Ng, C. P., & Henry, C. J. 2021. "Physicochemical, textural, and structural characteristics of wheat gluten-soy protein composited meat analogs prepared with the mechanical elongation method." *Food Structure* 28: 100183.

Devi, G. N., Padmavathi, G., Babu, V. R., & Waghray, K. 2015. "Proximate nutritional evaluation of rice (Oryza sativa L.)." *Journal of Rice Research* 8 (1): 23–32.

García Arteaga, V., Apestegui Guardia, M., Muranyi, I., Eisner, P., & Schweiggert- Weisz, U. 2020. "Effect of enzymatic hydrolysis on molecular weight distribution, techno-functional properties, and sensory perception of pea protein isolates." *Innovative Food Science & Emerging Technologies* 65: 102449.

Gharibzahedi, S. M. T., & Smith, B. 2021. "Effects of high hydrostatic pressure on the quality and functionality of protein isolates concentrate, and hydrolysates derived from pulse legumes: A review." *Trends in Food Science and Technology* 107: 466–479.

Gilbert, J. A., Bendsen, N. T., Tremblay, A., & Astrup, A. 2011. "Effect of proteins from different sources on body composition." *Nutrition, Metabolism, and Cardiovascular Diseases* 21: B16–B31.

Grahl, S., Strack, M., Mensching, A., & Mörlein, D. 2020. "Alternative protein sources in Western diets: Food product development and consumer acceptance of spirulina-filled pasta." *Food Quality and Preference*, Article 103933.

Hadidi, M., Ibarz, A., & Pouramin, S. 2021. "Optimization of extraction and deamidation of edible protein from evening primrose (Oenothera biennis L.) oil processing by-products and its effect on structural and techno-functional properties." *Food Chemistry* 334: 127613.

Hassan, B., Chatha, S. A. S., Hussain, A. I., Zia, K. M., & Akhtar, N. 2018. "Recent advances on polysaccharides, lipids, and protein-based edible films and coatings: A review International." *Journal of Biological Macromolecules* 109: 1095–1107.

Hinderink, E. B. A., Schröder, A., Sagis, L., Schroën, K., & Berton-Carabin, C. C. 2021. "Physical and oxidative stability of food emulsions prepared with pea protein fractions." *LWT* 146: 111424.

Jiang, H., Sheng, Y., & Ngai, T. 2020. "Pickering emulsions: Versatility of colloidal particles and recent applications." *Current Opinion in Colloid & Interface Science* 49: 1–15.

Jiménez-Munoz, L. M., Tavares, G. M., & Corredig, M. 2021. "Design future foods using plant protein blends for best nutritional and technological functionality." *Trends in Food Science & Technology* 113: 139–150.

Katyal, P., Mahmoudinobar, F., & Montclare, J. K. 2020. "Recent trends in peptide and protein-based hydrogels." *Current Opinion in Structural Biology* 63: 97–105.

Kumar, A., Nayak, R., Purohit, S. R., & Rao, P. S. 2020. "Impact of UV-C irradiation on the solubility of Osborne protein fractions in wheat flour." *Food Hydrocolloids* 105845.

Kumar, M., Tomar, M., Potkule, J., Verma, R., Punia, S., Mahapatra, A., Belwal, T., Dahuja, A., Joshi, S., & Berwal, M. K. 2021. "Advances in the plant protein extraction: Mechanism and recommendations." *Food Hydrocolloids* 115: 106595.

Kyriakopoulou, K., Keppler, J. K., & van der Goot, A. J. 2021. "The functionality of ingredients and additives in plant-based meat analogs." *Foods* 10: 600.

Lehikoinen, E., & Salonen, A. O. 2019. "Food preferences in Finland: Sustainable diets and their differences between groups." *Sustainability* 11 (5): 1259.

Liu, J., Zhou, H., Tan, Y., Mundo, J. L. M., & McClements, D. J. 2021. "Comparison of plant-based emulsifier performance in water-in-oil-in-water emulsions: Soy protein isolate, pectin and gum Arabic." *Journal of Food Engineering* 307: Article 110625.

Liu, S. X., Chen, D., & Xu, J. 2022. "Physiochemical properties of jet-cooked amaranth and improved rheological properties by processed oat bran." *Future Foods* 5: 100107.

Margolis, G., Myers, S., & Newbold, D. 2019. "Manufacturing of plant-based yogurt." *U.S. Patent* 16: 282, 293, 12 September.

Matsuyama, S., Kazuhiro, M., Nakauma, M., Funami, T., Nambu, Y., Matsumiya, K., & Matsumura, Y. 2021. "Stabilization of whey protein isolate-based emulsions via complexation with xanthan gum under acidic conditions." *Food Hydrocolloids* 111: 106365.

Mattila, P., Mäkinen, S., Eurola, M., Jalava, T., Pihlava, J.M., & Hellström, J. 2018. "Nutritional value of commercial protein-rich plant products." *Plant Foods for Human Nutrition* 73 (2): 108–115.

McClements, D. J., & Grossmann, L. 2021. "The science of plant-based foods: Constructing next-generation meat, fish, milk, and egg analogs." *Comprehensive Reviews in Food Science and Food Safety* 20: 4049–4100. https://doi.org/10.1111/1541-4337.12771.

Millward, D. J., Layman, D. K., Tomé, D., & Schaafsma, G. 2008. "Protein quality assessment: Impact of expanding understanding of protein and amino acid needs for optimal health." *American Journal of Clinical Nutrition* 87 (5): 1576S–1581S.

Mir, N. A., Riar, C. S., & Singh, S. 2020. "Structural modification in the album (Chenopodium album) protein isolates due to controlled thermal modification and its relationship with protein digestibility and functionality." *Food Hydrocolloids* 103: 105708.

Moghadam, M., Salami, M., Mohammadian, M., Khodadadi, M., & Emam-Djomeh, Z. 2020. "Development of antioxidant edible films based on mung bean protein enriched with pomegranate peel." *Food Hydrocolloids* 104: 105735.

Mohamed, S. A. A., El-Sakhawy, M., & El-Sakhawy, M. A.-M. 2020. "Polysaccharides, protein and lipid-based natural edible films in food packaging: A review." *Carbohydrate Polymers* 238: 116178. https://doi.org/10.1016/j.carbpol.2020.116178.

Muhialdin, B. J., Rani, N. F. A., & Hussin, A. S. M. 2020. "Identification of antioxidant and antibacterial activities for the bioactive peptides generated from bitter beans (Parkia speciosa) via boiling and fermentation processes." *LWT*, Article 109776. https://doi.org/10.1016/j.lwt.2020.109776.

Muneer, F., Johansson, E., Hedenqvist, M. S., Plivelic, T. S., & Kuktaite, R. 2019. "Impact of pH modification on protein polymerization and structure–function relationships in potato protein and wheat gluten composites." *International Journal of Molecular Sciences* 20 (1): 58.

Nasrabadi, M. N., Goli, S. A. H., Doost, A. S., Dewettinck, K., & Van der Meeren, P. 2019. "Bioparticles of flaxseed protein and mucilage enhance the physical and oxidative stability of flaxseed oil emulsions as a potential natural alternative for synthetic surfactants." *Colloids and Surfaces B: Biointerfaces* 184: 110489.

Nikbakht, M., Doost, A., &Mezzenga, Raffael. 2021. "Modification approaches of plant-based proteins to improve their techno-functionality and use in food products." *Food Hydrocolloids* 118: 106789. https://doi.org/10.1016/j.foodhyd.2021.106789.

Pojić, M., Mišan, A., & Tiwari, B. 2018. "Eco-innovative technologies for extraction of proteins for human consumption from renewable protein sources of plant origin." *Trends in Food Science and Technology* 75: 93–104.

Rahman, M. M., & Lamsal, B. P. 2020. "Ultrasound-assisted extraction and modification of plant-based proteins: Impact on physicochemical, functional, and nutritional properties." *Comprehensive Reviews in Food Science and Food Safety* 20: 1457–1480.

Satankar, V., Singh, M., Mageshwaran, V., Jhodkar, D., Changan, S., Kumar, M., & Mekhemar, M. 2021. "Cottonseed kernel powder as a natural health supplement: An approach to reduce the gossypol content and maximize the nutritional benefits." *Applied Sciences* 11 (9): 3901.

Sha, L., & Xiong, Y. L. 2020. "Plant protein-based alternatives of reconstructed meat: Science, technology, and challenges." *Trends in Food Science and Technology* 102: 51–61.

Shouqin, Z., Jun, X., & Changzheng, W. 2005. "High hydrostatic pressure extraction of flavonoids from propolis." *Journal of Chemical Technology & Biotechnology* 80: 50–54.

Sim, S. Y. J, Akila, S. R. V., Chiang, J. H., & Henry, C. J. 2021. "Plant proteins for future foods: A roadmap." *Foods* 10 (8): 1967. https://doi.org/10.3390/foods10081967.

Stephen, A. M., & Phillips, G. O. 2016. *Food polysaccharides and their applications*. CRC Press.

Taniya, M. S., Reshma, M. V., Shanimol, P. S., Krishnan, G., & Priya, S. 2020. "Bioactive peptides from amaranth seed protein hydrolysates induced apoptosis and anti-migratory effects in breast cancer cells." *Food Bioscience* 35: 100588. https://doi.org/10.1016/j.fbio.2020.100588.

Tomar, M., Bhardwaj, R., Kumar, M., Singh, S. P., Krishnan, V., Kansal, R., & Sachdev, A. 2021. "Nutritional composition patterns and application of multivariate analysis to evaluate indigenous Pearl millet ((Pennisetum glaucum (L.) R. Br.) germplasm." *Journal of Food Composition and Analysis* 103: 104086.

Torres-León, C., Ramírez-Guzman, N., Londoño-Hernandez, L., Martinez-Medina, G. A., Díaz-Herrera, R., Navarro-Macias, V., Alvarez-Pérez, O. B., Picazo, B., Villarreal-Vázquez, M., Ascacio-Valdes, J., & Aguilar, C. N. 2018. "Aguilar food waste, and by-products: An opportunity to minimize malnutrition and hunger in developing countries." *Frontiers in Sustainable Food Systems* 2: 52.

Van Ba, H., Touseef, A., Jeong, D., & Hwang, I. 2012. "Principle of meat aroma flavors and future prospects." In I. Akyar (Ed.), *Latest research into quality control*. Intech Open Access Publisher.

Varadavenkatesan, T., Pai, S., Vinayagam, R., Pugazhendhi, A., & Selvaraj, R. 2021. "Recovery of value-added products from wastewater using aqueous two-phase systems–a review." *Science of the Total Environment* 778: 146293.

Wen, C., Zhang, J., Qin, W., Gu, J., Zhang, H., & Duan, Y. 2020. "Structure and functional properties of soy protein isolate-lentinan conjugate obtained in Maillard reaction by slit divergent ultrasonic assisted wet heating and the stability of oil-in-water emulsions." *Food Chemistry* 331: 127374.

Willett, W., Rockström, J., Loken, B., Springmann, M., Lang, T., Vermeulen, S., Garnett, T., Tilman, D., DeClerck, F., & Wood, A. 2019. "Food in the anthropocene: The eat–lancet commission on healthy diets from sustainable food systems." *Lancet* 393: 447–492. https://doi.org/10.1016/S0140-6736(18)31788-4.

Xing, Q., Dekker, S., Kyriakopoulou, K., Boom, R. M., Smid, E. J., & Schutyser, M. A. I. 2020. "Enhanced nutritional value of chickpea protein concentrate by dry separation and solid-state fermentation." *Innovative Food Science & Emerging Technologies* 59: 102269.

Yan, W., Zhang, B., Yadav, M. P., Feng, L., Yan, J., Jia, X., & Jin, L. 2020. "Corn fiber gum-soybean protein isolates double network hydrogel as oral delivery vehicles for thermosensitive bioactive compounds." *Food Hydrocolloids*, Article 105865.

Yang, J., Liu, G., Zeng, H., & Chen, L. 2018. "Effects of high-pressure homogenization on Faba bean protein aggregation in relation to solubility and interfacial properties." *Food Hydrocolloids* 83: 275–286.

Yildiz, G., Andrade, J., Engeseth, N. E., & Feng, H. 2017. "Functionalizing soy protein nano-aggregates with pH-shifting and mano-thermo-sonication." *Journal of Colloid and Interface Science* 505: 836–846.

Zhang, J., Liu, L., Liu, H., Yoon, A., Rizvi, S. S. H., & Wang, Q. 2019. "Changes in conformation and quality of vegetable protein during the texturization process by extrusion." *Critical Reviews in Food Science and Nutrition* 59 (20): 3267–3280.

Zhang, J., Wen, C., Zhang, H., Duan, Y., & Ma, H. 2020. "Recent advances in the extraction of bioactive compounds with subcritical water: A review." *Trends in Food Science and Technology* 95: 183–195.

13 Packaging: A Legal Aspect of Plant-Based Protein

Nilushni Sivapragasam

13.1 INTRODUCTION

Packaging can be simply defined as the material that is used for protecting the contents from various damages and delivering the contents to the consumers, intact (Ivanković et al., 2017). Especially, food packaging warrants attention due to the growing global population and the complexity of food processing. A major part of the food packaging materials is derived from petroleum sources (Ivanković et al., 2017; Shaikh et al., 2021). These sources are nonrenewable and can cause threat to the environment due to non-biodegradability and bio-incompatibility (Ivanković et al., 2017). Thus, the last two decades have focused on biodegradable plastics for packaging.

The American Society for Testing and Materials (ASTM) defines the biodegradable plastics as "a plastic that undergoes degradation by biological processes during composting to yield carbon dioxide (CO_2), water (H_2O), inorganic compounds, and biomass at a rate consistent with other known compostable materials and that leaves no visible, distinguishable, or toxic residue." These biodegradable materials are called third-generation materials, which are extracted and/or isolated directly from the biomass, synthesized using bio-monomers, or extracted from naturally occurring or genetically modified microorganisms (Ivanković et al., 2017). In light of such fact, globally, 2.11 million tons of bioplastics were produced in the year 2019 and are anticipated to reach 2.43 million tons by the year 2024 (Shaikh et al., 2021). Unlike conventional packaging that uses the petroleum-based plastics, the biodegradable packaging carries many advantages; the conventional packaging mainly focuses on preventing the contents from physical damage, chemical contamination, and microbial degradation (Young et al., 2020). The biodegradable packaging comprises bio-active compounds, which adds multi-layer protection such as, but not limited to, oxygen scavenging, light barrier, gas barrier, moisture barrier, and microbial suppression (Panja et al., 2018). These properties of biodegradable packaging not only protect the contents from physical and chemical deterioration but also increase the shelf life and improve the quality of the products– which positively change the value equation of the freshness index and the safety of the food products (Young et al., 2020; Panja et al., 2018). Owing to the innate properties, the biodegradable packaging is currently called as active and intelligent packaging; the active packaging provides multiple barriers by interacting with the internal and the surrounding environment of the packaging while the intelligent packaging communicates the freshness and the quality of the packaged contents to the consumer (Panja et al., 2018; Dutta & Sit, 2022). Collectively, the biodegradable packaging benefits the end user and at the same time protects the environment from contamination.

13.2 PLANT-BASED PROTEIN IN PACKAGING

The biodegradable materials extracted from biomass include polysaccharides, proteins, and lipids (Chen et al., 2019). Proteins have attracted a lot of interest among the three macromolecules due to their unique structure, multiple functionalities, and the nutrition-value they could add to the contents (Mohanty et al., 2020). The proteins for biodegradable packaging can be derived from plant (cereals, legumes, oilseeds, algae) or animal sources (dairy and other animals) (Mihalca et al., 2021). Nevertheless, this chapter will focus, exclusively, on the proteins extracted from the plant

DOI: 10.1201/9781003369790-13

sources for synthesizing biodegradable packaging materials followed by their economic feasibility and legal aspects.

The protein-based packaging material can be synthesized as films or coatings. The films are stand-alone structures, which are synthesized through wet (solution casting) or dry process (extrusion, compression molding, injection molding) (Amin et al., 2021; Zink et al., 2016). On the other hand, unlike films, the coatings directly adhere on to the food products, and these are formed by either dipping the contents in the film-forming solution or spray dying the film-forming solution (Amin et al., 2021; Zink et al., 2016). Moreover, incorporating proteins as edible coatings can act as supplements because of the amino acids present in the coating. Besides, unlike other macromolecules, carbohydrates, and lipids, proteins can form better biodegradable packaging due to the enhanced mechanical stability and barrier properties (Senthilkumaran et al., 2022). This is owed to the unique structure of the proteins—which facilitates intermolecular bonding and hence increases the crosslinking; high degree of intermolecular interactions is critical to form intact films with better barrier properties (Bourtoom, 2009). The amino acids which are building blocks of proteins determine the structure-activity of the protein-based films. On the same token, the functional groups of amine, carboxyl, and sulfide contribute toward crosslinking in protein films (Senthilkumaran et al., 2022; Bourtoom, 2009). In addition, the proteins in nature exist as fibrous proteins or globular proteins; most of the plant proteins are globular while the animal-sourced proteins are fibrous (Said & Sarbon, 2019). Overall, the nature of the protein determines the physical and chemical properties of the films derived from protein sources. Figure 13.1 shows various plant-derived protein sources that can be utilized in biodegradable packaging.

Although protein-based films have better mechanical, optical barrier, and oxygen barrier properties, they do exhibit poor water vapor permeability due to its sensitivity to moisture (Zink et al., 2016). Thus, the protein-based films should be incorporated with hydrophobic materials to increase the stability toward water vapor (Rajeshkumar et al., 2022). In addition, the functionality of the proteins is influenced by pH, dielectric permittivity of the organic solvent, temperature, pressure, shear forces, ultrasound, and additives (Zink et al., 2016). Change in the structure and hence the functionality of the protein directly impacts the quality of the film (Hadidi et al., 2022). Therefore, synthesizing protein-based films with better properties warrants optimized conditions.

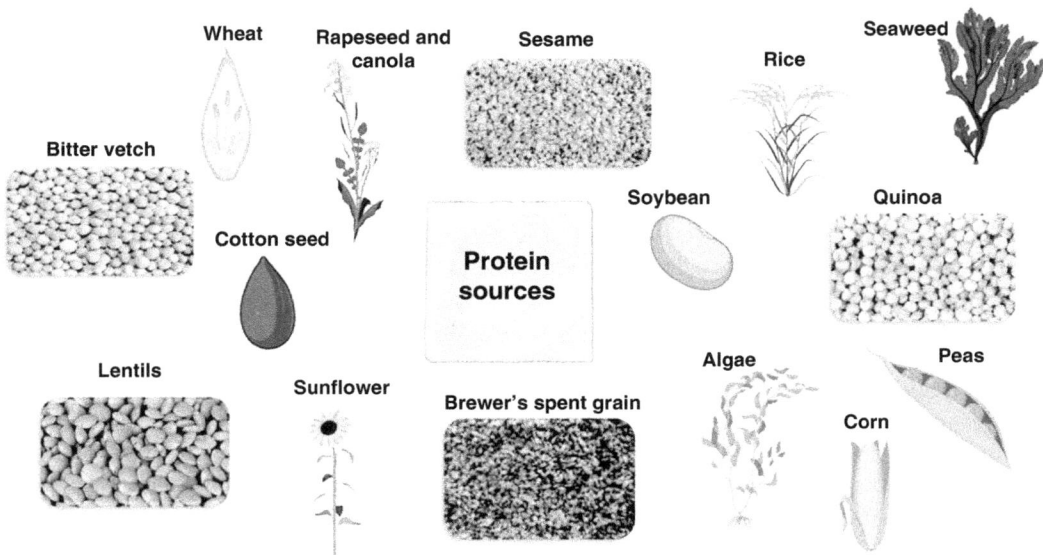

FIGURE 13.1 Various plant-based protein sources.

Effect of pH: Fabrication of protein-based packaging is highly influenced by the pH of the film-forming solution. The isoelectric point of proteins provides stability against denaturation, and hence the films formed at isoelectric pH of the particular protein contribute to minimal crosslinking (Sikorski, 2018; Zink et al., 2016). Lesser the availability of the functional groups, lesser the degree of crosslinking—resulting in a poor barrier property. Optimizing the pH changes the net charge on the protein, which destabilizes the intramolecular interaction to unfold the protein (Sikorski, 2018). At the unfolded/denatured state, the amino acid side chains become available for crosslinking (Sikorski, 2018). Therefore, the pH of the film-forming solution in protein packaging material directly impacts the stability and the barrier properties.

Effect of temperature: The temperature at which the protein-based packaging is synthesized determines the stability and barrier properties. The temperature-induced denaturation of the proteins destabilizes hydrogen bonding and electrostatic interaction while it stabilizes the hydrophobic interaction (Sikorski, 2018; Zink et al., 2016). Due to the change in the intramolecular interactions, the mobility of the peptide chains varies. In addition, if the packaging materials are fabricated under wet condition such as solution casting—the proteins can denature at lower temperature (Bourtoom, 2009). This is attributable to the increased mobility of the peptides in aqueous medium where the solvent can penetrate into the microenvironment of the proteins to cause destabilizing. In contrast, under dry condition such as extrusion process, the proteins maintain a static structure which does not facilitate the mobility of the peptide chains and hence the denaturation of the protein in minimal (Said & Sarbon, 2019). Overall, the temperature influences the protein structure and functionality, which in turn contribute toward the physical–chemical stability of the protein packaging material.

Effect of pressure: The protein-based packaging materials synthesized under hydrostatic condition denature the proteins mainly by decreasing the partial specific volume and stabilizing the hydrogen bonds (Sikorski, 2018; Zink et al., 2016). Plant-based proteins are mostly globular in nature with void spaces. Under hydrostatic pressure, the compressibility of these proteins increases due to the removal of voids—exposing the buried amino acid side chains for intermolecular interactions. Besides, the sulfhydryl reactivity increases under hydrostatic pressure (Sikorski, 2018). Collectively, these two factors augment the crosslinking and facilitate the formation of better protein-based packaging materials.

Effect of shear forces: Shear forces influence the fabrication of protein-based films by creating aggregates, de-aggregates, and protein denaturation. Depending on the strength of the shear forces, the proteins can collide with each other to form aggregates (Sikorski, 2018). The aggregates can also de-aggregate when the surface property of the aggerates and the pressure of the film-forming mixture change. Higher amounts of aggerates can cause poor mechanical stability and barrier properties of protein-based packaging materials because the aggregates can increase the intermolecular spacing of the films and coating (Chen et al., 2019). Thus, optimizing the shear forces is critical to obtain a favorable ratio of aggregates-to-de-aggregates. Besides, the shear force denatures the proteins by facilitating the absorption of proteins into the air pockets created during shearing (Sikorski, 2018; Zink et al., 2016). The denaturation allows proteins to unfold and expose the side chains of amino acids for intermolecular interactions (Sikorski, 2018). Together, the aggregates, de-aggregates, and protein denaturation impact the physicochemical properties of the protein-based packaging materials.

Effect of additives: Additives such as plasticizers (glycerol or sorbitol) or cross-linking agents (formaldehyde, glutaraldehyde, glyoxal) are used while synthesizing protein-based packaging materials (Chen et al., 2019; Sikorski, 2018). These additives can form strong intermolecular interactions to create packaging materials with better physical and chemical properties. The formaldehyde forms chemical crosslinking by creating methylene bridge between two ε-amino acids of protein. In contrast, glutaraldehyde forms intermolecular interactions specifically with lysine, cysteine, histidine, and tyrosine (Chen et al., 2019; Zink et al., 2016). Similarly, the glyoxal favors crosslinking with lysine and arginine (Chen et al., 2019). Moreover, the additives are small molecules, which can penetrate easily into the interlayer spacing of the packaging materials to form intact coatings or films (Senthilkumaran et al., 2022). Highly intact coatings or films provide good mechanical stability and enhanced barrier properties.

Effect of organic solvent: The wet process of fabricating protein-based packaging material mostly utilizes aqueous medium and occasionally organic solvents. The organic solvents can suppress the hydrophobic interaction due to weak polar—nonpolar interaction (Sikorski, 2018). Furthermore, the dielectric permittivity of the solvents can influence the intermolecular interaction. Destabilization of the covalent and the non-covalent interaction of the proteins can expose the side chains of amino acids and make them available for crosslinking (Sikorski, 2018).

Effect of ultrasound: Fabricating protein-based films under wet process can take place through ultrasound treatment. The ultrasound creates cavitation in the film-forming solution, which creates high shear forces in the solution—resulting in protein unfolding (Chen et al., 2019; Zink et al., 2016; Sikorski, 2018). Nevertheless, the degree of protein unfolding depends on the intensity of the ultrasound (Sikorski, 2018). Hence, the properties of protein-based packaging material can be influenced by ultrasound-assisted technique.

13.3   VARIOUS PLANT-BASED PROTEIN SOURCES USED AS PACKAGING MATERIALS

13.3.1   WHEAT GLUTEN

Wheat (*Triticum*) is a crop that is abundantly grown around the world. According to Foreign Agricultural Service of USDA, the global wheat production in the year 2021–2022 was 779.3 million metric tons, and it is anticipated to rise to 782.7 million metric tons by the end of 2023 (Mohammed et al., 2021). Protein and starch are the two major macromolecules found in the endosperm of the wheat grain. Depending on the variety of wheat, the protein can vary from 11.7% to 14.8% and the starch can be varying between 68% and 70% (Patni et al., 2014). The protein fraction of wheat is obtained after the isolation of starch—which is utilized majorly for bioethanol production (Mohammed et al., 2021). The protein fraction of wheat comprises albumin, globulin, gliadin, glutenin, and residual protein. The gliadin and glutenin together are called as "gluten protein," and this accounts for 70–80% of the total proteins found in wheat grain (Patni et al., 2014; Baghi et al., 2022). The gluten forms a complex network with densely packed structure due to its high proline content. The proline content in a gluten protein is around 11–29%, and this forms a dense packing of the gluten though "kinks" with secondary structure of protein (Wieser et al., 2023).

The wheat gluten is a complex network and exhibits better viscoelastic properties, mechanical stability, low solubility to water, and good barrier properties for gases (Zhang et al., 2007). The outstanding properties of wheat gluten are owed to its structure and composition. The gliadin of gluten fraction is around 40–50% whereas the glutenin is 35–40%. Depending on the molecular weight, gliadin can be classified as: α-Gliadin (31,800 Da), γ-gliadin (35,200 Da), ω1,2-gliadin (43,500 Da), and ω5-gliadin (50,900 Da). On the other hand, glutenin comprises both low and high molecular weight fractions, and this include: low molecular weight glutenin subunit (32,000 Da), high molecular weight glutenin subunit-y (68,700 Da), and high molecular weight glutenin subunit-x (86,800 Da) (Wieser et al., 2023; Patni et al., 2014; Baghi et al., 2022). A summary of the components in gluten protein is depicted in Figure 13.2.

Wheat gluten is one of the highly studied proteins to create packaging materials for food applications because of the complex nature of wheat gluten, biodegradability, abundance, and outstanding physical-and-chemical properties. Although wheat gluten alone can form packaging materials with good mechanical properties, due to the strain in the peptide chains—the packaging materials can experience cracking and brittleness (Zubeldía et al., 2015). Thus, to improve the properties of

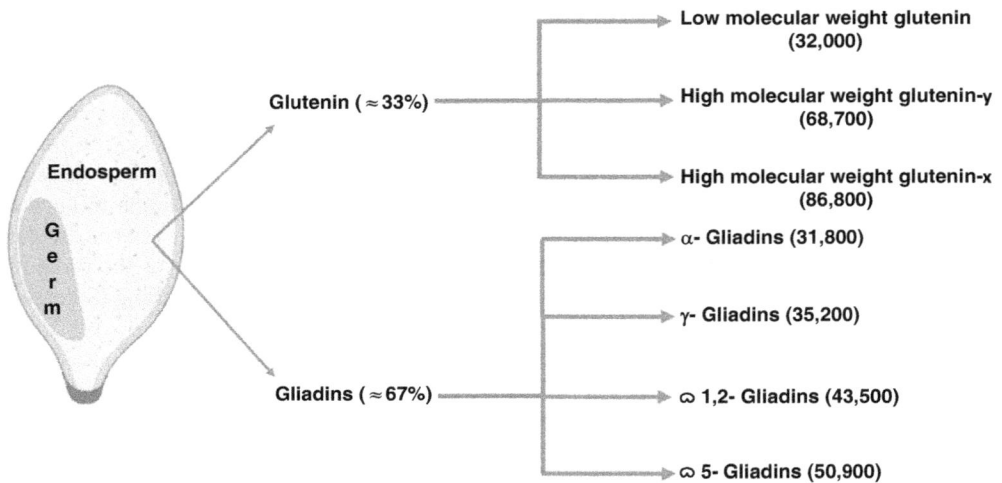

FIGURE 13.2 Components in gluten proteins found in wheat.

wheat gluten for packaging materials, plasticizers are usually added. Zubeldía et al. studied the influence of plasticizer on wheat gluten films prepared by compression molding. In this study, at 20% glycerol concentration in the presence and absence of water, the films showed different mechanical properties in terms of elastic modulus, maximum strength, and elongation at break (Zubeldía et al., 2015). Both elastic modulus and maximum strength were high in the presence of water. The elastic modulus and maximum strength in the presence of water were 95.2 MPa (versus 48.9 MPa without water) and 4.92 MPa (versus 4.37 MPa without water), respectively. In addition, the elongation at break was higher for the films without water compared to the films with water—which accounted for 53.2% and 28.2%, respectively (Zubeldía et al., 2015). This observation could be attributable to the effect of plasticizing effect of water. Water is an excellent plasticizer in wheat-gluten-based system because, comparatively, water has high melting peak and melting enthalpy compared to wheat gluten, which facilitates the penetration of water molecules into the microenvironment of the protein matrix (Zubeldía et al., 2015; Tanada-Palmu & Grosso, 2005). In the presence of, both, glycerol and water, both molecules can easily penetrate into the gluten protein network and reduce the strain on peptide chains along with reducing the effect of embrittlement. The crosslinking of water and glycerol would have aided in increasing elastic modules and maximum strength. On the other hand, in the absence of water, the mobility of the peptide chains would be limited, and hence the crosslinking of the peptide with glycerol would not be high enough, compared to a wet medium. This could also be translated to the observation with elongation at break where the glycerol alone decreased the strength of the film (as evidenced by elastic modulus and maximum strength) and increased the elongation at break. Besides, the effect of glycerol in terms of water vapor penetration, with increasing concentration of glycerol, was that the water vapor penetration increased. The water vapor permeability at 15% and 20% glycerol was 0.8 and 1.75 g mPa^{-1} s^{-1} m^{-2}, respectively. This could be due to the change caused by glycerol in gluten network. The crosslinking of glycerol could lower the strain in the peptide network to cause more interlayer spacing through which the water vapor can easily penetrate (Zubeldía et al., 2015). In a different study, similar observations were made with different concentrations of glycerol. With increasing concentration of glycerol from 3% to 7%, steady decrease in the tensile strength and puncture strength was observed (Kaushik et al., 2016). Furthermore, the water vapor transmission rate increased with increasing concentration of glycerol. Collectively, these observations can be attributable to the effect of glycerol in reducing the intermolecular interactions in the gluten network and increasing the intercalation—which results in decrease in tensile strength and increase in water vapor permeability. Thus, the concentration of plasticizer should be optimized to get better wheat-gluten-based packaging materials.

The processing conditions play a pivotal role in the synthesis of wheat gluten-based packaging materials. In terms of temperature, glutenin and gliadin have two different temperature regions for optimal crosslinking. The crosslinking for glutenin and gliadin happens at 60–70°C and approximately 90°C, respectively (Muneer et al., 2014). Thus, to obtain better crosslinking, the processing temperature should be adjusted. On the same note, wheat gluten protein film processed at 80°C and 100°C showed better elastic modulus and maximum strength at 100°C (Zubeldía et al., 2015). The elastic modulus for films produced at 80°C and 100°C was 15.08 and 22.58 MPa, respectively, while the maximum strength was 1.86 and 2 MPa at 80°C and 100°C, respectively. The temperature effect on mechanical property could be attributable to the unfolding of protein with increasing temperature, which allows the reformation of sulfhydryl (SH)–disulfide (SS) bonds—to enhance the crosslinking (Zubeldía et al., 2015). Similarly, pH is another crucial factor that affects the properties of the wheat gluten packaging materials. Packaging with better properties happens when the film-forming in solution is maintained at alkaline pH. For example, wheat gluten films synthesized at pH 9 showed enhanced maximum strength and strain compared to films formed at pH 6 (Jiménez-Rosado et al., 2019). This could be due to the ability of the alkaline pH to denature the gluten protein network better and increase the availability of the sulfhydryl (SH), hydrophobic, and other bonds to reform new bonds—resulting in films with better mechanical properties (Hemsri et al., 2012; Jiménez-Rosado et al., 2019). Thus, the processing conditions significantly affect the properties of the wheat gluten packaging materials.

Similar to processing conditions, the properties of wheat gluten packaging are influenced by reinforcing materials. Materials such as clays and fibers influence to property of the wheat gluten film. Wheat gluten reinforced by alkyl-ammonium-modified montmorillonite clay (MN-clay) showed increase in tensile strength and Young's modulus with increasing concentration from MN-clay from 1% to 3%. At 1% and 3% of MN-clay, the wheat gluten showed a tensile strength of 9 and 13.9 MPa, respectively, and the Young's modulus was 91.8 MPa for 1% MN-clay and 106.2 MPa for 3% MN-clay (Zhang et al., 2007). The increases in mechanical properties were further studied using transmission electron microscopy (TEM) for qualitative property and wide-angle X-ray diffraction (WAXD) for quantitative analysis. The TEM showed no significant difference for 1% and 3% MN-clay except for three to five particles of tactoids—which confirms the smoother surface morphology in the presence of MN-clay. The WAXD which showed a clear 2Θ peak at 5.2° for MN-clay decreased when it was incorporated into wheat gluten matrix. Moreover, the peak intensity decreased with increasing concentration of MN-clay in wheat gluten network. The WAXD observation demonstrates the exfoliation of MN-clay in wheat gluten matrix. In addition, unlike the other systems—the glass transition temperature (T_g) decreased when exfoliated with MN-clay in wheat gluten matrix. This decrease could be attributed to the suppressed mobility of the wheat gluten matrix which resulted in a decreased T_g (Zhang et al., 2007). Similarly, fibers obtained from coconut (Hemsri et al., 2012) and hemp (Muneer et al., 2014) showed better packaging properties when they were incorporated into wheat gluten. The coconut-fiber-reinforced wheat gluten films prepared by compression molding showed better first failure stress at 5% alkali treatment with NaOH, compared to 2.5% alkali treatment. In addition, when the coconut fibers were treated with silane followed by 5% alkali treatment, the first failure to stress was even better; the values of first failure stress for wheat gluten–2.5% alkali-treated coconut fiber, wheat gluten–5% alkali-treated coconut fiber, and wheat gluten silane–5% alkali-treated coconut fiber were 47.9, 51.7, and 58.6 MPa, respectively. The better tensile properties of silane–alkali-treated coconut fiber could be due to the enhanced interaction of silane with the fibers in coconut—enabling the wheat gluten films to withstand increased stress (Hemsri et al., 2012). Another study on compression moulding with wheat gluten and hemp fibers showed better film properties (Muneer et al., 2014). Individually, when hemp fiber was reinforced with gliadin, glutenin, and gluten—gliadins were able to withstand high stress–strain followed by gluten and glutenin. This could be due to the ability of the gliadin to unfold more and reform new bonds, making it stronger to withstand stress–strain. When randomly oriented hemp fibers were used for this study, the mechanical properties significantly improved with less pull-out of fibers (Muneer et al., 2014). This is because of the favorable orientation of the hydroxyl groups to form hydrogen bonding with the wheat gluten to create films with better mechanical properties.

Lignin is an important component of fiber that can be used to reinforce wheat gluten film. The abundance of diverse functional groups such as hydroxyl, carbonyl, carboxyl, and phenols has the potential to create intermolecular interactions with the complex wheat gluten network to form packaging materials with better properties (Yang et al., 2015). Nevertheless, the particle size of lignin influences the properties of the packing film. Thus, Yang et al. synthesized lignin nanoparticles and studied the properties of wheat gluten-based packing films fabricated using solution casting (Yang et al., 2015). The films reinforced with lignin nanoparticles showed better UV–visible barrier properties. At 550 nm, the transmittance of the film decreased with increasing concentration of lignin nanoparticles. The transmittance for film without lignin nanoparticle, 1% lignin nanoparticle, and 3% lignin nanoparticle was 89%, 72%, and 56%, respectively. Similarly, the transmittance decreased at 400 nm with increasing concentration of lignin nanoparticles. In terms of mechanical properties, Young's modulus and tensile strength significantly increased at high concentration of lignin nanoparticle. At 1% and 3% concentration of lignin nanoparticles, the Young's modulus was 243.2 and 553.2 MPa, respectively, while the tensile strength was 6.4 and 13.3 MPa at 1% and 3% of lignin nanoparticle, respectively. Although the Young's modulus and tensile strength was higher at 3% lignin nanoparticle, a drastic decrease in elongation at break was observed. The improved mechanical properties of the lignin nanoparticle reinforced films could be attributable to the strong intermolecular interaction between lignin and wheat gluten. Interestingly, the water barrier property also increased with increasing concentration of lignin. The moisture content for 1% and 3% lignin nanoparticle was 11.2% and 10.14%, respectively. In addition, the water uptake significantly decreased from 138.55% to 103.21% at 1% and 3% of lignin nanoparticle, respectively, in wheat gluten film. The lignin is a complex—highly branched network with an abundance of hydrophobic functional groups which facilitated the repulsion of water and decreased the water uptake and moisture content of the film (Yang et al., 2015).

Wheat gluten-based films are highly preferable in packing perishable foods because of the improved barrier properties and the nutrition values that could be obtained from the amino acids of wheat gluten. In light of such facts, the strawberries were dip coated with wheat gluten solution, wheat gluten–lipid composite, and wheat gluten–lipid bilayer (Tanada-Palmu & Grosso, 2005). In terms of visible decay, the films with composite and bilayer showed freshness until 12 days compared to the film dip coated with wheat gluten solution—which started decaying at 6 days. Firmness is an important parameter for perishable foods, and hence the firmness retention was calculated for control film (film with no gluten), film with wheat gluten solution, wheat gluten–lipid composite, and wheat gluten–lipid bilayer. The strawberries coated with control film showed 90% decrease of the strawberry firmness, and for the other films, the firmness decreased only up to 40%. In addition, compared to other films, the titratable acidity, total soluble solid, and reducing sugar were lower in wheat gluten–lipid bilayer film (Tanada-Palmu & Grosso, 2005). These observations reveal that wheat gluten-based films control the metabolism of the perishable food and facilitate the freshness index for increased shelf life. Furthermore, the wheat gluten–lipid bilayer showed better properties compared to other films due to the unique three-dimensional network formed within the bilayer films. Strikingly, wheat gluten packing materials are desirable for packing applications due to their biodegradability (Domenek et al., 2004). For example, wheat gluten films reinforced with hemp fibers showed 30–40% of biodegradation of the films in 90 days. After 180 days, the composite completely biodegraded (Muneer et al., 2014). In a different study, the wheat gluten films casted through solution casting, in the presence of glycerol as a plasticizer, biodegraded the films in 36 days in liquid medium through aerobic fermentation and in 50 days in a farmland soil (Patni et al., 2014).

Therefore, utilizing wheat gluten as packaging materials provides numerous benefits toward food safety, economy, and environmental concerns.

13.3.2 SOY PROTEIN

Soy (*Glycine max*) protein is obtained from soybean seeds. The soybean seeds comprise 35–40% protein, approximately 20% lipids, approximately 8.5% moisture, and approximately 9% dietary fiber (Friedman & Brandon, 2001). As depicted in Figure 13.3, depending on the protein content, the soy protein can be classified as soy flour, soy protein concentrate, and soy protein isolate (Koshy et al., 2015). The two major storage protein groups found in soy protein are albumin and globulin. The globulin is the abundant storage protein in soy beans and accounts for 30% of glycinin (11S) and 40% of β-conglycinin (7S) (Fukushima, 1991; Qin et al., 2022). In the native state, the soy proteins remain folded with hydrophobic subunits buried and hydrophilic subunits exposed to the external environment—making the soy proteins hydrophilic—at native state (Fukushima, 1991). In addition, due to the rich amino acid profile, the soy proteins have the highest Protein Digestibility Corrected Amino Acid Score (PDCAAS). The PDCAAS of soy protein is 0.92–1.00, which is superior to other plant-derived proteins such as pea (PDCAAS 0.66–0.91) and barley (PDCAAS 0.76–0.5) (Qin et al., 2022). Considering the rich amino acid profile, the soy proteins can be utilized as food packaging material. Besides, due to high protein content, soy protein isolates are commonly used in packaging materials. However, the hydrophilic nature of the soy protein limits its utilization in packaging material. Therefore, the native soy protein has to be modified to make it conducive for food packaging material.

When soy proteins are processed for packaging, either through wet or dry process, the native state of soy protein is denatured. During denaturation, the soy proteins become unfolded and expose sulfhydryl (SH) and hydrophobic units and reform disulfide (S–S) linkage (Koshy et al., 2015). Utilizing plasticizers such as glycerol or sorbitol in soy protein packaging process can increase the mobility of the soy protein films; the complex three-dimensional network of soy protein makes the soy protein-based packaging materials rigid and brittle (Koshy et al., 2015; Soliman et al., 2007). Penetration of the plasticizers can create voids to enable the peptide chain mobility. In a study with soy protein isolate, the effect of plasticizers glycerol, sorbitol, and PEG_{400} were studied (Soliman et al., 2007). The moisture content of the soy protein film significantly decreased with sorbitol compared to glycerol and PEG_{400}. Moreover, the moisture content of the film was less at 50% sorbitol (12.18%) compared to 60% sorbitol (12.64%). Similarly, better tensile strength and water vapor permeability were reported when sorbitol was used as a plasticizer, compared to glycerol and PEG_{400} (Soliman et al., 2007). These observations suggest that the plasticizers, which are small molecules, can reconstruct the soy protein network to form packaging materials with better properties.

In addition to plasticizers, the incorporation of crosslinkers such as formaldehyde, glutaraldehyde, or genipin (Gen) can enhance the physicochemical properties of the soy protein films

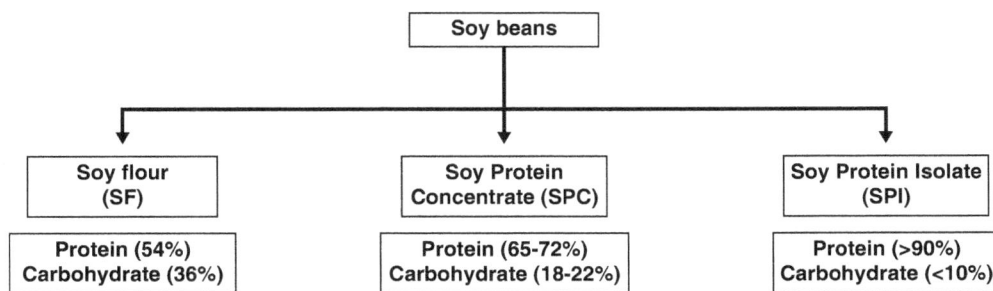

FIGURE 13.3 Classification of soy protein based on the protein content.

(Soliman et al., 2007; González et al., 2011). The crosslinking agents can form intermolecular interactions with the side chains of amino groups—which get exposed during film formation, as a result of denaturation. In light of such facts, the addition of genipin (Gen) to soy protein isolate in the presence of glycerol resulted in films with better physicochemical properties (González et al., 2011). The crosslinking between genipin (Gen) and soy protein isolate was confirmed by FTIR. The peak at 1668 cm^{-1} for soy protein isolate showed an increase in the absorption band when crosslinked with genipin (Gen). This clearly showed the amide linkage between soy protein isolate and genipin (Gen). Furthermore, the O–H stretching and N–H bending bands between 3000 and 3500 cm^{-1} widened with the addition of genipin (Gen) due to the modification in the hydroxyl and amine groups. In addition, the peaks at 1,048 cm^{-1} (C–O stretching) and 854 cm^{-1} (C–H bending of C=C) became sharper and defined. The crosslinking between soy protein isolate and genipin (Gen) also manifested better film properties. The tensile strength of the film increased significantly from 3.22 MPa for control film (film with no soy protein isolate and genipin [Gen]) to 4.6 MPa for film with 5% genipin (Gen). Increasing of genipin (Gen) more than 5% did not have a statistical significance in the tensile strength. Similarly, the water vapor permeability for control film and film with 2.5% genipin (Gen) was 2.41 × 10^{-10} and 1.72 × 10^{-10} g mPa^{-1} s^{-1} m^{-2}, respectively. Nevertheless, the percentage of crosslinking and the opacity increased with increasing concentration of genipin (Gen); the increased concentration of opacity blocks the light penetration and prevents the contents from oxidative damage. Thus, the insignificant improvement of the physicochemical property of the films could be due to the decrease in the mobility of the protein chain caused by the rigidity, upon adding high concentrations of genipin (Gen). Interestingly, the film with 1% genipin (Gen) showed biodegradability in 14 days compared to higher concentrations, which biodegraded only after 23 days for 2.5% genipin (Gen) and remained intact for 10% of genipin (Gen) (González et al., 2011). These findings suggest that optimal levels of natural crosslinkers such as genipin (Gen) can increase the feasibility of the soy protein isolate to use in food packaging applications due to intermolecular crosslinking.

The formaldehyde crosslinked soy protein isolate also improves mechanical properties of the film. For example, when the concentration of formaldehyde was increased from 0.1 to 0.5 mg mL^{-1} of soy protein isolate film-forming solution, the tensile strength significantly increased up to 0.4 mg mL^{-1} of film-forming solution (Soliman et al., 2007). In terms of water vapor permeability, a steady decrease was observed up to 0.3 mg mL^{-1} of soy protein isolate film-forming solution and increased slightly for 0.4 and 0.5 mg mL^{-1} of soy protein isolate film-forming solution. In contrast, the oxygen permeability decreased with increasing concentration of formaldehyde, up to 0.5 mg mL^{-1}. The improved properties of the films up to a certain concentration of formaldehyde show that the efficiency of crosslinking in manifesting better film properties was seen around 0.3–0.4 mg mL^{-1} of soy protein isolate film-forming solution. A similar trend was observed for films crosslinked with glutaraldehyde, except for the interlayer spacing that was higher in glutaraldehyde (Park et al., 2012). The higher interlayer spacing can be translated to the difference in molecular weight, where formaldehyde is a small molecule (MW 30.1) compared to glutaraldehyde (MW 100.12), which creates an increased interlayer spacing when crosslinked with soy protein isolate. In a separate study, the soy protein isolates crosslinked with glutaraldehyde showed that the molecular weight of the crosslinking was greater than 200 kDa, according to gel electrophoresis. The crosslinked films showed lowest solubility near the isoelectric point, and this point was slightly shifted (4.3–4.5) compared to soy protein isolate, which had an isoelectric point close to 4.8. In terms of surface hydrophobicity, which is an important parameter to prevent water vapor penetration and decrease moisture content of the film, the crosslinked showed decreased surface hydrophobicity (4.4) compared to non-crosslinked soy protein isolate (11.5). The decrease in surface hydrophobicity could be due to the increased exposure of hydrophilic groups on the surface while crosslinking—even though during films' formation, the proteins denature and reform disulfide and expose hydrophobic groups. In addition, the glutaraldehyde-crosslinked soy protein isolate showed better thermal stability. The mechanical properties steadily increased as the concentration of glutaraldehyde increased from 0.1 to 0.4 (w/w) protein solution. The tensile strength increased from 12.24 to 14.89 MPa, and the elongation at break increased from 43.75% to 71.25% while the non-crosslinked film had a tensile strength and elongation at break of 8.32 MPa and 38.71%, respectively. Furthermore, with increasing concentration of glutaraldehyde, the opacity increased, which shows the suppression of light penetration (Park et al., 2012). Collectively, cross-linking soy protein isolate with optimal concentration of crosslinkers exhibits excellent physical, chemical, and mechanical properties of the film.

Converting soy protein into packaging materials involves the addition of fillers. These fillers can reinforce the soy protein to demonstrate good barrier properties. The fillers for reinforcing soy protein can be materials such as, but not limited to, cellulose and its derivatives (carboxy methyl cellulose) (Han et al., 2015), polysaccharides (starch, lignocellulose, chitosan) (González & Alvarez Igarzabal, 2015), organic fillers (Mikus et al., 2021), or inorganic fillers (metal nano particles and clay) (Swain et al., 2012). Compared to cellulose, cellulose nanocrystals are widely used as fillers due to the enhanced interaction with the film matrix– owning to the decreased particle size. The addition of cellulose nanocrystals in a soy protein film solution showed a significant improvement in tensile strength (Yu et al., 2018). The tensile strength of soy protein isolate and cellulose nanocrystal-reinforced soy protein isolate was 3.8 and 5 MPa, respectively. In addition, the water vapor permeability did not have a statistical significance. However, when 10% pine needle extract was added, the water vapor permeability significantly dropped. The water vapor permeability for soy protein isolate, cellulose nanocrystal-reinforced soy protein isolate, and soy protein isolate–cellulose nanocrystal–pine needle extract was 2.1×10^{-10}, 2.0×10^{-10}, and 1.6×10^{-10} g m^{-1} s^{-1} Pa^{-1}, respectively. The significant decrease in water vapor permeability in the presence of pine needle could be due to the ordered crystalline structure that enabled an increased intermolecular interaction and reduced the free volume for the water vapor to penetrate (Yu et al., 2018). Similarly, in a different study when the soy protein isolate was reinforced with starch nanocrystal, the soy protein film matrix showed an increased crystallinity, according to X-ray diffraction (XRD) (González & Alvarez Igarzabal, 2015). The increased crystallinity reflected a steady decrease in water vapor permeability with increasing concentration of starch nanocrystals. Furthermore, the tensile strength and Young's modulus had a significant increase. The tensile strength for soy protein isolate and 40% starch nanocrystal-reinforced soy protein isolate was 1.1 and 5.08 MPa, respectively, while the Young's modulus for protein isolate and 40% starch nanocrystal-reinforced soy protein isolate was 26.89 and 310.34 MPa, respectively. The improved mechanical property of the starch nanocrystal-reinforced soy protein isolate is attributable to the increased intermolecular interaction, which enhances the mechanical properties (González & Alvarez Igarzabal, 2015). Not only the nanocrystals but also the addition of natural clay can be used to fabricate soy protein-based films with better properties. On the same token, addition of a natural clay, Cloisite 30B, showed better thermal stability and oxygen barrier properties. The oxygen barrier property steadily decreased when the concentration of the clay was increased from 1 to 8 w/w % (Swain et al., 2012). Similar to nanocrystals, the clay forms an ordered crystalline structure in the film matrix, which creates a tortious path for oxygen and reduces the permeability. Apart from these fillers, compounds such as hyperbranched polysiloxanes (HPS) and dendritic tannic acids (TA) can be used to improve the properties of the soy protein films (Li et al., 2021). The strain at failure, tensile strength, and Young's modulus were significantly improved by the addition of both HPS and TA. In addition, these molecules showed better UV-barrier properties. The soy protein isolate-HPS had 11.6% blockage of UV-A, while the addition of TA into soy protein isolate-HPS exhibited 98.6% blockage of UV-A. The improved UV-A barrier in the presence of TA is due to the abundant aromatic rings and phenolic hydroxyl, which can absorb the wavelength corresponding to UV-A (320–400 nm) (Li et al., 2021). In summary, the addition of fillers to soy protein-based food packaging properties can result in better packaging materials.

Soy protein-based packaging materials can be used as active packaging materials by adding antioxidant and antimicrobial compounds to film-forming solution. When mangosteen peel extract was added to soy protein isolate film-forming solution, the DPPH activity increased from 43.3%

(soy protein isolate) to 61.54% (soy protein isolate-mangosteen peel extract) (Huang et al., 2021). The antioxidant activity of soy protein isolate is due to the presence of amino acids cysteine, tyrosine, tryptophan, and histidine. This significant increase in antioxidant activity in the presence of mangosteen peel extract is owed to the abundant (poly)phenolic compounds in mangosteen peel—especially the xanthones. However, when ZnO-nano particles were added, the DPPH activity decreased with increasing concentration of ZnO-nano particles. The decrease in the antioxidant activity could be due to the adsorption of (poly)phenols of mangosteen peel extract on the ZnO-nano particles, which can reduce the bioavailability of the (poly)phenols to quench the free radicals. Unlike antioxidant activity, the antimicrobial activity was high in the presence of ZnO-nano particle. At the highest ZnO-nano particle concentration, the antimicrobial activity for *E. coli* and *S. aureus* was 100% and 97.4%, respectively. In the absence of ZnO-nano particle (soy protein isolate–mangosteen peel extract), the antimicrobial activity was 21.7% and 59.53% for *E. coli* and *S. aureus*, respectively. The pronounced antimicrobial activity of the ZnO-nano particle over mangosteen peel extract could be due to the induction of high oxidative stress in microbial cell wall, which results in the disintegration of the microbial cell wall and eventually cell death (Huang et al., 2021). In a separate study, when the soy protein isolate films were incorporated with either oregano oil or thyme oil to study the active packaging efficiency on beef patties, the oregano showed high antioxidant activity compared to thyme (Coskun et al., 2014). The IC_{50} of thyme and oregano was 7.38 and 6.64 g L^{-1}, respectively. Besides, both thyme- and oregano- incorporated films showed better lipid oxidation, free fatty acidity, and-peroxide values. Pine needle extract is another plant extract, which can be used as a potential material for active packaging. When pine needle extract was added to soy protein isolate–starch nanocrystal, the ABTS and DPPH activities steadily increased (González & Alvarez Igarzabal, 2015). The ABTS increased from 58% to 83% while DPPH increased from 25% to 55%. Similarly, when catechin was incorporated in soy protein isolate film, the antioxidant activity was 5–6% and 90% for soy protein isolate and soy protein isolate–catechin film (Han et al., 2015). The increased antioxidant activity in the presence of catechin is attributable to the abundant hydroxyl groups that can quench the free radicals by providing an electron or proton (Han et al., 2015). Not only antioxidant and antimicrobial activity, but β-cyclodextrin-added soy protein films were reported to also show anticholesterol activity (González & Alvarez Igarzabal, 2015). When the film soy protein isolate–starch nanocrystal–β-cyclodextrin was immersed in milk that contained 4.7 ppm cholesterol, the cholesterol decreased to 2.6 ppm when 20% of β-cyclodextrin was added to the soy protein isolate–starch nanocrystal film. The ability of the β-cyclodextrin to reduce the cholesterol in food products is due to the inclusion complex that can be formed between β-cyclodextrin and cholesterol (González & Alvarez Igarzabal, 2015). As seen from the aforementioned studies, the addition of bioactive compounds can convert soy protein isolate packaging material into active packaging materials with better physicochemical properties.

13.3.3 OIL CAKES

Oil seeds are widely used to extract oils for human consumption. The oil cakes are obtained once the oil is extracted from the oil seeds (Phang et al., 2022). These oil cakes are mainly used as animal feed and comprise rich amino acid profile (Grewell et al., 2014). Oils cakes especially from sunflower (*Helianthus annuus*) (Ayhllon-Meixueiro et al., 2000), canola, rapeseed, sesame, cottonseed, pumpkin seeds, and bitter vetch are currently being studied for their feasibility to use as food packaging materials.

The sunflower seeds are rich in lipids (≈65%) and protein (≈34%), and the oil cakes have 20–22% of proteins (Petraru et al., 2021). The major amino acids in sunflower oil cake protein are alanine, leucine, isoleucine, valine, tryptophan, phenylalanine, methionine, and cysteine (Petraru et al., 2021). Apart from proteins, the sunflower oil cakes have diverse (poly)phenolic compounds. These (poly)phenols can be bound to proteins through covalent or non-covalent linkages– making the sunflower oil cakes suitable for utilizing in packaging materials (González-Pérez, 2003). When the films were synthesized from sunflower protein isolates which were obtained through alkaline extraction (pH 12) from sunflower oil cake, the films did not show significant changes in the tensile strength with increasing concentration of sunflower protein isolate concentration (Ayhllon-Meixueiro et al., 2000). However, when different bases (at pH 12) were used to examine the mechanical properties, LiOH and NaOH showed higher tensile strength (3.9 MPa) compared to triethyl amine (1.8 MPa) and NH_4OH (1.3 MPa). The effect of bases on the tensile strength of sunflower protein isolate films can be attributed to the molecular structure and the volatility. For example, both LiOH and NaOH are small molecules which can easily penetrate into the protein chains and create bridges between protein chains. On the other hand, NH_4OH is a larger molecule with significant stearic hindrance, and this can keep the protein chains intact with high cohesive forces, more brittleness, and less tensile strength. In terms of triethyl amine, the chemical structure limits the formation of bridges, and thus reduces the tensile strength (Ayhllon-Meixueiro et al., 2000). The effect of (poly)phenols on the sunflower protein isolate films showed that the sunflower protein isolates with high (poly) phenolic content showed high opacity (Salgado et al., 2010). Sunflower protein isolates with 2.51% and 1.82% (poly)phenolic content had an opacity of 22.34 and 16.81 UA/mm, respectively. In addition, these protein isolates also showed maximum absorbance at 420 nm and 670 nm. The presence of (poly)phenols in sunflower protein isolates makes them suitable for food packaging materials because of the light barrier properties. The increased opacity with increasing (poly)phenolic content is due to the oxidation of the (poly)phenols under alkaline conditions, which oxidizes (poly)phenols to quinones—which in turn reacts with side chains of amino acids—to give intense color (Salgado et al., 2010). Similar to (poly)phenols that are naturally present in sunflower seeds, additives such as aldehydes, alcohols, plant tannins, and fatty acids can influence the sunflower-oil-cake-based packaging properties (Orliac et al., 2002). The sunflower protein isolate showed an increase in tensile strength up to 3% aldehyde and then a steady decline. The increase in tensile strength up to 3% is due to the increased crosslinking between aldehyde and the proteins, and the decline could be due to the increased interlayer spacing between protein chain at higher concentration of aldehyde. In terms of water uptake, the addition of plant tannins significantly reduced the water uptake due to the increased intermolecular interaction with sunflower seed protein and the hydrophobicity of the plant tannins (Orliac et al., 2002). Thus, studies have showed that sunflower oil cakes can be utilized for food packing materials with compatible additives.

Canola (*Brassica napus*) is another oil seed that is mainly cultivated for human consumption. Canola seeds have approximately 40% lipids and 17–36% protein while the oil cakes comprise up to 50% protein (Aider & Barbana, 2011; Zhang et al., 2018). The main amino acids found in canola oil seed are glutamine, glutamic acid, arginine, leucine, and aspartic acid. The canola oil seed has low concentration of sulfur-containing amino acids cysteine and methionine. Besides, the major storage proteins are napin and cruciferin, while oleosin acts as a structural protein (Aider & Barbana, 2011; Shi & Dumont, 2014). Canola protein isolate films fabricated with plasticizer (glycerol), co-plasticizer (stearic acid), and denaturant (sodium dodecyl sulfate- SDS) showed varying physicochemical properties of the films (Shi & Dumont, 2014). The tensile strength decreased and elongation at break increased with increasing concentration of glycerol—which could be due to the increased lubrication between the protein layers. When the amount of glycerol was kept constant in the film-forming solution and the concentration of stearic acid increased, the tensile strength increased from 1.5 MPa (at 25% glycerol and 5% stearic acid) to 2.1 MPa (at 25% glycerol and 15% stearic acid). The increase in stearic acid content could have decreased the chain–chain mobility between the protein layers, caused by glycerol—resulting in an increased tensile strength. Similarly, at constant glycerol concentration when the concentration of SDS was increased, the tensile strength increased from 2.3 MPa (at 25% glycerol and 5% SDS) to 3.8 MPa (at 25% glycerol and 15% SDS). Interestingly, the increase in tensile strength was high in the presence of SDS compared to stearic acid. This difference could be related to the ability of SDS to completely denature the canola protein and expose the buried amino acid groups which can participate in increased intermolecular interaction. Higher the intermolecular interactions, higher will be the tensile strength. In contrast, the water absorption capacity increased with increasing concentration of SDS. For example, the water absorption capacity for films with 30% glycerol, 30% glycerol + 15% stearic acid, and 30% glycerol + 15% SDS were 182%, 149%, and 1155%, respectively. The major increase in the water absorption in the presence of SDS is due to the denaturation of cruciferin, which could expose the hydrophilic amino acids and increase the water absorption capacity (Shi & Dumont, 2014). Cellulose nanocrystals are commonly used as fillers to reinforce the protein-based films. On the same token, when canola protein isolates were reinforced with cellulose nanocrystals, the films showed better mechanical and barrier properties. Increasing the concentration of cellulose nanocrystals formed a homogenous film-forming solution at pH 9.5 due to the negative charge on both canola protein and cellulose nanocrystal– facilitating the charge distribution and a better microstructure of the film (Osorio-Ruiz et al., 2019). The better films' morphology also translates to better mechanical strength. A steady increase in tensile strength was observed with increasing concentration of cellulose nanocrystal. This steady increase from 12% to 36% of cellulose nanocrystals in canola protein is because of the increased intermolecular interaction among glycerol–protein–cellulose nanocrystal. The sulfate (SO_4^{2-}) and hydroxyl (OH) with hydroxyl (OH) of glycerol and various functional groups of canola protein can create a rigid and intact structure through bridges and chelation to provide films with better mechanical strength. On the other hand, the mechanical properties and water vapor permeability in the absence of cellulose nanocrystal-like fillers can be compromised (Osorio-Ruiz et al., 2019). For a canola protein isolate solution when glycerol was added, the tensile strength decreased from 6.8 to 2.2 MPa with an increasing concentration of glycerol at 30% and 50%, respectively (Chang & Nickerson, 2015). Similar behavior was observed with water vapor permeability. The decrease in tensile strength and increase in water vapor permeability with increasing concentration of glycerol are due to the increased hydrophilicity of the films due to the OH group from glycerol and the ability of the glycerol to easily penetrate into protein layers and form void volume or free space (Chang & Nickerson, 2015). In summary, optimized conditions should be used for canola oil cakes to be used in food packaging.

Rapeseed, another member of family *Brassicaceae,* is also studied for its potential in food packaging applications. The oil cakes from rapeseed can consist of 40% protein with cruciferin and napin being the major proteins (Jia et al., 2021). The protein isolates of rapeseed oil cake can account for more than 90% protein (Jia et al., 2021). Sequential isoelectric precipitation of rapeseed oil cake showed better antioxidant properties, which is essential for synthesizing active packaging (Georgiev et al., 2022). In terms of water absorption capacity, the rapeseed protein isolate precipitated at pH 10.5–2.5 and showed higher water absorption capacity (2.36 g H_2O/g sample) compared to rapeseed protein isolate precipitated at pH 2.5–8.5 (2.11 g H_2O/g sample). The different water absorption capacity of these two fractions can be attributed to the change in the protein structure that occurs during the sequential precipitation. These two fractions also showed better antioxidant capacities with increasing concentration. Strikingly, the antioxidant capacity was high for pH 10.5–2.5 fraction compared to pH 2.5–8.5 fraction, and this again could be related to the structural change that takes place during sequential precipitation (Georgiev et al., 2022). In films with rapeseed oil cake, better properties for food packaging were obtained when the plasticizers sorbitol and sucrose were at 2% and 0.5% respectively while the polysorbate as an emulsifier was at 1.5% (Jang et al., 2011). When films with these optimized conditions were treated with gelatin and *Gelidium corneum,* the mechanical properties were further enhanced. The enhanced mechanical properties could be directly corelated with the increased intermolecular interaction of the film-forming substances (Jang et al., 2011).

Apart from aforementioned oil seed cakes, other seeds are studied for the suitability in food packaging. Sesame protein isolates obtained from sesame oil cakes showed good water vapor permeability when the films were formed at pH 12 compared to pH 9 (Sharma & Singh, 2016). The better permeability of water at pH 12 is due the complete denaturation of sesame oil cake protein that could expose the hydrophobic groups, suppressing the permeability of water vapor (Sharma &

Singh, 2016). For cottonseed oil cakes, increasing glycerol concentration showed a decrease in thermal stability due the increased void space and the mobility created by increased concentration of glycerol (Yue et al., 2012). The denaturation temperatures were 156.2, 139.3, and 136°C at 10%, 20%, and 30% glycerol concentration, respectively. Similar observations were reported with pumpkin oil cake (Popović et al., 2012). Besides, the bitter vetch (*Vicia ervilia*)-based films formed by the addition of zein protein showed better barrier properties due to the hydrophobicity and the structural modifications (Arabestani et al., 2016). Overall, the studies have showed the feasibility of using oil cake proteins in food packaging applications.

13.3.4 PULSE PROTEIN

Pulses (lentils, peas, and beans) are rich sources of protein, and these are potent starting materials to fabricate food packaging materials. Lentils (*Lens culinaris*) comprise 20–30% of protein, and the protein concentrates of lentils usually contain more than 60% of protein (Bamdad et al., 2006). Majority of the protein is globulin (70%), followed by albumin (16%), glutelin (11%), and prolamin (3%) (Jarpa-Parra, 2017). Since protein-based films are rigid and brittle due to high cohesive energy density, the addition of plasticizer is required (Apodaca et al., 2020; Khazaei et al., 2019). In light of such facts, the effects of plasticizers (glycerol, sorbitol, and polyethylene glycol- PEG 400) on lentil protein concentrates were studied (Apodaca et al., 2020). The water vapor permeability was high for glycerol (133.68 g m^{-2} day) compared to PEG 400 (74.35 g m^{-2} day) and sorbitol (14.13 g m^{-2} day). The increased water vapor permeability in glycerol could be attributed to the increased hydrophilicity and lubricant capacity compared to sorbitol and PEG 400. Moreover, PEG 400 had high water vapor permeability compared to sorbitol because of the stearic hindrance. Besides, the oxygen permeability was high for PEG 400 (58.46 cm^3 m^{-2} day) compared to glycerol (37.96 cm^3 m^{-2} day) and glycerol (0.62 cm^3 m^{-2} day)—which could be due to the increased free volume created by PEG 400 due to its bulk structure, compared to glycerol and sorbitol (Apodaca et al., 2020). In a different study with lentil protein concentrate and glycerol, the puncture strength was considerably high (1.552 N) but with less tensile strength compared to protein films from soy and pea (Bamdad et al., 2006). The findings of this study showed that although the lentil protein concentrate films with glycerol formed good cohesive films, the mechanical properties were not higher than other pulse-based protein films (Bamdad et al., 2006). Thus, lentil protein films require more reinforcement with fillers or additives to obtain good mechanical and barrier properties.

Similar to lentils, proteins from peas and beans can be used as food packaging materials. To study the effect of native and denatured state of pea protein on film-forming properties, yellow field peas (*Pisum sativum*) were used to form films (Choi & Han, 2001). In this study, the heat-denatured protein at 90°C for 25 min showed better tensile strength (2.896 MPa) compared to the native pea protein (1.105 MPa). The increased tensile strength in a denatured protein is due to the exposure of amino acids, which facilitates the intermolecular interactions to form rigid films. In a different study with pea protein, the effect of crosslinking on the film properties due to heat and UV treatment were assessed (Pérez Puyana et al., 2022). The degree of crosslinking increased with increasing heat treatment and UV treatment. The degree of crosslinking at 50°C for 24 h and 120°C for 24 h were 16.8% and 24.3%, respectively. Similarly, the UV treatment at 50 and 500 mJ cm^{-2} showed 3.6% and 18.1% degree of crosslinking, respectively. Increased degrees of crosslinking at different treatment conditions were also reflected on the Young's modulus, maximum stress, and strain at break. For example, the pea protein films subjected to 120°C for 24 h withstood high maximum stress (7.92 MPa) compared to 50°C for 24 h (2.6 MPa). In terms of UV treatment, the high strain at break was observed for 500 mJ cm^{-2} compared to 50 mJ cm^{-2} showed, and the values were 0.88 and 0.74 mm mm^{-1}, respectively (Pérez Puyana et al., 2022). These observations suggest that at high temperature, the proteins get completely unfolded and form new bonds and increase the intermolecular interaction—to enhance the rigidity and the mechanical properties of the films.

The UV-induced changes in the protein structure in proteins can be mainly due to the phenyl ala-nine and tyrosine in pea protein, which can facilitate the high-energy transitions and enhance the crosslinking (Pérez Puyana et al., 2022). In terms of antimicrobial properties, better antimicrobial properties were seen with UV treatment compared to heat treatment. This could be because of the ability of the UV to create new metabolites in pea protein with antimicrobial properties and also the possibility to generate high-energy electrons, which can deteriorate the cell walls of microbes (Pérez Puyana et al., 2022). Another study showed that incorporation of nisin—an antibiotic—into pea protein films not only inhibited the gram-positive bacteria but also enhanced the mechanical properties (Pérez Puyana et al., 2016). An enhanced Young's modulus was also observed when yellow pea protein was reinforced by whey protein isolate (Acquah et al., 2020). The Young's modulus of yellow pea protein and whey protein isolate-reinforced yellow pea protein was 6.65 and 28.64 MPa, respectively. The increased intermolecular interaction between the amino acid side chain of whey protein isolate and yellow pea protein could have contributed toward the signifi-cant difference in the Young's modulus. Although the mechanical property improved, the surface hydrophobicity decreased with the addition of whey protein isolate due to the less hydrophobic nature of whey protein isolate, compared to pea protein. This was also reflected in the contact angle measurement where the addition of whey protein isolate reduced the contact angle of pea protein and made it more hydrophilic. The contact angle of pea protein and whey protein isolate-reinforced yellow pea protein was 65 and 40°, respectively (Acquah et al., 2020). Overall, different types of protein treatment (heat or UV), the reinforcing agents, and additives can be used in pea protein-based films to enhance the packaging properties and make them suitable for food packag-ing applications.

Faba bean (*Vicia faba*), one of the highly consumed proteins in Asian continent, is another good source to synthesize packaging materials (Rojas-Lema et al., 2021). Faba bean protein films reinforced with increasing concentration of cellulose nanocrystals showed an increase in tensile modulus and tensile strength (Rojas-Lema et al., 2021). The tensile modulus of faba bean, faba bean–1% cellulose nanocrystal, faba bean–5% cellulose nanocrystal, and faba bean–7% cellulose nanocrystal was 3.8, 4.3, 5.6, and 7 MPa, respectively. In terms of tensile strength, faba bean, faba bean–1% cellulose nanocrystal, faba bean–5% cellulose nanocrystal, and faba bean–7% cellulose nanocrystal was 4.3, 4.2, 5.3, and 6.2 MPa, respectively. The increased mechanical properties due to increased concentration of cellulose nanocrystals are due to the more ordered crystalline struc-ture formed in the presence of cellulose nanocrystals due to increased intermolecular forces—that are facilitated by the functional groups and increased surface area. Simultaneously, the films started showing less hydrophilic character with increased concentration of cellulose nanocrystals in terms of contact angle measurements, and this also reflected in less water vapor permeability. Besides, the oxygen transfer rate decreased with increased concentration of cellulose nanocrys-tal—because a more rigid structure with less void volume was formed—hindering the penetra-tion of oxygen (Rojas-Lema et al., 2021). Not only the fillers, but also the change in pH can significantly affect the physicochemical properties of the film. The puncture strength increased with increasing pH values. At pH 7, 8.5, and 10, the puncture strength of faba protein was 14, 17, and 18 MPa, respectively (Montalvo-Paquini et al., 2014). The better puncture strength at alka-line condition could be corelated to the high solubility of the protein under alkaline condition—forming homogenous films with high puncture strength. Similar increase in tensile strength was observed in a different faba protein film. At pH 7, 9, and 12, the tensile strength was 0.6, 1.2, and 3 MPa, respectively (Saremnezhad et al., 2011). In terms of water vapor permeability, the faba films showed less permeability to films at alkaline condition; at pH 7, 9, and 12, the water vapor permeability was 3, 2.3, and 1.7 g m^{-1} Pa^{-1} s^{-1}, respectively. The increased solubility of the proteins at alkaline condition creates a well-ordered film– which suppresses the permeability of water vapor (Montalvo-Paquini et al., 2014; Saremnezhad et al., 2011). In summary, several studies have proved the feasibility of utilizing pulse proteins—lentils, peas, and beans for food packaging applications.

13.3.5 Other Protein Sources

Apart from aforementioned sources of protein, there are several other potential protein sources that could be considered for food packaging applications. Recently, algae—which is the main producer of ocean—is being investigated for its suitability in food packaging applications (Yap et al., 2022). In light of such facts, algae-based packaging is used as edible packaging, non-edible food packaging, and biodegradable food packaging which accounts for 60%, 23%, and 17%, respectively (Yap et al., 2022). Moreover, microalgae are considered for algal-protein-based packaging since the microalgae comprise high protein content (>30%) compared to macroalgae, which have only 7–15% of protein (Yap et al., 2022; Benelhadj et al., 2016). Nevertheless, the protein extraction from algae is a cumbersome process, and hence the studies on algae protein-based packaging are scarcely explored. Films produced using *Undaria pinnatifida*—a seaweed widely used as a food source in Japan and Korea—showed good film properties (Yang et al., 2016). These films were formed with gelatin to enhance the physicochemical properties of the film. The films with 3% (w/w) *Undaria pinnatifida* and 2% (w/w) gelatin showed better tensile strength, Young's modulus, and water vapor permeability and the values were 26.63 MPa (versus 1.72 MPa for films with no gelatin), 379.2 MPa (versus 21.04 MPa for films with no gelatin), and 1.49×10^{-9} g m m^{-2} s^{-1} Pa^{-1} (versus 1.78×10^{-9} g m m^{-2} s^{-1} Pa^{-1}), respectively. In addition, the films showed better antimicrobial properties with the addition of vanillin. Increasing concentration of vanillin from 0% to 0.5% reduced the microbial population from 6 log CFU g^{-1} to 3.3 log CFU g^{-1}. When these films were wrapped around smoked chicken breast, the *E. coli* population decreased from 5.12 log CFU g^{-1} on the zeroth day to 3.26 log CFU g^{-1} on the twentieth day (Yang et al., 2016a). In a different study, the protein isolated from *Spirulina* coupled with lysozyme showed statistical insignificance in opacity but enhanced Young's modulus (Benelhadj et al., 2016). These studies on algae have proved the feasibility of utilizing algae in food packaging applications.

Corn (*Zea mays*) is a crop that is widely used for bioethanol production. The major by-product of corn ethanol production is distillers dried grains with soluble (DDGS), which has around 27% protein (Yang et al., 2016b). When DDGS protein was complexed with green tea, oolong tea, and black tea extracts, the tensile strength was compromised. In contrast, low water vapor permeability was observed for films with black tea extract compared to green tea and oolong tea (Yang et al., 2016b). For example, for DDGS films with no tea extract, the water vapor permeability was 2.46×10^{-9} g m m^{-2} s^{-1} Pa^{-1} and for 0.5% incorporation of green tea, oolong tea, and black tea, the values were 2.34 and 2.27×10^{-9} g m m^{-2} s^{-1} Pa^{-1}, respectively. Interestingly, the black tea extracts had low phenolic compounds compared to green tea and oolong tea, and the less water vapor permeability could be due to the complex molecular structure that is formed between DDGS protein and black tea (poly)phenols. In terms of antioxidant activity, films with 0.5% green tea showed high antioxidant activity in terms of ABBTS and DPPH compared to oolong tea and black tea. This observation can be related to the high phenolic content of green tea, which was 50 mg GAE g^{-1} compared to oolong tea (48 mg GAE g^{-1}) and black tea (35 mg GAE g^{-1}). The high antioxidant activity of DDGS protein-green tea film was also translated to the low thiobarbituric acid reactive substances (TBARS) values reported when these films were wrapped around the pork for ten days. Compared to control film, film with oolong tea, and films with black tea, the TBARS was low for films with green tea and the values were 1.6 (control film), 1.3 (film with black tea), 1.2 (film with oolong tea), and 0.8 (film with green tea) mg malonaldehyde (MDA) per kilogram of sample (Yang et al., 2016b). When corn zein proteins were used for fabricating films with low-density polyethylene (LLDPE), the tensile strength decreased and the water vapor permeability increased, but the antioxidant property increased with the addition of antioxidant agents thymol, carvacrol, and eugenol (Park et al., 2012). When these films were wrapped around beef patties, the films with eugenol showed high DPPH activity in terms of ascorbic acid equivalent. When corn zein proteins were complexed with tung oil and coated or laminated and coated, significant changes were observed in terms of tensile strength, Young's modulus, toughness, water vapor permeability, and gas permeability. The laminated and coated films with tung oil had significantly high tensile strength and toughness of 5.87 and 0.82 MPa, respectively, compared to the untreated film, which had tensile strength and toughness of 2.08 and 0.25 MPa, respectively (Rakotonirainy et al., 2001). In terms of barrier properties, the water vapor permeability was significantly low for aminated and coated films (1.14 pg Pa^{-1} s^{-1} m^{-1}) compared to untreated films (17.9 pg Pa^{-1} s^{-1} m^{-1}). Similarly, the oxygen and carbon dioxide permeabilities were low for laminated and coated films compared to untreated films. The better mechanical and barrier properties of the laminated and coated film could be attributable to the complex structure that is formed during the film-forming process. When the laminated and coated film was used to pack the broccoli, the freshness of the broccoli was retained over 160 h, compared to other films. The broccoli wrapped with laminated and coated films also did not exhaust oxygen, unlike other films, and contained high amounts of carbon dioxide. The increased carbon dioxide with well-balanced oxygen levels would have contributed toward the freshness of the broccoli over time (Rakotonirainy et al., 2001).

Brewer's spent grain, a major by-product of beer production, is another potential source for protein-based packaging materials (Lee et al., 2015; Proaño et al., 2020). The spent grains comprise 150–250 g kg^{-1} of proteins—which could be translated to 15–25% of protein (Proaño et al., 2020). Composite films formed with brewer's spent grain protein and chitosan showed increase in tensile strength and decrease in elongation at break with increasing concentration of chitosan (Lee et al., 2015). For example, film with no chitosan and film with a ratio of 30:70 of brewer's spent grain protein to chitosan showed a tensile strength of 4.32 and 26.2 MPa, respectively. Moreover, the elongation at break decreased from 36.38% to 28.54% for film with no chitosan and film with 30:70 brewer's spent grain protein: chitosan, respectively. The findings from this study showed that the addition of chitosan increased the intermolecular interaction between the amino functional group of chitosan and the various functional groups of amino acids in brewer's spent grain protein (Lee et al., 2015). Similarly, the increased intermolecular interaction also contributed toward reduced water vapor permeability—with increasing concentration of chitosan. In contrast, with increasing concentration of chitosan, the opacity decreased—suppressing the light barrier property. In terms of antimicrobial activity, addition of chitosan showed a good inhibitory property against *Listeria monocytogenes* (ATCC 19111), *Staphylococcus aureus* (KCTC 1621), *Salmonella Typhimurium* (KCTC2514), and *Escherichia coli* O157:H7, with enhanced inhibitory effects on gram-positive bacteria, which is due to the difference in the structure of the gram-positive and gram-negative cell walls. The inhibition zone of *L. monocytogenes*, *S. aureus*, *E. coli*, and *S. Typhimurium* were 18.63, 19.09, 17.15, and 17.22 mm, respectively, at 30:70 brewer's spent grain protein: chitosan (Lee et al., 2015). When brewer's spent grain protein films were formed by adding polyethylene glycol (PEG), the tensile strength and elastic modulus increased when 0.1% PEG was added (Proaño et al., 2020). The tensile strength and elastic modulus were 1.2 MPa (versus 1 MPa for control film) and 0.45 MPa (versus 0.4 MPa for control film), respectively. The films with same formulation also showed better antioxidant property—which is essential for the prevention of oxidative stress and the oxygen-induced deterioration of the contents (Proaño et al., 2020).

Rice (*Oryza sativa*), a major cereal used across the globe as a staple, and quinoa (*Chenopodium quinoa*), a pseudocereal highly consumed due to the rich and balanced amino acid profile, are also feasible sources for protein-based packaging materials. The wild and Pasankalla quinoa crosslinked with transglutaminase was studied using SDS-PAGE, and the change in molecular weights was significant (Escamilla-García et al., 2019). The transglutaminase crosslinking was high for Pasankalla quinoa with greater than 35 kDa, compared to the wild-type, which could be due to the increased availability of free amino acid side chains in the Pasankalla quinoa. The change in the degree of crosslinking reflected when these protein films were further crosslinked with chitosan. The degree of crosslinking for Pasankalla quinoa with chitosan varied between 26% and 68%, while for the wild-type quinoa, the degree of crosslinking was between 12% and 44%. The increased degree of crosslinking with chitosan also contributed toward less water vapor permeability for Pasankalla quinoa (2.42 to 3.23 × 10^{-13} g cm Pa^{-1} cm^{-2} s^{-1}) compared to wild-type quinoa

(3.4 to 4.69×10^{-13} g cm Pa^{-1} cm^{-2} s^{-1}) (Escamilla-García et al., 2019). A different film with chitosan-cross-linked quinoa protein showed high tensile strength when the ratio of chitosan: quinoa protein was 90:10, and the tensile strength was 12.8 MPa (Caro et al., 2015). Similarly, the water content of this film was lower and accounted for 0.61 g H_2O per gram of dry film as opposed to greater than 0.7 g H_2O per gram of dry films with lower chitosan content (Caro et al., 2015). Thus, the quinoa protein-based films show promising materials for food packaging. Similar to quinoa, rice also can be used for food packaging materials. Especially, the protein from rice husk accounts for 8% and is mainly used as animal feed (Felix et al., 2015). Utilizing such by-products contributes toward sustainable food system. On the same token, when proteins from rice husks were molded into films with glycerol as a plasticizer, sodium bisulfite as a reducing agent, and glyoxal and L-cysteine as crosslinkers, the films showed better stress–strain curve with L-cysteine as an additive. This film with 70:30 rice husk protein: glycerol and 1% (w/w) showed maximum strain at break of 180 mm mm^{-1} × 10 Young's modulus (MPa) compared to films other films: 70:30 rice husk protein: glycerol, 70:30 rice husk protein: glycerol + 0.3% (w/w) sodium bisulfite, and 70:30 rice husk protein: glycerol + 3% (w/w) glyoxal with strain at break 40, 105, and 120 mm mm^{-1} × 10 Young's modulus (MPa), respectively. The increased strain at break in the presence of L-cysteine could be attributed to the increased intermolecular crosslinking between rice husk protein and L-cysteine (Felix et al., 2015). The studies on quinoa and rice have shown that the most abundant crops can be utilized as sustainable food packaging materials.

13.4 LEGAL ASPECTS OF PLANT-BASED PROTEIN FOR FOOD PACKAGING

The technological innovations of food system include developing novel food sources with novel techniques as well as innovating novel food packaging system. Biodegradable food packaging materials is a technological innovation, which leaves less carbon footprints (Krzywonos & Piwowar-Sulej, 2022). These biodegradable packaging systems contribute to circular economy and sustainability of food system.

Plant-based packaging materials, as biodegradable sources, are intensely investigated due to the economic and environmental issues caused by synthetic packaging materials. In light of such facts, proteins, carbohydrates, and lipid-based packaging materials from plant-sources, over animal sources, are considered because plant sources contribute toward the equilibrium and the stability of the ecosystem (Chen et al., 2019). Carbohydrates, lipids, and proteins can be extracted from plant biomass. Plant proteins are utilized and highly investigated for packaging applications due to their innate stability, which is owed to the high cohesive energy—compared to lipids and carbohydrates. Moreover, when the protein-based packaging degrades, it provides nitrogen for soil to enhance soil fertility (Chen et al., 2019; van der Spiegel et al., 2013). Such resource-efficiency in food innovation is gearing toward developing plant protein-based packaging materials.

The protein-based packaging is gradually entering the market, and thus the safety of using such packaging materials is becoming a concern in terms of consumer adaptability and acceptability. Compared to terrestrial environment, the protein obtained from aqueous environments such as algae and duckweeds is vulnerable to heavy metal toxicity. In terms of terrestrial environment, the plant proteins from seeds, pulses, cereal, and nuts carry considerable amount of pesticide, toxins, and allergens (van der Spiegel et al., 2013). Thus, when utilizing such sources for food packaging applications, strict rules and regulations need to be in place. The Food and Drug Administration (FDA) of the United States uses "Generally Recognized as Safe (GRAS)" for any substances that is either added to the food and/or comes in close contact with the food. In addition, the FDA also implements the Hazard Analysis of Critical Control Point (HACCP) to ensure the biological, chemical, and physical hazard from start to end of the product processing. In terms of Europe legislation, the European Food Safety Authority (EFSA) has introduced regulation (EC) 396/2005 for hazardous residues of plant protection products, regulation (EC) 396/2005 and Directive 2002/32/EC for contaminants such as heavy metals, mycotoxins, environmental contaminants, pesticidal residues, endogenous substances, and toxic weeds in food products (Lähteenmäki-Uutela et al., 2021). Apart from the United States and Europe, other countries also regulate safety aspects of food system through stringent laws. Therefore, such local and global regulations ensure the safety of the end products and drift consumer preferences by increasing the consumer adaptability and acceptability of the novel food packaging techniques that are based on plant proteins.

REFERENCES

Acquah, C., Zhang, Y., Dubé, M., & Udenigwe, C. (2020). Formation and Characterization of Protein-Based Films from Yellow Pea (Pisum sativum) Protein Isolate and Concentrate for Edible Applications. *Current Research in Food Science*, 2, 61–69. https://doi.org/10.1016/j.crfs.2019.11.008.

Aider, M., & Barbana, C. (2011). Canola Proteins: Composition, extraction, Functional Properties, Bioactivity, Applications as a Food Ingredient and Allergenicity—A Practical and Critical Review. *Trends in Food Science & Technology*, 22, 21–39. https://doi.org/10.1016/j.tifs.2010.11.002.

Amin, U., Khan, M. U., Majeed, Y., Rebezov, M., Khayrullin, M., Bobkova, E., & Thiruvengadam, M. (2021). Potentials of polysaccharides, lipids and proteins in biodegradable food packaging applications. *International Journal of Biological Macromolecules*, 183, 2184–2198. https://doi.org/https://doi.org/10.1016/j.ijbiomac.2021.05.182.

Apodaca, E., Montanari, A., Castro, L., Umilta, E., Arroyo, L., Zurlini, C., & Villaran, M. C. (2020). Lentil By-Products as a Source of Protein for Food Packaging Applications. *American Journal of Food Technology*, 15, 1–10. https://doi.org/10.3923/ajft.2020.1.10.

Arabestani, A., Kadivar, M., Amoresano, A., Illiano, A., Di Pierro, P., & Porta, R. (2016). Bitter Vetch (Vicia ervilia) Seed Protein Concentrate as Possible Source for Production of Bilayered Films and Biodegradable Containers. *Food Hydrocolloids*, 60. https://doi.org/10.1016/j.foodhyd.2016.03.029.

Ayhllon-Meixueiro, F., Vaca-Garcia, C., & Silvestre, F. (2000). Biodegradable Films from Isolate of Sunflower (Helianthus annuus) Proteins. *Journal of Agricultural and Food Chemistry*, 48(7), 3032–3036. https://doi.org/10.1021/jf9907485.

Baghi, F., Gharsallaoui, A., Dumas, E., & Ghnimi, S. (2022). Advancements in Biodegradable Active Films for Food Packaging: Effects of Nano/Microcapsule Incorporation. *Foods*, 11. https://doi.org/10.3390/foods11050760.

Bamdad, F., Goli, S., & Kadivar, M. (2006). Preparation and Characterization of Proteinous Film from Lentil (Lens culinaris): Edible Film from Lentil (Lens culinaris). *Food Research International*, 39, 106–111. https://doi.org/10.1016/j.foodres.2005.06.006.

Benelhadj, S., Fejji, N., Degraeve, P., Attia, H., Ghorbel, D., & Gharsallaoui, A. (2016). Properties of Lysozyme/Arthrospira Platensis (Spirulina) Protein Complexes for Antimicrobial Edible Food Packaging. *Algal Research*, 15, 43–49. https://doi.org/https://doi.org/10.1016/j.algal.2016.02.003.

Bourtoom, T. (2009). Edible Protein Films: Properties Enhancement. *International Food Research Journal*, 16, 1–9.

Caro, N., Quiñonez, E., Diaz-Dosque, M., Lopez, L., Abugoch, L., & Tapia, C. (2015). Novel Active Packaging Based on Films of Chitosan and Chitosan/Quinoa Protein Printed with Chitosan-Tripolyphosphate-Thymol Nanoparticles via Thermal Ink-Jet Printing. *Food Hydrocolloids*, 52. https://doi.org/10.1016/j.foodhyd.2015.07.028.

Chang, C., & Nickerson, M. T. (2015). Effect of Protein and Glycerol Concentration on the Mechanical, Optical, and Water Vapor Barrier Properties of Canola Protein Isolate-Based Edible Films. *Food Science and Technology International*, 21(1), 33–44. https://doi.org/10.1177/1082013213503645.

Chen, H., Wang, J., Cheng, Y., Wang, C., Liu, H., Bian, H., & Han, W. (2019). Application of Protein-Based Films and Coatings for Food Packaging: A Review. *Polymers (Basel)*, 11(12). https://doi.org/10.3390/polym11122039.

Choi, W.-S., & Han, J. (2001). Physical and Mechanical Properties of Pea-Protein-based Edible Films. *Journal of Food Science*, 66, 319–322. https://doi.org/10.1111/j.1365-2621.2001.tb11339.x.

Coskun, B., Calikoglu, E., Emiroglu, Z., & Candogan, K. (2014). Antioxidant Active Packaging with Soy Edible Films and Oregano or Thyme Essential Oils for Oxidative Stability of Ground Beef Patties. *Journal of Food Quality*, 37. https://doi.org/10.1111/jfq.12089.

Domenek, S., Feuilloley, P., Gratraud, J., Morel, M. H., & Guilbert, S. (2004). Biodegradability of Wheat Gluten Based Bioplastics. *Chemosphere*, 54(4), 551–559. https://doi.org/10.1016/s0045-6535(03)00760-4.

Dutta, D., & Sit, N. (2022). Application of Natural Extracts as Active Ingredient in Biopolymer Based Packaging Systems. *Journal of Food Science and Technology*. https://doi.org/10.1007/s13197-022-05474-5.

Escamilla-García, M., Delgado-Sánchez, L., Ríos-Romo, R., García-Almendárez, B., Méndez-Méndez, J., Amaro Reyes, A., & Regalado, C. (2019). Effect of Transglutaminase Cross-Linking in Protein Isolates from a Mixture of Two Quinoa Varieties with Chitosan on the Physicochemical Properties of Edible Films. *Coatings*, 9, 736. https://doi.org/10.3390/coatings9110736.

Felix, M., Lucio-Villegas, A., Romero, A., & Guerrero, A. (2015). Development of Rice Protein Bio-Based Plastic Materials Processed by Injection Molding. *Industrial Crops and Products*, 79, 152–159. https://doi.org/10.1016/j.indcrop.2015.11.028.

Friedman, M., & Brandon, D. L. (2001). Nutritional and Health Benefits of Soy Proteins. *Journal of Agricultural and Food Chemistry*, 49(3), 1069–1086. https://doi.org/10.1021/jf0009246.

Fukushima, D. (1991). Recent Progress of Soybean Protein Foods: Chemistry, Technology, and Nutrition. *Food Reviews International*, 7(3), 323–351. https://doi.org/10.1080/87559129109540915.

Georgiev, R., Kalaydzhiev, H., Ivanova, P., Silva, C. L. M., & Chalova, V. I. (2022). Multifunctionality of Rapeseed Meal Protein Isolates Prepared by Sequential Isoelectric Precipitation. *Foods*, 11(4). https://doi.org/10.3390/foods11040541.

González, A., & Alvarez Igarzabal, C. I. (2015). Nanocrystal-Reinforced Soy Protein Films and Their Application as Active Packaging. *Food Hydrocolloids*, 43, 777–784. https://doi.org/10.1016/j.foodhyd.2014.08.008.

González, A., Strumia, M. C., & Alvarez Igarzabal, C. I. (2011). Cross-Linked Soy Protein as Material for Biodegradable Films: Synthesis, Characterization and Biodegradation. *Journal of Food Engineering*, 106(4), 331–338. https://doi.org/10.1016/j.jfoodeng.2011.05.030.

González-Pérez, S. (2003). *Physico-Chemical and Functional Properties of Sunflower Proteins*. Wageningen University and Research.

Grewell, D., Schrader, J., & Srinivasan, G. (2014). Developing Protein-Based Plastics. In *Soy-Based Chemicals and Materials* (Vol. 1178, pp. 357–370). American Chemical Society. https://doi.org/10.1021/bk-2014-1178.ch015.

Hadidi, M., Jafarzadeh, S., Forough, M., Garavand, F., Alizadeh, S., Salehabadi, A., & Jafari, S. M. (2022). Plant Protein-Based Food Packaging Films; Recent Advances in Fabrication, Characterization, and Applications. *Trends in Food Science & Technology*, 120, 154–173. https://doi.org/10.1016/j.tifs.2022.01.013.

Han, J., Shin, S.-H., Park, K.-M., & Kim, K. M. (2015). Characterization of Physical, Mechanical, and Antioxidant Properties of Soy Protein-Based Bioplastic Films Containing Carboxymethylcellulose and Catechin. *Food Science and Biotechnology*, 24(3), 939–945. https://doi.org/10.1007/s10068-015-0121-0.

Hemsri, S., Grieco, K., Asandei, A. D., & Parnas, R. S. (2012). Wheat Gluten Composites Reinforced with Coconut Fiber. *Composites Part A: Applied Science and Manufacturing*, 43(7), 1160–1168. https://doi.org/10.1016/j.compositesa.2012.02.011.

Huang, X., Zhou, X., Dai, Q., & Qin, Z. (2021). Antibacterial, Antioxidation, UV-Blocking, and Biodegradable Soy Protein Isolate Food Packaging Film with Mangosteen Peel Extract and ZnO Nanoparticles. *Nanomaterials (Basel)*, 11(12). https://doi.org/10.3390/nano11123337.

Ivanković, A., Zeljko, K., Talić, S., bevanda, a., & Lasić, M. (2017). Biodegradable Packaging in the Food Industry. *Archiv für lebensmittelhygiene*, 68, 23–52. https://doi.org/10.2376/0003-925X-68-26.

Jang, S. A., Lim, G. O., & Song, K. B. (2011). Preparation and Mechanical Properties of Edible Rapeseed Protein Films. *Journal of Food Science*, 76(2), C218–C223. https://doi.org/10.1111/j.1750-3841.2010.02026.x.

Jarpa-Parra, M. (2017). Lentil Protein: A Review of Functional Properties and Food Application. An Overview of Lentil Protein Functionality. *International Journal of Food Science & Technology*, 53. https://doi.org/10.1111/ijfs.13685.

Jia, W., Rodriguez-Alonso, E., Bianeis, M., Keppler, J. K., & van der Goot, A. J. (2021). Assessing Functional Properties of Rapeseed Protein Concentrate Versus Isolate for Food Applications. *Innovative Food Science & Emerging Technologies*, 68, 102636. https://doi.org/10.1016/j.ifset.2021.102636.

Jiménez-Rosado, M., Zarate-Ramírez, L. S., Romero, A., Bengoechea, C., Partal, P., & Guerrero, A. (2019). Bioplastics Based on Wheat Gluten Processed by Extrusion. *Journal of Cleaner Production*, 239, 117994. https://doi.org/10.1016/j.jclepro.2019.117994.

Kaushik, R., Sharma, N., Khatkar, B. S., Sharma, P., & Sharma, R. (2016). Isolation and Development of Wheat Based Gluten Edible Film and Its Physico-Chemical Properties. *International Food Research Journal*, 24.

Khazaei, H., Subedi, M., Nickerson, M., Martínez-Villaluenga, C., Frias, J., & Vandenberg, A. (2019). Seed Protein of Lentils: Current Status, Progress, and Food Applications. *Foods*, 8(9). https://doi.org/10.3390/foods8090391.

Koshy, R., Mary, S., Pothan, L., & Thomas, S. (2015). Environment Friendly Green Composites Based on Soy Protein Isolate: A Review. *Advanced Structured Materials*, 75, 433–467. https://doi.org/10.1007/978-81-322-2470-9_14.

Krzywonos, M., & Piwowar-Sulej, K. (2022). Plant-Based Innovations for the Transition to Sustainability: A Bibliometric and in-Depth Content Analysis. *Foods*, 11(19). https://doi.org/10.3390/foods11193137.

Lähteenmäki-Uutela, A., Rahikainen, M., Lonkila, A., & Yang, B. (2021). Alternative Proteins and EU Food Law. *Food Control*, 130, 108336. https://doi.org/10.1016/j.foodcont.2021.108336.

Lee, J.-H., Lee, J.-H., Yang, H.-J., & Song, K. (2015). Preparation and Characterization of Brewer's Spent Grain Protein-Chitosan Composite Films. *Journal of Food Science and Technology*, 52. https://doi.org/10.1007/s13197-015-1941-x.

Li, J., Jiang, S., Wei, Y., Li, X., Shi, S. Q., Zhang, W., & Li, J. (2021). Facile Fabrication of Tough, Strong, and Biodegradable Soy Protein-Based Composite Films with Excellent UV-Blocking Performance. *Composites Part B: Engineering*, 211, 108645. https://doi.org/10.1016/j.compositesb.2021.108645.

Mihalca, V., Kerezsi, A., Weber, A., Gruber-Traub, C., Schmucker, J., Vodnar, D., & Pop, O. L. (2021). Protein-Based Films and Coatings for Food Industry Applications. *Polymers*, 13, 769. https://doi.org/10.3390/polym13050769.

Mikus, M., Galus, S., Ciurzyńska, A., & Janowicz, M. (2021). Development and Characterization of Novel Composite Films Based on Soy Protein Isolate and Oilseed Flours. *Molecules*, 26(12). https://doi.org/10.3390/molecules26123738.

Mohammed, A., Omran, A. A. B., Hasan, Z., Ilyas, R. A., & Sapuan, S. M. (2021). Wheat Biocomposite Extraction, Structure, Properties and Characterization: A Review. *Polymers (Basel)*, 13(21). https://doi.org/10.3390/polym13213624.

Mohanty, B., & Hauzoukim, S. S. (2020). Functionality of Protein-Based Edible Coating- Review. *Journal of Entomology and Zoology Studies*, 8, 1432–1440.

Montalvo-Paquini, C., Rangel-Marrón, M., Palou, E., & López-Malo, A. (2014). Physical and Chemical Properties of Edible Films from Faba Bean Protein. *International Journal of Biology and Biomedical Engineering*, 8, 125–131.

Muneer, F., Johansson, E., Hedenqvist, M., Gällstedt, M., & Newson, W. (2014). Preparation, Properties, Protein Cross-Linking and Biodegradability of Plasticizer-Solvent Free Hemp Fibre Reinforced Wheat Gluten, Glutenin, and Gliadin Composites. *BioResources*, 9. https://doi.org/10.15376/biores.9.3.5246-5261.

Orliac, O., Rouilly, A., Silvestre, F., & Rigal, L. (2002). Effects of Additives on the Mechanical Properties, Hydrophobicity and Water Uptake of Thermo-Moulded Films Produced from Sunflower Protein Isolate. *Polymer*, 43(20), 5417–5425. https://doi.org/10.1016/S0032-3861(02)00434-2.

Osorio-Ruiz, A., Avena-Bustillos, R. J., Chiou, B.-S., Rodríguez-González, F., & Martinez-Ayala, A.-L. (2019). Mechanical and Thermal Behavior of Canola Protein Isolate Films as Improved by Cellulose Nanocrystals. *ACS Omega*, 4(21), 19172–19176. https://doi.org/10.1021/acsomega.9b02460.

Panja, P., Mani, A., & Thakur, P. K. (2018). Current Status of Active and Intelligent Packaging in Food Technologies. *Trends and Prospects in Postharvest Management of Horticultural Crops*, 299–320.

Park, H.-Y., Kim, S.-J., Kim, K. M., You, A.-S., Kim, S. Y., & Han, J. (2012). Development of Antioxidant Packaging Material by Applying Corn-Zein to LLDPE Film in Combination with Phenolic Compounds. *Journal of Food Science*, 77, E273–E279. https://doi.org/10.1111/j.1750-3841.2012.02906.x.

Park, S., Bae, D., & Rhee, K. (2012). Soy Protein Biopolymers Cross-Linked with Glutaraldehyde. *Journal of Oil & Fat Industries*, 77, 879–884. https://doi.org/10.1007/s11746-000-0140-3.

Patni, N., Yadava, P., Agarwal, A., & Maroo, V. (2014). An Overview on the Role of Wheat Gluten as a Viable Substitute for Biodegradable Plastics. *Reviews in Chemical Engineering*, 30. https://doi.org/10.1515/revce-2013-0039.

Pérez Puyana, V., Cuartero, P., Jiménez Rosado, M., Martínez, I., & Romero, A. (2022). Physical Crosslinking of Pea Protein-Based Bioplastics: Effect of Heat and UV Treatments. *Food Packaging and Shelf Life*, 32, 100836. https://doi.org/10.1016/j.fpsl.2022.100836.

Pérez Puyana, V., Felix, M., Romero, A., & Guerrero, A. (2016). Development of Pea Protein-Based Bioplastics with Antimicrobial Properties. *Journal of the Science of Food and Agriculture*, 97. https://doi.org/10.1002/jsfa.8051.

Petraru, A., Ursachi, F., & Amariei, S. (2021). Nutritional Characteristics Assessment of Sunflower Seeds, Oil and Cake. Perspective of Using Sunflower Oilcakes as a Functional Ingredient. *Plants*, 10, 2487. https://doi.org/10.3390/plants10112487.

Phang, L. Y., Mohammadi, M., Jenol, M. A., & Gozan, M. (2022). Bioplastic Production from Oil Producing Plants. *Biorefinery of Oil Producing Plants for Value-Added Products*, 543–562. https://doi.org/10.1002/9783527830756.ch27.

Popović, S., Peričin, D., Vaštag, Ž., Lazić, V., & Popović, L. (2012). Pumpkin Oil Cake Protein Isolate Films as Potential Gas Barrier Coating. *Journal of Food Engineering*, 110, 374–379. https://doi.org/10.1016/j.jfoodeng.2011.12.035.

Proaño, J., Salgado, P., Cian, R., Mauri, A., & Drago, S. (2020). Physical, Structural and Antioxidant Properties of Brewer's Spent Grain Protein Films. *Journal of the Science of Food and Agriculture*, 100. https://doi.org/10.1002/jsfa.10597.

Qin, P., Wang, T., & Luo, Y. (2022). A Review on Plant-Based Proteins from Soybean: Health Benefits and Soy Product Development. *Journal of Agriculture and Food Research*, 7, 100265. https://doi.org/10.1016/j.jafr.2021.100265.

Rajeshkumar, G., Seshadri, S. A., Richelle, R. R., Mitha, K. M., & Abinaya, V. (2022). 24, Production of Biodegradable Films and Blends from Proteins. In S. Mavinkere Rangappa, J. Parameswaranpillai, S. Siengchin, & M. Ramesh (Eds.), *Biodegradable Polymers, Blends and Composites* (pp. 681–692). Woodhead Publishing. https://doi.org/10.1016/B978-0-12-823791-5.00003-X.

Rakotonirainy, A. M., Wang, Q., & Padua, G. W. (2001). Evaluation of Zein Films as Modified Atmosphere Packaging for Fresh Broccoli. *Journal of Food Science*, 66, 1108–1111. https://doi.org/10.1111/j.1365-2621.2001.tb16089.x.

Rojas-Lema, S., Nilsson, K., Trifol, J., Langton, M., Gomez-Caturla, J., Balart, R., & Moriana, R. (2021). Faba Bean Protein Films Reinforced with Cellulose Nanocrystals as Edible Food Packaging Material. *Food Hydrocolloids*, 121, 107019. https://doi.org/10.1016/j.foodhyd.2021.107019.

Said, N., & Sarbon, N. (2019). Protein-Based Active Film as Antimicrobial Food Packaging: A Review. *Active Antimicrobial Food Packaging*, 53. https://doi.org/10.5772/intechopen.80774.

Salgado, P. R., Molina Ortiz, S. E., Petruccelli, S., & Mauri, A. N. (2010). Biodegradable Sunflower Protein Films Naturally Activated with Antioxidant Compounds. *Food Hydrocolloids*, 24(5), 525–533. https://doi.org/10.1016/j.foodhyd.2009.12.002.

Saremnezhad, S., Azizi, M., Barzegar, M., Abbasi, S., & Ahmadi, E. (2011). Properties of a New Edible Film Made of Faba Bean Protein Isolate. *Journal of Agricultural Science and Technology*, 13.

Senthilkumaran, A., Babaei-Ghazvini, A., Nickerson, M. T., & Acharya, B. (2022). Comparison of Protein Content, Availability, and Different Properties of Plant Protein Sources with Their Application in Packaging. *Polymers (Basel)*, 14(5). https://doi.org/10.3390/polym14051065.

Shaikh, S., Yaqoob, M., & Aggarwal, P. (2021). An Overview of Biodegradable Packaging in Food Industry. *Current Research in Food Science*, 4, 503–520. https://doi.org/10.1016/j.crfs.2021.07.005.

Sharma, L., & Singh, C. (2016). Sesame Protein Based Edible Films: Development and Characterization. *Food Hydrocolloids*, 61, 139–147. https://doi.org/10.1016/j.foodhyd.2016.05.007.

Shi, W., & Dumont, M.-J. (2014). Processing and Physical Properties of Canola Protein Isolate-Based Films. *Industrial Crops and Products*, 52, 269–277. https://doi.org/10.1016/j.indcrop.2013.10.037.

Sikorski, Z. (2018). Fennema's Food Chemistry (Fifth Edition)—Edited by Srinivasan Damodaran and Kirk L. Parkin. *Journal of Food Biochemistry*, 42. https://doi.org/10.1111/jfbc.12483.

Soliman, E., Tawfik, M., Hosni, E.-S., & Moharram, Y. (2007). Preparation and Characterization of Soy Protein Based Edible/Biodegradable Films. *American Journal of Food Technology*, 2. https://doi.org/10.3923/ajft.2007.462.476.

Swain, S. K., Priyadarshini, P. P., & Patra, S. K. (2012). Soy Protein/Clay Bionanocomposites as Ideal Packaging Materials. *Polymer-Plastics Technology and Engineering*, 51(12), 1282–1287. https://doi.org/10.1080/03602559.2012.700542.

Tanada-Palmu, P. S., & Grosso, C. R. F. (2005). Effect of Edible Wheat Gluten-Based Films and Coatings on Refrigerated Strawberry (Fragaria ananassa) Quality. *Postharvest Biology and Technology*, 36(2), 199–208. https://doi.org/10.1016/j.postharvbio.2004.12.003.

van der Spiegel, M., Noordam, M. Y., & van der Fels-Klerx, H. J. (2013). Safety of Novel Protein Sources (Insects, Microalgae, Seaweed, Duckweed, and Rapeseed) and Legislative Aspects for Their Application in Food and Feed Production. *Comprehensive Reviews in Food Science and Food Safety*, 12(6), 662–678. https://doi.org/10.1111/1541-4337.12032.

Wieser, H., Koehler, P., & Scherf, K. A. (2023). Chemistry of Wheat Gluten Proteins: Qualitative Composition. *Cereal Chemistry*, 100(1), 23–35. https://doi.org/10.1002/cche.10572.

Yang, H. J., Lee, J. H., Lee, K. Y., & Song, K. (2016a). Antimicrobial Effect of an Undaria Pinnatifida Composite Film Containing Vanillin Against Escherichia Coli and Its Application in the Packaging of Smoked Chicken Breast. *International Journal of Food Science & Technology*, 52. https://doi.org/10.1111/ijfs.13294.

Yang, H.-J., Lee, J.-H., Won, M., & Song, K. B. (2016b). Antioxidant Activities of Distiller Dried Grains with Solubles as Protein Films Containing Tea Extracts and Their Application in the Packaging of Pork Meat. *Food Chemistry*, 196, 174–179. https://doi.org/10.1016/j.foodchem.2015.09.020.

Yang, W., Kenny, J., & Puglia, D. (2015). Structure and Properties of Biodegradable Wheat Gluten Bionanocomposites Containing Lignin Nanoparticles. *Industrial Crops and Products*, 348–356. https://doi.org/10.1016/j.indcrop.2015.05.032

Yap, X., Gew, L. T., Khalid, M., & Yow, Y.-Y. (2022). Algae-Based Bioplastic for Packaging: A Decade of Development and Challenges (2010–2020). *Journal of Polymers and the Environment*. https://doi. org/10.1007/s10924-022-02620-0.

Young, E., Mirosa, M., & Bremer, P. (2020). A Systematic Review of Consumer Perceptions of Smart Packaging Technologies for Food. *Frontiers in Sustainable Food Systems*, 4, 63. https://doi.org/10.3389/ fsufs.2020.00063.

Yu, Z., Sun, L., Wang, W., Zeng, W., Mustapha, A., & Lin, M. (2018). Soy Protein-Based Films Incorporated with Cellulose Nanocrystals and Pine Needle Extract for Active Packaging. *Industrial Crops and Products*, 112, 412–419. https://doi.org/10.1016/j.indcrop.2017.12.031.

Yue, H., Cui, Y. D., Shuttleworth, P., & Clark, J. (2012). Preparation and Characterisation of Bioplastics Made from Cottonseed Protein. *Green Chemistry*, 14, 2009–2016. https://doi.org/10.1039/C2GC35509D.

Zhang, X., Do, M. D., Dean, K., Hoobin, P., & Burgar, I. M. (2007). Wheat-Gluten-Based Natural Polymer Nanoparticle Composites. *Biomacromolecules*, 8(2), 345–353. https://doi.org/10.1021/bm060929x.

Zhang, Y., Liu, Q., & Rempel, C. (2018). Processing and Characteristics of Canola Protein-Based Biodegradable Packaging: A Review. *Critical Reviews in Food Science and Nutrition*, 58(3), 475–485. https://doi. org/10.1080/10408398.2016.1193463.

Zink, J., Wyrobnik, T., Prinz, T., & Schmid, M. (2016). Physical, Chemical and Biochemical Modifications of Protein-Based Films and Coatings: An Extensive Review. *International Journal of Molecular Sciences*, 17(9). https://doi.org/10.3390/ijms17091376.

Zubeldía, F., Ansorena, M. R., & Marcovich, N. E. (2015). Wheat Gluten Films Obtained by Compression Molding. *Polymer Testing*, 43, 68–77. https://doi.org/10.1016/j.polymertesting.2015.02.001.

14 Plant Protein-Based Foods, Trend from a Business Perspective

Market, Consumers' Challenges, and Opportunities in Future

Fatma Boukid, Supriya Kumari and Zakir Showkat Khan

14.1 INTRODUCTION

Plant-based diets are trending upward as healthier and more sustainable lifestyles. Despite the relevance of animal-derived products in the human diet, their overconsumption might be related to serious health issues due to their content of cholesterol and saturated fatty acids. Zoonotic diseases, the use of antibiotics and hormones in animal production, and foodborne diseases are negatively impacting consumers' perception of animal products (Sattar et al. 2021; Kang et al. 2021). On the other hand, plant-based foods are cholesterol-free, and there are a variety of sources that might provide fats/oils with low content of saturated fatty acids such as sunflower oil and rapeseed oil. Thus, plant-based diets were found to be associated with reduced risk of cardiovascular disease, blood pressure, diabetes, and mortality (Desmond et al. 2021). Furthermore, animal productions were reported to have higher carbon and water footprints and greenhouse gas production (Kumar et al. 2022). Plant sources contribute to soil fertility preservation and conservation of biodiversity in alignment with the sustainability goals (Mariutti et al. 2021; Boukid et al. 2019). Population growth is alarming, and thus the increased use of plant sources is required to ensure food security in the future. Growing niches, i.e., vegetarians, vegans, and flexitarians, are contributing to the increase of plant-based products labeled vegan and vegetarian (Mintel 2020). Ethical concerns over animal welfare are among the boosters to shift toward more or only plant products (Boukid 2021a).

Alternative protein sources are diverse and include plant proteins, insects, algae, and duckweed (Siddiqui et al. 2022; Boukid, Sogari, and Rosell 2022). Plant proteins are particularly gaining attention due to their long history of use, availability, and familiarity (Boukid et al. 2021b; Boukid and Pasqualone 2021). Furthermore, plant proteins can be manufactured as mainstreams or side streams to starch and oil extraction (Boukid 2021; Martinez and Boukid 2021b). From a manufacturer's perspective, these proteins are versatile ingredients, acting as an essential source of nutrition as well as improving techno-functionality of food formulations to prepare a wide variety of both conventional and novel food products. Recent market trends are revolving around the successful creation and manufacture of plant protein-based products that can act as alternatives to meat, dairy, egg, fermented dairy, and seafoods (Mefleh, Pasqualone, Caponio, and Faccia 2022; Castro-Alba et al. 2019). Scientific research and large-scale manufacturers are aiming to create plant-based alternatives that mimic the appearance, texture, umami flavor, and protein amino acid profile of various foods from animal origin (Kumar et al. 2017; Joshi and Kumar 2015). The market of plant protein-based foods is growing fast, and new launches are continuously available on supermarket shelves to offer consumers more options. In this light, this chapter intends to provide insights about the global market landscape of alternative plant-based foods, consumer challenges/acceptance, and potential prospects for creating new plant protein-based foods.

DOI: 10.1201/9781003369790-14

14.2 PLANT-BASED MARKET LANDSCAPE

14.2.1 MAIN SOURCES

The global plant-based protein market was estimated to be USD 12.2 billion in 2022 and is expected to reach USD 17.4 billion by 2027 (Markets and Markets 2022). Soy proteins lead the market of plant proteins followed by wheat and pea proteins. This is due to their high content (40% per seed), affordability, availability, high nutritional quality (PDCAAS = 1) and high functionality (e.g., solubility, emulsification, and gelling). Wheat proteins have a long history of use in bakery, snacks, and nutritional supplements. Large-scale wheat production, government support, and cheap labor costs are the major factors driving market growth (MeticulousResearch® 2019). Legumes, specially pulses, are gaining lot of interest as they are gluten-free rich sources of proteins. The market of pea protein is expected to reach USD 176.03 million by 2025, and it is fueled by the increasing demand of GMO-free, allergen-free, gluten-free, and lactose-free products (Meticulous Research® 2019).

Cereal-derived proteins (rice protein and zein) and vegetable proteins such as potato protein are currently marketed and used chiefly in gluten-free foods and sports drinks and supplements (Boukid et al. 2021b). The market is fueled by an increasing demand for healthy diet options with lactose and gluten-free sources of protein (Transparency market Research 2019). Other plant sources such as faba bean, oat, chickpea, lupine, and mung bean protein are new entries to the market (Boukid and Castellari 2022; Boukid and Pasqualone 2021; Boukid 2021c). These proteins come to respond to the increased demand of proteins and to diversify the portfolio of ingredients and thus enable a versatility of options to consumers (Boukid 2021a, 2021b, 2021c).

14.2.2 MAIN FORMS

Plant proteins are available in the market in different forms (Figure 14.1). Plant protein flours are obtained by size reduction/separation and typically contain a protein content ranging from 12% to 40% depending on the source. Plant proteins in their purified forms are available in different forms. Based on their purity, protein concentrates contain around 54–65% protein, and they are produced by dry processing (milling and air classification or using a wet process). The process includes milling and elimination of soluble components from the flour. Protein isolates contain the highest content of protein (≥90% protein). Commonly, they are produced using an alkaline extraction followed by an isoelectric precipitation. Purity has a crucial role to play in the price of protein ingredients; isolates have a higher price compared to concentrates as they have higher protein content and usually lighter color. Concentrates' market is the largest due to the diversity of their applications including food, feed, and aquaculture. Protein isolates can be further processed to obtain functional ingredients. Hydrolyzed proteins are gaining interest, and they are obtained by subjecting isolates to an enzymatic treatment to obtain shorter peptides with interesting functional properties (e.g., higher solubility) and nutritional value (e.g., higher digestibility). Plant proteins are available in dry and liquid forms. Protein in powder form is governing the market due to its stability and easiness for transportation/storage. Liquid forms are moving fast in the market due to the easy workability and fast blending with different food components (Grand View Research 2019). Biopeptides are further purified to obtain peptides with bioactive effects (e.g., antioxidant, antihypertensive, immunomodulatory, antimicrobial, and antidiabetic) (Wen et al. 2020). Beside health benefits, peptides have desirable functional properties (high solubility, emulsifying and foaming properties, and water/oil retention capacities) owning to the hydrophilic/lipophilic property, conformational flexibility, and chain length of peptides (Fan et al. 2022). Textured vegetable proteins (TVPs) have a long history of use in their dry forms, which are obtained through low moisture extrusion (LM-TVP). LM-TVPs have a porous structure and a stable shelf life. Prior to their use, they go through rehydration, and consequently their texture becomes chewy and juicy, suiting meat-substitute applications (Maningat, Jeradechachai, and Buttshaw 2022). High moisture extrudates (HM-TVP) have a fibrous

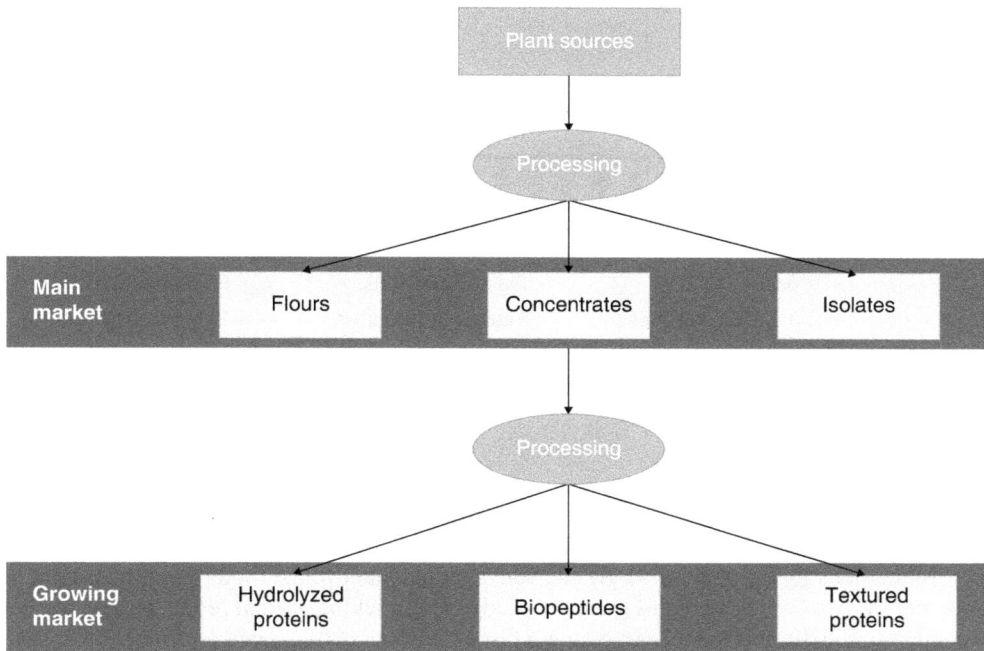

FIGURE 14.1 Plant protein ingredients available in the market.

structure like muscle meat and thus are able to recreate meat-like bite and texture (Baune et al. 2022). HM-TVP are currently sold business to business as semi-finished products and not available in the supermarket. They are characterized by high moisture content and thus have a short shelf life compared to those that are dry.

14.3 GEOGRAPHICAL DISTRIBUTION

Figure 14.2 shows that Europe dominates the market of plant-based alternatives followed by North America and Asia Pacific, Latin America and Middle East and Africa. The *plant-based* foods market is well established in Europe, and it is expected to reach USD 16.70 billion by 2029. In 2022, Germany was reported to have the major share in Europe (Business Wire 2022). North American market is estimated to grow at *8.60% of compound annual growth rate (CAGR)* from 2021 to 2028. The Asian market was valued at USD 17,473 million in 2020 and is anticipated to reach a CAGR of 10.8% by 2026 (Industry ARC 2022). Plant-based foods are increasingly being adopted by Asians as they are fiber-rich foods with low saturated fats, and this would help against chronic diseases including diabetes, obesity, and cardiovascular disease (Bryant et al. 2019). The Latin American market is also growing fast with a CAGR of 8.28% over the period from 2021 to 2028 (Research and Markets 2021). Brazil is the largest market due to growing investments and startups (Gómez-Luciano et al. 2019). Middle East and Africa market holds the smallest share and is projected to grow at a CAGR of 6.01% by 2027 (Morder Intelligence 2021). The shift in consumer behavior is still a recent concept in this region (Owusu-Kwarteng et al. 2022).

14.4 MAIN APPLICATIONS

Plant proteins are used in different applications, namely foods and beverages (e.g., meat, poultry, seafood, bakery, meat analog, dairy and dairy alternatives, cereals and snacks, beverage), animal

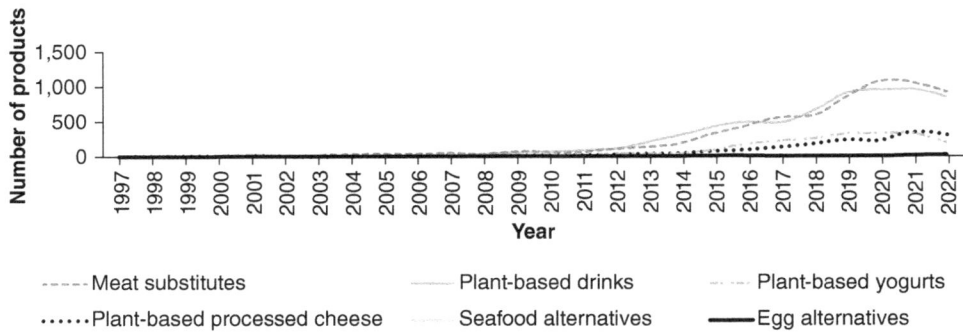

FIGURE 14.2 Evolution of plant protein-based alternative foods.

feed, nutrition and health supplements, cosmetics, and pharmaceuticals (Meticulous Research® 2019). Food and beverage segment holds the largest share of the overall plant protein-based food market (Meticulous Research® 2019). Beside conventional plant-based foods, plant-based products are boosting the use of plant proteins to create products mimicking animal products. The market of plant protein-based alternatives is witnessing a fast expansion as illustrated by Figure 14.2. A total of 17,817 plant-based alternative products have been launched since 1997. In terms of market share, meat substitutes hold the largest share of alternative products, followed by plant-based drinks (alternatives to milk), plant-based yogurt, and cheese [Figure 14.3(a) and (b)]. Meat substitutes have a longer history of consumption, and this might explain their faster and higher growth compared to the other categories (Boukid and Gagaoua 2022). The first meat substitute product registered in Mintel GNPD goes back to 1997. Dairy alternatives also hold an important position as they are highly used by people having cows' milk allergy and lactose intolerance (Akin and Ozcan 2017). Just recently, egg and seafoods' alternatives started to be commercialized, and this explains the limited number of products (Fresán et al. 2019; Sprague et al. 2017). For all products, a pick of launches was observed in 2019–2020. This might be attributed in part to changes in food habit during the COVID-19 outbreak due to consumers' belief that plant-based foods consumption is healthier options (Siddiqui et al. 2022).

14.4.1 MEAT SUBSTITUTES

Plant-based meat market was valued at USD 7.9 billion in 2022 and is projected to reach USD 15.7 billion by 2027 (Markets and Markets 2022). Plant-based meats are available in different forms such as nuggets, burgers, sausages, and meatballs (Saget et al. 2021). These products are currently available in supermarkets and franchises. Plant-based meats are cholesterol-free and usually with low amount of fat unlike conventional meat products (Tso et al. 2021; Modlinska et al. 2020). Advances in technologies are boosting the organoleptic features improvement of these products (Boukid et al. 2021b). Structuring technologies are allowing to transform the globular structure of plant-based proteins into fibrous to suit meat products (Cornet et al. 2022). Beside the texture, flavors and colorant ingredients are being developed to mimic meat features using innovative technologies such precision fermentation (Razavizadeh et al. 2022). Meat alternatives are improving to include higher amounts of proteins and complete essential amino acids profile, low salt, and low saturated fats (Guo et al. 2020; Boukid and Castellari 2021). Beyond Meat, Impossible Meat, Boulder Brands, Hain Celestia, Nestlé, Garden Protein International, Vivera, Lightlife Foods, Woolworths, Naturli' Foods, and Sainsbury's are some examples of producers (Mintel 2020).

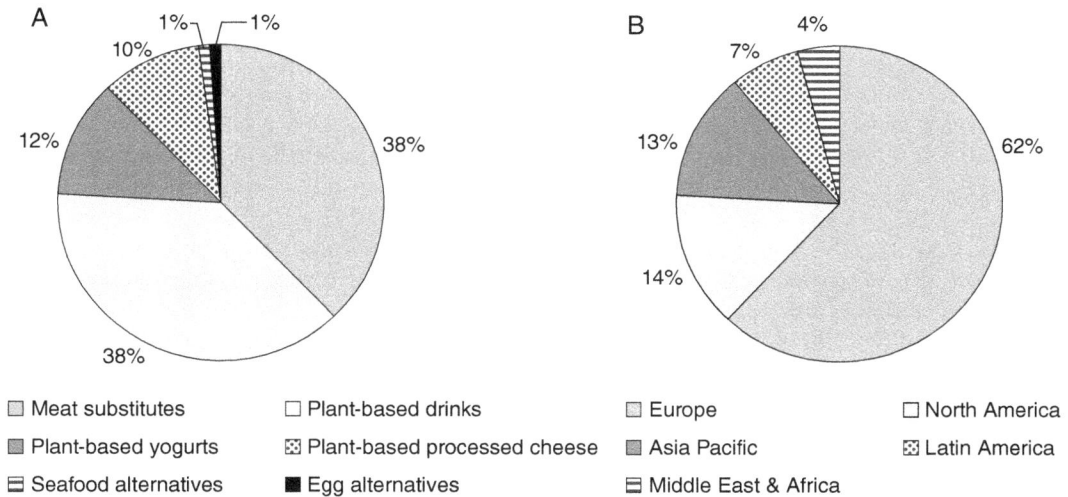

FIGURE 14.3 Market segmentation of plant protein-based foods. (a) Food applications, (b) geographical distribution.

14.4.2 PLANT-BASED DAIRY DRINKS (ALTERNATIVES TO MILK)

The global market of dairy milk alternatives reached a value of USD 22.6 billion in 2020 and is projected to reach USD 40.6 billion by 2026 (Markets and Markets 2021). This can be related to rising prevalence toward cow's milk allergenicity and lactose intolerance (Munekata et al. 2020; Boukid et al. 2021b). The main sources for plant-based dairy are almond, rice, soy, and oat (McClements, Newman, and McClements 2019). Compared to cow's milk, plant-based milk alternatives had in some cases lower protein, calcium, and vitamin D (McClements, Newman, and McClements 2019; A. R. A. Silva, Silva, and Ribeiro 2020). Plant-based milk alternatives were reported to be less digestible due to the globular structure of plant-based proteins (J. G. S. Silva et al. 2020; Chalupa-Krebzdak, Long, and Bohrer 2018). Plant-based dairy are also being used for feeding infants, specially soy-based products as they contain a complete amino acids profile and high protein digestibility (Gasparre, Mefleh, and Boukid 2022). Nevertheless, fortification is required to provide sufficient micronutrients including Vitamin D, iron, and zinc necessary for infant growth (Verduci et al. 2019).

14.4.3 PLANT-BASED YOGURT

Plant-based fermented alternatives to yogurt are gaining interest among consumers, adults, and children (Manus et al. 2021; Masiá et al. 2022). Plant-based alternative to yogurts market was estimated at USD 2.2 billion in 2022 and expected to reach USD 5.1 billion by 2032 (Globe News Wire 2022). The main types are soymilk yogurt, coconut milk yogurt, almond milk yogurt, and rice milk yogurt. These products can be fermented using lactic acid bacteria to improve taste and texture (Brückner-Gühmann, Benthin, and Drusch 2019). Otherwise, hydrocolloids are used to create the gel-like structure of yogurt (Jiang et al. 2020). Nuritionally, they provide less protein but higher fat, carbohydrates, and energy, but in most cases they lack vitamins and minerals present in animal-derived products (Boukid et al. 2021a). Manufacturers still encounter nutritional and organoleptic challenges to meet the quality of conventional products (Grasso, Alonso-Miravalles, and O'Mahony 2020). Danone is one of the leading company of plant-based yogurts with products such as Silk® and Vega™.

14.4.4 PLANT-BASED CHEESE

Plant-based cheese market was valued at USD 2.43 billion in 2021 and is expected to grow at a CAGR of 12.6% by 2030 (Grand View Research 2022). Plant-based cheese are made using emulsions of oil-in-water, containing protein, fat/oils, emulsifiers, flavors, colors, and preservatives (Wei and Yano 2020). These ingredients can go through fermentation to simulate cheese taste and flavor. However, mostly the use of hydrocolloids, fats, and starches contributes into reaching the desired consistency (Mefleh, Pasqualone, Caponio, and Faccia 2022; Mefleh, Pasqualone, Caponio, De Angelis, et al. 2022). Nutritionally, plant-based cheese have higher total fats, saturated fatty acids, and carbohydrates but lower protein, salt, and sugar than dairy cheese (Boukid et al. 2021a). Some of the main producers are Treeline Cheese, Follow Your Heart, Daiya Foods Inc., Kite Hill, Violife, Dr-Cow Tree Nut Cheese, Tofutti Brands, Inc., Tyne Chease Limited, Miyoko's Creamery, and Parmela Creamery.

14.4.5 SEAFOOD ALTERNATIVES

Seafood alternatives was valued *at* USD 767 million in 2022 and is expected to keep growing owing to serious concerns over overfishing and the environmental impact of industrial fishery (Future Market Insights 2022). Based on a recent market search, a total of 149 seafood alternatives were launched from 2002 to 2021 (Boukid et al. 2022). These products are tuna, shrimps, calamari, fish fingers, fish sticks, salmon, caviar, and fillet alternatives. The nutritional quality varied among the commercial products (Brand Essece Research 2020). Remarkably, calamari, fish fingers, fish sticks, and salmon alternatives had amounts similar to their conventional counterparts. Micronutrients, omega-3 fatty acids, and vitamins (A, B, and D) are added to improve the nutritional quality of plant-based products (Ran et al. 2022). The main producers are Ahimsa Foods, Good Catch Foods, Atlantic Natural Foods, Sophie's Kitchen Inc, New Wave Foods, Gardein, Ocean Hugger Foods, Impossible Foods Inc, Qishan Food Limited Company, SoFine Foods, Tofuna Fysh, and Vivera.

14.4.6 EGG ALTERNATIVES

The global egg alternatives market was valued at USD 1.5 billion in 2021 and is expected grow at a CAGR of 8.3% by 2031 (FACTMR 2021). Alternative eggs are available mainly as powders, but also as a liquid and in egg-shaped forms (Mintel 2021). Commercial plant-based eggs are made using a variety of ingredients including starches, plant-based proteins, algae flours, and fibers to suit different applications (Boukid and Gagaoua 2022). The top applications are bakery products and mayonnaise (Mustafa et al. 2018; Serventi, Yang, and Bian 2020). Cargill Inc., Kerry, Corbion NV, Glanbia Plc, Tate & Lyle Plc, Ingredion Incorporated, and Puratos Group are among the important manufacturers leading this market (Future Market Insights 2019).

14.5 OPPORTUNITIES AND CHALLENGES IN THE PLANT-BASED MARKET

Plant-sourced trends have been focusing on new ingredients and innovative technologies to provide plant-based alternatives with features similar to animal products (Rincon, Botelho, and de Alencar 2020). The main challenge is matching the nutritional aspects and quality characteristics of animal products. Rising awareness about the nutritional profiles of plant-based products in comparison to those of conventional animal products might help consumers making a rational decision of purchase and consumption. To emulate the organoleptic characteristics, a large array of proteins sources (e.g., legume proteins, oilseeds, cereals) and purified and refined fats and oils (coconut, cocoa, sunflower seeds, and rapeseed) as well as additives were used (Sridhar et al. 2022). There is still room for innovation in terms of technological and nutritional quality to provide high-value products to the consumers.

14.5.1 Plant-Sourced Meat Substitutes

Plant-based meat is made from plant proteins, oils, grains, vitamins, and minerals (Sridhar et al. 2022). Nutritionally, plant-based meat substitutes are found to be high in carbohydrates, sugar, and dietary fiber and low in calories, fat, and saturated fat (Curtain and Grafenauer 2019). Plant-based meat has less minerals in comparison with the natural animal-sourced meat products because of the iron present in animal-sourced meat as a heme form (Vliet et al. 2021). Biofortification such as iron and vitamin B12 contributes to improving the nutritional quality of the plant-based products. Another study revealed that the risk of obesity and cardiovascular disease was decreased when plant-based meat substitutes were consumed (Crimarco et al. 2020). Furthermore, combining proteins would contribute to developing plant-based meat substitutes with balanced amino acids profile. These substitutes can then be enriched with important elements to mimic the functioning and texture of meat from animals. Although there has been improvement in the texture and flavour of items made from plants instead of meat, there are still issues with creating a satisfying sensory experience and raising the nutritional value of these products (Tyndall et al. 2022). The sensory qualities of plant-based meat-substitute products will benefit from the influence of new plant proteins and processing technologies, which will help them reach desired meat-like characteristics and enhance the eating experience (Fiorentini, Kinchla, and Nolden 2020)

14.5.2 Plant-Sourced Dairy Substitutes

Although plant-based dairy substitutes have been taken and enjoyed for decades, their popularity has exploded in the last ten years. To improve their texture profile and sensory characteristics including mouthfeel and nutritional profile, a variety of plant-sourced components, emulsifiers, and flavorings agents are currently being used to make nutrient-rich plant-based dairy products. Plant-based milk prepared using a combination of coconut and chickpea extracts had more calcium content in comparison with animal-based milk (Rincon, Botelho, and de Alencar 2020). There are a number of plant-based milk variants which were developed such as chickpea and coconut milk blend (Rincon, Botelho, and de Alencar 2020), soy–almond milk blend (Dhankhar and Kundu 2021), grains-based milk (Clegg et al. 2021), quinoa-fermented beverage (Grasso et al. 2021), and soft cheese-, soybeans-, coconut-, cashew-, almond-, and hemp-based yogurts (Atalar 2019). Numerous studies have emphasized that the use of fermentation technology enhanced the flavor, texture, and nutritional quality of plant-based dairy products. Making plant-based soft and spreadable cheese is currently feasible; yet, hard cheese making is still challenging. Mimicking the hard texture requires more research to identify new ingredients and/or processing to reach texture similar to that of dairy cheese. Plant-based yogurts are more advanced because consumers appreciate thickness and creaminess (Greis et al. 2023). However, the main disadvantages are the high use of carbohydrates and/oils and low protein content. Focusing on developing dairy products, relying on more proteins is still challenging, but the use of lactic bacteria and precision fermentation might contribute into improving the nutritional and organoleptic quality.

14.5.3 Plant-Sourced Seafood Substitutes

Plant-sourced fish and seafoods are prepared using different ingredients such as vegetable/plant-sourced oils, flavorings agents, preservatives, stabilizers, and thickeners. These nutrients serve as the foundation for their high nutritional value, protein content, lack of cholesterol and mercury, and energy content. The use of a variety of plant ingredients creates the same texture and flavour as traditional fish products, demonstrating the resemblance of plant-sourced fish foods to traditional fish food products with better sustainability by easing the burden on the decreasing ocean fisheries. On the contrary, increased consumer demand sparked the creation of many sustainable seafood solutions. For instance, numerous international companies that produce plant-sourced seafood have introduced plant-sourced seafood substitutes like fish burgers, crab cakes, smoked salmon, and tuna chunks. However, these seafood industries might not aim to replicate the structure and texture of conventional seafood, which should be researched through novel manufacturing techniques and an understanding of chemical and nutritional aspects to improve the final product's structural and sensory qualities and meet consumer expectations by simulating the flavor, aroma, appearance, and texture of the real thing (Kazir and Livney 2021).

14.5.4 PLANT-SOURCED EGGS

As an alternative to traditional eggs, food professionals have created plant-based eggs. Plant-based eggs, which represent a healthier version and more sustainable and versatile product compared to conventional eggs, will be more in demand in upcoming years. The "cage-free by 2025" initiative, promoted by several supermarket chains throughout Europe, aims to sell only cage-free eggs (such as a barn, free-range, etc.) within the next five years. Countries like the United Kingdom and Italy have demonstrated a growing demand for higher animal welfare and sustainability standards in egg production (Rondoni, Millan, and Asioli 2022). Two of the key ingredients for mimicking plant-based eggs are emulsified canola oil and mung bean protein. During cooking, the proteins spread out and assemble, creating a structure akin to gel. The canola oil droplets add to the final product's opaque look, textural qualities, flavour profile, and mouthfeel. Additionally, these products include transglutaminase, an enzyme that crosslinks proteins to increase gel strength and water-holding ability, better simulating an actual egg (Gharibzahedi et al. 2018).

Aquafaba is obtained in the leftover liquid as the byproduct of the pulses when boiled. Commercially manufactured aquafaba is made by drying this water under the proper processing conditions, and it is then shipped to suppliers. The quality of aquafaba is influenced by a variety of processing factors. Aquafaba is used an egg replacement (Raikos, Hayes, and Ni 2020). In vegan mayonnaises, the proteins in aquafaba have been showing success. Aquafaba, therefore, has great economic product potential for vegan consumers. Additionally, pulse-cooking foods like chickpeas typically don't have a strong flavor or aroma. As a result, in addition to technological options, it can be an effective organoleptic choice for using eggs in a variety of recipes for people.

14.5.5 NEW INGREDIENTS

Different food product categories meet the demands of consumers in various ways. There are some natural ingredients, which could be utilized directly without applying any processing methods. These ingredients for the development of plant-sourced product development could be categorized as natural ingredients and semi-refined ingredients (Loveday 2020).

14.5.6 NATURAL INGREDIENTS

Natural ingredients consist of various natural plant sources such as cereals, pseudocereals, edible seeds, nuts, oilseeds, algae, tubers, and roots. They include products like almonds, nuts, flaxseeds, and vegetables, which either can be consumed directly or with minimal processing (Loveday 2020). Microalgae can be used as a food colorant and intriguing novel chemicals with biological activity that could be exploited as functional components can be found naturally in microalgae which are generally cultivated for the extraction of high-value components such as pigments or proteins (Priyadarshani and Rath 2012). Microbial proteins, usually referred to as single-cell proteins, are frequently obtained from bacteria, fungus, or microalgae (Hadi and Brightwell 2021). According to the research done thus far, algal biomass exhibits potential qualities as a new source of protein; on average, most of the algae tested have protein values that are comparable to or often even better than those of traditional plant sources (Becker 2007). Duckweed is another example of naturally grown plant in fresh water such as ponds and lakes. In terms of its chemical composition, duckweed has enormous potential as a future food source. When taken as food, it can give the human body energy and macro- and micronutrients, preserving health and supporting the body's structure. In contrast to most cereal grains, duckweed includes larger amounts of fiber, lipids, and minerals. It also has a protein level that is comparable to or even higher than that of most cereal grains (Xu et al. 2022).

14.5.7 SEMI-REFINED INGREDIENT

Non-protein ingredients in semi-refined foods, like fiber and phytochemicals, might add textures and sensorial characteristics that negatively restrict the variety of food applications. For instance, it can be challenging to include algae granules in goods like yogurt or ice cream because they often have a savory "umami" flavor and are green in color. This is because consumers typically anticipate these products to be light in color and to have a sweet or neutral flavor. Semi-refined ingredients may need pricey reinvention work to ensure that food quality is maintained throughout and macronutrient levels comply with legal requirements because they will fluctuate depending on the growing season and region. It is becoming more and more obvious that preserving natural food microstructures and phytochemicals through less extensive processing has health benefits. Whole grains and fresh or dried turmeric are examples. On the grounds of sustainability and health, it is a challenge worth pursuing to use semi-refined ingredients (Loveday 2020).

14.5.8 NEW TECHNOLOGIES

Microwave-assisted extraction technique, high-voltage electrical discharge, radiofrequency of proteins, and supercritical fluid extraction technique are some of the novel methods that are already been utilized for the development of plant-sourced food products. These technologies are very much popular in the extraction of plant proteins as well as in making them nutritionally rich. There are some nutritional differences in foods developed using plant sources such as meat, dairy, and seafood alternatives, which should be overcome through the application of an advanced version of processing techniques such as fermentation, fortification, and encapsulation. Although many plant-sourced alternatives mimic the sensorial profile like mouthfeel, umami taste and flavor, and texture of animal-based foods, there are nutritional differences in plant-based alternatives to meat, dairy, and seafood (Sridhar et al. 2022).

Apart from these technologies, there are a few novel biotechnological techniques that can be utilized for the development of upcoming plant-based food products. Cell culture technique is one of the techniques, which could be used for fermentation-derived animal protein as well as for plant cell culture (Loveday 2020; Nordlund et al. 2018)—for example, cultured berry cells contain around 19% protein and can have enhanced bioactive compounds (Nordlund et al. 2018). Gene-editing techniques like CRISPR/Cas9 are also being applied to increase the degree of yields of crops and enhance resistance to drought and disease situations (Brandt and Barrangou 2019). Reduced food waste may result from resistant varieties of apples, mushrooms, and potatoes, and gene editing may make antinutrients less prevalent or perhaps impossible to detect (Arora and Narula 2017).

14.6 PLANT PROTEIN: WAYS OF EXTRACTION, ISOLATION, AND CHARACTERIZATION

Scientists have focused on utilizing new plant protein extraction techniques, which have been found to enhance the protein yield while maintaining both functional and nutritional values. A number of studies have shown higher protein yields when using combined conventional as well as unconventional techniques of protein extraction simultaneously (Gorgüç, Bircan, and Yılmaz 2019). However, the nonconventional protein extraction techniques are reported to have low energy consumption, higher extraction yield, and lower consumption of solvents. Besides, these nonconventional methods are environment friendly and green ways of protein extraction.

Enzyme-assisted extraction is one of the nonconventional methods for the commercial recovery of good-quality proteins from plant sources (Liu et al. 2016; Ochoa-Rivas et al. 2017). Usually, a hard and stiff cellular wall acts as a barrier to the cell's protein extraction. Essentially, the non-conventional technique, i.e., enzyme-based extraction, focuses on disrupting cell wall integrity by enzyme-based breakdown of cell wall components, viz. complex carbohydrates (cellulose, hemicel-lulose, and pectin) (Jung et al. 2006). The specific action of enzymes like pectinase and amylases during cell wall degradation contributes to the systematic liberation of cellular proteins from plant foods such as cereals, pulses, and oilseeds (Rommi et al. 2014). Protease increases the yield of plant protein by separating protein from the bound polysaccharide matrix. Disruption of the cell wall facilitates the recovery of the cellular proteins. Subsequently, the enzyme protease also facilitates the breakdown of the higher molecular weight proteins into smaller, more soluble fractions, hence keeping promising extraction parameters. In addition, protease works at optimal pH, thereby pre-venting the denaturation of plant-based protein.

Another nonconventional technique for protein extraction from plant sources is the use of ultrasound-based extraction. In ultrasound-based extraction, the ultrasound waves produce acoustic cavitation, creating hotspots of greater temperature as well as pressure to extract constituents from the plant-based matrix (Belwal et al. 2018; Chemat et al. 2017; Fu et al. 2020; Soria and Villamiel 2010). The rapid development and then the collapse of bubbles produced through ultrasound on the cellular surface of the plant food sample, micro-flow and shock-wave generate greater shear as well as mechanical forces that disrupt membranes and cell walls (Lupatini et al. 2017). Two different kinds of instruments are employed for ultrasound-based extraction of plant proteins, such as ultra-sound probes and ultrasound bath. Additionally, the bath-type ultrasound arrangement is used for the direct as well as in direct contact with the food matrix all through the extraction. Ultrasound bath comprises a chamber encompassing the liquid phase for the transmission of ultrasound waves, ultrasound transducer, and ultrasound generator. However, the ultrasound probe allows direct con-tact with food source at varying altitudes. There are different kinds of probes varying in sizes and specifications (Kumar et al. 2022). Different researchers had utilized both the ultrasound bath and probe for the extraction of proteins from the plant-based foods like sunflower and rapeseed. The influence of greater-intensity ultrasound waves on the physicochemical and functional attributes of sunflower meal was evaluated (Malik et al. 2017). The lower intensity ultrasound waves are com-monly employed as nondestructive method, whereas high-intensity ultrasound waves are usually used for diffusion, emulsification, and extraction processes (Chemat et al. 2017).

Pulsed electric field treatment is a new nonconventional technology basically employed for the preservation of foodstuffs, inactivation of microbes and enzymes, and the extraction of protein from cellular matrix. However, this technique is relatively ineffective in accomplishing a greater yield of protein in comparison to other nonconventional techniques. The recovery of protein, using pulsed electric field at low temperature with long pulse exposure time as well as higher electric field inten-sity, increases. Nonetheless, the pulsed field input attributes must be optimized to extract proteins in their intuitive state. Protein quality is minimally compromised during processing and storage, making it a favorable method in comparison to conventional heat treatment (Kumar et al. 2021).

Another method of protein extraction from plant sources is the use of microwaves. Microwave extraction is using nonionizing electromagnetic radiation with frequencies between 300 MHz and 300 GHz. Microwaves sample through the linked action of dipole rotation as well as ionic conduc-tion, breaking the hydrogen bonds observed in the cell walls of the plant matrix. This exposure of microwave enhances the cell wall porosity, facilitates solvent penetration into the cell, and sim-plifies the effective release of intracellular constituents into the solvent system. Microwave pro-cessing creates a large quantity of heat energy, thereby initiating the thermally unstable bioactive compounds to decompose, causing it inappropriate for the extraction of proteins. Generally, the minimal extraction time and less solvent requirements are the main advantages of this microwave extraction method. Sequential use of microwaves with additional physical or biochemical methods can enhance protein extraction efficiency (Kumar et al. 2021).

High-pressure-assisted extraction is another way for protein extraction. This extraction technique is performed in three steps. First, the food is blended with the extraction medium and positioned within a pressure vessel. Furthermore, the system pressure is heightened from room temperature to the expected level in a brief period. The pressure of fluid typically varies from 100 to 1,000 MPa. As the pressure of the system increases, the pressure difference between the inside of the plant cell and the environment increases, initiating cell deformation as well as cellular wall damage. Solvents penetrate inside the damaged cell walls and membranes into the cell interior, enhancing the mass transfer of some soluble constituents. In one of the recent research, scientists prepared a response model to foresee the concentration of obtained proteins through high hydrostatic pressure, depending on the pressure applied ranging from 100 and 300 MPa and the type of solvent, for example, phosphate-buffered salts, trichloroacetic acid, and Tris-HCl (Altuner 2016). In another study, the extraction of protein from rice bran was said to be higher at greater pressures of 600–800 MPa while being lower at 200 MPa (Tang et al. 2002).

14.7 CONCLUSION

Plant-based alternatives to meat, dairy, and seafood are in great demand due to rising demands for wellness and sustainability. As a result, both start-ups and established food firms are making major investments to develop the next generation of plant-based alternatives. There are nutritional differences in plant-based meat, dairy, and seafood alternatives, which should be overcome through the application of advanced techniques like fermentation, fortification, and encapsulation. Although many plant-based alternatives mimic sensorial characteristics such as mouthfeel, umami meat taste and flavor, and texture profile of animal-sourced foods, there are nutritional differences in plant-sourced alternatives to meat, dairy, and seafood to be addressed in new product development.

REFERENCES

Akin, Zeynep, and Tulay Ozcan. 2017. "Functional Properties of Fermented Milk Produced with Plant Proteins." *LWT—Food Science and Technology* 86 (December): 25–30.

Altuner, E. M. 2016. "A Predictive Modelling Study for Using High Hydrostatic Pressure, a Food Processing Technology, for Protein Extraction." *Procedia Food Science* 7: 121–124.

Arora, L., and A. Narula. 2017. "Gene Editing and Crop Improvement Using CRISPR-Cas9 System." *Frontiers in Plant Science* 8: 1932.

Atalar, I. 2019. "Functional Kefir Production from High Pressure Homogenized Hazelnut Milk." *LWT* 107: 256–263.

Baune, M.-C., N. Terjung, M. Çağlar Tülbek, and F. Boukid. 2022. "Textured Vegetable Proteins (TVP): Future Foods Standing on Their Merits as Meat Alternatives." *Future Foods* (September): 100181.

Becker, E. W. 2007. "Micro-Algae as a Source of Protein." *Biotechnology Advances* 25, no. 2: 207–210.

Belwal, T., S. M. Ezzat, L. Rastrelli, I. D. Bhatt, M. Daglia, A. Baldi, H. P. Devkota, I. E. Orhan, J. K. Patra, G. Das, C. Anandharamakrishnan, and A. G. Atanasov. 2018. "A Critical Analysis of Extraction Techniques Used for Botanicals: Trends, Priorities, Industrial Uses and Optimization Strategies. *TrAC Trends in Analytical Chemistry* 100: 82–102.

Boukid, F. 2021a. "Oat Proteins as Emerging Ingredients for Food Formulation: Where We Stand?" *European Food Research and Technology* 247, no. 3: 535–544.

Boukid, F. 2021b. "The Realm of Plant Proteins with Focus on Their Application in Developing New Bakery Products." In *Advances in Food and Nutrition Research*. Academic Press.

Boukid, F. 2021c. "Peanut Protein—an Underutilised by-Product with Great Potential: A Review." *International Journal of Food Science & Technology* 57 (December).

Boukid, F., M.-C. Baune, M. Gagaoua, and M. Castellari. 2022. "Seafood Alternatives: Assessing the Nutritional Profile of Products Sold in the Global Market." *European Food Research and Technology* 1 (March): 1–10. https://doi.org/10.1007/S00217-022-04004- Z.

Boukid, F., and M. Castellari. 2021. "Veggie Burgers in the EU Market: A Nutritional Challenge?" *European Food Research and Technology* 247.

Boukid, F., and M. Castellari. 2022. "How Can Processing Technologies Boost the Application of Faba Bean (Vicia Faba L.) Proteins in Food Production?" *E-Food* 3, no. 3: e18.

Boukid, F., and M. Gagaoua. 2022a. "Meat Alternatives: A Proofed Commodity?" *Advances in Food and Nutrition Research* 101 (January): 213–236.

Boukid, F., and M. Gagaoua. 2022b. "Vegan Egg: A Future-Proof Food Ingredient?" *Foods* 11, no. 2: 161.

Boukid, F., M. Lamri, B. N. Dar, M. Garron, and M. Castellari. 2021a. "Vegan Alternatives to Processed Cheese and Yogurt Launched in the European Market During 2020: A Nutritional Challenge?" *Foods* 10, no. 11: 2782.

Boukid, F., and A. Pasqualone. 2021. "Lupine (Lupinus spp.) Proteins: Characteristics, Safety and Food Applications." *European Food Research and Technology* 1: 3. https://doi.org/10.1007/s00217-021-03909-5.

Boukid, F., C. M. Rosell, S. Rosene, S. Bover-Cid, and M. Castellari. 2021b. "Non-Animal Proteins as Cutting-Edge Ingredients to Reformulate Animal-Free Foodstuffs: Present Status and Future Perspectives." *Critical Reviews in Food Science and Nutrition* 137 (March): 1–31.

Boukid, F., G. Sogari, and C. M. Rosell. 2022. "Edible Insects as Foods: Mapping Scientific Publications and Product Launches in the Global Market (1996–2021)." *European Food Research and Technology* 1, no. 10 (August 1–16): 1–9. https://doi.org/10.3920/JIFF2022.0060.

Boukid, F., E. Zannini, E. Carini, and E. Vittadini. 2019. "Pulses for Bread Fortification: A Necessity or a Choice?" *Trends in Food Science and Technology* 88 (June): 416–428.

Brand Essece Research. 2020. "Plant-Based Seafood Alternatives Market Size Future Scenario, and Industrial Opportunities to 2027 | Report Analysis 2021–2027." https://brandessenceresearch.com/food-and-beverage/plant-based-seafood-alternatives-market-size?MND_Priyanka.

Brandt, K., and R. Barrangou. 2019. "Applications of CRISPR Technologies Across the Food Supply Chain." *Annual Review of Food Science and Technology* 10: 133–150.

Brückner-Gühmann, M., A. Benthin, and S. Drusch. 2019. "Enrichment of Yoghurt with Oat Protein Fractions: Structure Formation, Textural Properties and Sensory Evaluation." *Food Hydrocolloids* 86 (January): 146–153.

Bryant, C., K. Szejda, N. Parekh, V. Deshpande, and B. Tse. 2019. "A Survey of Consumer Perceptions of Plant-Based and Clean Meat in the USA, India, and China." *Frontiers in Sustainable Food Systems* 3 (February): 11.

Business Wire. 2022. "Europe Plant-Based Food Market Report 2022: Product Launches by Plant-Based Foods & Protein Alternatives Manufacturers Presents Opportunities—ResearchAndMarkets.Com." *Business Wire.* www.businesswire.com/news/home/20221013005739/en/Europe-Plant-based-Food-Market-Report-2022-Product-Launches-by-Plant-based-Foods-Protein-Alternatives-Manufacturers-Presents-Opportunities-ResearchAndMarkets.com.

Castro-Alba, V., C. E. Lazarte, D. Perez-Rea, N. G. Carlsson, A. Almgren, B. Bergenståhl, and Y. Granfeldt. 2019. "Fermentation of Pseudocereals Quinoa, Canihua, and Amaranth to Improve Mineral Accessibility Through Degradation of Phytate." *Journal of the Science of Food and Agriculture* 99, no. 11: 5239–5248.

Chalupa-Krebzdak, S., C. J. Long, and B. M. Bohrer. 2018. "Nutrient Density and Nutritional Value of Milk and Plant-Based Milk Alternatives." *International Dairy Journal* 87. Elsevier Ltd.

Chemat, F., N. Rombaut, A. G. Sicaire, A. Meullemiestre, A. S. Fabiano-Tixier, and M. Abert-Vian. 2017. "Ultrasound Assisted Extraction of Food and Natural Products: Mechanisms, Techniques, Combinations, Protocols and Applications. A Review." *Ultrasonics Sonochemistry* 34: 540–560.

Clegg, M. E., A. T. Ribes, R. Reynolds, K. Kliem, and S. Stergiadis. 2021. "A Comparative Assessment of the Nutritional Composition of Dairy and Plant-Based Dairy Alternatives Available for Sale in the UK and the Implications for Consumers' Dietary Intakes." *Food Research International* 148: 110586.

Cornet, Steven H. V., Silvia J. E. Snel, Floor K. G. Schreuders, Ruud G. M. van der Sman, Michael Beyrer, and Atze Jan van der Goot. 2022. "Thermo-Mechanical Processing of Plant Proteins Using Shear Cell and High-Moisture Extrusion Cooking." *Critical Reviews in Food Science and Nutrition* 62, no. 12: 3264–3280.

Crimarco, A., S. Springfield, C. Petlura, T. Streaty, K. Cunanan, J. Lee, . . . C. D. Gardner. 2020. "A Randomized Crossover Trial on the Effect of Plant-Based Compared with Animal-Based Meat on Trimethylamine-N-Oxide and Cardiovascular Disease Risk Factors in Generally Healthy Adults: Study with Appetizing Plantfood—Meat Eating Alternative Trial (SWAP-MEAT)." *The American Journal of Clinical Nutrition* 112, no. 5: 1188–1199.

Curtain, F., and S. Grafenauer. 2019. "Plant-Based Meat Substitutes in the Flexitarian Age: An Audit of Products on Supermarket Shelves." *Nutrients* 11, no. 11: 2603.

Desmond, M. A., J. G. Sobiecki, M. Jaworski, P. Płudowski, J. Antoniewicz, M. K. Shirley, S. Eaton, J. Książyk, M. Cortina-Borja, B. De Stavola, and M. Fewtrell. 2021. "Growth, Body Composition, and Cardiovascular and Nutritional Risk of 5- to 10-y-Old Children Consuming Vegetarian, Vegan, or Omnivore Diets." *The American Journal of Clinical Nutrition* 113, no. 6: 1565.

Dhankhar, J., and P. Kundu. 2021. "Stability Aspects of Non-Dairy Milk Alternatives." *Milk Substitutes-Selected Aspects*, 1–28.

FACTMR. 2021. "Vegan Egg Market Size, Share, Trends & Forecast, 2021–2031." www.factmr.com/report/vegan-eggs-market.

Fan, H., H. Liu, Y. Zhang, S. Zhang, T. Liu, and D. Wang. 2022. "Review on Plant-Derived Bioactive Peptides: Biological Activities, Mechanism of Action and Utilizations in Food Development." *Journal of Future Foods* 2, no. 2: 143–159.

Fiorentini, M., A. J. Kinchla, and A. A. Nolden. 2020. "Role of Sensory Evaluation in Consumer Acceptance of Plant-Based Meat Analogs and Meat Extenders: A Scoping Review." *Foods* 9, no. 9: 1334.

Fresán, Ujué, M. A. Mejia, W. J. Craig, K. Jaceldo-Siegl, and J. Sabaté. 2019. "Meat Analogs from Different Protein Sources: A Comparison of Their Sustainability and Nutritional Content." *Sustainability (Switzerland)* 11, no. 12.

Fu, X., T. Belwal, G. Cravotto, and Z. Luo. 2020. "Sono-Physical and Sono-Chemical Effects of Ultrasound: Primary Applications in Extraction and Freezing Operations and Influence on Food Components. *Ultrasonics Sonochemistry* 60: 104726.

Future Market Insights. 2019. "Egg Replacement Ingredient Market Analysis and Review 2019–2026 | Future Market Insights (FMI)." www.futuremarketinsights.com/reports/egg-replacement-ingredient-market.

Future Market Insights. 2022. "Plant-Based Fish Market Size, Share, Trends & Forecast—2032." www.futuremarketinsights.com/reports/plant-based-fish-market.

Gasparre, N., M. Mefleh, and F. Boukid. 2022. "Nutritional Facts and Health/Nutrition Claims of Commercial Plant-Based Infant Foods: Where Do We Stand?" *Plants* 11, no. 19: 2531.

Gharibzahedi, S. M. T., S. Roohinejad, S. George, F. J. Barba, R. Greiner, G. V. Barbosa-Cánovas, and K. Mallikarjunan. 2018. "Innovative Food Processing Technologies on the Transglutaminase Functionality in Protein-Based Food Products: Trends, Opportunities and Drawbacks." *Trends in Food Science & Technology* 75: 194–205.

Globe News Wire. 2022. "Vegan Yogurt Market to Reach $5.1 Billion, Globally, by." www.globenewswire.com/en/news-release/2022/03/16/2404570/0/en/Vegan-Yogurt-Market-to-Reach-5-1-Billion-Globally-By-2032-At-8-9-CAGR-Future-Market-Insights.html.

Gómez-Luciano, C. A., L. K. de Aguiar, F. Vriesekoop, and B. Urbano. 2019. "Consumers' Willingness to Purchase Three Alternatives to Meat Proteins in the United Kingdom, Spain, Brazil and the Dominican Republic." *Food Quality and Preference* 78 (December).

Görgüç, A., C. Bircan, and F. M. Yılmaz. 2019. "Sesame Bran as an Unexploited By-Product: Effect of Enzyme and Ultrasound-Assisted Extraction on the Recovery of Protein and Antioxidant Compounds." *Food Chemistry* 283: 637–645.

Grand View Research. 2019. "Pea Protein Market Size & Share | Industry Growth Report, 2019–2025." www.grandviewresearch.com/industry-analysis/pea-protein-market.

Grand View Research. 2022. "Vegan Cheese Market Size & Share Report, 2022–2030." www.grandviewresearch.com/industry-analysis/vegan-cheese-market.

Grasso, M., A. Remani, A. Dickins, B. M. Colosimo, and R. K. Leach. 2021. "In-situ Measurement and Monitoring Methods for Metal Powder Bed Fusion: An Updated Review." *Measurement Science and Technology* 32, no. 11: 112001.

Grasso, N., L. Alonso-Miravalles, and J. A. O'Mahony. 2020. "Composition, Physicochemical and Sensorial Properties of Commercial Plant-Based Yogurts." *Foods* 9, no. 3: 252.

Greis, M., A. A. Nolden, A. J. Kinchla, S. Puputti, L. Seppä, and M. Sandell. 2023. "What if Plant-Based Yogurts Were Like Dairy Yogurts? Texture Perception and Liking of Plant-Based Yogurts Among US and Finnish Consumers." *Food Quality and Preference* 104848.

Guo, Z., F. Teng, Z. Huang, B. Lv, X. Lv, O. Babich, W. Yu, Y. Li, Z. Wang, and L. Jiang. 2020. "Effects of Material Characteristics on the Structural Characteristics and Flavor Substances Retention of Meat Analogs." *Food Hydrocolloids* 105 (August).

Hadi, J., and G. Brightwell. 2021. "Safety of Alternative Proteins: Technological, Environmental and Regulatory Aspects of Cultured Meat, Plant-Based Meat, Insect Protein and Single-Cell Protein." *Foods* 10, no. 6: 1226.

Industry ARC. 2022. "Asia Plant Based Food Market Share, Size and Industry Growth Analysis 2021–2026." www.industryarc.com/Report/19784/asia-plant-based-food-market.html.

Jiang, Z. Q., J. Wang, F. Stoddard, H. Salovaara, and T. Sontag-Strohm. 2020. "Preparation and Characterization of Emulsion Gels from Whole Faba Bean Flour." *Foods* 9, no. 6.

Joshi, V. K., and S. Kumar. 2015. "Meat Analogues: Plant Based Alternatives to Meat Products: A Review." *International Journal of Food and Fermentation Technology* 5, no. 2: 107.

Jung, S., B. P. Lamsal, V. Stepien, L. A. Johnson, and P. A. Murphy. 2006. "Functionality of Soy Protein Produced by Enzyme-Assisted Extraction." *Journal of the American Oil Chemists' Society* 83, no. 1: 71–78.

Kang, ‚J., Md A. Hossain, H. C. Park, O. M. Jeong, S. W. Park, and M. Her. 2021. "Cross-Contamination of Enrofloxacin in Veterinary Medicinal and Nutritional Products in Korea." *Antibiotics* 10, no. 2: 1–6.

Kazir, M., and Y. D. Livney. 2021. "Plant-Based Seafood Analogs." *Molecules* 26, no. 6: 1559.

Kumar, G., S. Upadhyay, D. K. Yadav, S. Malakar, P. Dhurve, and S. Suri. 2022. "Application of Ultrasound Technology for Extraction of Color Pigments from Plant Sources and Their Potential Bio-Functional Properties: A Review." *Journal of Food Process Engineering* e14238.

Kumar, P., M. K. Chatli, N. Mehta, P. Singh, O. P. Malav, and A. K. Verma. 2017. "Meat Analogues: Health Promising Sustainable Meat Substitutes." *Critical Reviews in Food Science and Nutrition* 57, no. 5: 923–932.

Kumar, P., N. Mehta, A. Abubakar, A. K. Verma, U. Kaka, N. Sharma, A. Q. Sazili, M. Pateiro, M. Kumar, and J. M. Lorenzo. 2022. "Potential Alternatives of Animal Proteins for Sustainability in the Food Sector." *Food Reviews International* 1–26.

Kumar, M., M. Tomar, J. Potkule, Reetu Verma, Sneh Punia, Archana Mahapatra, Tarun Belwal, J. F. Kennedy, and V. Satankar. 2021. "Advances in the Plant Protein Extraction: Mechanism and Recommendations." *Food Hydrocolloids* 115: 106595.

Liu, J. J., M. A. A. Gasmalla, P. Li, and R. Yang. 2016. "Enzyme-Assisted Extraction Processing from Oilseeds: Principle, Processing and Application." *Innovative Food Science & Emerging Technologies* 35: 184–193.

Loveday, S. M. 2020. "Plant Protein Ingredients with Food Functionality Potential." *Nutrition Bulletin* 45, no. 3: 321–327.

Lupatini, A. L., L. de Oliveira Bispo, L. M. Colla, J. A. V. Costa, C. Canan, and E. Colla. 2017. "Protein and Carbohydrate Extraction from S. Platensis Biomass by Ultrasound and Mechanical Agitation." *Food Research International* 99: 1028–1035.

Malik, M. A., H. K. Sharma, and C. S. Saini. 2017. "High Intensity Ultrasound Treatment of Protein Isolate Extracted from Dephenolized Sunflower Meal: Effect on Physicochemical and Functional Properties." *Ultrasonics Sonochemistry* 39: 511–519.

Maningat, C., C., T. Jeradechachai, and M. R. Buttshaw. 2022. "Textured Wheat and Pea Proteins for Meat Alternative Applications." *Cereal Chemistry* 99, no. 1: 37–66. https://doi.org/10.1002/CCHE.10503.

Manus, J., M. Millette, C. Dridi, S. Salmieri, B. R. Aguilar Uscanga, and M. Lacroix. 2021. "Protein Quality of a Probiotic Beverage Enriched with Pea and Rice Protein." *Journal of Food Science* 86, no. 8: 3698–3706.

Mariutti, L. R. B., K. S. Rebelo, A. Bisconsin-Junior, J. S. de Morais, M. Magnani, I. R. Maldonade, N. R. Madeira, A. Tiengo, M. R. Maróstica, and C. B. B. Cazarin. 2021. "The Use of Alternative Food Sources to Improve Health and Guarantee Access and Food Intake." *Food Research International* 149 (November).

Markets and Markets. 2021. "Dairy Alternatives Market Size | Trends—Forecasts to 2026 | COVID-19 Impact on Dairy Alternatives Market | Markets and Markets." https://www.marketsandmarkets.com/Market-Reports/dairy-alternatives-market-677.html.

Markets and Markets. 2022. "Plant-Based Protein Market Size, Share | 2022–2027." https://www.marketsandmarkets.com/Market-Reports/plant-based-protein-market-14715651.html.

Martinez, M. M., and F. Boukid. 2021. "Future-Proofing Dietary Pea Starch." *ACS Food Science & Technology* 8: 1371–1372.

Masiá, C., P. E. Jensen, I. L. Petersen, and P. Buldo. 2022. "Design of a Functional Pea Protein Matrix for Fermented Plant-Based Cheese." *Foods* 11, no. 2. https://doi.org/10.3390/FOODS11020178.

McClements, D. J., E. Newman, and I. F. McClements. 2019. "Plant-Based Milks: A Review of the Science Underpinning Their Design, Fabrication, and Performance." *Comprehensive Reviews in Food Science and Food Safety.* Blackwell Publishing Inc. https://doi.org/10.1111/1541-4337.12505.

Mefleh,M., A. Pasqualone, F. Caponio, D. De Angelis, G. Natrella, C. Summo, and M. Faccia. 2022. "Spreadable Plant-Based Cheese Analogue with Dry- Fractioned Pea Protein and Inulin–Olive Oil Emulsion-Filled Gel." *Journal of the Science of Food and Agriculture* 102, no. 12: 5478–5487. https://doi.org/10.1002/JSFA.11902.

Mefleh, M., Antonella Pasqualone, Francesco Caponio, and Michele Faccia. 2022. "Legumes as Basic Ingredients in the Production of Dairy-Free Cheese Alternatives: A Review." *Journal of the Science of Food and Agriculture* 102, no. 1: 8–18. https://doi.org/10.1002/JSFA.11502.

Meticulous Research®. 2019. "Plant Based Protein Market Worth $14.32 Billion by 2025- Exclusive Report by Meticulous Research®." www.globenewswire.com/news-release/2019/08/20/1904339/0/en/Plant-Based-Protein-Market-worth-14-32-billion-by-2025-Exclusive-Report-by-Meticulous-Research.html.

Mintel. 2020. "GNPD—Plant Proteins in Meat Substitutes." https://portal.mintel.com/portal/login?next= https%3A%2F%2Fwww.gnpd.com%2Fsinatra%2Fanalysis%2Fchart_results%2Fsearch%2FFlM f6yv1YN%2F%2F%3Fanalysis_id%3Dcc8808dc-ef15-429.

Mintel. 2021. "How Plant-Based Eggs Will Crack into Mainstream Food." https://clients.mintel.com/.

Modlinska, K., D. Adamczyk, D. Maison, and W. Pisula. 2020. "Gender Differences in Attitudes to Vegans/Vegetarians and Their Food Preferences, and Their Implications for Promoting Sustainable Dietary Patterns: A Systematic Review." *Sustainability (Switzerland)* 12, no. 16. https://doi.org/10.3390/SU12166292.

Morder Intelligence. 2021. "Middle East & Africa Plant-Based Meat and Dairy Products Market—Growth|Trends| Forecast." www.mordorintelligence.com/industry-reports/middle- east-and-africa-plant-based-meat-and-dairy-products-industry.

Munekata, P. E. S., R. Domínguez, S. Budaraju, E. Roselló-Soto, F. J. Barba, K. Mallikarjunan, S. Roohinejad, and J. M. Lorenzo. 2020. "Effect of Innovative Food Processing Technologies on the Physicochemical and Nutritional Properties and Quality of Non- Dairy Plant-Based Beverages." *Foods*. MDPI Multidisciplinary Digital Publishing Institute. https://doi.org/10.3390/foods9030288.

Mustafa, R., Y. He, Y. Y. Shim, and M. J. T. Reaney. 2018. "Aquafaba, Wastewater from Chickpea Canning, Functions as an Egg Replacer in Sponge Cake." *International Journal of Food Science and Technology* 53, no. 10: 2247–2255. https://doi.org/10.1111/IJFS.13813.

Nordlund, E., M. Lille, P. Silventoinen, H. Nygren, T. Seppänen-Laakso, A. Mikkelson, . . . H. Rischer. 2018. "Plant Cells as Food–A Concept Taking Shape." *Food Research International* 107: 297–305.

Ochoa-Rivas, A., Y. Nava-Valdez, S. O. Serna-Saldívar, and C. Chuck-Hernández. 2017. "Microwave and Ultrasound to Enhance Protein Extraction from Peanut Flour Under Alkaline Conditions: Effects in Yield and Functional Properties of Protein Isolates." *Food and Bioprocess Technology* 10: 543–555.

Owusu-Kwarteng, J., D. Agyei, F. Akabanda, R. A. Atuna, and F. K. Amagloh. 2022. "Plant-Based Alkaline Fermented Foods as Sustainable Sources of Nutrients and Health-Promoting Bioactive Compounds." *Frontiers in Sustainable Food Systems* 6 (June): 197.

Priyadarshani, I., and B. Rath. 2012. "Commercial and Industrial Applications of Micro Algae–A Review." *Journal of Algal Biomass Utilization* 3, no. 4: 89–100.

Raikos, V., H. Hayes, and H. Ni. 2020. "Aquafaba from Commercially Canned Chickpeas as Potential Egg Replacer for the Development of Vegan Mayonnaise: Recipe Optimisation and Storage Stability." *International Journal of Food Science & Technology* 55, no. 5: 1935–1942.

Ran, X., X. Lou, H. Zheng, Q. Gu, and H. Yang. 2022. "Improving the Texture and Rheological Qualities of a Plant-Based Fishball Analogue by Using Konjac Glucomannan to Enhance Crosslinks with Soy Protein." *Innovative Food Science & Emerging Technologies* 75 (January): 102910.

Razavizadeh, S., G. Alencikiene, L. Vaiciulyte-Funk, P. Ertbjerg, and A. Salaseviciene. 2022. "Utilization of Fermented and Enzymatically Hydrolyzed Soy Press Cake as Ingredient for Meat Analogues." *LWT* 165 (August): 113736.

Research and Markets. 2021. "Latin America Plant-Based Food and Beverage Market 2021–2028." www.researchandmarkets.com/reports/5360089/latin-america-plant-based-food-and-beverage.

Rincon, L., R. B. A. Botelho, and E. R. de Alencar. 2020. "Development of Novel Plant-Based Milk Based on Chickpea and Coconut." *LWT* 128: 109479.

Rommi, K., T. K. Hakala, U. Holopainen, E. Nordlund, K. Poutanen, and R. Lantto. 2014. "Effect of Enzyme-Aided Cell Wall Disintegration on Protein Extractability from Intact and Dehulled Rapeseed (Brassica Rapa L. and Brassica Napus L.) Press Cakes." *Journal of Agricultural and Food Chemistry* 62, no. 32: 7989–7997.

Rondoni, A., E. Millan, and D. Asioli. 2022. "Plant-Based Eggs: Views of Industry Practitioners and Experts." *Journal of International Food & Agribusiness Marketing* 34, no. 5: 564–587.

Saget, S., M. Costa, C. S. Santos, M. W. Vasconcelos, J. Gibbons, D. Styles, and M. Williams. 2021. "Substitution of Beef with Pea Protein Reduces the Environmental Footprint of Meat Balls Whilst Supporting Health and Climate Stabilisation Goals." *Journal of Cleaner Production* 297 (May).

Sattar, A. A., R. Mahmud, M. A. S. Mohsin, N. N. Chisty, lM. H. Uddin, N. Irin, T. Barnett, G. Fournie, E. Houghton, and M. A. Hoque. 2021. "COVID-19 Impact on Poultry Production and Distribution Networks in Bangladesh." *Frontiers in Sustainable Food Systems* 5 (August): 306.

Serventi, L., Y. Yang, and Y. Bian. 2020. "Cooking Water Applications." *Upcycling Legume Water: From Wastewater to Food Ingredients* (January): 105–120.

Siddiqui, S. A., N. Z. Bahmid, Chayan M. M. Mahmud, F. Boukid, Melisa Lamri, and Mohammed Gagaoua. 2022. "Consumer Acceptability of Plant-, Seaweed-, and Insect-Based Foods as Alternatives to Meat: A Critical Compilation of a Decade of Research." *Critical Reviews in Food Science and Nutrition* (February): 1–22. https://doi.org/10.1080/10408398.2022.2036096.

Silva, A. R. A., Marselle M. N. Silva, and B. D. Ribeiro. 2020. "Health Issues and Technological Aspects of Plant-Based Alternative Milk." *Food Research International* 131. Elsevier Ltd.

Silva, J. G. S., A. P. Rebellato, E. T, dos Santos Caramês, R, Greiner, and J. A. L. Pallone. 2020. "In Vitro Digestion Effect on Mineral Bioaccessibility and Antioxidant Bioactive Compounds of Plant-Based Beverages." *Food Research International* 130 (April).

Soria, A. C., and M. Villamiel. 2010. "Effect of Ultrasound on the Technological Properties and Bioactivity of Food: A Review." *Trends in Food Science & Technology* 21, no. 7: 323–331.

Sprague, M., M. B. Betancor, J. R. Dick, and D. R. Tocher. 2017. "Nutritional Evaluation of Seafood, with Respect to Long-Chain Omega-3 Fatty Acids, Available to UK Consumers." *Proceedings of the Nutrition Society* 76 (OCE2).

Sridhar, K., S. Bouhallab, T. Croguennec, D. Renard, and V. Lechevalier. 2022. "Recent Trends in Design of Healthier Plant-Based Alternatives: Nutritional Profile, Gastrointestinal Digestion, and Consumer Perception." *Critical Reviews in Food Science and Nutrition*, 1–16.

Tang, S., N. S. Hettiarachchy, and T. H. Shellhammer. 2002. "Protein Extraction from Heat-Stabilized Defatted Rice Bran. 1. Physical Processing and Enzyme Treatments." *Journal of Agricultural and Food Chemistry* 50, no. 25: 7444–7448.

Transparencymarketresearch. 2019. "Rice Protein Market—Global Industry Analysis and Forecast 2027." www.transparencymarketresearch.com/rice-protein-market.html.

Tso, R., A. J. Lim, and C. G. Forde. 2021. "A Critical Appraisal of the Evidence Supporting Consumer Motivations for Alternative Proteins." *Foods* 10, no. 1: 24. https://doi.org/10.3390/foods10010024.

Tyndall, S. M., G. R. Maloney, M. B. Cole, N. G. Hazell, and M. A. Augustin. 2022. "Critical Food and Nutrition Science Challenges for Plant-Based Meat Alternative Products." *Critical Reviews in Food Science and Nutrition* 1–16.

Verduci,E., S. D'elios, L. Cerrato, P. Comberiati, M. Calvani, S. Palazzo, A. Martelli, M. Landi, T. Trikamjee, and D. G. Peroni. 2019. "Cow's Milk Substitutes for Children: Nutritional Aspects of Milk from Different Mammalian Species, Special Formula and Plant-Based Beverages." *Nutrients* 11, no. 8: 1739.

van Vliet, S., J. R. Bain, M. J. Muehlbauer, F. D. Provenza, S. L. Kronberg, C. F. Pieper, and K. M. Huffman. 2021. "A Metabolomics Comparison of Plant-Based Meat and Grass-Fed Meat Indicates Large Nutritional Differences Despite Comparable Nutrition Facts Panels." *Scientific Reports* 11, no. 1: 1–13.

Wei, W., and H. Yano. 2020. "Development of 'New' Bread and Cheese." *Processes* 8, no. 12: 1541.

Wen, C., J. Zhang, H. Zhang, Y. Duan, and H. Ma. 2020. "Plant Protein-Derived Antioxidant Peptides: Isolation, Identification, Mechanism of Action and Application in Food Systems: A Review." *Trends in Food Science & Technology* 105 (November): 308–322.

Xu, J., Y. Shen, Y. Zheng, G. Smith, X. S. Sun, D. Wang, Y. Zhao, W. Zhang, and Y. Li. 2022. "Duckweed (Lemnaceae) for Potentially Nutritious Human Food: A Review." *Food Reviews International* 1–15.

Index

Note: Page numbers in **bold** indicate a table and page numbers in *italics* indicate a figure on the corresponding page.

For Product Safety Concerns and Information please contact our EU
representative GPSR@taylorandfrancis.com
Taylor & Francis Verlag GmbH, Kaufingerstraße 24, 80331 München, Germany